World Geomorphological Landscapes

Series editor

Piotr Migoń, Wroclaw, Poland

More information about this series at http://www.springer.com/series/10852

Tomáš Pánek · Jan Hradecký
Editors

Landscapes and Landforms of the Czech Republic

Springer

Editors
Tomáš Pánek
Department of Physical Geography
and Geoecology
University of Ostrava
Ostrava
Czech Republic

Jan Hradecký
Department of Physical Geography
and Geoecology
University of Ostrava
Ostrava
Czech Republic

ISSN 2213-2090 ISSN 2213-2104 (electronic)
World Geomorphological Landscapes
ISBN 978-3-319-27536-9 ISBN 978-3-319-27537-6 (eBook)
DOI 10.1007/978-3-319-27537-6

Library of Congress Control Number: 2015956361

Printed on acid-free paper

This Springer imprint is published by SpringerNature
The registered company is Springer International Publishing AG Switzerland

Series Editor Preface

Landforms and landscapes vary enormously across the Earth, from high mountains to endless plains. At a smaller scale, nature often surprises us creating shapes which look improbable. Many physical landscapes are so immensely beautiful that they receive the highest possible recognition—they hold the status of World Heritage Sites. Apart from often being immensely scenic, landscapes tell stories which not uncommonly can be traced back in time for tens of million years and include unique geological events such as meteorite impacts. In addition, many landscapes owe their appearance and harmony not solely to the natural forces. For centuries, and even millennia, they have been shaped by humans, who have modified hill slopes, river courses and coastlines, and erected structures which often blend with the natural landforms to form inseparable entities.

These landscapes are studied by geomorphology—'the science of scenery'—a part of earth sciences that focuses on landforms, their assemblages, surface and subsurface processes that moulded them in the past and that change them today. To show the importance of geomorphology in understanding the landscape, and to present the beauty and diversity of the geomorphological sceneries across the world, we have launched a book series called *World Geomorphological Landscapes*. It aims to be a scientific library of monographs that present and explain physical landscapes, focusing on both representative and uniquely spectacular examples. Each book will contain details on geomorphology of a particular country or a geographically coherent region. This volume presents the impressive geodiversity of the Czech Republic. This Central European country may seem small but it hosts a very wide range of landscapes and landforms, the origin of which can be traced back to the Mesozoic. Among geomorphic highlights of the Czech Republic are block-faulted mountains with elevated planation surfaces and the evidence of past mountain glaciation, karst plateaus, deep fluvial canyons, astounding 'rock cities' in sandstone, flysch mountain ranges affected by huge landslides, and many others. They are presented and illustrated through carefully selected 25 examples from the entire country.

The World Geomorphological Landscapes series is produced under the scientific patronage of the International Association of Geomorphologists (IAG)—a society that brings together geomorphologists from all around the world. The IAG was established in 1989 and is an independent scientific association affiliated with the International Geographical Union (IGU) and the International Union of Geological Sciences (IUGS). Among its main aims are to promote geomorphology and to foster dissemination of geomorphological knowledge. I believe that this lavishly illustrated series, which keeps to the scientific rigour, is the most appropriate means to fulfil these aims and to serve the geoscientific community. To this end, my great thanks go to Tomáš Pánek and Jan Hradecký for agreeing to coordinate this volume. I am also very grateful to all individual authors who accepted invitations to contribute and delivered fine contributions which collectively show how varied and geomorphologically rich even a relatively small country can be.

For me, to write the preface to the Czech Republic volume is of particular pleasure. Living just across the border, I consider this country—within an easy reach of a day trip—as part of my homeland. I was fortunate to be able to see many geomorphological landscapes of this country myself and even to carry out some research. A little evidence of this involvement is my own modest contribution to this volume, regarding the granite landscape of Jizerské hory, in the northern part of the country.

I hope that the book will convince the readers that a little geomorphological paradise is located right in the heart of Europe.

Piotr Migoń

Contents

Short Biodata of Authors

Jiří Adamovič is a researcher at the Institute of Geology of the CAS, v.v.i. in Prague, Czech Republic. His subjects of study include sedimentology, stratigraphy and petrology of detrital sedimentary rocks. He has been involved in projects dealing with sandstone tectonics, lithological controls on the origin of sandstone relief, and hydrothermal alteration of sandstone upon its interaction with magma. He is the prime author of the recently published Atlas of sandstone rock cities of the Czech Republic and Slovakia.

Břetislav Balatka is a senior research member of the Department of Physical Geography and Geoecology at Faculty of Science, Charles University in Prague. He is known as an excellent teacher and tutor in the field of geography. His research interests are focused on physical geography and regional geomorphology of Central Europe, including the origin and morphostratigraphy of fluvial accumulation terraces, landform evolution during the Quaternary, recent and present-day geomorphologic processes and their impacts on the environment.

Pavel Bosák is Professor of Earth Science at the Institute of Geology of the CAS, v.v.i. in Prague, Czech Republic, and Associate Researcher at the Institute of Karst Research ZRC SAZU, Postojna, Slovenia. He has conducted a number of projects at home and abroad (Europe, North Africa, Asia, Caribbean region). He is a specialist in palaeokarstology and karstology; recently he has been working on cave and karst sediments and their sedimentology, paleogeographic significance and dating by a combination of methods, especially in the region of Central–Eastern Europe.

Rudolf Brázdil is Professor of Physical Geography at the Institute of Geography, Masaryk University in Brno. His research is focused on climate variability and climate change, with particular attention to hydrometeorological extremes during the past millennium based on instrumental, documentary and dendrochronological data in the Czech Lands and Central Europe. His scientific projects in the past years have been concentrated on historical climatology and historical hydrology.

Jiří Bruthans is a hydrogeologist and geologist at the Faculty of Science, Charles University in Prague. His research focuses on karst hydrogeology and evolution, using tracers and groundwater dating tools and study of erosion and weathering processes and feedback. Concerning geomorphology, he studied evolution of spectacular salt caves and landscapes of salt diapirs in Iran and recently focused his attention on the effect of gravity loading stress on evolution of sandstone landforms in the Czech Republic and USA.

Vladimír Cajz worked for the Institute of Geology and is currently a researcher at Institute of Geophysics, Czech Academy of Sciences. He has been involved mostly in research projects on Tertiary volcanism within the Central European Volcanic Province, focusing on its structural aspects, geochemistry and magnetostratigraphy.

Monika Chudaničová is a Ph.D. student of Environmental Geography at the Faculty of Science, University of Ostrava. Her main research interests lie in the fields of fluvial geomorphology, fluvial sedimentology and application of magnetic susceptibility.

Radek Dušek has a Ph.D. in Geodesy. He is a cartographer and geodesist at the Faculty of Science, University of Ostrava. His main scientific interest is connected with the use of DEM in geomorphological studies and analyses of old maps in the riverine landscape evolution. He is the author and co-author of many papers dealing with landform evolution in the Outer Western Carpathians.

Radomír Grygar is Associate Professor in Geology at the Faculty of Mining and Geology, VŠB—Technical University of Ostrava. He has been Head of the Department of Geological Engineering from 1999 till 2014. His principal field of interest is regional geology of Bohemian Massif, structure geology and morphotectonic analysis. He is author or co-author of many papers on structure and tectonic development of individual regions of Bohemian Massif and above all these on territory of the contact between Bohemian Massif and Outer West Carpathians.

Jan Hradecký is Associate Professor in Physical Geography and Geoecology at the Faculty of Science, University of Ostrava. He has been Head of the Department of Physical Geography and Geoecology from 2011 till 2015. From 2015 he is Dean of the Faculty of Science. His principal field of interest is slope and fluvial geomorphology, Quaternary landscape evolution and interdisciplinary approaches used in geomorphological and geoecological research. He is author or co-author of many papers on landforms evolution of flysch Outer Western Carpathians. He is the editor of journal of Geoenvironmental Disasters and reviewer of Catena, Geomorphology, ESPL, Geografiska Annaler, Progress in Physical Geography, etc.

Radim Jarošek is a senior specialist of the Landscape Protected Area Poodří. He is specialized in riverine landscape management and river restoration.

Jaroslav Kadlec is a geologist working at the Institute of Geophysics CAS v.v.i. and Associate Professor in Geology at Charles University in Prague, focused on Quaternary environmental history. His main fields of interest cover reconstruction of fluvial, lacustrine and karst processes using sedimentological and environmental magnetic techniques. He is a pioneer in multidisciplinary approach including palaeomagnetic dating applied in studies of cave deposits in the Czech Republic.

Jan Kalvoda is Professor of Physical Geography at the Charles University in Prague, Department of Physical Geography and Geoecology (Faculty of Science). He is a member of the Quaternary Palaeoenvironment Group at the University of Cambridge. His research activities are concentrated on dynamic geomorphology of tectonic active regions, Quaternary landform evolution in Europe and Asia, recent geodynamic processes, natural hazards and risks. He has worked in high-mountain ranges such as the Himalayas, Karakoram, Tian-Shan and Pamir. He has also examined the physical geography and regional geomorphology of the Czech Republic.

Karel Kirchner is Associate Professor of Physical Geography and Geoecology and Head of the Department of Environmental Geography (Institute of Geonics of the Czech Academy of Sciences). He is external lecturer at the Department of Geography of Faculty of Science, Masaryk University Brno and a member of the Working group on Geomorphosites of the International Association of Geomorphologists. His research topics are geomorphologic problems of the eastern part of Bohemian Massif and detailed geomorphological mapping, present-day geomorphologic processes in Outer Western Carpathians, as well as studies on geomorphosites in Moravia.

Marek Křížek is senior lecturer in the Department of Physical Geography at Charles University in Prague. His research interests in geomorphology focus on periglacial and glacial geomorphology and Quaternary evolution of environment, morphometrical analysis, mathematical and statistical methods in geomorphology, and exoscopy.

Lucie Kubalíková is a geomorphologist at Institute of Geonics, Czech Academy of Science, and a member of the Working group on Geomorphosites of the International Association of Geomorphologists. Her research interests focus especially on geoconservation, geotourism and geoeducation. She is the co-author of several regional studies on geosites and geomorphosites within the Czech Republic.

Jan Lenart is Assistant Professor of Geomorphology and Speleology at the University of Ostrava. He is working in crevice-type caves connected with evolution of deep-seated gravitational slope deformations in Central and Eastern Europe. His interests also focus on artificial and karst cavities.

Jan Lipina is a professional photographer. His work is focused on studio and landscape photos. He is an expert in the contemporary and historical Poodří Region.

Zdeněk Máčka is a geomorphologist in the Department of Geography of the Masaryk University in Brno. His main scientific interest is in the field of fluvial geomorphology. He is engaged in the research of interactions between in-channel large wood and hydrogeomorphological processes, historical changes of alluvial channels and hydromorphological assessment of rivers.

Pavel Mentlík is Associate Professor of Physical Geography and Geoecology at the University of West Bohemia in Plzeň where he belongs to the Centre of Biology, Geosciences and Environmental Education. His main scientific interest is glacial chronology of Central European mountainous areas and environmental changes during the Late Pleistocene. He focuses on the Bohemian Forest and the Tatra Mountains. Additionally, he deals with regional geomorphology and geodiversity of Western Bohemia.

Jan Mertlík is a geomorphologist and conservationist who worked for the Nature Conservation Agency of the Czech Republic for many years. He also promotes the area of the Bohemian Paradise in a local development agency based at Turnov. He regularly organizes field seminars called "Klokočky" focused on the spread of geological and geomorphological knowledge of this area.

Piotr Migoń is Professor of Geography at the University of Wrocław, Poland, and Director of Institute of Geography and Regional Development. His main subjects of research are rock control in geomorphology, especially in granite and sandstone areas, weathering, mass movement in mountain terrains and long-term landform evolution. The Sudetes Mountains in Central Europe is his main research area, but he was also involved in projects in other European countries (Czech Republic, Sweden, Great Britain, Portugal), in the United States, Mexico, China, Jordan and Namibia. During 1997–2001 he was the Secretary of the International Association of Geomorphologists, and its Vice-President in 2009–2013.

Jan Miklín is Assistant Professor at the Department of Physical Geography and Geoecology of University of Ostrava, with specialization on cartography, geoinformatics, landscape ecology and nature conservation.

Jan Mrlina is a geophysicist in the Department of Tectonics and Geodynamics of the Institute of Geophysics at the Academy of Sciences in Prague, Czech Republic. His research is focused on various applications of gravity surveying in geodynamics, mineral resources exploration, archaeology, geoengineering, etc. His special interest is related to volcanic structures in the Czech Republic and Greece, with constraining geophysical and geomorphological data. Besides discovering the only Quaternary maar in the Bohemian Massif, and studying the internal structure of volcanoes, he also performs gravity monitoring with the aim of detecting subsurface mass movements under the volcanic landforms.

Monika Mulková is Assistant Professor in the Department of Physical Geography and Geoecology, Faculty of Science, University of Ostrava. Her research focuses especially on using remote sensing in detecting landscape changes and landscape development. She is interested in land use/land cover changes in the coal mining regions.

Petr Neruda is an archaeologist working at the Anthropos Institute of Moravian Museum in Brno. He is interested in Palaeolithic technology, raw material distribution and settlement dynamics with special focus on Middle Palaeolithic Neanderthals. He carried out excavations of the Middle/Upper Palaeolithic Moravský Krumlov IV open-air site, the Magdalenian site in Loštice, Brno-Štýřice III and rescue excavations in the Balcarka Cave. Recently, he coordinates geoarchaeological research in the Kůlna and Výpustek caves in the Moravian Karst.

Tomáš Pánek is Associate Professor in Physical Geography and Geoecology at the Faculty of Science, University of Ostrava. His research interests involve geomorphology and Quaternary geology with a special focus on mass movements, neotectonics and palaeoenvironmental reconstructions in the Western Carpathians, Bohemian Massif, Crimean peninsula, Taurus Mountains in Turkey, Caucasus, Caspian Sea region of Kazakhstan, etc. He has published more than 40 peer-reviewed papers and acts as a reviewer in leading journals in the field of earth sciences.

Vlastimil Pilous studied Geography at the Charles University in Prague and worked in the Krkonoše National Park Management Headquarters. In his studies he deals with mesoforms and microforms (e.g. evorsion potholes) concentrating on the Krkonoše-Jizera Pluton and Crystalline as well as mid-mountain landforms affected by geomorphological processes (especially water and glacial erosion and waterfalls) and slope movements (debris flows). Having created a typology of travertine and tufa landforms, he also continuously studies historical anthropogenous (particularly mining-related) landforms in the Krkonoše Mountains and their role in mountain landscape evolution.

Petr Popelka is Associate Professor of Czech and Czechoslovak History in the Department of History, Faculty of Arts, University of Ostrava (since 2013 the head of this department) and research fellow at the Centre for Economic and Social History University of Ostrava. In his research work, he focuses mainly on the history of entrepreneurs and enterprise in the era of industrialisation, on the genesis of modern transport infrastructure in Central Europe in the eighteenth and nineteenth centuries and on landscape changes in the nineteenth and twentieth centuries.

Renata Popelková is Assistant Professor in the Department of Physical Geography and Geoecology, Faculty of Science, University of Ostrava. In her research work, she focuses mainly on landscape structure and landscape development using geoinformation technologies. In the last 10 years she has specialized in land use/land cover changes in the coal mining regions.

Pavel Raška is senior lecturer in the Department of Geography and Head of the Landscape Synthesis Research Unit at the J.E. Purkyně University in Ústí nad Labem. His research interests focus on natural hazards and risks, and landscape dynamics under human impacts. Methodologically, much of his work is devoted to the development of techniques used to analyse and interpret documentary proxies.

Václav Škarpich has a Ph.D. in Environmental Geography. He is a fluvial geomorphologist and hydrologist at the Faculty of Science, University of Ostrava. His main scientific interest is linked with geomorphology of fluvial systems, changes of river systems, restoration of gravel-bed rivers and water management of gravel-bed rivers. He is the author and co-author of several papers dealing with fluvial system evolution in the Outer Western Carpathians.

Veronika Kapustová is a geomorphologist in the Department of Physical Geography and Geoecology, University of Ostrava. Her research interests focus on hill slope-channel coupling and past and contemporary controls of landslide activity. She is also interested in the record of landscape changes in the sedimentary archives. She participated in research on landslide phenomena in the Outer Western Carpathians (Czech Republic) and the Crimean Mountains (Ukraine).

Jakub Stemberk is currently a Ph.D. student in the Department of Physical Geography and Geoecology, Faculty of Sciences, Charles University in Prague. His research topics are tectonic geomorphology, long-term morphotectonic relief evolution, especially evolution of river basins and valley forms in the Sudetes Mountains.

Petra Štěpančíková is a geomorphologist in the Department of Neotectonics and Thermochronology in the Institute of Rock Structure and Mechanics of Czech Academy of Sciences in Prague. Her research topics are tectonic geomorphology, paleoseismology, long-term morphotectonic relief evolution, and study of active faults in intraplate regions as well as within active plate boundaries (Czech Republic, Spain, Mexico, California, Israel).

Zuzana Vařilová is a geologist with focus on regional geology—especially Cretaceous sandstones and sandstone landscapes, weathering processes, rock-slope instability and landslides. She has worked more than 10 years in the Bohemian Switzerland NP Administration, where an integrated rock-fall risk management has been developed (including risk mapping and evaluation, monitoring system and remedial works). Now she works as a Curator of Geological Collections and Scientific Secretary at the Museum of the city of Ústí nad Labem. She has studied at the Faculty of Science, Charles University in Prague. A comprehensive research of sandstone rock form deterioration—case study Pravčická brána Arch was carried out within her Ph.D. study. She has participated in research projects in the Czech Republic and abroad (e.g. Ethiopian Highlands).

Marián Velešík is Ph.D. student of Environmental Geography at the Faculty of Science, University of Ostrava. His principal field of interest is riverine landscape and its ecosystem services and historical geography.

Vít Vilímek is Associate Professor at the Charles University in Prague (Department of Physical Geography and Geoecology). His research interests focus on natural hazards (e.g. GLOFs, landslides) and geomorphological aspects of neotectonics. He works currently as head of the Czech Geomorphological Association. His areas of expertise are the Peruvian Andes (Cordillera Blanca, Machu Picchu), Ethiopian Highlands and selected regions in Europe. Recently a new cooperation in the Patagonian Andes has been opened. He initiated creation of GLOFs Database running under ICL (International Consortium on Landslides) in the frame of World Centre of Excellence on Landslide Disaster Reduction.

Jan Vítek is Associate Professor of Physical Geography at the Faculty of Science, University of Hradec Králové, Czech Republic. His research focuses on mesoforms and microforms on the rock surface, incurred by weathering of sandstones, marlites, granitoids, metamorphic rocks and others. He designed typology of pseudokarst forms in the Czech republic.

Karel Žák is a geologist and geochemist at the Institute of Geology of the Czech Academy of Sciences. His main scientific interests are in the application of isotope geochemistry in environmental studies, including study of karst processes. During the last several years, his studies focused on processes occurring in caves during glacials. He is also interested in hydrothermal processes and ore deposit formation.

1 Introduction

Tomáš Pánek and Jan Hradecký

Although the territory of the Czech Republic lacks numerous attractive types of landscapes which are common in the neighbouring European countries (e.g. rugged alpine mountains, floodplains of large rivers or scenic coastlines), it is a country of exceptionally diverse landforms. Owing to its position in Central Europe, landscapes and landforms of the Czech Republic bear imprints of the intersection of major Eurasian geotectonic and bioclimatic domains. Strongly denuded Proterozoic and Paleozoic basement of the Variscan Bohemian Massif meets here with the young Cenozoic fold-and-thrust belt of the Western Carpathians. Sudetic Mountains with tundra-like landscapes and periglacial phenomena on watersheds are in contrast to vineyard-covered limestone hills in Southern Moravia resembling the Mediterranean realm. Due to complex geotectonic and geomorphological evolution, many landscapes in the Czech Republic can be classified as geo-regions of first-order importance. The area of Late Cretaceous transgression in the Central and Eastern Bohemia includes spectacular "rock cities" which are without any doubt most impressive in Europe and some of the largest on the world-wide scale. Some of the regions in the Czech Republic obtained their credit due to the traditional history of scientific research. It pays e.g. for loess sections in Southern Moravia, which are among the most investigated and important natural archives of Late Quaternary climate changes in Eurasia. Other examples are small young volcanoes in Western Bohemia (especially Komorní hůrka/Kammerberg volcano) which were fascinating for J.W. Goethe at the beginning of the nineteenth century and inspired him to contribute to the theories of neptunism and plutonism, i.e. great geological paradigms of that time. A possibility to expand long-term research in a geologically and geomorphologically rich country is still highly attractive for geoscientists from all around the world.

The Czech Republic is located in the contact zone of four important European geomorphological provinces—Bohemian Highlands, Central Polish Lowland, Western Carpathians and West-Pannonian Basin (Czudek 2005). The Bohemian Highlands occupy 84 % of the total area of the Czech Republic and their origin is related to the Variscan orogeny at the end of the Paleozoic era. Thereby, they represent the largest and at the same time the oldest geomorphological province of the Czech Republic containing the highest peak of the CR—Sněžka Mt (1603 m) as well as other high mountain ranges of the CR—Hrubý Jeseník Mts (Praděd Mt 1491 m), Králický Sněžník Mts (Králický Sněžník 1423 m) and Šumava Mts (Plechý Mt 1378 m). The altitude of the CR territory ranges from 115 m (Labe/Elbe River valley near Hřensko) up to 1602 m (Sněžka Mt in the Krkonoše Mts). Western Carpathians (c. 9 % of the CR total area) are built partly by a system of hilly lands, highlands and uplands in the easternmost part of the republic and partly by a system of lowlands (the so-called Carpathian Foredeep) in the mountain range foreground. West-Pannonian Basin occupies 6.5 % of the total CR area. It represents lowland (exceptionally hilly) relief overreaching into the Czech territory from the territory of Austria and Slovakia in a form of Dolnomoravský úval and Dyjskosvratecký úval basins. Central Polish Lowland occupies as little as about 0.5 % of the total state territory.

Czech geomorphologists (Demek et al. 2006) proposed a very detailed geomorphological division of the country. This approach is based on the delineation of morphostructures, morphography and general morphometric properties. The scheme of the geomorphological division is presented in Fig. 1.1 where localisation of geomorphological provinces and geomorphological units is shown.

This book is not intended to act as a textbook whose aim is to systematically describe all regions in the Czech Republic. A lot of basic regional-geomorphological units displayed in Fig. 1.1 are not included in the text. Rather,

T. Pánek (✉) · J. Hradecký
Department of Physical Geography and Geoecology,
Faculty of Science University of Ostrava, Ostrava, Czech Republic
e-mail: tomas.panek@osu.cz

J. Hradecký
e-mail: jan.hradecky@osu.cz

T. Pánek and J. Hradecký (eds.), *Landscapes and Landforms of the Czech Republic*,
World Geomorphological Landscapes, DOI 10.1007/978-3-319-27537-6_1

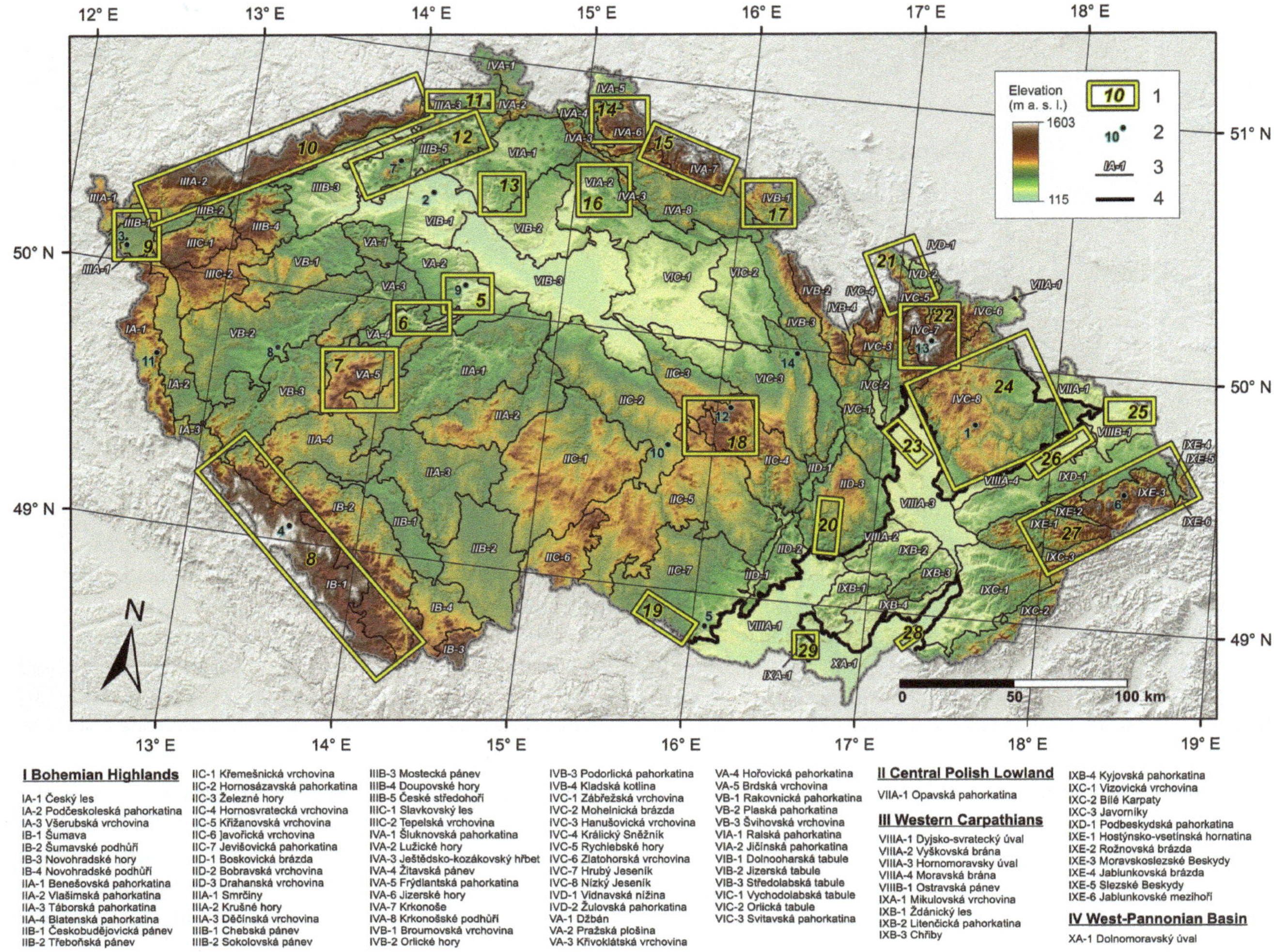

Fig. 1.1 Geomorphological division of the Czech Republic based on Demek et al. (2006): *1* localisation of the studied geomorphological landscapes in this book—numbers represent the number of a given chapter; *2* climatological stations—for details see Chap. 3 and Table 3.1; *3* borders of geomorphological units; *4* border of European geomorphological provinces within the Czech Republic)

the book covers those landscapes which best represent both geomorphological particularities and diversity of the country. It involves areas with magnificent landscapes forming natural background, famous historical sites and scientifically valuable regions representing geosites that are of crucial importance for the understanding of geomorphological evolution of Central Europe. A special intention was to include geomorphological regions from where recent intensive investigations have brought a new insight into the geodynamics of the region. In such a way, the book emphasizes internationally well-known geomorphological highlights of the country (e.g. sandstone "rock cities"; Chaps. 11, 13, 16 and 17), but it also tends to describe so far less popular landform assemblages (e.g. abundant cryogenic landforms in the Brdy area which have been hidden to the public for decades due to the presence of the army; Chap. 7).

The book is divided into three sections. In the first section, it introduces the main aspects of geology and tectonic development (Chap. 2). Then, it deals with the past and recent climatic pattern of the country (Chap. 3) and finally, it summarizes long-term geomorphological history of the Czech Republic as an interplay of complex morphotectonic evolution, rock control and Cenozoic climatic oscillations (Chap. 4).

The core of the book is the second section, which represents 25 geomorphologically most spectacular regions of the Czech Republic. Starting in the historical heart of the Czech Republic and one of the most impressive towns in Europe, Prague (Chap. 5), it describes the most interesting geomorphological landscapes of the Central Bohemia such as Bohemian Karst (Chap. 6) and Brdy Highland (Chap. 7). From the point of view of tectonics, these regions belong to the most stable regions in Central Europe. Thanks to the presence of planation surfaces, unique karst landforms and complete assemblage of Quaternary river terraces, they host some of the most valuable imprints of the long-term landscape evolution in the Czech Republic. Long-term landscape evolution is also characteristic of the Bohemian Forest (Chap. 8), the most extensive forest landscape in Central Europe situated in the borderland with Bavaria (Germany). The area is strictly protected on both sides of the border via the Bohemian Forest National Park in the Czech Republic and Bavarian Forest in Germany. Due to considerable neotectonic uplift and survival of extensive summit plateaus reaching far above the Pleistocene snow line, this mountain range is one of the three regions in the Czech Republic that bear clear geomorphological markers of the existence of Pleistocene mountain glaciers.

Moving to the west, we can study geodynamically most active landscape in the country. It is evidenced by the occurrence of the youngest Pleistocene volcanoes and current seismicity in the area surrounding the Ohře/Eger rift (Chap. 9). Another morphostructural expression of this dramatic geodynamic setting is the area of the Krušné hory Mts, the longest mountain range with some of the most expressive fault scarps in the Czech Republic (Chap. 10). In the NE continuation of the Krušné hory Mts, we can observe a highly diverse landscape which originated as a result of intensive neotectonic movements and multi-stage Tertiary volcanic activity. It is a geological domain with pronounced rock control and presence of Late Cretaceous sandstones. One of the most valuable geomorphological landscapes of the Czech Republic is the area of the Elbe Sandstones (Chap. 11) involving the deeply incised valley of the Labe/Elbe River and sandstone "rock cities" in the Bohemian Switzerland National Park with some iconic sandstone landforms (e.g. Pravčická brána Arch). The nearby neovolcanic terrain of the České Středohoří Mts (Chap. 12) is another excellent demonstration of rock control in the landscape, but in this case it is connected with the exhumation of subvolcanic bodies, recently shaped by numerous large landslides. The Kokořín area (Chap. 13) is a much smaller unit, but it is a unique demonstration of how hydrothermal activity related to Tertiary volcanism affected lithological properties of Cretaceous sandstones and lead to the evolution of specific microforms such as mushroom rocks.

Jizerské hory Mts (Chap. 14) represent a true Sudetic landscape, typical of the northern borderland of the Czech Republic. High-elevation planation surfaces with numerous granite tors alternate with steep fault slopes hosting plenty of waterfalls and traces of active debris flow processes. Such features are even more noticeable in the Krkonoše Mts, the highest and formerly most glaciated mountain range of the Czech Republic (Chap. 15). Due to their extraordinary natural heritage, the Krkonoše Mts were established to be the first national park in the country in 1963. To the south of the Jizerské hory and Krkonoše Mts, a large exceptional landscape with several sandstone rock cities and neovolcanic bodies can be visited. Known as Bohemian Paradise (Chap. 16), it was included (as the only region in the Czech Republic) into the European Geopark Network due to its outstanding geological, geomorphological and historical heritage. The last region containing the phenomena of sandstone rock cities is the area of Adršpach–Teplice Rocks and Broumov Cliffs (Chap. 17), a system of elevated tablelands with characteristic mesas and cuestas, escarpments dissected by numerous rock columns, pillars, gorges and pseudokarst caves.

With the geomorphological landscape of the Žďárské vrchy Highland (Chap. 18), our book enters the territory of Moravia. Žďárské vrchy Highland represents a spectacular region at the main watershed between the Northern and Black Sea and gives a proof that in specific circumstances "rock cities" could develop even in crystalline rocks. Another landscape with a strong geomorphological imprint of crystalline rocks is the Dyje canyon-like valley, included in the Podyjí National Park just at the border with Austria (Chap. 19). It is one of famous deeply incised river valleys at the SE marginal slope of the Bohemian Massif with rock slopes and mysterious pseudokarst ("ice") caves. Continuing to the NE, adjacent to the northern suburb of the Brno city (historical capital of Moravia), there is the most extensive karst landscape in the Czech Republic—the Moravian Karst (Chap. 20). With a complete list of exo- and endokarst phenomena on Devonian limestones, numerous examples of Early Cretaceous paleokarst and long tradition of geo-archaeological investigation connected especially with the name of K. Absolon (1877–1960), this territory is on the top from the point of view of the significance for the understanding of the long-term geomorphological history of the Czech Republic.

As for the Rychlebské hory Mts (Chap. 21), we have to move to the Sudetic north again. This landscape is spectacular for a high diversity of granite landforms, but what makes this region particularly valuable from the scientific point of view is the recent discovery of Late Pleistocene/Holocene tectonics along the Sudetic Marginal Fault. In this respect, tectonic landforms along the foot of the Rychlebské hory Mts are among the youngest ones in Central Europe and they demonstrate ongoing crustal deformations even in the domain of the old cratonised Bohemian Massif. The mountain region of the Hrubý Jeseník Mts (Chap. 22) represents the second highest range in the Czech Republic and despite its limited glaciation during the Pleistocene; by far, the most conspicuous geomorphological phenomena of this area are numerous periglacial features such as cryoplanation terraces, thufurs, sorted polygons and other microforms. Although nearby to the Hrubý Jeseník Mts, the region of Litovelské Pomoraví (Chap. 23) represents a completely different picture—alluvial landscape close to the outflow of the Morava River from a mountainous catchment. It is the only example of an anastomosing river pattern in the Czech Republic with excellent riparian habitats. Further to the east, the Nízký Jeseník upland (Chap. 24) forms a vast elevated plateau at the margin of the Bohemian Massif. It is marked by several Quaternary volcanoes and in many sites "penetrated" by large underground mines giving an insight into the interior of the rock massif formed by folded Carboniferous shales.

The remaining chapters describe the Western Carpathians—the youngest geological domain of the Czech Republic. The dirty mining landscape of the Ostrava-Karviná region (Chap. 25) is an example of drastic anthropogenic transformation of an originally agricultural landscape of the Carpathian Foredeep. This process lasted less than 200 years, but fortunately recent revitalisation of the devastated area has gradually been returning the nature back to this landscape. In comparison with this "black land", the stretches of the Odra River SW of the city of Ostrava (Chap. 26) looks like a pristine landscape. It is perhaps the best example of a meandering river belt in the Czech Republic as it survived decades of river training. The highest area of the Western Carpathians in the Czech Republic is represented by the Moravskoslezské Beskydy Mts and surrounding mountains (Chap. 27). Formed by folded and thrust flysch rocks, this region is especially prone to landslide activity, which represents a natural hazard, but it also tends to increase geodiversity of the landscape with numerous rock slopes, block fields and long pseudokarst/crevice-type caves. The southernmost part of the Czech Republic (Southern Moravia) is a domain of vast alluvial plains and rolling hills with vineyards, a landscape resembling Southern Europe. The most impressive alluvial plain here is situated along the lower course of the Morava River (Chap. 28). This is a semi-natural riverine landscape characterised by recent dynamic changes of channel morphology, but it also contains an exceptional sedimentary record of the Late Pleistocene-Holocene floodplain evolution. Nearby lowland covered by numerous Late Glacial sand dunes (often called "Moravian Sahara") is a unique example of aeolian landforms in the Czech Republic. Southernmost hilly landscape belonging to the Western Carpathians in the Czech Republic is represented by the Pavlov Hills (Chap. 29). These Jurassic limestone hills are a part of the "bridge" between Eastern Alps and Western Carpathians. The region provides excellent manifestations of rock control dominated by rigid limestone beds, but it also reveals good examples of tectonic landforms and contains some of the most important loess sequences in Europe.

The final chapters (Chaps. 30 and 31) provide to the readers the main aspects of geo-conservation and geotourism in the Czech Republic. They emphasize a long tradition of landscape protection in the country and also recent tendencies to prepare inventories of especially valuable landforms ("geomorphosites") as well as their inclusion in the network of protected areas. Some of the regions make use of their geoheritage in establishing geoparks which promote geodiversity and geoheritage to the public and accelerate tourism in particular areas.

We would like to express many thanks to Prof Piotr Migoń for offering us to participate in the editorial work of this volume and for his excellent supervision of this book series. We acknowledge the technical support of Dr. Veronika Kapustová and Dr. Jan Miklín. Many thanks belong to Monika Hradecká for her help with English corrections. We would like to warmly thank all contributors. It was a real pleasure to cooperate on such an exceptional book describing geomorphological and landscape beauties of our country.

References

Czudek T (2005) Vývoj reliéfu krajiny České republiky v kvartéru. Moravské zemské muzeum, Brno, 240 pp

Demek J, Mackovčín P, Balatka B, Buček A, Cibulková P, Culek M, Čermák P, Dobiáš D, Havlíček M, Hrádek M, Kirchner K, Lacina J, Pánek T, Slavík P, Vašátko J (2006) Hory a nížiny. Zeměpisný lexicon ČR. AOPK ČR, Brno, 582 pp

Part I

Physical Environment

Geology and Tectonic Development of the Czech Republic

2

Radomír Grygar

Abstract

Even if the Czech Republic occupies a small area in Central Europe, it is unique by the very interesting and varied geological and tectonic development that is recorded in the structure of the present-day Earth's crust, especially in the case of the Bohemian Massif. The Bohemian Massif can be interpreted as a heterogeneous unit composed of four separate regional domains. Each of them is defined especially by a specific stratigraphic content, tectomagmatic development and tectonic limitation in relation to its surroundings. The history of its development involves a long time period from the Paleoproterozoic to the recent period, i.e. about 2.1×10^9 years. Basic features of the Earth's crust structure, reflecting in geological maps, were however impressed on the area of the country only by relatively younger phases of Variscan orogeny and, to a lesser extent, Alpine orogeny that affected the eastern part of the country—the Western Carpathians. At the beginning of the Westphalian, the Bohemian Massif became part of the stabilised Variscan crust of the West European Platform, which in consequence meant that it began to act as a single unit, in which any mutual lateral displacement of units, metamorphosis and associated ductile deformation took place no longer. The Western Carpathians are one of partial branches of the vast orogenic belt of the Alpides created from the former Tethys Ocean. The development of the Western Carpathians already begins shortly after terminating the Variscan orogeny. At present, the Carpathians are divided from south to north into the Inner, Central and Outer Western Carpathians. The Central as well as the Inner Carpathians do not occur in the territory of the Czech Republic. The younger accretionary complex in the area of Moravia and Silesia is composed of the Pouzdřany, Ždánice, Subsilesian, Silesian and Fore-Magura Units.

Keywords

Bohemian Massif • Variscan orogeny • Epi-Variscan Platform • Alpine orogeny • Western Carpathians • Tectonic development

2.1 Introduction

The Czech Republic occupies a comparatively small area in Central Europe. In addition to other particularities, it is unique by the very interesting and varied geological development that is recorded in the structure of the present-day Earth's crust, especially in the central and western part of the territory—the Bohemian Massif. According to current knowledge, the history of its development involves a long time period from the Paleoproterozoic to the recent period, i.e. about 2.1×10^9 years. Basic features of the Earth's crust structure, shown on geological maps (Fig. 2.1), were however impressed on the area only by

R. Grygar (✉)
Institute of Geological Engineering, Technical University of Ostrava, 70833 Ostrava, Czech Republic
e-mail: radomir.grygar@vsb.cz

T. Pánek and J. Hradecký (eds.), *Landscapes and Landforms of the Czech Republic*, World Geomorphological Landscapes, DOI 10.1007/978-3-319-27537-6_2

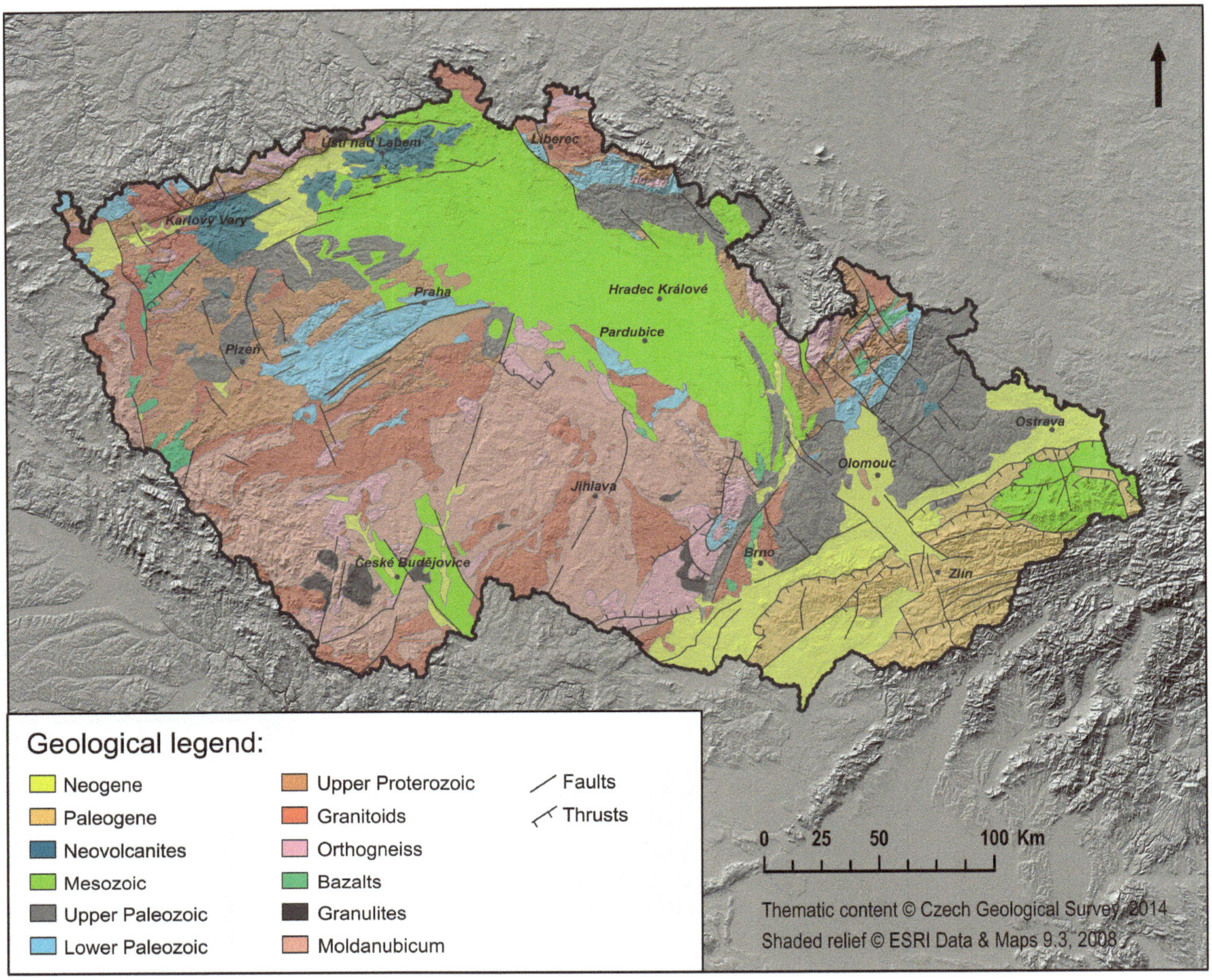

Fig. 2.1 Simplified geological map of Czech Republic draped over digital elevation model (Original made by Czech Geological Survey, courtesy of Lucie Kondrová)

relatively younger phases of geological development—Variscan orogeny and, to a lesser extent, Alpine orogeny that affected primarily the eastern part of the country—the Western Carpathians.

The territory of the Czech Republic is, besides its geographical and geopolitical position, a significant area in the geological pattern of Europe. In Moravia, two parts of Europe vastly different in age, geological development and geophysical parameters of the Earth's crust, meet. Bohemia and part of western Moravia and Silesia are a portion of the Bohemian Massif, one of the most significant and extensive fragments of the Variscan orogen formed during the Devonian and the Carboniferous (over the interval c. 380–320 Ma) by collision of the peri-Gondwana microcontinents (i.e. microcontinents situated in the early Paleozoic originally along the northern edge of the Gondwana continent) with Avalonia and Baltica (East European Platform).

The eastern part of Moravia and Silesia belong to the Western Carpathian orogen, which is one of sub-parts of the Alpine orogen—a vast mountain system of southern Europe. It was formed by collisions of continental fragments situated between the northern edge of Africa and the so-called Epi-Variscan Platform of Western Europe during the Mesozoic and the Tertiary.

The aim of the following chapter is thus to characterise the specific features of development and detailed division of both the above-mentioned different geological units constituting the territory of the Czech Republic.

2.2 The Bohemian Massif

The Bohemian Massif is one of the largest continuously outcropping fragments of the originally vast Variscan orogen that crops out from the basement of younger Epi-Variscan Platform sediments. The comparatively extensive Variscan orogen was formed gradually in the course of joining the peri-Gondwana fragments to (see Franke 1989; Franke et al. 2000; Matte et al. Laurussia1986, 1990, 1991; Winchester 2002; Ziegler 1982, 1984), i.e. to the more northerly situated continent created as the result of Caledonian convergence between Laurentia and Baltica.

Based on the current concepts of development of continents (Condie 1989), which start from the application of principles of plate tectonics, the Bohemian Massif can be interpreted as a heterogeneous unit composed of four separate regional domains. Each of them is defined especially by a specific stratigraphic content, tectomagmatic development and tectonic boundaries in relation to its surroundings.

According to the binding regional geological division of the Bohemian Massif (Commission 1994), the Bohemian Massif can be divided, based on the differences in structure and geological development, into four autonomous regions: Moldanubian Unit, Teplá-Barrandian Unit, Saxo-Thuringian Unit (subdivided by the younger Elbe Fault Zone into the Krušné hory Mountains Zone and the Lugicum = West Sudetic Zone) and Moravo–Silesian Unit (Fig. 2.2). This basic division reflects the existence of four independent

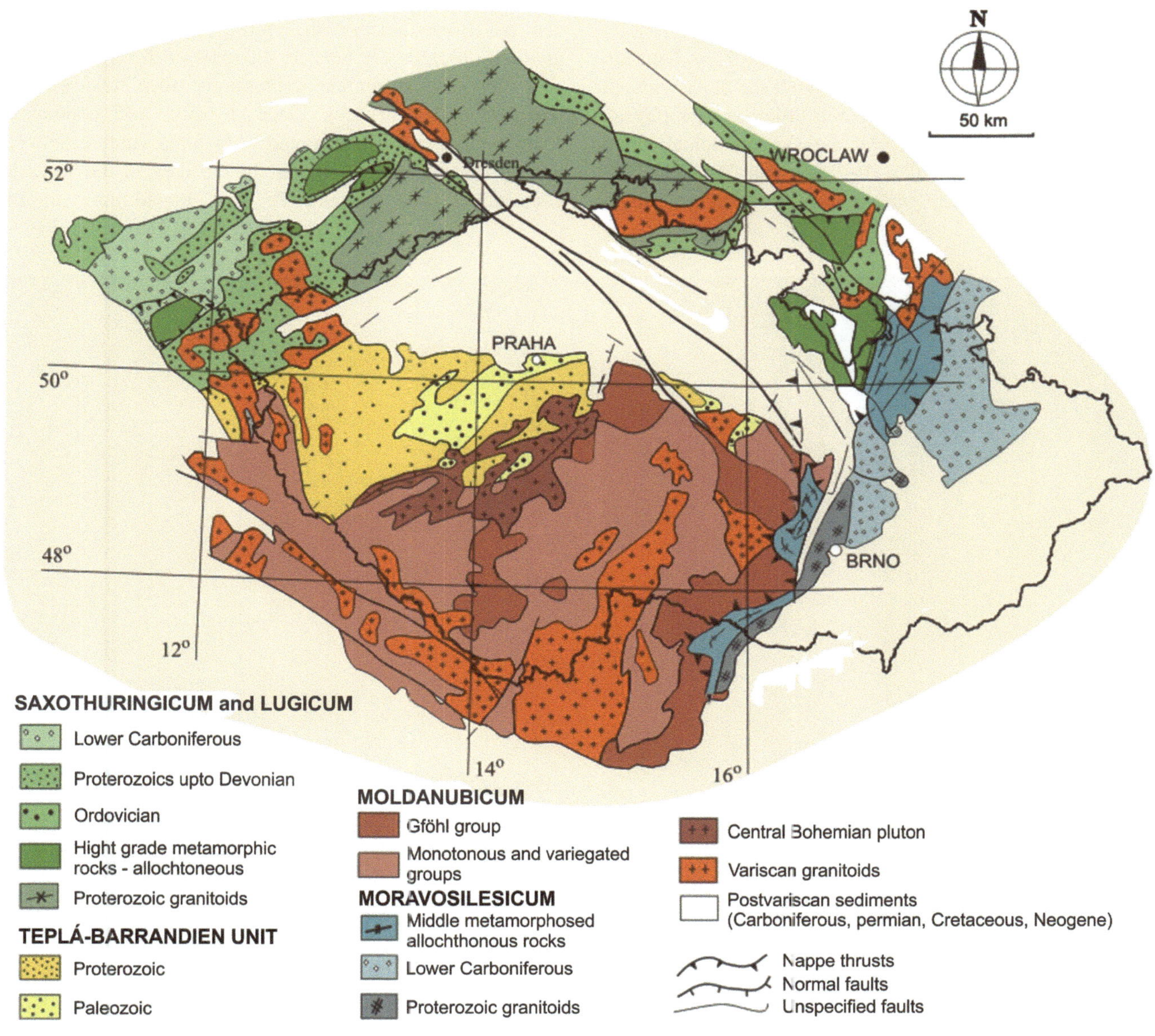

Fig. 2.2 Main Variscan regional units of the Bohemian Massif (Simplified according to Schulmann et al. 2005)

crustal fragments separated originally by oceanic domains, the traces of which are indicated today by the occurrences of ophiolite complexes and/or belts with high-pressure and mantle rocks (Mariánské Lázně complex, Letovice ophiolite complex, blueschists of Železný Brod crystalline complex and Rýchory ridge, high-temperature and high-pressure rocks of the Gföhl Unit). Paleomagnetic data (Krs et al. 2001; Tait et al. 2000), a number of common features in the development of the Cadomian basement, the presence of Cadomian calc–alkaline granitoids, formed by melting rocks above a subduction zone and flysch sequences deformed during the Cadomian orogeny, document that these units were part of a belt of island arcs and perhaps accretionary complexes on the northern edge of Gondwana. In the boundary period between the Proterozoic and the Paleozoic, this belt was situated in the area of low southern latitudes (Fig. 2.3).

It is the independence of the Moldanubian Unit that remains problematic. In comparison with the other units, the Moldanubian Unit has a different lithology, geophysical characteristics of the crust and the subcontinental mantle (Beránek and Dudek 1981; Beránek and Zátopek 1981; Babuška and Plomerová 2001), tectonic boundaries (with the Moravo–Silesian Unit and the Teplá-Barrandian Unit are quite evident) and above all different metamorphic development, given by the substantially deeper present-day denudation level.

In spite of the fact that the above-mentioned units are separated by significant sutures and tectonic zones, they have a number of common features related to the Neoproterozoic and, in part of them, also in the Cambro–Ordovician development; on the contrary, they differ markedly as far as the early stage of Paleozoic development during the Variscan orogeny is concerned.

After the end of the Variscan orogeny, the Bohemian Massif was gradually transformed into a platform unit. Paleomagnetic data for the Early Permian document that all western, central and northern Europe behaved as a single unit designated the North European Platform. It also included the Bohemian Massif. During the Carboniferous, it gradually became dry land. In the most of the area, with the exception of intramontane depressions, deep erosion of the Variscan basement took place. Erosion and continental sedimentation were interrupted only for a short time by a marine transgression over part of the area in the Jurassic, Cretaceous and Neogene Periods. In addition to the

Fig. 2.3 Proterozoic biotitic gneiss of Gföhl group of Moldanubicum. Locality Náměšť nad Oslavou (*Photo* R. Grygar)

deposition of sediments in the depressions, various types of volcanic bodies of Cretaceous to Quaternary age provided the finishing touches to the surface of the platform cover of the Bohemian Massif.

The basis of regional division of the Bohemian Massif is natural geological boundaries represented by significant sutures, and also other types of tectonic boundaries separating microcontinents (also smaller units, so-called terranes) with different paleogeographical provenances, lithologies and tectonometamorphic effects, ages of rock complexes, and maybe with different characters of magmatic manifestations.

The pre-Variscan geodynamic development of the units can be divided into two phases: Neoproterozoic and Paleoproterozoic. The Neoproterozoic development is evidenced best in the Teplá-Barrandian Unit (Fig. 2.4) and in the Saxo-Thuringian Unit reworked slightly by the Variscan orogeny. In a lithological record, a transition to the regime of active subduction in the upper part of the Neoproterozoic (Kralupy-Zbraslav Group), which was accompanied by the formation of island arcs and subsequently of an accretionary wedge of flysch sediments above the subducting oceanic lithosphere, is evident. During the Cambrian, the subduction died away and the active margin was transformed to a passive margin. Parts consolidated by the pan-African orogeny began separating from the mother continent, Gondwana, in the Cambrian. A system of rifts was formed; along them, the comparatively continuous Avalonian-Cadomian belt of microcontinents was broken up.

Fig. 2.4 Outcrops of Proterozoic silicites from Teplá-Barrandien region. Lokality Hudlice (*Photo* R. Grygar)

The paleomagnetic, paleobiogeographical data and analysis of clastic micas and zircons show that East Avalonia and fragments outcropping on the eastern periphery of the Bohemian Massif separated first and at the fastest rate (Bruno-Vistulicum, Malopolska Massif and also Lysa Gora Unit of Holy Cross Mountains (Góry Swietokrzyskie)—e.g. Belka et al. 2002). The Malopolska Massif and the Bruno-Vistulicum came to direct contact already during the Cambrian.

Later, at the Cambrian–Ordovician boundary, the remaining Armorican microcontinents began separating as well. The separation of these microcontinents is also indicated, in addition to the paleomagnetic records, by extensive bimodal rift volcanism and magmatic activity, the beginning of which falls within the period from 520 to 480 Ma (i.e. Cambrian–Ordovician boundary). This magmatism is observable above all along the rims of gradually separating blocks in the whole area of the Armorican group of continents. Changes in tectonic regime at the Cambrian–Ordovician boundary caused that the Cambrian sedimentation cycle, connected partially with the Cadomian development, was sharply separated from the Ordovician–Devonian cycle in the Teplá-Barrandian Unit so as in the Saxo-Thuringian Unit. An acceleration of the expansion of the originally continental rifts then gradually led to the formation of the Rheic Ocean and, on the contrary, to the closure of the Tornquist Ocean.

From the paleomagnetic data and paleoclimatic indicators (Krs et al. 2001) it is obvious that, e.g. the Teplá-Barrandian Unit moved gradually from lower southern latitudes (about 40° South Latitude in the Ordovician, 20° South Latitude in the Silurian) to the northern hemisphere. In the Devonian, it was located in the tropical equatorial region; during the Early Devonian, it crossed the equator. The extensional regime in these continental fragments continued till the Early Devonian, when the tectonic regime began to change, with the onset of the Variscan orogeny, into compressional one. For this reason, the early Paleozoic sequences that deposited along the passive margins of these fragments in the majority of the units are largely continuous.

The Variscan orogeny was a result of collisions of the Armorican microcontinents and their final amalgamation to Avalonia and Bruno-Vistulicum. In the Bohemian Massif, it is a case of the collisions between the Teplá-Barrandian plate and the Moldanubicum and Saxo-Thuringicum, followed with the final amalgamation of the above-mentioned units to the Variscan foreland formed in N and NW by the block of

East Avalonia consolidated by the Caledonian orogeny and in NE by the Bruno-Vistulicum. Differences in the Variscan development of the four basic units of the Bohemian Massif are given by their different positions in the Armorican group of microcontinents, which entered to the processes of Variscan collisions at considerable intervals. The geometry of subduction zones determined, to a certain extent, even later processes of continental collisions, especially the vergence of overthrusting movements on the boundaries of colliding fragments. Sutures that controlled the processes of Variscan orogeny were as follows: Gföhl suture and its equivalents in the French Massif Central (south Brittany) and the Iberian Peninsula (Galician), Teplá suture (suture between the Teplá-Barrandian Unit and the Saxo-Thuringicum) and Rheic and Rheno-Hercynian sutures, between Avalonia and the northern margin of the Saxo-Thuringicum. First the Gföhl suture between the Moldanubian microblock and the Teplá-Barrandian microblock was closed (based on an analogy with the French Massif Central, this happened probably in the period from the Silurian till the Early Devonian). From the suture, metamorphic complexes mostly of Precambrian to Early Paleozoic age were thrust out towards the south and in the case of the Bohemian Massif towards the south-east. The thickened Moldanubian crust was strongly heated, which led to the origin of extensive granitoid bodies in the Early Carboniferous. A rapid exhumation of the thickened orogenic root caused its deep erosion up to the level of the middle crust. In consequence, less metamorphosed and unmetamorphosed supracrustal units are missing in the Moldanubicum.

The Teplá suture, forming the present-day geological boundary between the Teplá-Barrandian microblock and the Saxo-Thuringicum, was also closed in the Devonian; an obduction of high-pressure rocks took place at the end of the Middle Devonian (c. 380–370 Ma). Rocks from rather deep parts of the Saxo-Thuringian Ocean and both continental margins were thrust out towards NW over the Saxo-Thuringian autochthon. The outermost sutures of the Variscides are the Rheic and Rheno-Hercynian sutures. The older Rheic suture, which was closed already during the Devonian, is indicated by calc-alkaline volcanism and high-pressure low-temperature (HP–LT) metamorphism in the area of so-called Northern Phyllite Zone on the boundary of the Saxo-Thuringicum and the Rheno-Hercynicum. The equivalent of the Rheno-Hercynian suture in the Bohemian Massif is probably complexes on the boundary of the Bruno-Vistulicum and the Lugodanubicum, of which the Devonian–Carboniferous, mostly flysch complexes were thrust out towards E over the Bruno-Vistulian foreland.

The gradual migration of tectonic deformation in time and space from south to north together with different geometries of the main zones of shortening created the characteristic fan-like zonal structure of the Variscan orogen as defined already in classic papers (Suess 1926; Kossmat 1927; Stille 1951). Based on the age of protoliths of rocks of the basement and the Variscan mantle, main tectonic deformation phases, intensity of metamorphism, pre- and post-Variscan magmatic manifestations, the following zones can be defined from south to north: Moldanubian Zone, Saxo-Thuringian Unit, Rheno-Hercynian Zone and Subvariscan Foredeep; they can be observed in all the European Variscides (Cháb et al. 2010).

The Moldanubian Zone is characterised by an inverted internal metamorphic structure, high intensity of metamorphism, presence of HP–HT rocks (Fig. 2.3), which differ from similar rocks in other zones by higher temperatures and pressures of equilibration of high-pressure parageneses. The Teplá-Barrandian Unit, which represents together with the Armorican Massif the best preserved relic of the Cadomian crust, covered partly with discordantly laid unmetamorphosed Early Paleozoic sequences, was earlier regarded either as part of the Moldanubian Region (Kossmat 1927; Franke 1989), or as part of the Saxo-Thuringian Unit (Ellenberger and Tamain 1980; Mísař et al. 1983). It follows from the paleomagnetic data and well-documented suture lines, which delimit it, that these units have, in the framework of the orogen, independent positions. The termination of sedimentation in the Devonian and the main phase of deformation between the Givetian and the Fammenian make it different from the surrounding units, likewise the presence of the fundamental complex slightly reworked by the Variscan orogeny. The metamorphic development, in contrast with the Saxo-Thuringian Unit and the Moldanubian Region, is caused by Early Carboniferous subsidence along the extensive West Bohemian and Central Bohemian shear zones (Zulauf 1994).

The Saxo-Thuringian Unit is characterised, in comparison with the Teplá-Barrandinian Unit, by higher intensity of Variscan reworking of the Cadomian basement, by largely continuous unmetamorphosed to weakly metamorphosed sequences of the Paleozoic in the period from the Cambrian to the Lower Carboniferous (Fig. 2.5) and by Devonian to Carboniferous extension accompanied by intraplate volcanism. A characteristic feature is the presence of allochthonous relics that were thrust out from the Teplá suture and occupy the highest structural position, and the presence of granulite complexes underlying the Lower Carboniferous flysch units.

The Rheno-Hercynian Zone represents a predominantly Devonian–Carboniferous accretionary complex thrust out from the Rheno-Hercynian suture between Avalonia and the Saxo-Thuringicum. Older rock complexes crop out at the surface only rarely. The Zone is characterised by weak metamorphism and fold–thrust structure.

The Variscan Foredeep (Subvariscicum) represents a classical foreland basin formed by lithospheric flexure before the fronts of nappes of the Rheno-Hercynicum

Fig. 2.5 Outcrop of Ordovician quartzite belongs to allochtonian sequences of Krkonoše–Jizera units in the Lugicum region. Locality under the Ještěd Mt. (*Photo* R. Grygar)

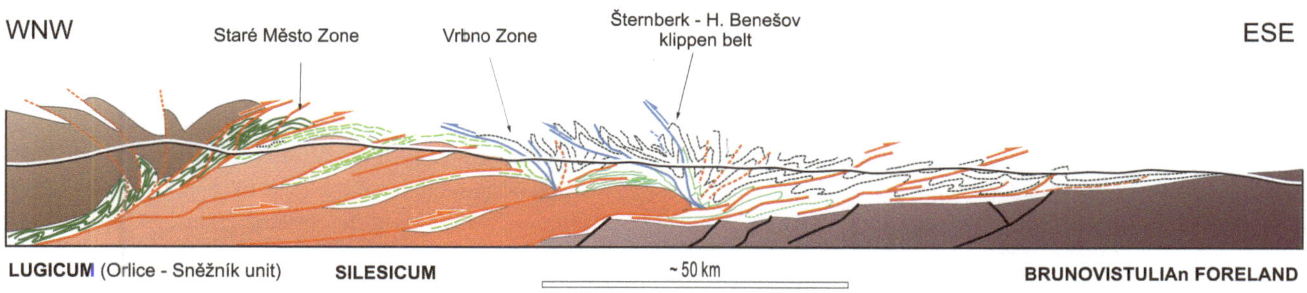

Fig. 2.6 Schematic structural cross section of the Variscan accretion wedge of the Moravo-Silesian unit. According to Grygar and Vavro 1995

finishing their thrust over the Avalon-Bruno-Vistulicum foreland (Figs. 2.6 and 2.7). The stratigraphic range of a molasse, first marine and later continental, is from the Namurian to the Westphalian. Variscan folds disappeared in the course of its filling. The character of fauna and flora shows that at that time oceanic barriers no longer existed in Europe and that newly created Variscan Europe formed one unit with the Gondwana.

2.3 *The Post-Orogenic—Platform Development of the Bohemian Massif*

Basic features of the structure of basement of the Bohemian Massif were formed during the Variscan orogeny. At the beginning of the Westphalian, the Bohemian Massif became part of the stabilized Variscan crust of the West European

Fig. 2.7 Fold and thrust structures of the Variscan flysh foredeep with eastward vergency. Locality Stará Ves near Bílovec (*Photo* R. Grygar)

Platform, which in consequence meant that it began to act as a single unit, in which any mutual lateral displacement of units, metamorphosis and associated ductile deformation no longer were taking place. The majority of later deformations are brittle, when vertical (largely in the order of several hundred metres to several kilometres) and/or lateral movements (in the order of kilometres to several tens of kilometres as a maximum) occured. Mostly it is the case of faults and tectonic zones, created by the Variscan orogeny and later reactivated, which reacted to changes in the stress regime in the lithosphere during the Mesozoic and the Tertiary in the course of so-called Saxonian tectonics due to deformations in the foreland of the Alpine orogen. The most significant lines are NW–SE faults (Sudetic direction), parallel to the Tornquist line, NE–SW faults (Krušné hory Mts. direction) and NNE–SSW faults (originated at the end of the Variscan phase as so-called furrows—Boskovice, Blansko, Jihlava). The Bohemian Massif is segmented by these faults into a series of blocks that show different characters of dominant movements in different phases. Platform sediments are only exceptionally affected by large-wavelength flat-lying folds, such as folds in the Cretaceous in the surroundings of the Orlice Basin, Hořice Ridge, etc.

A transition from the orogenic phase to the post-orogenic phase took place during the Westphalian (Late Carboniferous), in the course of which ductile deformations in the area of Variscan foreland basins terminated. After thickening the Variscan crust in the compressional phases of the Variscan orogeny, a gravitational collapse of the orogen happened. It was accompanied by the formation of asymmetrically bounded inner continental molasse basins (Fig. 2.8). The basins are often created by crustal subsidence along the originally compressional structures (Mattern 2001).

The Variscan molasse basins can be divided into two groups: the older group of intramontane basins of Namurian–Westphalian age is largely parallel to the major zones of the Variscan orogen. After a change in tectonic regime, when especially horizontal movements along the fault systems of Sudetic and NNE-SSW direction began to play their role, the other (i.e. younger) group of Stephanian–Permian basins, having often the character of narrow and deep asymmetric tectonic trenches, was formed. Already in the Early Permian (Saxonian) we can observe that subsidence slowed down and the area of the basins gradually decreased. In the Lugicum area, sedimentation of the Variscan molasse was terminated as late as in the Triassic.

The synconvergent granitoid magmatism, the culmination of which was recorded in the internal zones of the Variscides in the period from the Late Viséan to the Namurian (345–325 Ma), continued by intrusions of post-tectonic, mostly geochemically strongly differentiated granitoids till the Early Permian. As well, manifestations of the volcanic activity passed without interruption from the orogenic period to the post-orogenic molasse stage. In the Westphalian to the Lower Stephanian, explosive acid magmatism extended especially in the area of Central Bohemian and Western Bohemian basins. The final phase of subsequent intraplate magmatism falls to the period from the Stephanian to the Autunian. During this phase, in addition to the acidic members, alkaline members were also formed.

The termination of Triassic sedimentation can be regarded as the beginning of platform development of the Bohemian Massif. Almost for the whole remaining period of the Triassic and the Jurassic, it was exposed to extensive erosion and peneplanisation. In the Late Jurassic, only a narrow strip of the Massif along the Elbe Fault Zone was

Fig. 2.8 Tectonic contact in the zone of the Hronov–Poříčí thrust fault of Upper Cretaceous marine sediments (*left side*) with Permian continental red coloured sandstones (*right side*). Locality Malé Svatoňovice in the SW limb of the Lower Silesian basin (*Photo* R. Grygar)

flooded by the sea to form a channel connecting the North German Basin with the Tethyan area. After a short period, the sea again retreated from this area. The cover, which is more significant from the point of view of thickness, occurs in the south-east slopes of the Bohemian Massif, periodically reached by transgressions from the area of the western Tethys.

The long period of prevailing denudation of the Bohemian Massif was replaced, on a larger scale, by sedimentation only during the eustatic rise in the level of world's oceans in the Late Cretaceous, when part of the Bohemian Massif subsided along the faults of the Elbe Fault Zone and became a site of at first continental and later marine sedimentation in the Bohemian Cretaceous Basin. At the end of the Cretaceous and in the Paleogene, inversion of the Bohemian Cretaceous Basin occurred as a result of folding in the Alpine area. Some NW–SE faults that acted as normal faults or strike-slip faults in the course of deposition of Cretaceous sediments were used for shortening the basin in this stage. The best-known example of an inverted fault is the Lusatian Fault (reverse fault) (Adamovič and Coubal 1999).

In the Tertiary, continental basins of rather small extent were formed in the area of the Ohře/Eger Rift and in southern Bohemia. In the pre-rift stage, older depressions in relief were filled; the rift stage is connected with an increased rate of subsidence of the basin bottom and with sedimentation of several hundred metres of Miocene sediments (Fig. 2.9). In the course of sedimentation, extensive volcanic activity took place along the faults, especially those limiting the south-eastern margin of the rift (Doupov Mountains, Central Bohemian Uplands).

The Quaternary is a period when the Bohemian Massif was, after the regression of the sea of the Carpathian Foredeep in the Tertiary, solely dry land. It is a very short period in comparison with the duration of the other geological units (c. 1.6–1.8 Ma). In the Quaternary, the character of geological, especially exogenous processes was affected by the existence of vast ice sheets that covered considerable part of northern Europe.

Fig. 2.9 Open pit mine Družba in the Neogene Sokolov basin located in the Ohře/Eger graben. Coal seam outcropping along boundary normal falt. In the footwall (*right side*) weathered metamorphic rock of the Saxo-Thuringian zone of the Krušné Hory Mountains cropping out (*Photo* R. Grygar)

2.4 The Western Carpathians

The Western Carpathians are one of partial branches of the vast orogenic belt of the Alpides created from the former Tethys Ocean that extends from Spain to south-eastern Asia. In the territory of the Czech Republic, they occur merely in the easternmost areas of Moravia and Silesia. The development of the Western Carpathians already begins shortly after terminating the Variscan orogeny that gave rise to a huge supercontinent called Pangaea.

The beginnings of the origin of narrow rift basins, meaning the beginning of disintegration of Pangaea, are evident as early as in the Triassic. During the Jurassic and the Cretaceous, the broadening and differentiation of the basins occurred between Africa and Europe. However, at the end of the Jurassic, some of them began to close again, which led later even to continental collisions of partial microblocks that took place in the European area in three waves in the course of the Jurassic to the Early Cretaceous (c. 160–120 Ma), Late Cretaceous (110–80 Ma) and Paleogene to Neogene (45–12 Ma). The basement of the Mesozoic and Tertiary units, later folded during the Alpine cycle, with the exception of oceanic domains, is formed, both in the Alps and in the Carpathians, mostly by various parts of the crust formed by the Variscan orogeny.

In the Western Carpathians, migration of orogenic processes towards the north manifested itself in characteristic orogenic zonation that became the base for the inner zonation of the orogen. At present, the Carpathians are divided from south to north into the Inner, Central and Outer Western Carpathians. The Central as well as the Inner Carpathians do not occur in the territory of the Czech Republic.

In the area of eastern Moravia and Silesia, the Outer Carpathians are represented by two accretionary flysch complexes and the Carpathian Foredeep. The older accretionary complex is composed of Cretaceous but largely Paleogene siliciclastic complexes of the Magura Group of nappes immediately adjacent to the klippen zone interpreted earlier as part of the Outer Carpathians; at present, especially its inner parts are interpreted as part of the Late Cretaceous–Early Tertiary accretionary complex that occurs in the place of an assumed suture after the oceanic domain, the so-called Vahicum.

In the Magura Group of nappes in rhythmically bedded units, which are characteristic of flysch basins, sandy members predominate over claystones, siltstones and coarser-grained clastics. The total thickness of the sediments is several kilometres. The frontal parts of the nappes of this group reached comparatively far, as far as the Moravo–Silesian boundary, approximately the Hodonín–Valašské Meziříčí–Třinec line. They are partly overlain by sediments of the Vienna Basin and Late Miocene and Pliocene sediments of the filling of the Hornomoravský úval basin. This complex was shortened already during the Paleogene, but the thrust over the external group of nappes took place during the Miocene at the end of closure of flysch basins (42–23 Ma). The amplitude of overthrust is estimated at several tens of kilometres.

The younger accretionary complex in the area of Moravia and Silesia is composed of the Pouzdřany, Ždánice, Subsilesian, Silesian and Fore-Magura Units. In the Polish and the Slovakia area, the Skole and Dukla Units belong to this group of nappes as well. In contrast to the above-mentioned Magura Group, they contain sediments of a broader stratigraphic range, from the Jurassic to the Middle Miocene. In addition to the flysch siliciclastic sequences of Jurassic to Early Miocene age (alternation of sandstone, claystones, conglomerates), carbonates of Jurassic and Cretaceous age are present, mainly in the Silesian Unit. The carbonates outcrop as olistoliths and tectonic shreds—klippen in the vicinity of overthrust lines of sub-nappe units. During the Miocene, the units of this younger accretionary complex were thrust over the Carpathian Foredeep, formed by a deflection of the foreland composed of the Cadomian basement of the Bruno-Vistulicum and its Paleozoic and Mesozoic cover.

The Foredeep began to be created already during the Oligocene/Miocene transition, whilst sedimentation continued to the Badenian. Marine sedimentation predominated. The Foredeep was formed by a series of sub-basins created simultaneously with the overthrust of flysch nappes. That is why the sediments of the Foredeep occur in the foreland of the nappes, on them and also far under their fronts. As a result of changes in tectonic regime, rapid changes in the extent of the basin, stratigraphic hiatuses and erosion of older sediments developed unevenly along the longitudinal axis of the basin occurred. Thicknesses of largely clayey and sandy sediments can be even more than 2 km.

In the Carpathian, the Vienna Basin with a very complicated history of tectonic development began to be formed in the depressions of flysch nappes in the area between Vienna and Uherské Hradiště. It belongs to a group of basins of the pull-apart type that, in the Western Carpathians, opened as a result of transtension caused by rotation of the Carpathians in relation to the northern European foreland. During the Late Miocene, marine sandy and clayey sediments passed gradually to brackish and later lacustrine and river sediments of up to Pliocene age. Marine Miocene sediments contain rather limited deposits of hydrocarbons; in continental sediments lignite seams were formed. The total thickness of the sediments is up to 5 km.

A still younger basin of this type is the Hornomoravský úval Basin; it was created by the rejuvenation of movements along the faults of the Elbe Fault Zone. It originated at the end of the Miocene and sedimentation of continental sediments several hundred metres in thickness continued till the Pliocene.

The extension of the crust of the Central and Southern Carpathians in the area above the oceanic lithosphere subducting to the south enabled the ascent of andesitic and basaltic magmas. Volcanic activity culminated in the Middle Miocene, but it continued, in a limited degree, to the Pliocene. To this epoch, minor occurrences of trachyandesite and trachybasalt veins in the surroundings of Uherský Brod belong, whose age of 16 Ma determined by the K–Ar method corresponds to the Late Badenian, (Přichystal et al. 1988).

2.5 Conclusion

The Bohemian Massif is one of the largest continuously outcropping fragments of the originally vast Variscan orogen that crops out from the basement of younger Epi-Variscan Platform sediments. The comparatively extensive Variscan orogen was formed gradually in the course of joining the peri-Gondwana fragments to Laurussia, i.e. to the more northerly situated continent created as the result of Caledonian convergence between Laurentia and Baltica.

Based on the current concepts of development of continents, which start from the application of principles of plate tectonics, the Bohemian Massif can be interpreted as a heterogeneous unit composed of separate regional domains. Each of them is defined especially by a specific stratigraphic content, tectomagmatic development and tectonic limitation in relation to its surroundings. In spite of the fact that the above-mentioned units are separated by significant sutures and tectonic zones, they have a number of common features especially regarding the Neoproterozoic and, in part of them, also the Cambro–Ordovician development; on the contrary, they differ markedly in terms of the early stage of Paleozoic development during the Variscan orogeny.

After the end of the Variscan orogeny, the Bohemian Massif was gradually transformed into a platform unit. During the Carboniferous, it gradually became dry land. In the most of the area, with the exception of intramontane depressions, deep erosion of the Variscan basement took place. Erosion and continental sedimentation were interrupted only for a short time by a marine transgression over part of the area in the Jurassic, Cretaceous and Neogene Periods. In addition to the deposition of sediments in the depressions, various types of volcanic bodies of Cretaceous to Quaternary age provided the finishing touches to the surface of the platform cover of the Bohemian Massif.

The Western Carpathians are one of partial branches of the vast orogenic belt of the Alpides created from the former Tethys Ocean that extends from the area of Spain to south-eastern Asia. In the territory of the Czech Republic, they occur merely in the easternmost areas of Moravia and Silesia. The development of the Western Carpathians already began shortly after termination of the Variscan orogeny. At present, the Carpathians are divided from south to north into the Inner, Central and Outer Western Carpathians. Only Outer Western Carpathians occur in the territory of the Czech Republic.

References

Adamovič J, Coubal M (1999) Instrusive geometries and Cenozoic stress history of the northern part of Bohemian Massif. Geolines 9:5–14

Babuška V, Plomerová J (2001) Subcrustallitosphere around the Saxothuringian-Moldanubian Suture Zone—a model derived from anisotropy of seismic wawe velocities. Tectonophysics 332:185–199

Belka Z, Valverde-Vaquero P, Dörr W, Ahrendt H, Wemmer K, Franke W, Schäfer J (2002) Accretion of first Gondwana-derived terranes at the margin of Baltica. Geol Soc Lond Spec Publ 201:19–36

Beránek B, Dudek A (1981) Geologický výklad transformovaných polí v Českém masivu a Západních Karpatech. Sbor geo Věd, UG 17:47–60

Beránek B, Zátopek A (1981) Earth's crust structure in Czechoslovakia and in Central Europe by methods of explosion seismology. In: Zátopek A (ed) Geophysical synthesis in Czechoslovakia. Veda, Bratislava, pp 243–264

Cháb J, Breiter K, Fatka O, Hladil J, Kalvoda J, Šimůnek Z, Štorch P, Vašíček Z, Zajíc J, Zapletal J (2010) Outline of the Geolog of the Bohemian Massif: the Basement Rocks and their Carboniferous and Permian Cover. ČGS Praha. pp 295

Commission, R.o.t.WG.f.R.G.C.o.t.B.M.a.t.f.C.S. (1994) Regional gelogical subdivision of the Bohemian Massif on the territory of the Czech Republic. J Czech Geol Soc 39(1): 127–144

Condie KC (1989) Plate Tectonic and Crustal Evolution. Pergamon Press

Ellenberger F, Tamain AG (1980) Hercynian Europe. Episodes 22–27

Franke W (1989) Tectonostratigraphic units in the Variscan Belt of Central Europe. Geol Soc Amer Spec Paper 230: 67–90

Franke W, Haak V, Oncken O, Tanner D, Editors (2000) Orogenic processes: quantification and Modelling in the Variscan Belt. Geol Soc Lond Spec pap 179: p 464

Grygar R, Vavro M (1995) Evolution of lugosilesian orocline (north-eastern periphery of the Bohemian Massif): kinematics of variscan deformation. J Czech Geol Soc 40(1–2):65–90

Kossmat F (1927) Gliederung der varistischen Gebirgsbaues. Abhandlungen des Sachsischen Geologischen Landesamts 1:1–39

Krs M, Pruner P, Man O (2001) Tectonic and paleogeographie interpretation of the paleomagnetism of Variscan and preVariscan formations of the Bohemian Massif, with special reference to the Barrandian terrane. Tectonophysics 332(1–2):93–114

Matte P (1986) Tectonics and plate tectonics model for the Variscan belt of Europe. Tectonophysics 126:329–374

Matte P (1991) Accretionary History and Crustal Evolution of the Variscan Belt in Western-Europe. Tectonophysics 196(3–4):309–337

Matte P, Maluski H, Rajlich P, Franke W (1990) Terrane boundaries in the Bohemian Massif: result of large-scale Variscan shearing. Tectonophysics 177:150–170

Mattern F (2001) Permo-Silesian movements between Baltica and Western Europe: tectonic and.basin families. Terra 13(5):368–375

Mísař Z, Dudek A, Havlena V, Weiss J (1983) Geologie ČSSR I. SPN Praha, Český masiv, p 333

Přichystal A, Repčok I, Krejčí Z (1988) Radiometrické datování trachyandezitu u Uherského Brodu (magurská skupina flyšového pásma). Geol výzk Mor Slez v r 1997(5):33–34

Schulmann K, Kröner A, Hegner A, Wendt E, Konopásek J, Lexa O, Štípská P (2005) Chronological constraints on the pre-orogenic history, burial and exhumation of deep-seated rocks along the eastern margin of the Variscan orogen. Bohemian Massif, Czech Republic 305(5):407–448

Stille H (1951) Das mitteleuropaische variszische Grundgebirge im Bilde des gesamteuropaischen. Geol Jb Beih 2:138

Suess E (1926) Intrusionstektonik und Wandertektonik im variszischen Grundgebirge. Gebruder Borntrager, Leipzig, p 138

Tait J, Schatz M, Bachtadse V, Soeffel H (2000) Palaeomagnetism and palaeozoic palaeogeography of gondwana and european terranes. In: Franke W, Haak V, Oncken O, Tanner D (eds) Orogenic Processes: quantification and Modelling in the Variscan Belt. Special Publication of the Geological Society of London, London, pp 21–35

Winchester JA (2002) Palaeozoic amalgamation of Central Europe: new results from recent geological and geophysical investigations. Tectonophysics 360:5–21

Ziegler PA (1982) Geological Atlas of Western and Central Euprope. Shell Internationale Petroleoum, Maatschappij B.Y., Amsterodam

Ziegler PA (1984) caledonian and hercynian consolidation of western and Central Europe, a working hypothesis. Geol Mijnbouw 63:93–108

Zulauf G (1994) Ductile normal faulting along the West Bohemian Shear Zone (Moldanubian/Tepla-Barrandian boundary): evidence for late variscan extensional collapse in the variscan internides. Geol Rundsch 83:276–292

Climate in the Past and Present in the Czech Lands in the Central European Context

3

Jan Hradecký and Rudolf Brázdil

Abstract

Central Europe and Czech Lands (recent Czech Republic) itself have recently represented an area with a transitional type between the oceanic and continental types of temperate climate. The climate changed during geological history and various climates played an important role in the evolution of landforms due to changes in type and intensity of weathering and earth surface processes. This chapter describes general trends in climate oscillations during the Tertiary and the Quaternary within the Czech Lands and Central Europe. Climatological and hydrological extremes and fluctuations during the last centuries are along with human activity fundamental drivers of recent changes in the landscape evolution.

Keywords

Climate • Tertiary • Quaternary • Climate oscillations • Climatological and hydrological extremes

3.1 Introduction

The relief of the Czech Republic shows distinct signs of polygenesis and its evolution was strongly controlled by climate that underwent considerable changes during the Cenozoic Era. Different climatic conditions influenced the evolution of landforms that represent a parallel to the relief development in contemporary morphoclimatic zones of the Earth. Geomorphological legacy of the oldest Cenozoic periods has greatly been changed in the subsequent phases, particularly cold phases of the Quaternary. Climate reconstructions are based on the modern analysis of stable isotopes in marine sediments and ice cores (e.g. Svensson et al. 2008; Vinther et al. 2010). From the point of view of the region, however, key analyses are mainly paleobotanical (pollen analyses) (e.g. Davis et al. 2003). It is particularly paleobotanical data that make it possible to derive mean annual temperatures and mean precipitation totals for the territory of Central Europe (Mosbrugger et al. 2005). Individual parts of the following text describe climatic conditions in different periods of Cenozoic landscape evolution, mainly in the period of the late Tertiary and the Quaternary, characterised by significant climate oscillation with the displays of the alternation of cold and dry glacial periods and warm and humid interglacial periods. From the point of view of landscape and relief evolution, special phenomena are extreme events whose existence has been recorded but their frequency decreases if we go deeper into the past due to the incompleteness of datasets. The reconstruction of natural extremes that formed the relief of Czech landscape and that comprise particularly hydrometeorological events was derived mainly from documentary and instrumental data (e.g. Brázdil et al. 2005, 2012). The final part of this chapter focuses on contemporary climatic conditions.

J. Hradecký (✉)
Department of Physical Geography and Geoecology, Faculty of Science, University of Ostrava, Ostrava, Czech Republic
e-mail: jan.hradecky@osu.cz

R. Brázdil
Institute of Geography, Faculty of Science, Masaryk University, Brno, Czech Republic
e-mail: brazdil@sci.muni.cz

R. Brázdil
Global Change Research Centre, Czech Academy of Sciences, Brno, Czech Republic

T. Pánek and J. Hradecký (eds.), *Landscapes and Landforms of the Czech Republic*, World Geomorphological Landscapes, DOI 10.1007/978-3-319-27537-6_3

3.2 Tertiary Climates of Central Europe

Unlike the subsequent period of the Quaternary, climatic conditions of the Tertiary were greatly different than those of today and led to the origin of different landform assemblages that can be identified in the landscape of the Czech Republic even nowadays. Sediments that fill basin structures of the Czech Republic, namely organogenic sediments of the rank of coal (e.g. in Mostecká pánev Basin), are other important evidence of different climatic conditions of the Tertiary. From the point of view of paleogeography, very important sediments are Miocene formations of the Carpathian Foredeep.

Tertiary climate was characterised by the alternation of warmer and colder oscillations with a tendency towards gradual cooling (Chlupáč et al. 2002). Climate has been reconstructed in the Wiess Elster Basin in the vicinity of the border between Bohemia and Germany. Paleobotanical analyses show that climate in Central Europe in the period from the Middle Eocene to Lower Oligocene was tropical, with mean annual temperature ranging from 23 to 25 °C, mean annual precipitation from 1,000 mm to 1,600 mm and coldest month mean (CMM) from 17 to 21 °C (Mosbrugger et al. 2005). Lower temperatures were associated with a majority of the Oligocene period with CMM around 5 °C, while the latest Chattian was marked by a temperature peak which was recorded by Mosbrugger et al. (2005) from the Lower Rhine Basin. This peak corresponds to the Late Oligocene Warming known from isotope records (Zachos et al. 2001). The warmest period of the Neogene was the Miocene (Chlupáč et al. 2002) in which the trend of warming continued up to the Middle Miocene. This warming seems to be rather stepwise, while the curves show several short-term variations. In the Weisselster Basin record there is evidence of short-term cooling at the base of the Aquitanian (Mosbrugger et al. 2005). The Middle Miocene temperature peak in Central Europe corresponds to the Middle Miocene Climatic Optimum that is observed globally. After Mosbrugger et al. (2005), the Miocene cooling seems to be between 14 and 13 Ma when considering all different records and analysed climate variables. In the Molasse Basin, CMM decreased more rapidly than in both other regions, and, at the end of the Middle Miocene, CMM dropped below 4 °C. The transition between the Miocene and the Pliocene shows a gradual trend in climate cooling. During the Late Pliocene the cooling intensified and CMM fell below the freezing point (Mosbrugger et al. 2005).

3.3 Quaternary Climatic Cycle

Quaternary evolution in Central Europe is connected with fundamental changes in environmental conditions and essential paleogeographic changes that were related to the transgression of the continental ice sheets, temperature drop and changes in morphogenesis.

A comprehensive overview of landscape evolution is brought by Quaternary climate and sediment model compiled by Ložek (1973, 1999a, b, 2007). The author characterises the evolution using four phases (Fig. 3.1): early glacial period, pleniglacial period, late glacial and interglacial. It is evident that global trends of the climate system oscillation were reflected in regional cycles. On the basis of an extensive set of data about Quaternary sediments Ložek (1999a, b) was able to derive a rather general model that characterises not only climate parameters but also points to the conditions of the evolution of soils and vegetation and the processes of weathering and material deposition. The four phases of Ložek's model are described below.

The phase of early glacial is characterised by the onset of cold climate. Mean annual temperatures range between +3 and −1 °C, depending on the location. Cooling brings a distinct decline in precipitation (mean annual precipitation totals are estimated to 200–400 mm). The landscape undergoes gradual aridization, which becomes evident in pedogenesis and vegetation composition. Interglacial forests are divided into smaller units whose species composition changes towards a boreal forest (taiga). Conifers start to appear, while the species of Central European temperate forest are in recession. Very dry periods bring forth chernozem steppes. The transformation of ecosystems gradually gives rise to cold continental steppes in which grasses and chernozems prevail. Temperatures and precipitation totals continue to decrease, while the cycle passes into the phase of the so-called pleniglacial.

Pleniglacial phase is characterised by the transgression of the continental glacier and conditions of periglacial climate in a great part of the territory. Glaciers start to appear in the topmost areas of mountain ranges and the ice sheet expands into the northernmost parts of the recent Czech Republic territory. Mean annual temperatures drop to −3 to −5 °C, which leads to the occurrence of permafrost. Tree vegetation recedes considerably, while groups of trees only survive in protected areas or vanish totally. The development of vegetation is limited by very low precipitation totals, ranging between 100 and 200 mm/year. Cold and dry climate is

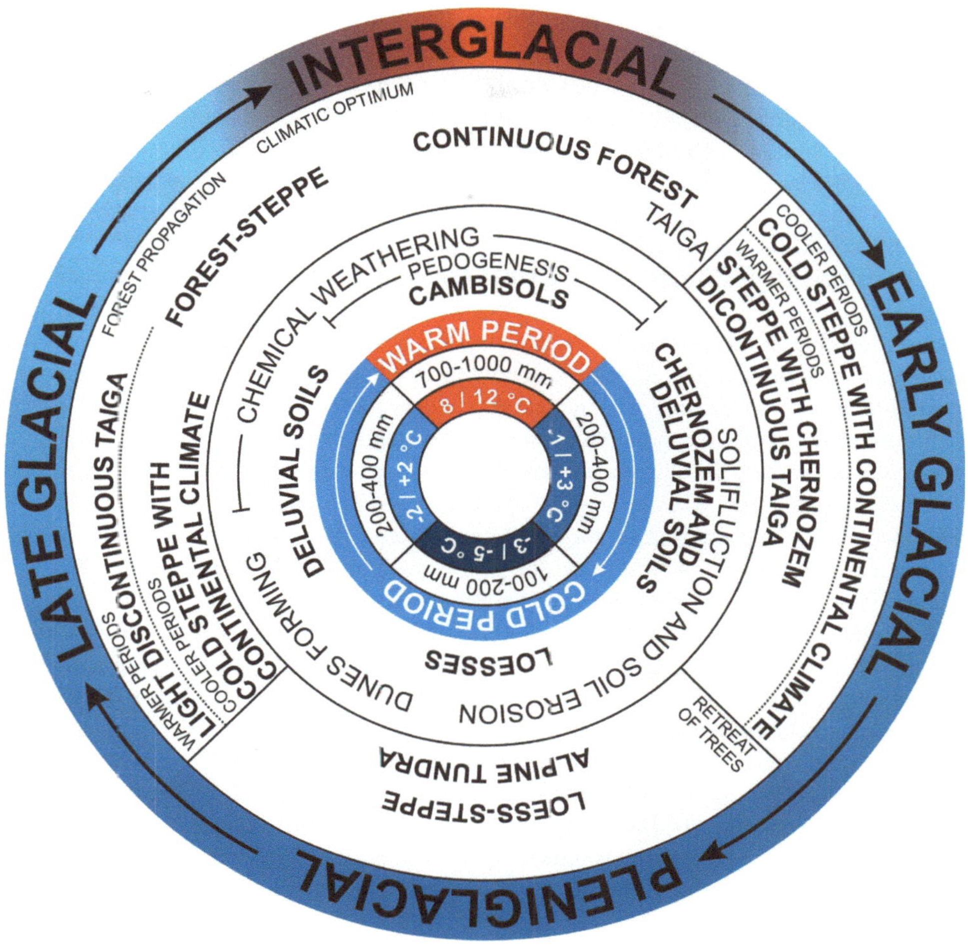

Fig. 3.1 Simplified Quaternary climate and sediment model compiled by Ložek (2007)

marked by strong atmospheric flow and loess deposition. This leads to the formation of cold loess steppes. However, the foreground of the ice sheet and higher locations witness the formation of tundra or sub-alpine ecosystems with cryogenic soils. Soil-forming substrates are very rich in salts and calcium carbonate, which leads to the spreading of halophile and calciphile species. The character of non-glaciated parts of the landscape is significantly affected by intensive congelifraction, presence of permafrost and gelifluction. The period of low temperature is replaced by gradual warming of climate, which leads to temperature oscillation. The landscape starts to enter the period of late glacial.

Mean annual temperatures in *late glacial phase* are still relatively low, ranging between –2 and +2 °C, however, the warming trend and increasing humidity are evident (200–400 mm). The climate is characterised by significant instability, which is reflected by the fact that cold continental steppes are preserved at many places, while the onset of thermophilic vegetation is very slow. Degradation of permafrost makes itself felt both by the increase in the thickness of its active layer and gradual decomposition of continuous permafrost areas into isolated permafrost patches. The retreat of the continental ice sheet along with the deglaciation of the highest mountain ranges brings fundamental changes in environmental conditions. Periods of warm oscillations make conditions for light discontinuous taiga with birch, pine and sea-buckthorn. Colder phases still witness the occurrence of cold continental steppe. Large accumulations of material weathered during cold periods start to be influenced by chemical weathering. Towards the end of the late glacial, forest species start to appear and the area covered by forests gradually extends.

Interglacial phase is a phase of warm climate. Annual temperature means increase greatly (8–12 °C) and the climate becomes more humid (700–1000 mm/year). The landscape changes fundamentally. With the onset of climatic optimum the open landscape is gradually closed by the Central European forest that replaces forest-steppe communities. Soils rich in calcium enable intensive spreading of basophilic species and high temperatures predetermine the spreading of xerophilic communities. Intensive chemical weathering and leaching alkali out of soil gradually gives rise to cambisols. Forest formations change regarding the species and gradual acidification of the surroundings facilitates the spreading of acidophilic species. The continuous forest reaches the phase of climax. However, the climatic system undergoes further development and enters the phase of cooling, which involves the spreading of cold-loving species, acidification intensifies and taiga spreads in the landscape.

3.4 Quaternary Climate in Central Europe

Climate cooling at the end of the Tertiary led to the Pleistocene, characterised by the alternation of cold (glacials) and warm (interglacials) periods. Using marine isotope stages (MIS) as a framework, it is possible to identify 104 stages of cool (52) or warm (52) climate periods during the whole Quaternary (and 103 MIS within the Pleistocene). Early Pleistocene was characterised by the mean annual temperature below 0 °C; however, this very old period in Central Europe is not covered well by precise data. On the basis of geomorphological proxy data, Czudek (2005) estimates that during cold phases of the Early Pleistocene mean annual temperatures dropped to −3 to −4 °C. The formation of cryogenic structures in the southern Moravia can indicate mean monthly temperatures of the coldest month to –20 °C, which would point to the occurrence of continuous permafrost (Vandenberghe 2001b). With respect to the climate of our territory, there is relatively little information on the Middle Pleistocene. Our conclusions are again drawn from proxy data (e.g. ice wedges and a range of pseudomorphoses). Mean annual temperature is estimated for −5 °C. Temperatures of the coldest months were on average around −20 °C or even lower. The Late Pleistocene was characterised by the peak in periglacial landform-shaping processes in the territory of the recent Czech Republic (Czudek 2005). In the Eemian interglacial period mean annual temperatures were around 13 °C and the climate was very humid (Czudek 2005). Subsequent cooling in the Weichsellian glacial period again brought mean annual temperatures below the freezing point (−2 and −5 °C). The greatest drop in temperatures came in the pleniglacial (73–13 ka BP) when in the phase of the Last Glacial Maximum (LGM) mean annual temperatures were −6 to −8 °C; mean January temperatures ranged between −18 and −20 °C. The warmest summer months reached temperatures between 5 and 6 °C (2005). Ložek (1999a) states that climate had continental character, with long and cold winters, short springs, but relatively warm summers. He further mentioned that annual precipitation totals ranged between 100 and 200 mm and they occurred particularly in the warm part of the year. An interesting approach in the reconstruction of environmental conditions during the LGM is brought by the study of Alvaradoa et al. (2011) in which a drop in temperatures by 5–7 °C was confirmed by the analysis of dissolved noble gases in groundwater of the Bohemian Cretaceous Basin. The end of the Pleistocene (Late Pleistocene) was characterised by a distinct increase in temperatures; however, with considerable oscillation between interstadials (Bölling and Alleröd) and stadials (Older and Younger Dryas). Mean annual temperatures in interstadials ranged between 2 and 5 °C, while during stadials they were around −2 to −3 °C (Czudek 2005).

Warming at the end of the Younger Dryas brought radical changes to the environment. Considerable retreat of the glaciers led to the onset of the Holocene interglacial. Climate warming was accompanied by increased precipitation that accelerated vegetation growth and changes in the pedogenetic conditions, weathering and relief evolution. Individual chronozones of the Holocene landscape evolution are shown in Fig. 3.2 including reconstructed temperatures and precipitation totals after Ložek (2007) and Starkel (1990a). In the Preboreal (10,300–9,300 BP) mean annual temperature was by c. 3 °C lower than nowadays. Continuing continental climate was warmed in the course of the Preboreal (9300–8400 BP); mean annual temperature was by 2–3 °C higher than nowadays (Czudek 2005).

Climatic optimum was reached in the Atlantic (8400–5100 BP) in which the mean annual temperature was up to 3 °C higher than today (Czudek, 2005). An important characteristic of the Atlantic climate was significantly higher precipitation, namely by 100 %, if compared with nowadays (Ložek 1999c). The beginnings of the Subboreal (5100–2400 BP) were by 1 °C warmer than the present-day mean. The main feature of the Subboreal period was ambivalence and drier periods alternating with more humid ones and warmer periods alternating with colder ones (Czudek 2005). The period of the Subatlantic (2400 BP—the present day) brought cooling and, at the same time, increased precipitation (Ložek 1999c). Humid climatic phases are reflected in the landslide activity phases in the Carpathian part of the Czech Republic (Pánek et al. 2010) that also correspond with the Polish landslide activity chronology (Fig. 3.2).

The latest pollen data proxies and the relationship of pollen and the climate changes during Holocene are brought by the research of Veron et al. (2014) from the peat bog of Boží Dar (Krušné hory Mts). The authors confirm very cold Late Glacial dominated by *Cyperaceae* grass land (12.5–11.0 kyr BP), the Early Holocene warming and the onset of *Pinus* (11.0–9.0 kyr BP). During the Boreal the temperature increased with an increase in the shade-intolerant *Corylus* and a concurrent decrease in *Pinus* (9.0–8.1 kyr BP). The Atlantic chronozone is classified as the warmest and wettest period of the Holocene characterised by the species of *Alnus* and *Fraxinus* (8.1–4.3 kyr BP). The following Subboreal chronozone was detected as drier and possibly colder and characterised by the decline in temperature-sensitive species (< 4.3 kyr BP). Similar results of temperature trends were

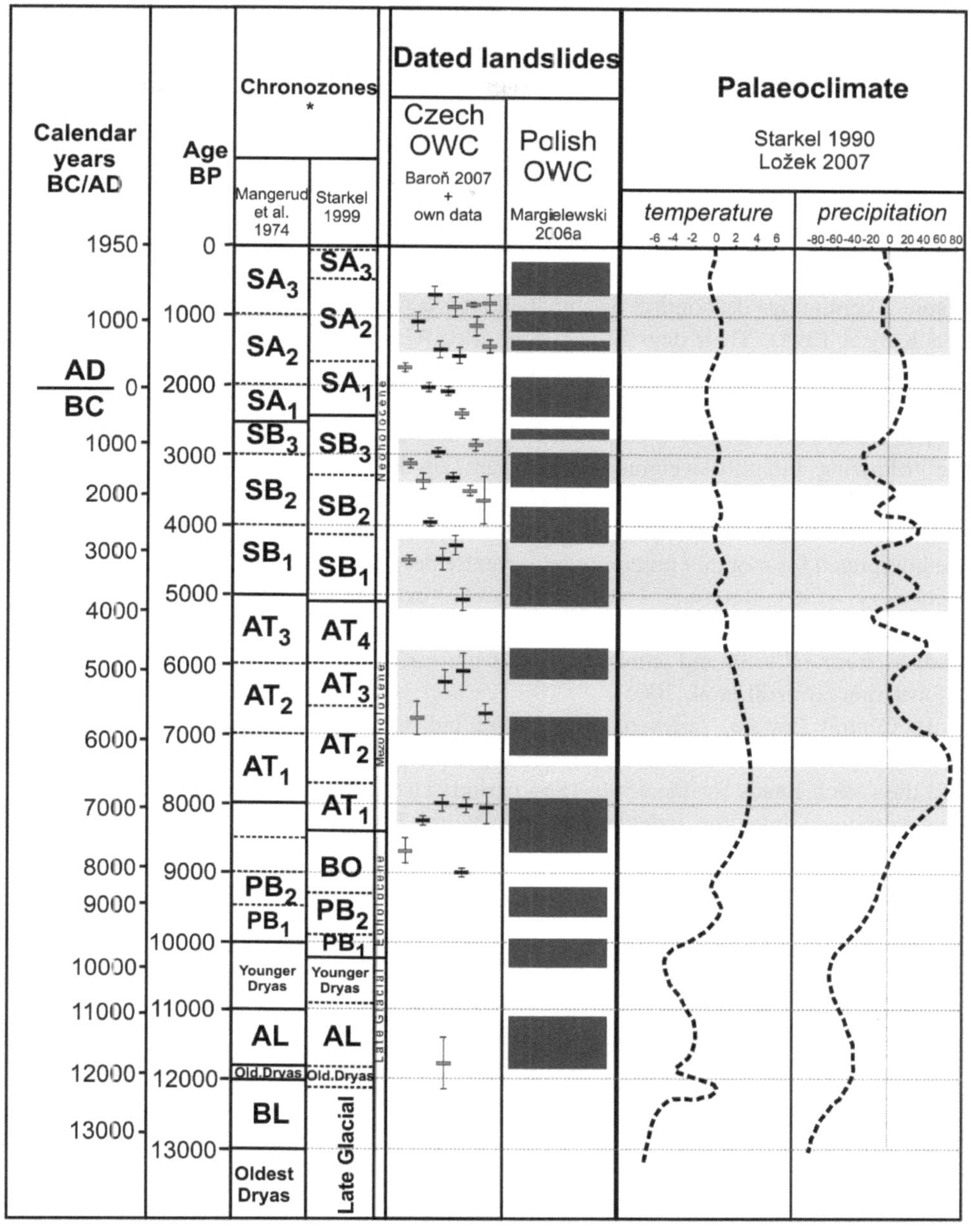

Fig. 3.2 Correlations of dated landslides (both in the Czech and Polish parts of the Flysch Carpathians) with paleoclimate (Pánek et al. 2010). The scheme is based on a diagram performed by Margielewski (2006); time-span of individual chronozones after Mangerud et al. (1974) and Starkel (1999); dated landslides in the Czech part of the Outer Western Carpathians (OWC) after Baroň (2007) (20 cases—*black boxes*) and dating performed by the authors of this study (15 cases—*gray boxes*); dated landslides (landslide phases derived from the dating of 67 landslides) in the western part of the Polish Outern Western Carpathians after Margielewski (2006); paleotemperature and paleoprecipitation after Starkel (1990). Despite the fact that landslides occurred in the Czech part of the OWC throughout the entire Late Glacial and Holocene, significant landslide aktivity clusters (*horizontal grey bars*) are correlated to periods characterised by high precipitation/low temperature

brought by the synthesis of pollen data proxies of all Central Europe made by Davis et al. (2003).

3.5 Climate and Floods of the Past 500 Years in the Czech Lands

Climate of the past millennium is usually divided into Medieval Warm Anomaly (MWA), Little Ice Age (c. 1300–1860) and recent global warming (e.g. Grove 2004; Matthews and Briffa, 2005; Xoplaki et al. 2011; Stocker et al. 2013). High-resolution climatic data in the Czech Lands are related to the beginnings of systematic instrumental meteorological observations. The longest series are available from the Prague-Klementinum station (temperatures from 1775 and precipitation from 1804) and the Brno station (temperatures from 1800, precipitation from 1803) (Brázdil et al. 2012). The data from the period before the instrumental period can be extended with dendrochronological and documentary data. The tree-ring data of fir (*Abies alba* Mill.) were compiled from different places in South Moravia and used for March–July precipitation reconstruction in the 1376–1996 period (Brázdil et al. 2002). Recently this series has been complemented with other samples and used to reconstruct May–June Z-index as a drought indicator from AD 1500 (Büntgen et al. 2011). Because of restrictions of tree-ring-based reconstructions to only a few months of the year (usually of the vegetation period), documentary data related to weather and climate are more promising. They describe directly weather and/or human activities or phenomena with a direct relation to the weather. Their sources are very diverse: annals, memories, chronicles, diaries, letters, economic records, pictures, etc. Weather-related documentary data in the Czech Lands between the eleventh and fifteenth century are relatively scarce and do not allow to obtain a continuous description of climatic patterns (Brázdil and Kotyza 1995). Their density increases after AD 1500. Because of qualitative information in documentary data, the series of temperature indices has to be interpreted in an ordinal scale. For example, in case of the seven degree scale, the following monthly weighted indices are applied: –3 extremely cold, –2 very cold, –1 cold, 0 normal, 1 warm, 2 very warm, 3 extremely warm. Analogously, similar indices are interpreted for monthly precipitation: –3 extremely dry, –2 very dry, –1 dry, 0 normal, 1 wet, 2 very wet, 3 extremely wet. Monthly indices are then summarised to obtain seasonal (indices from −9 to 9) and annual (indices from −36 to 36) information (Brázdil et al. 2005).

In Central Europe, series of temperature indices were created separately for the territory of Germany, Switzerland and the Czech Lands for the 1500–1854 period. They were then combined into one index series for Central Europe, which is fully representative also for the Czech Lands with respect to high spatial temperature correlations. In addition, a mean series of air temperature was calculated from the measurements at 11 Central European stations in South Germany, Switzerland, Austria and Bohemia (Prague-Klementinum) that date back to the year 1760. Finally, these two series were statistically analysed (using standard paleoclimatological method) for the common period of 1760–1854 in order to identify inter-relations and ultimately to reconstruct the temperature from before the year 1760 using several statistical techniques. Based on this stepwise analysis, temperature series of Central Europe could be produced for the seasons and the year for the entire 1500–2007 period (Dobrovolný et al. 2010). The reconstruction is very good and accounts for 81 % of the corresponding annual temperature variability (for seasons from 73 % in autumn to 83 % in winter). The Central European temperatures exhibit a great interannual and interdecadal variability and an increasing trend from the nineteenth century that has been particularly pronounced since the 1970s, which is in agreement with the observed global warming (Fig. 3.3a). The coldest periods occurred in the last three decades of the sixteenth century and in the late seventeenth century. As for seasons, remarkable periods are the coldest 30-year periods in the late sixteenth century (winter 1572–1601, summer 1569–1598) corresponding to *Little-Ice-Age*

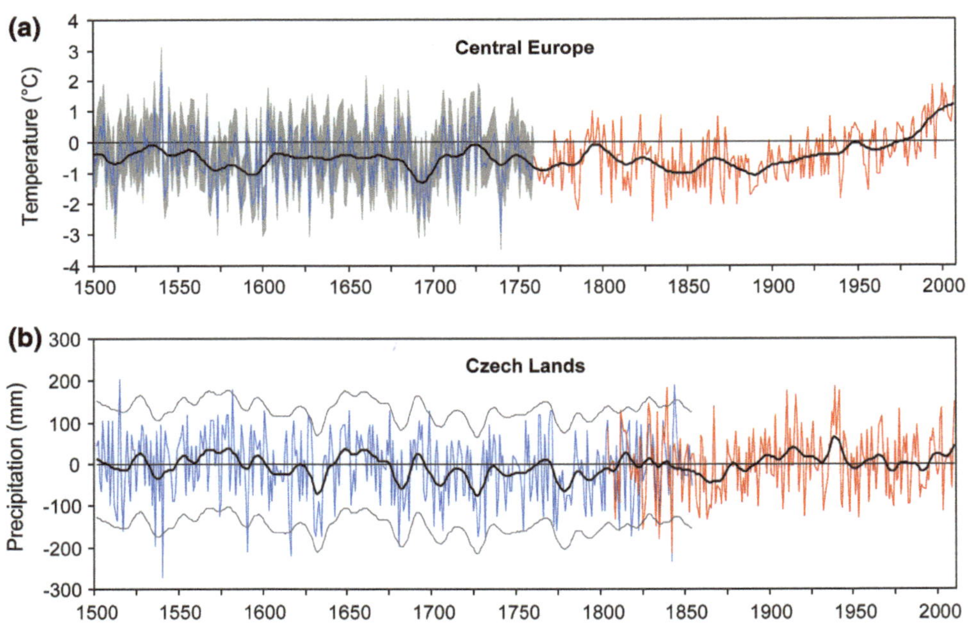

Fig. 3.3 (**a**) Annual temperature fluctuations in Central Europe in the 1500–2007 period derived from documentary and instrumental data and expressed as deviations from the 1961–1990 reference period. Uncertainty limits of reconstructed values are expressed by grey colour (Dobrovolný et al. 2010); (**b**) annual precipitation fluctuations in the Czech Lands in the 1501–2010 period expressed as deviations from the 1961–1990 reference period. Uncertainty limits are given by a 95 % confidence interval (Dobrovolný et al. 2014). Values in both graphs are smoothed by 30-year Gaussian filter; deviations for the pre-instrumental period are in *blue colour*, for the instrumental period in *red colour*

type event sensu Wanner (2000) (see also Matthews and Briffa, 2005). The other existing temperature reconstruction for the Czech Lands is based on winter wheat harvest dates and gives March–June temperatures for the 1501–2008 period (Možný et al. 2012).

The same standard paleoclimatological approach as for temperatures was applied to reconstruct precipitation totals from the series of precipitation indices for the Czech Lands. The 1803–1854 period was used as an overlap period between the series of documentary-based precipitation indices and the mean Czech series calculated from homogenised series of precipitation totals from 14 stations (Dobrovolný et al. 2015). With respect to higher spatial variability of precipitation, the reconstruction of annual totals explains only 36 % of corresponding precipitation variability (for seasons from 26 % in winter to 36 % in autumn). Fluctuations in annual precipitation totals in the last 500 years are characterised by great inter-annual and inter-decadal variability, but generally with missing long-term trends (Fig. 3.3b). The wettest 30-year periods were recorded analogously as for temperatures, in the second half of the sixteenth century (winter 1555–1584, summer 1568–1597).

Fluctuations in floods, which are the most disastrous natural events in the Czech Lands, are another important feature of the climate. Based on meteorological causes of their origin, floods can be divided into winter and summer floods. Winter floods are related to snow melt due to sudden warming (accompanied by rain) or ice jams in rivers and they usually occur from November to April. Summer floods,

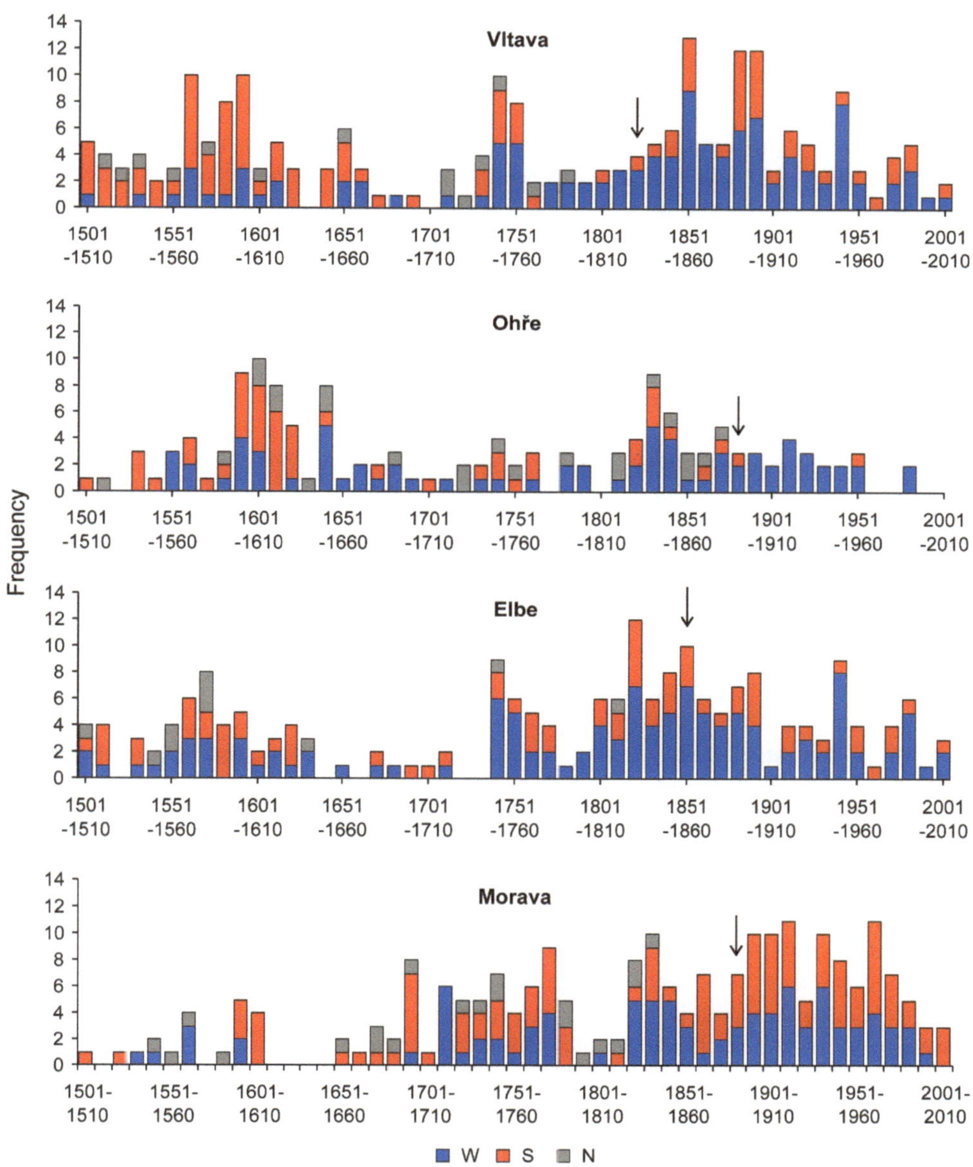

Fig. 3.4 Decadal frequencies of floods in the Czech Lands in the 1501–2010 period with differentiation according to the synoptic type of the flood (*W* winter, *S* summer, *N* not specified): the River Vltava (from České Budějovice to its mouth into the Labe River near Mělník), the Ohře River (from Kadaň to its mouth into the Labe River at Litoměřice), the Labe/Elbe River (from Brandýs nad Labem to Děčín), the Morava River (from Olomouc to Strážnice) (Brázdil et al. 2005, 2011). Arrows mark the beginning of systematic hydrological measurements

Table 3.1 Selected annual climatological characteristics of the geomorphological units of the Czech Republic 1961–1990, (Source: www.chmi.cz)

Station (Geomorphological Unit)[a]	Altitude (m a.s.l.)	Mean temperature (°C)	Mean precipitation (mm)
Červená (Nízký Jeseník)	749	5.5	739
Doksany (Dolnooharská tabule)	158	8.5	454
Cheb (Chebská pánev)	483	7.2	560
Churáňov (Šumava)	1118	4.2	1091
Kuchařovice (Znojemská pahorkatina)	334	8.5	471
Lysá hora (Moravskoslezské Beskydy)	1322	2.6	1390
Milešovka (České středohoří)	837	5.2	555
Plzeň (Plzeňská kotlina)	360	7.3	533
Praha—Karlov (Pražská plošina)	261	9.4	447
Přibyslav (Českomoravská vrchovina)	532	6.6	677
Přimda (Český les)	742	5.8	684
Svratouch (Žďárské vrchy)	733	5.7	762
Praděd (Hrubý Jeseník)	1490	1.1	1129
Ústí nad Orlicí (Svitavská pahorkatina)	402	7.1	763

[a]For the localization of geomorphological units see Fig. 1.1

related to heavy precipitation lasting for several days, are typical of May to October (flash floods do not occur in the studied rivers). Quite rich documentary evidence together with water level and discharge measurements starting in the ninteenth century allow to compile long series of flood frequency for the most important rivers in the Czech Lands (the Labe/Elbe River and its tributaries Vltava and Ohře Rivers in Bohemia, the Morava River in Moravia) from AD 1500 (Fig. 3.4). Despite differences between individual rivers, remarkable maxima of flood frequency are typical of the nineteenth century (mainly winter floods) and of the second half of the sixteenth century (mainly summer floods) (Brázdil et al. 2005).

3.6 Contemporary Climate

The location of the Czech Lands in Central Europe together with their relief and position in relation to the main pressure systems in the Atlantic-European area, influencing atmospheric circulations patterns, are the main factors deciding about spatial and temporal variability of climate in the Czech Republic. This territory falls within the transitory temperate climate zone and therefore the area of interest experiences the influence of both the oceanic and continental climate. Temperature conditions largely reflect location and elevation above the sea level. The highest mean temperatures (more than 10 °C) are measured in lowlands and southern areas of the country and thus Southern Moravia, Osoblažsko area and middle and lower portions of the Labe River belong to the warmest locations (Tolasz et al. 2007). On the other hand, the lowest mean air temperatures are measured in mountainous regions, namely in the highest locations (less than 3 °C) of the Krušné hory Mts., Krkonoše Mts., Šumava Mts., Jizerské and Orlické hory Mts., Králický Sněžník Mts., Hrubý Jeseník Mts. and Moravskoslezské Beskydy Mts. Maximum precipitation is reached in summer half-year. Generally, about 40 % of precipitation falls in summer, 25 % in spring, 20 % in autumn and 15 % in winter. The rainiest locations are found in the highest elevations of the Czech and Moravian ranges where mean annual precipitation considerably exceeds 1200 mm (Krušné hory Mts., Šumava Mts., Jizerské hory Mts., Krkonoše Mts., Orlické hory Mts., Hrubý Jeseník Mts. and Moravskoslezské Beskydy Mts.) (Tolasz et al. 2007). The lowest annual precipitation is measured in Žatec Basin behind the Krušné hory Mts., in the rain shadow area. Other locations with low precipitation throughout the year are Southern and Central Moravia (area of Moravian basins), Polabí area and Opava region. Žatec region, Polabí Lowland and Southern Moravia are also characterised by the highest occurrence of dry periods given by precipitation deficit (Tolasz et al. 2007). Selected climatological parameters for some geomorphological units are presented in Table 3.1.

3.7 Conclusion

The Czech Republic is located in the transition area whose climate results from interaction of both maritime and continental air masses. The geographical position was crucial during the geological history and therefore the paleoclimate was influenced by transgressions and regressions of the Scandinavian ice sheets. The paleoclimate oscillations and historical fluctuations were reconstructed with the use of various proxy data. It is evident that a very important role in the evolution of landforms was played by warm and wet climate of the Tertiary. A complete change started at the end of the Tertiary when the climate pattern changed into Quaternary oscillations between colder and drier periods of the glacials and warmer and wetter periods of the interglacials. Climate changes created conditions for the evolution of periglacial landforms and limited areas were affected by glacier action (the highest mountains were glaciated and the northernmost areas of the recent Czech Republic were covered by masses of the Scandinavian ice sheet (Elsterian and Saalian glacials). Extreme temperatures during the whole Pleistocene are connected with the Last Glacial Maximum ($\sim$26 ka BP) when mean annual air temperatures oscillated between −6 and −8 °C. The Holocene started with higher temperatures and higher precipitation and this led to the evolution of vegetation and changes in landform evolution. Human activities during the Post-Atlantic period created new conditions for the acceleration of alluviation under more humid climate. With the use of the methods of historical climatology and instrumental measurements in the last five centuries it was possible to identify climate and flood fluctuations which are very important in recent landform evolution.

Acknowledgements Rudolf Brázdil acknowledges support by the Grant Agency of the Czech Republic for the project no. P209/11/0956. Petr Dobrovolný is acknowledged for providing data for Fig. 3.1, L. Řezníčková for the elaboration of Figs. 3.1 and 3.2 and Václav Škarpich for the elaboration of Fig. 3.1. Jan Hradecký acknowledges support by the Student Grant Competition of the University of Ostrava (SGS18/PřF/2015-2016).

References

Baroň I (2007) Výsledky datování hlubokých svahových deformací v oblasti Vsetínska a Frýdeckomístecka. Geologické výzkumy na Moravě a ve Slezsku v roce 2006:10–12

Brázdil R, Kotyza O (1995) History of Weather and Climate in the Czech Lands I. Period 1000–1500. Zürcher Geographische Schriften 62, Zürich, 260 p

Brázdil R, Štěpánková P, Kyncl T, Kyncl J (2002) Fir tree-ring reconstruction of March–July precipitation in southern Moravia (Czech Republic), 1376–1996. Clim Res 20:223–239

Brázdil R, Dobrovolný P, Elleder L, Kakos V, Kotyza O, Květoň V, Macková J, Müller M, Štekl J, Tolasz R, Valášek H (2005a) Historical and Recent Floods in the Czech Republic. Masaryk University, Czech Hydrometeorological Institute, Brno, Praha, 370 p

Brázdil R, Pfister C, Wanner H, von Storch H, Luterbacher J (2005b) Historical climatology in Europe—the state of the art. Clim Change 70:363–430

Brázdil R, Řezníčková L, Valášek H, Havlíček M, Dobrovolný P, Soukalová E, Řehánek T, Skokanová H (2011) Fluctuations of floods of the River Morava (Czech Republic) in the 1691–2009 period: interactions of natural and anthropogenic factors. Hydrol Sci J 56:468–485

Brázdil R, Bělínová M, Dobrovolný P, Mikšovský J, Pišoft P, Řezníčková L, Štěpánek P, Valášek H, Zahradníček P (2012) Temperature and Precipitation Fluctuations in the Czech Lands During the Instrumental Period. Masaryk University, Brno, 236 p

Büntgen U, Brázdil R, Dobrovolný P, Trnka M, Kyncl T (2011) Five centuries of Southern Moravian drought variations revealed from living and historic tree rings. Theor Appl Climatol 105:167–180

Chlupáč I, Brzobohatý R, Kovanda J, Stráník Z (2002) Geologická minulost České republiky. Academia, Praha, 438 p

Corcho Alvaradoa JA, Leuenberger M, Kipfer R, Pacese T, Purtschert R (2011) Reconstruction of past climate conditions over central Europe from groundwater data. Quat Sci Rev 30:3423–3429

Czudek T (2005) Vývoj reliéfu krajiny České republiky v kvartéru. Moravské zemské muzeum, Brno, 240 p

Davis BAS, Brewer S, Stevenson AC, Guiot J, Contributors Data (2003) The temperature of Europe during the Holocene reconstructed from pollen data. Quat Sci Revs 22:1701–1716

Dobrovolný P, Moberg A, Brázdil R, Pfister C, Glaser R, Wilson R, van Engelen A, Limanówka D, Kiss A, Halíčková M, Macková J, Riemann D, Luterbacher J, Böhm R (2010) Monthly and seasonal temperature reconstructions for Central Europe derived from documentary evidence and instrumental records since AD 1500. Clim Change 101:69–107

Dobrovolný P, Brázdil R, Trnka M, Kotyza O, Valášek H (2015) Precipitation reconstruction for the czech lands, AD 1501–2010. Int J Climatol 35:1–14

Grove JM (2004) Little Ice Age: Ancient and Modern. Routledge, London, 718 p

Ložek V (1973) Příroda ve čtvrtohorách. Academia, Praha, 372 p

Ložek V (1999a) Ochranářské otázky ve světle vývoje přírody, 1. Část—Okno do minulosti—klíč k problémům současnosti. Ochrana přírody 54:7–12

Ložek V (1999b) Ochranářské otázky ve světle vývoje přírody, 2. Část—Vývoj současných ekosystémů. Ochrana přírody 54:35–40

Ložek V (1999c) Ochranářské otázky ve světle vývoje přírody, 5. Část—Holocén a jeho problematika. Ochrana přírody 54:131–136

Ložek V (2007) Zrcadlo minulosti—Česká a slovenská krajina v kvartéru. Dokořán, Praha, 200 p

Mangerud J, Andersen ST, Berglund B, Donner JJ (1974) Quaternary stratigraphy of Norden, a proposal for terminology and classification. Boreas 3:109–126

Margielewski W (2006) Records of the Late Glacial—Holocene palaeoenvironmental changes in landslide forms and deposits of the Beskid Makowski and Beskid Wyspowy Mts. Area (Polish Outer Carpathians). Folia Quaternaria 76:1–149

Matthews JA, Briffa KR (2005) The 'Little Ice Age': re-evaluation of an evolving concept. Geogr Ann 87A:17–36

Mosbrugger V, Utescher T, Dilcher DL (2005) Cenozoic continental climatic evolution of Central Europe. Proc Natl Acad Sci U S A 102:14964–14969

Možný M, Brázdil R, Dobrovolný P, Trnka M (2012) Cereal harvest dates in the Czech Republic between 1501 and 2008 as a proxy for March–June temperature reconstruction. Clim Change 110:801–821

Pánek T, Hradecký J, Minár J, Šilhán K (2010) Recurrent landslides predisposed by fault-induced weathering of flysch in the Western Carpathians. In: Calcaterra D, Parise M (eds) Weathering as a Predisposing Factor to Slope Movements. Geological Society, London, pp 183–199

Starkel L (1990) Stratygrafia holocenu jako interglacjału. Przegląd Geologiczny 38:13–16

Starkel L (ed) (1999) Geografia Polski—środowisko przyrodnicze. Polske Wydawnictwo Naukowe, Warszawa, 669 p

Stocker TF, Qin D, Plattner G-K, Tignor MMB, Allen SK, Boschung J, Nauels A, Xia Y, Bex V, Midgley PM (eds) (2013) Climate Change 2013: the Physical Science Basis. Working Group I Contribution to the Fifth Assessment Report of the Intergovernmental Panel on Climate Change. Cambridge University Press, Cambridge, UK, 1535 p

Svensson A, Andersen KK, Bigler M, Clausen HB, Dahl-Jensen D, Davies SM, Johnsen SJ, Muscheler R, Parrenin F, Rasmussen SO, Roethlisberger R, Seierstad I, Steffensen JP, Vinther BM (2008) A 60 000 year Greenland stratigraphic ice core chronology. Clim Past 4:47–57

Tolasz, R, Brázdil R, Bulíř O, Dobrovolný P, Dubrovský M, Hájková L, Halásová O, Hostýnek J, Janouch M, Kohut M, Krška K, Křivancová S, Květoň V, Lepka Z, Lipina P, Macková J, Metelka L, Míková T, Mrkbica Z, Možný M, Nekovář J, Němec L, Pokorný J, Reitschläger JD, Richterová D, Rožnovský J, Řepka M, Semerádová D, Sosna V, Stříž M, Šercl P, Škáchová H, Štěpánek P, Štěpánková P, Trnka M, Valeriánová A, Valter J, Vaníček K, Vavruška F, Voženílek V, Vráblík T, Vysoudil M, Zahradníček J, Zusková I, Žák M, Žalud Z. (2007) Climate Atlas of Czechia. Český hydrometeorologický ústav, Universita Palackého, Praha, Olomouc, 256 p

Vandenberghe J (2001) Permafrost during the Pleistocene in the north west and central Europe. In: Paepe R, Melnikov VP, van Overloop E, Gorokhov VD (eds) Permafrost response on economic development. Kluwer, Dordrecht, Environmental security nad natural resources, pp 185–194

Veron A, Novak M, Brizova E, Stepanova M (2014) Environmental imprints of climate changes and anthropogenic activities in the Ore Mountains of Bohemia (Central Europe) since 13 cal. kyr BP. Holocene 24:919–931

Vinther BM, Jones PD, Briffa KR, Clausen HB, Andersen KK, Dahl-Jensen D, Johnsen SJ (2010) Climatic signals in multiple highly resolved stable isotope records from Greenland. Quat Sci Rev 29:522–538

Wanner H (2000) Vom Ende der letzten Eiszeit zum mittelalterlichen Klimaoptimum. In: Wanner H, Gyalistras D, Luterbacher J, Rickli R, Salvisberg E, Schmutz C (eds) Klimawandel im Schweizer Alpenraum. Vdf Hochschulverlag AG an der ETH Zürich, Zürich, pp 73–78

Web Site of Czech Hydromeorological Institute. www.chmi.cz. Accessed 12 Feb 2010

Xoplaki E, Fleitmann D, Diaz H, von Gunten L, Kiefer T (eds) (2011) Medieval Climate Anomaly. Pages News 19: 4–32

Zachos J, Pagani M, Sloan L, Thomas E, Billups K (2001) Trends, rhythms, and aberrations in global climate 65 Ma to present. Science 292:686–693

Long-Term Geomorphological History of the Czech Republic

4

Tomáš Pánek and Veronika Kapustová

Abstract

Consisting of two distinct geotectonic domains (Bohemian Massif and Western Carpathians), the territory of the Czech Republic reveals exceptionally high diversity of landforms. Major geomorphological features of the Czech Republic are the result of Late Paleozoic Variscan orogenetic processes, Late Cretaceous marine transgression, Alpine tectonics since the Late Cretaceous and Cenozoic neovolcanic processes. Geomorphological history of the Bohemian Massif can be traced back to the late Permian ($\sim$260 Ma) but its major landforms (e.g. basic planation surfaces and river valleys) emerged after the exhumation of Cretaceous deposits during the Paleogene and Neogene. Gross land surface of the Western Carpathians is much younger and its subaerial evolution spans the last $\sim$15 Ma since the Miocene. Quaternary landscape evolution of the country was dictated by to this day continuing uplift of the majority of its area and changing climatic pattern resulting in the evolution of a staircase of river terraces, continental and mountain glaciation, loess accumulation and the development of various periglacial landforms. Holocene landscape evolution is under the dominance of fluvial and slope processes, which have been strongly modified by human activity since the Atlantic chronozone and especially the Bronze Age.

Keywords

Bohemian Massif • Western Carpathians • Morphostructures • Low-temperature thermochronology • River terraces • Denudation chronology

4.1 Introduction—Milestones of Geomorphological Evolution

The contact of two distinct geotectonic and geomorphic domains—Bohemian Massif and Western Carpathians—makes the Czech Republic a unique territory with exceptionally high diversity of landforms which originated throughout completely different time scales. Only a few countries in Europe offer a possibility to study such different morphostructures in a very limited space. The landforms of the Czech Republic are a product of complex tectonic evolution, lithological diversity and climate changes during the Cenozoic. However, the long-term evolution of relief of the Czech Republic needs to be considered in relation to four chronological milestones: (1) Variscan orogenetic processes, (2) Late Cretaceous transgression and deposition of a thick sedimentary pile across a vast part of the Bohemian Massif, (3) collision of Alpine and Carpathian orogenic wedges with a European passive margin leading to the thrusting of the Outer Western Carpathians and inducing several stages of neotectonic activity in the Bohemian Massif since the Late Cretaceous and (4) volcanic activity in the Eocene–Pleistocene time span, bringing forth numerous neovolcanic terrains, especially in the northern part of the Bohemian Massif.

T. Pánek (✉) · V. Kapustová
Department of Physical Geography and Geoecology, Faculty of Science, University of Ostrava, Ostrava, Czech Republic
e-mail: tomas.panek@osu.cz

V. Kapustová
e-mail: veronika.smolkova@osu.cz

T. Pánek and J. Hradecký (eds.), *Landscapes and Landforms of the Czech Republic*,
World Geomorphological Landscapes, DOI 10.1007/978-3-319-27537-6_4

Active and passive morphostructures which originated during these stages were sculptured in diverse climatic conditions involving predominantly humid tropical and subtropical climates in the Mesozoic and Tertiary and cold periglacial and temperate conditions during the Quaternary. The territory of the Czech Republic also shows an exceptionally long (>25 ka) history of human activity which has significantly modified its landforms and landscapes.

In this chapter, we briefly summarize main features of the geomorphological history of the Czech Republic, which started as early as the late Permian period. Its aim is not to provide an exhaustive review of landscape evolution since it has recently been a subject of other comprehensive studies (e.g. Czudek 2005). It rather tends to put together evidence from traditional geomorphic studies along with the results of modern research which are supported by geochronological data. Despite the fact that radiometric data related to various types of landforms and sediments are still relatively scarce in the Czech Republic, they bring a new insight into some major geomorphic problems including the origin of planation surfaces in the Bohemian Massif (Danišík et al. 2010) or chronology of the last glaciation in the highest mountains of the country (Mentlík et al. 2013; Engel et al. 2014).

4.2 Pre-Quaternary Evolution

Due to morphostructural distinctiveness of both domains and different time scales of their geomorphological evolution, the Bohemian Massif and Western Carpathians are described separately. Traditional evidence in pre-Quaternary landscape reconstruction—planation surfaces, correlative sediments and weathering mantles—has recently been complemented with radiometric and thermochronological data that allow better determining age constraints of landforms genesis. In such a context, an insight into the long-term evolution of both the Bohemian Massif and the Western Carpathians has experienced a great progress in the last decade.

4.2.1 Bohemian Massif

Variscan orogenetic processes terminated during the early Permian period having produced a rugged mountainous terrain, evidence of which is given by coarse-grained clastics (e.g. conglomerates) that were deposited at the foot of elevations during the Permian (especially Saxonian) (Chlupáč et al. 2002; Migoń and Danišík 2012). Intensive denudation processes reduced the mountain terrain to a planation surface (sometimes called late Permian peneplain or post-Hercynian planation surface), which most likely occupied a majority of the Bohemian Massif already in the late Permian (Balatka and Kalvoda 2008). Much of the Mesozoic including Late Triassic–Early/Late Cretaceous is only poorly documented from the point of view of geomorphology; however, thick kaolinitic and lateritic weathering mantles situated under Late Cretaceous sedimentary rocks point to weathering and denudation of the area under humid tropical conditions (Engel and Kalvoda 2002).

Of major importance for the genesis of contemporary landforms of the Bohemian Massif was the Late Cretaceous epoch when marine transgression (especially between the Cenomanian and Coniacian) caused burial of basement rocks by a thick pile of clastic deposits. It was especially sandstone facies of the Bohemian Cretaceous Basin that were later sculptured to the famous "rock cities". Nowadays, such sandstone landscapes represent first-order geosites of the Czech Republic (Cílek 2010; see Chaps. 11, 13, 16 and 17 in this book). However, the overall role and extent of the Late Cretaceous transgression has recently been significantly re-evaluated, mainly in the context of new thermochronological (AFT and (U-Th)/He) data (Danišík et al. 2010, 2012; Migoń and Danišík 2012). It is clear that the extent of Late Cretaceous sedimentation was much greater than previously assumed (e.g. Chlupáč et al. 2002) as it even covered recently uplifted areas such as Rychlebské hory Mts in the NE part of the Bohemian Massif where a > 6.5-km-long burial of basement rocks can be inferred from thermochronological data (Danišík et al. 2012). Thermochronological dating and thermal modelling have crucial implications for timing and origin of planation surfaces, which are traditionally interpreted as Paleogene (Eocene) etchplains (Czudek 2005 and references therein). A combination of zircon (U-Th)/He (ZHe), apatite fission track (AFT) and apatite (U-Th-[Sm])/He (AHe) thermochronology performed for samples from the planation surfaces in the main ridge of the Krkonoše Mts shows that such surfaces most likely originated by the exhumation of a late Permian planation surface from a several-kilometre-thick pile of Cretaceous sediments and that later processes (especially during the Paleogene) only completed its evolution (Danišík et al. 2010). Further research is needed to resolve whether a similar scenario of the evolution of planation surfaces is applicable to other areas of the Bohemian Massif, namely the Šumava Mts, Krušné hory Mts and Českomoravská vrchovina Upland (Fig. 4.1).

Orogenetic processes in the Alps and Carpathians induced major tectonic rejuvenation in the Bohemain Massif. The first major impulse in the form of rapid cooling and uplift is dated back to the Late Cretaceous and recorded via apatite fission-track data, tectonic deformations of Cretaceous deposits and sedimentary evidence (Danišík et al. 2010, 2012). Thermal history of the NE part of the Bohemian Massif implying burial, erosion and uplift is shown in Fig. 4.2. Several stages of neotectonic activity during the Tertiary led to the formation of the characteristic

Fig. 4.1 Planation surface (suggested by majority authors as "etchplain") in the ridge of the Králický Sněžník Mt. (*Photo* T. Pánek)

horst-and-graben structure of the Bohemian Massif with an elevated rim of mountains enclosing a rather stable or slightly uplifted Bohemian Basin (Ivan 1999). The Tertiary in Central Europe was characteristic by a predominantly warm and humid climate (Mosbrugger et al. 2005) that especially during the Paleogene paved the way for deep weathering and formation of kaolinitic mantles (e.g. granites of the Western Bohemia or SE marginal slope of the Bohemian Massif), reaching locally several tens of metres in thickness (Kužvart 1965). The most important stage in the evolution of caves in the Bohemian Massif occurred between the Paleogene and the Lower Miocene (Bosák 1985). Final shaping of the "main" planation surface is supposed to have taken place during the Eocene (e.g. Czudek 2005). An important neotectonic pulse took place starting from the Oligocene when massive volcanic activity formed the Ohře/Eger rift that affected the NW and N parts of the Bohemian Massif (Chlupáč et al. 2002). Although a majority of the volcanic products were removed by erosion, recent landforms of selectively denuded sub-volcanic bodies form characteristic landscapes in the NW and N parts of the Bohemian Massif (e.g. České Středohoří Mts and Doupovské hory Mts; see Chap. 12 in this book). Flat surfaces cutting neovolcanic rocks (Balatka and Kalvoda 2008) and associated weathering mantles (Migoń and Lidmar-Bergström 2001) serve as an evidence of post-volcanic planation that most likely took place in the Lower Miocene.

Main features of recent major valley networks were established in the Late Tertiary (especially the Miocene and Pliocene) (Malkovský 1975; Tyráček and Havlíček 2009). However, the reconstructed Miocene drainage pattern significantly differs from the recent one (Fig. 4.3). In comparison with the modern situation, major differences involve drainage of the upper part of the Labe catchment eastward to the Paratethys Basin (i.e. recent Carpathian Foredeep), outflow of the southern part of the Vltava catchment southward to the Alpine foreland and termination of the rivers from Western Bohemia (e.g. Berounka) in Neogene freshwater lakes that were situated along the Eger rift below the Krušné

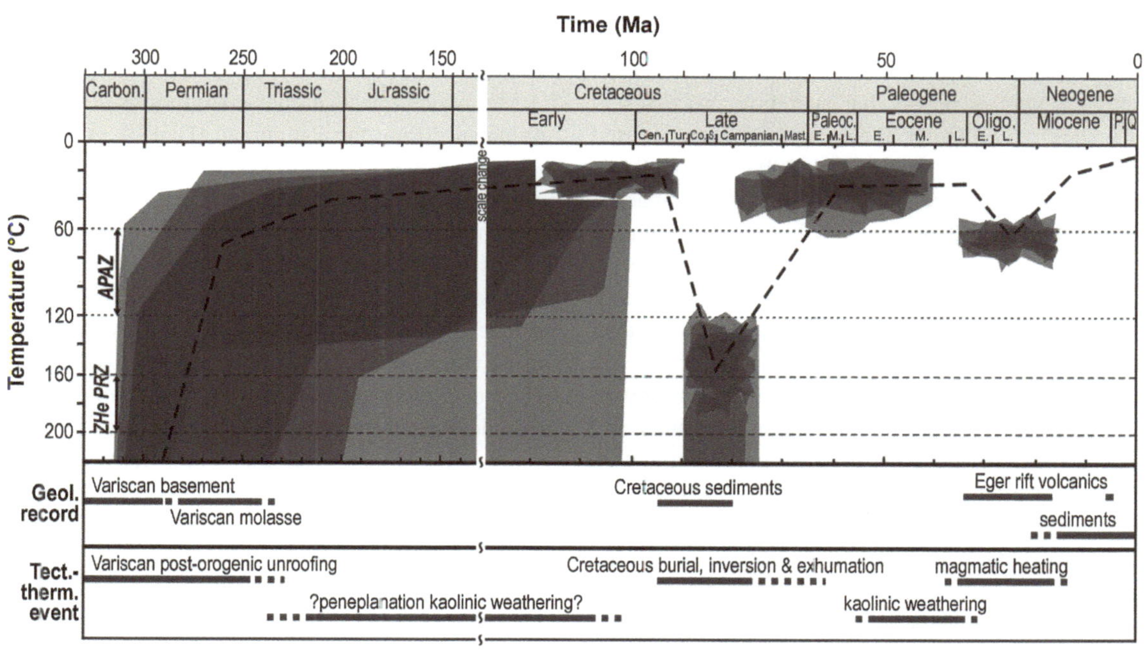

Fig. 4.2 Thermal history of the NE part of the Bohemian Massif (Rychlebské hory Mts.) revealing main stages of uplift, denudation and the burial (Reprinted from Tectonics, 31, Danišík et al. (2012) by the American Geophysical Union, with permission from John Wiley and Sons)

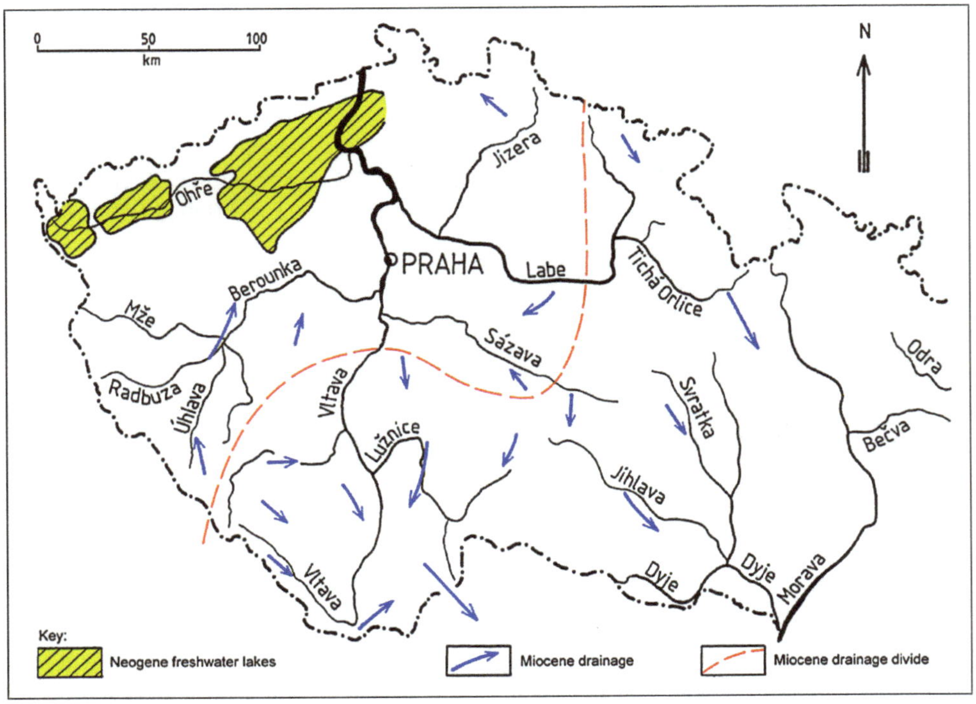

Fig. 4.3 Reconstructed Miocene drainage pattern in comparison with the present river network (Reprinted from Global and Planetary Change, 68/4, Tyráček and Havlíček (2009), with permission from Elsevier)

hory Mts (Tyráček and Havlíček 2009). An interesting problem is the existence of deeply incised valleys buried by Neogene deposits, which are frequent along the SE margin of the Bohemian Massif (Czudek 2005 and references therein).

4.2.2 Western Carpathians

Recent low-temperature AFT and (U-Th)/He thermochronological studies reveal that accretion processes in the flysch wedge of the Outer Western Carpathians started already in the Eocene times (Botor et al. 2006; Danišík et al. 2008) but the final emplacement of nappes and subaerial landscape evolution of the thrust-and-fold belt started in the Miocene between the Karpatian and Badenian (Menčík et al. 1983; Stráník et al. 1993). Therefore, gross land surface of the flysch Outer Western Carpathians is hardly older than ~15 Ma.

Post-thrusting Neogene evolution of the topography of Outer Western Carpathians is still a matter of doubts and exact evidence of the neotectonic evolution and denudation chronology of this area is rather fragmentary. Earlier ideas (e.g. Demek 1976; Buzek 1969) assumed three levels of post-Badenian planation surfaces that originated during cyclic phases of planation interrupted by tectonic activity. However, recent detailed studies reveal that the surfaces originally considered as "planation surfaces" are in fact structural features predisposed by sub-horizontally-inclined flysch beds (Ivan et al. 2000; Pánek et al. 2009). New radiometric dating performed in various parts of the Outer Western Carpathians suggest that planation surfaces, if existing, are generally younger than Pannonian (i.e. <11.6–7.1 Ma). In the area of Bílé Karpaty Nappe (Magura Unit), 1.4 km of post-Sarmatian denudation is estimated according to stratigraphic methods and dating of neovolcanic rocks (Bíl et al. 2004). Thermochronology based on (U-Th)/He dating shows that planation surfaces in the Podbeskydská pahorkatina Upland (Silesian Unit) are much younger than 7.1 Ma, which rejects earlier assumptions that these surfaces originated during the Pannonian (Danišík et al. 2008).

It seems that planation surfaces of various genesis (e.g. pediments) are common in the Czech part of the Western Carpathians, but they probably do not represent denudation levels which can be correlated across large territories (e.g. between the Outer and Inner Carpathians). Besides other evidence, a few million years did not provide sufficient time for the origin of regional-scale planation surfaces. They are rather local erosion surfaces which originated in site-specific geological environments and in relation to local base-level conditions. A classic example is represented by pronounced flats cutting anticlines of the Pavlov Hills in Southern Moravia (e.g. Stolová hora Mt; Fig. 4.4), which probably originated partly due to abrasion of the Upper Badenian Sea (Ivan 1973).

Specific pre-Quaternary evolution is attributed to Neogene basins situated at the front of flysch nappes. They involve both Carpathian Foredeep basins (e.g. Ostrava

Fig. 4.4 Planation surface (probably of the Upper Badenian/Miocene age) in the top of the Stolová hora Mt., Pavlov Hills in Southern Moravia (*Photo* J. Vítek)

Basin, Moravian Gate or Hornomoravský úval Basin) and the northern promontory of the Vienna Basin (part of the Pannonian Basin). Up to the Badenian times, these basins provided important accumulation areas covered by the Paratethys Sea. Some of these basins remained to be important accumulation centres up to the end of the Miocene, for example the Vienna Basin (herein presented as the Dolnomoravský úval Basin) which intensively subsided in the pull-apart regime (Ivan et al. 2000). However, post-Miocene basin inversion caused tectonic uplift of many previous accumulation areas and their Miocene deposits were included in adjacent upland units. Examples of uplifted Neogene deposits can be found in the Kyjovská pahorkatina Upland along the western rim of the Vienna Basin or in the Litenčická pahorkatina Upland in the vicinity of the Vyškov Gate foredeep basin (Fig. 4.5).

Fig. 4.5 Neogene deposits of the Karpatian age (Miocene) were uplifted and contemporary form landslide-prone hilly landscape along the southern rim of the Vyškov Gate (Litenčická pahorkatina Hilly land-Větrník hill) (*Photo* T. Pánek)

4.3 Pleistocene Geomorphological Evolution

It is assumed that tectonic activity in the Pleistocene continued from the Neogene both in the Bohemian Massif and the Western Carpathians (Czudek 2005). At the beginning of the Quaternary, the overall topography of the Czech Republic was largely similar to the current one (Czudek 2005). A somewhat different picture was related to the central part of the country, the Bohemian Basin, characterized by very flat landscape dominated by meandering rivers (Balatka and Kalvoda 2008; Tyráček and Havlíček 2009). Early and Middle Pleistocene intensification of uplift in this area caused river incision and formation of the characteristic landscape for central Bohemia with canyon-like valleys (e.g. Vltava, Sázava or Berounka Rivers; see Chap. 5 in this book) incised into a pre-Quaternary planation surface (Balatka and Kalvoda 2008). Similar tectonic movements (so-called Drahany phase) affected the eastern part of the Bohemian Massif and the Carpathian Foredeep (Tyráček and Havlíček 2009).

Tectonic movements are assumed for many faults and the most prominent structures in the Bohemian Massif seem to be the Krušné hory Fault, Mariánské Lázně Fault, Hronov-Poříčí Fault and the Sudetic Marginal Fault. It should be emphasized that the NE part of the Bohemian Massif (Bruntál volcanic field in the Nízký Jeseník Upland) experienced significant volcanic activity during the Pleistocene, with the youngest volcanic activity dated back to the Late Pleistocene (Cajz et al. 2013) (see Chap. 24 in this book).

4.3.1 Evolution of River Valleys

Pleistocene evolution of rivers in the Czech Republic is well documented by a staircase of river terraces that developed along a majority of river systems in the Bohemian Massif and the Western Carpathians. Although the research of river terraces has a long tradition in the country (Balatka and Sládek 1962; Tyráček and Havlíček 2009 and references therein), radiometric data exactly determining the age of individual accumulations are almost missing (Homolová et al. 2012).

There is a major difference between the evolution of river valleys in the western (Bohemia) and eastern (Moravia and Silesia) parts of the Czech Republic. It is reflected in the number of river terraces, thickness of their alluvial component and vertical spacing of individual terrace risers (Tyráček and Havlíček 2009). Terrace systems of the largest rivers in the Bohemian Massif, including drainage basins of the Labe and Vltava Rivers, usually consist of seven main terrace accumulations with several secondary levels (see Chap. 5 in this book). Their evolution started in the Early

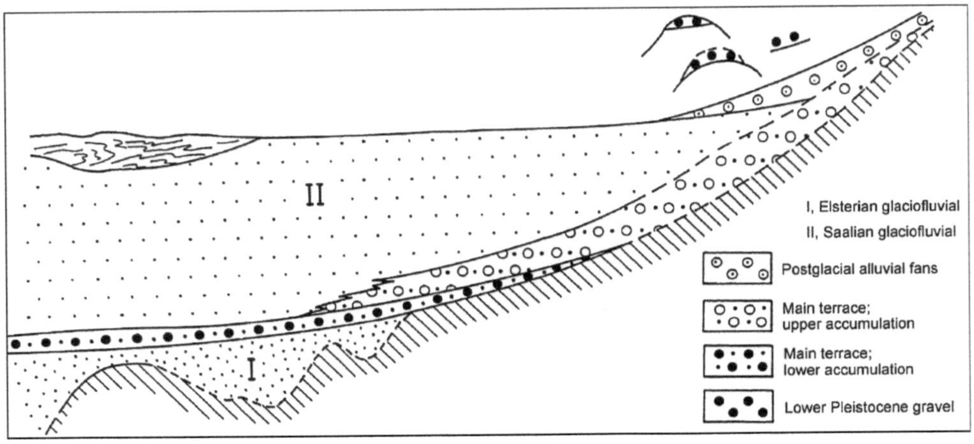

Fig. 4.6 Morphostratigraphical scheme of the fluvial and glacial sediments in the Odra River catchment of the Northern Moravia and Silesia (Reprinted from global and planetary change, 68/4, Tyráček and Havlíček (2009), with permission from Elsevier)

Pleistocene, following neotectonic uplift of the central part of the Bohemian Massif and cyclic climate-driven changes of the environment. According to traditional models (which have recently been refined and modified by absolute dating, e.g. Homolová et al. 2012), sandy and gravely accumulations of river terraces were results of braided rivers activity during cold glacial or stadial phases, whereas humid interglacial/interstadial conditions were favourable for river incision. Total Pleistocene river incision in the central Bohemian Massif is about 100 m, which is inferred from the altitudinal position of Late Neogene terrace remnants (Balatka and Kalvoda 2008). However, the rate of vertical erosion was uneven throughout the Pleistocene, with a few centimetres per ka during the Early and Late Pleistocene and up to 60–80 cm/ka during the younger part of the Middle Pleistocene (Balatka and Kalvoda 2008). Such fast Middle Pleistocene incision was responsible for the formation of canyon-like valleys of the Labe, Vltava, Berounka and Sázava Rivers. Middle Pleistocene (Holstein) incision of the Labe River in the Děčínská vrchovina Hilly Land (between Děčín town and Hřensko area) caused overdeepening of the valley by 90 m (Balatka and Kalvoda 1995). Specific Pleistocene evolution of river terraces was documented for the Ohře River in the NW part of the Bohemian Massif that follows the tectonically active Ohře/Eger rift system. Balatka and Sládek (1975) reported a stretch of 23 terrace steps in this valley that most likely reflect pronounced Quaternary tectonic mobility of this area.

The eastern part of the Czech Republic, including the eastern margin of the Bohemian Massif and the Western Carpathians, reveals different aspects of the Pleistocene river valley evolution. The area of the Central and Southern Moravia (Morava River catchment) contains a well-developed sequence of Pleistocene river terraces including five terraces along the Morava River and six terraces in the Dyjskosvratecký úval Basin (Czudek 2005; Tyráček and Havlíček 2009). These terraces reflect progressive eastward tilting of the eastern margin of the Bohemian Massif towards the Carpathian Foredeep (Tyráček and Havlíček 2009). A characteristic feature of this area was a major change in river courses, exemplified, e.g. by Early Pleistocene SW-oriented drainage of the Morava River through the Vyškov Gate. Recent southward-oriented course of the Morava River through the Napajedla Gorge was induced by tectonic subsidence at the Lower/Middle Pleistocene transition (Tyráček and Havlíček 2009). The NE part of Moravia and Silesia within the catchment of the Odra River reveals an incomplete sequence of river terraces with only 2–3 steps (Fig. 4.6). This is partly result of two Middle Pleistocene advances of continental glaciers which destroyed the majority of Early Pleistocene terrace remnants. Of major importance in the Odra River basin is the Main Terrace which originated during the early Saalian and can be correlated with similar river terraces in the Morava River basin (Tyráček and Havlíček 2009).

4.3.2 Continental and Mountain Glaciation

Knowledge of the extent and chronology of the Pleistocene continental and mountain glaciation in the Czech Republic has recently experienced great progress (Nývlt et al. 2011). Three Northern European ice sheets advanced to Northern Bohemia and Northern Moravia and Silesia during both the Elsterian (marine isotope stages /MIS/16 and 12) and older Saalian (MIS6) glaciations (Nývlt et al. 2011). Southward advance of glaciers was blocked by the northern slopes of the Sudetes and the Beskydy Mountains. Repeated ice sheet advances formed conspicuous terminal moraines with complicated internal structure and frequent glacitectonic features, typical example of the Hlučín terminal moraine NW from the Ostrava City (Macoun et al. 1965). Northern Bohemia and Silesia experienced maximum glaciation during the Elsterian, whereas the southernmost strike of the continental glacier in the Ostrava Basin and the Moravian Gate took place during the first stage of the Saalian glaciation when the

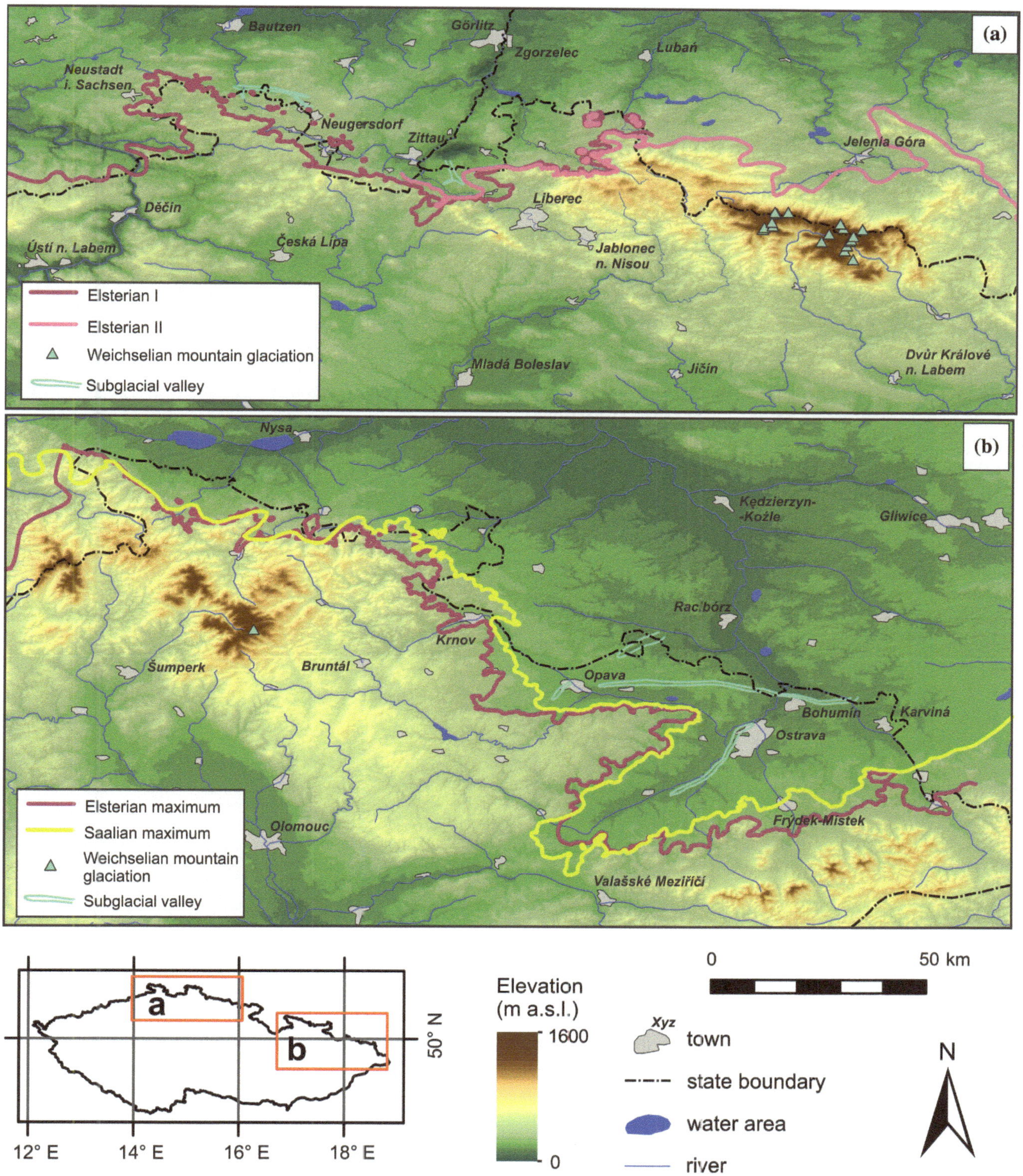

Fig. 4.7 Extent and chronology of the Pleistocene continental and mountain glaciation in the northern part of the Czech Republic (*Data sources* elevation—SRTM3, base map data—digital chart of the world, glaciation—Nývlt et al. 2011)

ice sheet and its meltwater crossed the main European watershed between the Odra and Danube Rivers (Fig. 4.7; Macoun et al. 1965; Tyráček and Havlíček 2009; Nývlt et al. 2011).

Mountain glaciation was restricted to the highest mountains of the Czech Republic. The existence of rather small mountain glaciers is only verified in the Krkonoše Mts, Šumava Mts (Bohemian Forest) and Hrubý Jeseník Mts. Nevertheless, the existence of small glaciers in the highest parts of the Krušné hory Mts, Jizerské hory Mts, Králický Sněžník Mt and the Moravskoslezské Beskydy Mts cannot be completely ruled out. Although it is supposed that the glaciation of the highest mountains repeated several times during the Pleistocene (Nývlt et al. 2011), recent radiometric dating (e.g. Mentlík et al. 2013; Engel et al. 2014) reveals that morphologically distinct moraine ridges are not older than the Weichselian. Absolute chronology of the last glaciation in the Šumava and Krkonoše Mts has recently been determined by radiometric dating involving cosmogenic nuclide dating (^{10}Be) of moraine boulders and polished bedrock surfaces as well as radiocarbon dating of basal sequences of peat-bogs filling glacial cirques (Mentlík et al. 2010, 2013; Engel et al. 2014). Sedimentary record from the glacier cirque of the Labský důl valley in the Krkonoše Mts reveals that this area was ice-free at the end of MIS3 (~30 ka BP) before the onset of the last glaciation during MIS2 (Engel et al. 2010). Maximum extent of glaciers in the highest part of the Krkonoše Mts seems to be synchronous with the Last Glacial Maximum and is ^{10}Be-dated to ~21 ka BP, whereas the last small glaciers on the slopes with cold (N to NE) orientation readvanced during the Younger Dryas cold event ~13 ka BP (Fig. 4.8; Engel et al. 2014). The glacial chronology of the Šumava Mts (Bohemian Forest) is in agreement with the local glaciation chronology in the Krkonoše Mts, with maximum ^{10}Be-dated advance of mountain glaciers ~24–20 ka BP (Mentlík et al. 2013) and the final retreat of glaciers from the cirques occurring at ~14 ka BP (Mentlík et al. 2010, 2013; see Chap. 8 in this book).

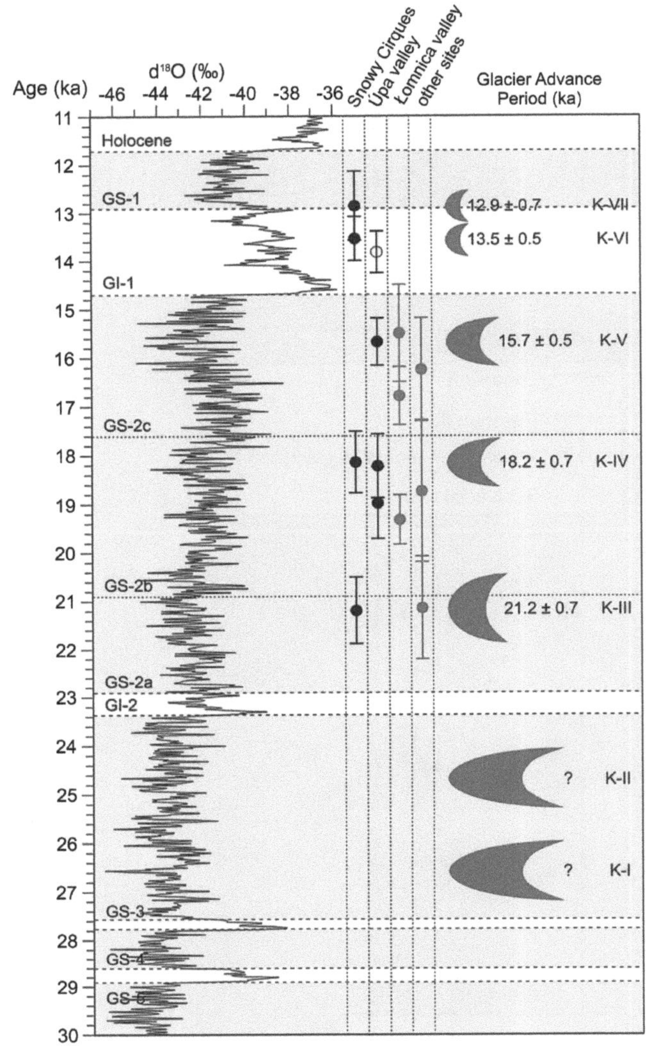

Fig. 4.8 Chronology of the last glaciation in the Krkonoše Mts. (reprinted from geomorphology, 206, Engel et al. (2014), with permission from Elsevier)

4.3.3 Periglacial Phenomena

Situated between two large European ice sheets (e.g. Alps and Northern Europe), during the cold stages of the Pleistocene, a majority of the area of the Czech Republic was covered by tundra and/or taiga landscape with the predominance of periglacial processes. Good evidence of periglacial landscape evolution comes especially from the Weichselian Pleniglacial when the thickness of the permafrost locally exceeded 200 m (Czudek 2005). A recent study on cryogenic cave carbonates suggests that a minimum permafrost thickness in the Bohemian Karst was 65 m during the second half of the Weichselian (Žák et al. 2012). Landform assemblages related to Pleistocene periglacial zone are common both in the mountains (especially the Sudetes) where they are represented by patterned ground, cryoplanation terraces, blockfields, rock glaciers, nivation hollows and in the case of less elevated landscapes rich in dells, by asymmetric valleys, solifluction sheets and ice wedge casts (Czudek 2005; Treml et al. 2010; see Chap. 22 in this book). High intensity of periglacial processes and slope retreat in less resistant rocks (such as flysch or Neogene clastics in Central and Southern Moravia) is documented by the evolution of cryopediments, which are represented by gently inclined piedmont surfaces truncating often intensively folded bedrock (Vandenberghe and Czudek 2008). Recent investigation of Pánek et al. (2014) revealed that the originally rugged surface morphology of a large landslide (~55 ka old) protruding to the foothills of the Beskydy Mountains was significantly smoothed and bevelled in maximally several tens of thousands of years during the second half of the Weichselian Glacial (Fig. 4.9).

Fig. 4.9 Pediment-like surface at the northern margin of the Outer Western Carpathians (Podbeskydská pahorkatina hilly land in the vicinity of Frýdek-Místek town) originated due to the periglacial transformation of landslide body dated to ~55 ka BP (*Photo* T. Pánek)

Open periglacial landscape abundant in weathered material and sandy deposits along the floodplains of braided rivers was highly susceptible to wind-blown sediment transport processes that led to the accumulation of both sand dunes and loess across the country (Czudek 2005). Cold stadial conditions with limited vegetation cover were particularly suitable for aeolian processes. Recent morphologically distinct sand dunes (up to 19 m high) along the Labe, Lužnice and Morava Rivers probably originated during the Upper Pleniglacial and Younger Dryas (Havlíček 2007; see Chap. 28 in this book). Loess deposits (forming often loess-paleosol sequences) are common both in the lowlands and hilly landscapes. Some of the loess outcrops are among the most famous and studied ones in the world. For instance, the loess outcrop called "Calendar of Ages" in Southern Moravia (Pavlov Hills) reflects landscape changes throughout the whole Weichselian glacial almost continuously (Antoine et al. 2013; see Chap. 29 in this book).

4.4 Holocene Geomorphological Evolution

Recent paleoseismic investigation on selected faults (e.g. Sudetic Marginal Fault or the Mariánské Lázně Fault) reveals the ongoing crustal deformations in the Holocene. Štěpančíková et al. (2010) stated that the Sudetic Marginal Fault at the foot of the Rychlebské hory Mts was reactivated as a normal fault after the early Holocene colluvial sedimentation but before the occurrence of buried soil dated to 430 ± 120 cal yrs BP (see Chap. 21 in this book). Paleoseismic trenching within the Mariánské Lázně fault in Western Bohemia revealed fault-related warping of deposits dated to ca 4.8 ka BP (Štěpančíková and Fischer 2012). Such information emphasizes the fact that the Holocene/recent activity of many faults in the Bohemian Massif and Western Carpathians should be re-evaluated.

Main geomorphological agents acting in the territory of the Czech Republic during the humid and temperate conditions of the Holocene were fluvial and slope processes that started to be significantly modified by human activities since the Atlantic chronozone (i.e. Neolithic). Czech rivers witnessed the development of floodplains, which have recently been organized into 2–3 low terrace steps (Czudek 2005). They usually consist of two sequences: coarse-grained channel facies (gravels and sands) overlain by finer overbank deposits (sands, silty sands). The overall thickness of floodplain deposits is up to ~10 m in the case of the largest rivers in the country (Czudek 2005). Gravels forming the basal sediments of floodplains started to accumulate already during the Late Glacial. For instance, gravels in the lower part of the Dyje River floodplain (Southern Moravia) were deposited as early as in the Late Pleniglacial (~19.5 cal ka BP) and were episodically re-sedimented up to the early Holocene (Havlíček 2007). The re-sedimentation of older floodplain sequences was often related to major catastrophic floods. For example, in the case of the Odra River floodplain (Northern Moravia), Czudek and Hiller (2001) described complete re-sedimentation of >2 m thick gravely sand layer during floods in the second half of the thirteenth century. The onset of the overbank accumulation was often related to human impact (deforestation, agricultural activities, etc.) and therefore not considered synchronous across the country. On the other hand, since overbank deposits of the rivers in Bohemia and Central and Southern Moravia started to accumulate most likely already during the Neolithic (with acceleration during the "Medieval Warm Period"; Kadlec et al. 2009), the accumulation onset in the mountains of the Outer Western Carpathians was probably not earlier than during the sixteenth to seventeenth centuries when substantial parts of mountain slopes were deforested during the so-called "Wallachian colonisation" (Fig. 4.10; Stacke et al. 2014). More radiocarbon data on floodplain sequences are needed to establish a reliable chronology of the Holocene evolution of Czech rivers. It is especially relevant for rivers in the Bohemian part of the country where radiometric data related to floodplains are almost missing.

Recent investigations and dating of Holocene slope processes reveal that especially neovolcanic terrains of the Northern Bohemia, elevated areas of the Bohemian Cretaceous Basin and the flysch Outer Western Carpathians experienced several stages of mass movement activity. Landslides and earthflows with length up to >1 km originated usually during more humid phases of the Holocene (e.g. Atlantic, beginning of Subboreal and Subatlantic, etc.; Pánek et al. 2013). It is assumed that frequency of shallow

Fig. 4.10 Overbank deposits of the Bečva River (thickness ~2 m) in the Western Carpathians were deposited since the deforestation of slopes in the basin during the "Wallachian Colonization" from sixteenth to seventeenth centuries (*Photo* T. Pánek)

landslides, together with sheet and gully erosion, fluctuated in accordance with human impact during the Holocene.

Some landscapes in the Czech Republic have been transformed by direct anthropogenic impact. One of the major changes is evidenced by the shortening of river channels by up to 30 % in some areas (e.g. lower course of the Svratka River, Southern Moravia) in the last 250 years (Demek et al. 2008). In the last 200 years, mining-affected regions such as the Ostrava Basin (see Chap. 25 in this book) or lignite basins in the NW Bohemia (e.g. Cheb, Sokolov and Most Basins) have undergone nearly complete transformation of the landscape accompanied by the emergence of new topographical features.

4.5 Conclusion

Major geomorphological features of the Czech Republic are the result of Late Paleozoic Variscan orogenetic processes, Late Cretaceous marine transgression, Alpine tectonics starting from the Late Cretaceous and Cenozoic neovolcanic processes. Geomorphological history of the Bohemian Massif can be traced back to the late Permian (~260 Ma), but its major landforms (e.g. basic planation surfaces or river valleys) emerged after the removal of Cretaceous deposits during the Paleogene and Neogene. Gross land surface of the Western Carpathians is much younger and its subaerial evolution has been taking place for the last ~15 Ma since the Miocene. The morphostructures of both the Bohemian Massif and Western Carpathians were significantly reshaped by processes conditioned by climate changes that involved humid and mostly warm conditions in the Mesozoic and Tertiary and a highly fluctuating climate pattern in the Quaternary. Although both the continental and mountain glaciation only affected a small fraction of the country (~10 %), periglacial processes in the extraglacial zone became significantly imprinted in the topography of the Czech Republic across all altitudinal zones. It is of great importance for the geodiversity of the Czech Republic that both various types of tectonic and structural landforms (e.g. SW marginal fault scarp of the Krušné hory Mts, or sandstone "rock cities" in the Bohemian Paradise) and climate-related generations of landforms (e.g. glacial cirques in the Krkonoše Mts) can be demonstrated by numerous examples which are accessible for tourists and open to ongoing scientific research.

References

Antoine P, Rousseau DD, Degeai JP, Moine O, Lagroix F, Kreutzer S, Fuch M, Hatté C, Gauthier C, Svoboda J, Lisá L (2013) High-resolution record of the environmental response to climatic variations during the Last Interglacial-Glacial cycle in Central Europe: the loess-palaeosol sequence of Dolní Věstonice (Czech Republic). Quatern Sci Rev 67:17–38

Balatka B, Sládek J (1962) Říční terasy v českých zemích. Geofond v nakl. ČSAV, Praha, 578 pp

Balatka B, Sládek J (1975) Geomorfologický vývoj dolního Poohří. Rozpravy ČSAV, ř. mat. a přír. věd 87: 98 (In Czech)

Balatka B, Kalvoda J (1995) Vývoj údolí Labe v Děčínské vrchovině. Sborník ČGS 100:173–192 (In Czech)

Balatka B, Kalvoda J (2008) Evolution of Quaternary river terraces related to the uplift of the central part of the Bohemian Massif. Geografie-Sborník ČGS 113:205–222

Bíl M, Krejč O, Franců J, Hrouda F, Přichystal A (2004) Approximation of missing eroded sediments in the Bílé Karpaty unit (Magura group of the nappes, Outer West Carpathians). Studia Geomorphologica Carpatho-Balcanica 38:59–66

Bosák P (1985) Periody a fáze krasovění v Českém krasu. Český kras 11:36–55 (In Czech)

Botor D, Dunkl I, Rauch-Wlodarska M, von Eynatten H (2006) Attempt to Dating of Accretion in the West Carpathian Flysch Belt: Apatite Fission Track Thermochronology of Tuff Layers. Geolines 20:21–23

Buzek L (1969) Geomorfologie Štramberské vrchoviny. Spisy Pedagogické fakulty v Ostravě 11, 91 pp (In Czech)

Cajz V, Skácelová Z, Schnabl P, Radoň M (2013) Svrchněkenozoický severomoravský vulkanismus: rekonstrukce činnosti, paleomagnetismus, geofyzikální obraz, návrh litostratigrafie. Zprávy o geologických výzkumech 2012:20–25 (In Czech)

Chlupáč I, Brzobohatý R, Kovanda J, Stráník Z (2002) Geologická minulost České republiky. Academia, Praha, 436 pp (In Czech)

Cílek V (2010) Saxon-Bohemian Switzerland: sandstone rock cities and fascination in a romantic landscape. In: Migoń P (ed) Geomorphological Landscapes of the World. Springer, pp 201–209

Czudek T, Hiller A (2001) Vývoj údolní nivy řeky Odry v Ostravské pánvi. Geografie—Sborník ČGS 106: 94–99

Czudek T (2005) Vývoj reliéfu krajiny České republiky v kvartéru. Moravské zemské muzeum, Brno, 238 pp (In Czech)

Danišík M, Pánek T, Matýsek D, Dunkl I, Frisch W (2008) Apatite fission track and (U-Th)/He dating of teschenite intrusions gives time constraints on accretionary processes and development of planation surfaces in the Outer Western Carpathians. Zeitschrift für Geomorphologie, NF Hauptbänd 52:273–289

Danišík M, Migoń P, Kuhlemann J, Evans NJ, Dunkl I, Frisch W (2010) Thermochronological constraints on the long-term erosional history of the Karkonosze Mts. Central Europe. Geomorphology 117:78–89

Danišík M, Štěpančíková P, Evans NJ (2012) Constraining long-term denudation and faulting history in intraplate regions by multisystem thermochronology: An example of the Sudetic marginal Fault (Bohemian Massif, central Europe). Tectonics 31. doi: 10.1029/2011TC003012

Demek J (1976) Planation surfaces of the Moravian Carpathians (Czechoslovakia). Sborník ČSSZ 81:9–15

Demek J, Havlíček M, Chrudina Z, Mackovčin P (2008) Changes in land-use and the river network of the Graben Dyjsko-svratecký úval (Czech Republic) in the last 242 years. J Landscape Ecol 2:22–51

Engel Z, Kalvoda J (2002) Morphostructural development of the sandstone relief in the bohemian cretaceous basin. In: Přikryl R, Viles H (eds) Understanding and Managing stone decay. Karolinum Press, Praha, SWAPNET, pp 225–231

Engel Z, Nývlt D, Křížek M, Treml V, Jankovská V, Lisá L (2010) Sedimentary evidence of landscape and climate history since the end of MIS 3 in the Krkonoše Mountains, Czech Republic. Quatern Sci Rev 29:913–927

Engel Z, Braucher R, Traczyk A, Léanni L, Team Aster (2014) ^{10}Be exposure age chronology of the last glaciation in the krkonoše mountains, Central Europe. Geomorphology 206:107–121

Havlíček P (2007) Kvartérně—geologický výzkum a vývoj údolní nivy v přírodním parku 'Niva Dyje mezi Břeclaví a Lednicí. Zprávy o geologických výzkumech 2006:58–59

Homolová D, Lomax J, Špaček P, Decker K (2012) Pleistocene terraces of the Vltava River in the Budějovice basin (Southern Bohemian Massif): New insight into sedimentary history constrained by luminescence data. Geomorphology 161–162:58–72

Ivan A (1973) Outline of denudation chronology of the Mikulovská vrchovina (Highland). Folia, Facultatis Scientiarum Naturalium Universitatis Purkynianea Brunensis—Geographia 13: 35–43

Ivan A (1999) Geomorphological aspects of late Saxonian epiplatform orogeny of the Bohemian Massif. Moravian Geogr Rep 7:18–33

Ivan A, Kirchner K, Krejčí O (2000) K poznání morfostrukturních rysu reliéfu moravské části Západních Karpat a Panonské pánve. Geografický časopis 52:221–230

Kadlec J, Grygar T, Světlík I, Ettler V, Mihaljevič M, Diehl JF, Beske-Diehl S, Svitavská-Svobodová H (2009) Morava River floodplain development during the last millennium, Strážnické Pomoraví, Czech Republic. Holocene 19:499–509

Kužvart M (1965) Geologické poměry moravskoslezských kaolínů. Sborník Geologických Věd 6:87–146 (In Czech)

Macoun J, Šibrava V, Tyráček J, Kneblová-Vodičková V (1965) Kvartér Ostravska a Moravské brány. Archive Manuscript, Ústřední ústav geologický, Praha. 419 pp (in Czech with German summary)

Malkovský M (1975) Palaeogeography of the Miocene of the Bohemian Massif. Věstník Ústředního ústavu geologického 50:27–31

Menčík E, Adamová M, Dvořák J, Dudek A, Jetel J, Jurková A, Hanzlíková E, Houša V, Peslová H, Rybářová L, Šmíd B, Šebesta J, Tyráček J, Vašíček Z (1983) Geologie Moravskoslezských Beskyd a Podbeskydské pahorkatiny (Geology of the Moravskoslezské Beskydy Mountains and Podbeskydská pahorkatina hilly country). Ústřední Ústav Geologický, Praha (in Czech, with English Summary)

Mentlík P, Minár J, Břízová E, Lisá L, Tábořík P, Stacke V (2010) Glaciation in the surroundings of Prášilské Lake (Bohemian Forest, Czech Republic). Geomorphology 117:181–194

Mentlík P, Engel Z, Braucher R, Léanni K, Team Aster (2013) Chronology of the Late Weichselian glaciation in the Bohemian Forest in Central Europe. Quatern Sci Rev 65:120–128

Migoń P, Lidmar-Bergström K (2001) Weathering mantles and their significance for geomorphological evolution of central and northern Europe since the Mesozoic. Earth-Sci Rev 56:285–324

Migoń P, Danišík M (2012) Erosional history of the Karkonosze Granite Massif—constraints from adjacent sedimentary basins and thermochronology. Geol Q 56:441–456

Mosbrugger V, Utescher T, Dilcher DL (2005) Cenozoic continental climatic evolution of Central Europe. Proc Natl Acad Sci 102:14964–14969

Nývlt D, Engel Z, Tyráček J (2011) Pleistocene glaciations of Czechia. In: Ehlers J, Gibbard PL, Hughes PD (eds) Quaternary Glaciations—Extent and Chronology. Elsevier, Amsterdam, pp 37–46

Pánek T, Hradecký J, Minár J, Hungr O, Dušek R (2009) Late Holocene catastrophic slope collapse affected by deep-seated gravitational deformation in flysch: Ropice Mountain, Czech Republic. Geomorphology 103:414–429

Pánek T, Smolková V, Hradecký J, Baroň I, Šilhán K (2013) Holocene reactivations of catastrophic complex flow-like landslides in the Flysch Carpathians (Czech Republic/Slovakia). Quatern Res 80:33–46

Pánek T, Hartvich F, Jankovská V, Klimeš J, Tábořík P, Bubík M, Smolková V, Hradecký J, (2014). Large Late Pleistocene landslides from the marginal slope of the Flysch Carpathians. Landslides 11:981–992

Stacke V, Pánek T, Sedláček J (2014) Late Holocene evolution of the Bečva River floodplain (Outer Western Carpathians, Czech Republic). Geomorphology 206:440–451

Stráník Z, Menčík E, Eliáš M, Adámek J (1993) Flyšové pásmo Západních Karpat, Autochtonní mesozoikum a paleogén na Moravě a ve Slezsku. In: Přichystal A, Obstová V, Suk M (eds) Geologie Moravy a Slezska. Moravské zemské muzeum a Sekce geologických věd PřF MU, Brno, pp 107–122 (In Czech)

Štěpančíková P, Hók J, Nývlt D, Dohnal J, Sýkorová I, Stemberk J (2010) Active tectonics research using trenching technique on the south-eastern section of the Sudetic Marginal Fault (NE Bohemian Massif, central Europe). Tectonophysics 485:269–282

Štěpančíková P, Fischer (2012) Late Quaternary activity within the Mariánské Lázně Fault zone as revealed by trenching survey; Cheb Basin, Kopanina site In: Blahůt J, Klimeš J, Štěpančíková P, Hartvich F (eds) Sborník abstraktů 12. mezinárodní konference Stav geomorfologických výzkumů v roce 2012 Sokolov 18.-20.4.2012, pp 51–52

Treml V, Křížek M, Engel Z (2010) Classification of Patterned Ground Based on Morphometry and Site Characteristics: A Case Study from the High Sudetes, Central Europe. Permafrost Periglac Process 21:67–77

Tyráček J, Havlíček P (2009) The fluvial record in the Czech Republic: A review in the context of IGCP 518. Glob Planet Change 68:311–325

Vandenberghe J, Czudek T (2008) Pleistocene cryopediments on variable terrain. Permafrost Periglac Process 19:71–83

Žák K, Richter DK, Filippi M, Živor R, Deininger M, Mangini A, Scholz D (2012) Coarsely crystalline cryogenic cave carbonate—a new archive to estimate the Last Glacial minimum permafrost depth in Central Europe. Clim Past 8:1821–1837

Part II

Landscapes and Landforms

The Geomorphological Evolution and Environmental Hazards of the Prague Area

5

Jan Kalvoda and Břetislav Balatka

Abstract

Landform evolution of the Prague area in the central part of the Bohemian Massif was controlled by the coupled occurrence of episodic tectonic uplift and variable climato-morphogenetic processes during the Cenozoic. Much older geological history of the region commenced in the Precambrian times and was very diverse in terms of transformations of the natural environment. Present-day landform patterns of the Prague area are determined by epigenetic and antecedent deepening of canyon-like valleys of the Vltava River and its tributaries to large planation surfaces during the Quaternary. These dynamic processes have led to the origin of river accumulation terraces as well as erosion and denudation slopes with weathered mantle of deposits. The extraordinary geodiversity and biodiversity of the landscape in the Prague area is associated with geomorphic hazards, including devastating floods and landslides. Prague is also faced to severe impact of modern urban development and related human activities on the architectural heritage.

Keywords

Prague • Landform evolution • Geomorphological processes • Environmental hazards

5.1 Introduction

Historical location of Prague has been substantially influenced by favourable natural conditions, including its extraordinary efficient geographical position in the central part of the Bohemian Massif. Archaeological findings give evidence that the Prague area has been occupied since 5 000 years B.P. and variable cultures of the Neolitic and Bronze Ages are also documented (Fridrichová et al. 1995; Fridrich 1997). The history of settlements continued in the pre-Christian centuries (Celts, Slavonic tribes, etc.) and thanks to an attractive combination of environmental, especially relief features, climatic and hydrological conditions (Hrdlička 1984; Kubíková et al. 2005), was not interrupted up to now. Even the present-day heritage evidences of multi-cultural urban patterns of ancient Prague (Fig. 5.1), represented, e.g. by variable architectural styles (especially Gothic, Renaissance and Baroque), have grown around a thousand years. Geological and geomorphic factors played an important role in the specific location and development of Prague, and historical evidence also shows many ways in which human activities have modified landform and environmental characteristics (Fig. 5.2). The aim of this study is to explain when and how the main rock assemblages and landform patterns of the Prague area have been evolved. Principal geomorphic events in paleogeographical history of the area of Prague are emphasized, including the evolution of the Vltava River valley and its accumulation terraces during the Quaternary. Main recent geomorphic hazards are also illustrated as the topical evidence of relationships between natural and human processes in the environment of the Prague area.

J. Kalvoda (✉) · B. Balatka
Faculty of Science, Charles University in Prague, Albertov 6, 128 43 Prague 2, Czech Republic
e-mail: kalvoda@natur.cuni.cz

T. Pánek and J. Hradecký (eds.), *Landscapes and Landforms of the Czech Republic*, World Geomorphological Landscapes, DOI 10.1007/978-3-319-27537-6_5

Fig. 5.1 Landscape of Prague in the middle of the nineteenth century drawn by an anonymous artist 150 years ago. The Prague Kettle with the oldest districts of the town is surrounded by flat plains and hills of the Prague Plateau. Painting shows the Vltava valley by a bird-eye view from the south to the north that is downstream. The Prague Castle (Hradčany) and the Lesser Town are situated on the left side and the Old Town and the New Town on the right side of the Vltava River valley (Reproduction of the lithography from the archives of the Map Collection of the Charles University in Prague.)

The geomorphological unit of the Prague Plateau and the adjacent areas (Figs. 5.1 and 5.3) which includes planation surfaces, slightly inclined (mainly) denudational slopes of different age and, by contrast, deeply incised fluvial valleys, including the Prague Kettle depression, has been formed since the beginning of the Tertiary (Balatka 1985; Chlupáč 1999). The degree of uplift of the central part of the Bohemian Massif has been "masked" since that time by the concurrent action of differential tectonic movements and intensive erosion, denudation and transport of solid rock fragments and its weathered counterparts (Kalvoda and Balatka 2006). The current substantial height differences and relief contrasts have nevertheless developed only recently as in the late Cenozoic. Examples of maximal height differences in the Prague area are: (a) 225 m of total difference (a plateau at 400 m a.s.l. westwards from Zličín and the Vltava level below Prague at 175 m a.s.l.), (b) the Bílá hora Hill (380 m a.s.l.) is situated only 6.5 km from the Vltava river, i.e. a difference is about 200 m, (c) the Na Vidouli Plateau (371 m a.s.l.) is situated at 4 km and the Petřín Hill (318 m) only at 750 m from the Vltava river valley floor at an altitude of 188 m (Fig. 5.2), the difference of relative heights being thus 183 and 130 m, respectively. Valley meanders and bends (Fig. 5.4), characteristic of the middle course of the Vltava River, were formed as bends on the bottom of the Pliocene wide valley with a low longitudinal channel gradient. The contemporary landforms appeared during the phase of valley deepening during the Quaternary, mainly by the development of larger bends with flights of river terraces inside the bends.

5.2 Principal Geomorphic Events in Paleogeographical History of the Prague Area

Landform evolution in the Prague area was determined by neotectonic and climato-morphogenetic processes during the Cenozoic. However, main events in the geological history of the region are much older and very diverse in terms of

Fig. 5.2 Contemporary majestic view of the Prague city and its bridges (a view upstream) also shows susceptibility of the canyon-like Vltava valley to environmental problems associated with road traffic or other anthropogenic pollutions, floods and mass movements (*Photo* Michal Němejc)

transformations of the natural environment. The oldest crystalline rocks of the central part of the Bohemian Massif (Fig. 5.5) have a complex past. The process of their origin commenced with sedimentation of transported material from the weathered mantle of the Precambrian continent into the epicontinental sea. Marine transgression penetrated the region during the Late Proterozoic and early Paleozoic. Then, these marine sediments were metamorphosed to a different degree already during the early Paleozoic (Chlupáč 1999; Kříž 1999). Strong uplift occurred during the Cadomian tectogenesis, whilst weaker uplift also occurred in the Early Ordovician and was followed by a very strong uplift in the Carboniferous. These uplift episodes were accompanied by erosion and denudation, which were particularly severe during the early Variscan times.

The position of this foundation of the central part of the Bohemian Massif at the end of the Ordovician was over 60° of southern latitude (Chlupáč et al. 2002). The Bohemian Massif, as the margin of the Gondwana ancient continent, shifted to that place from the northern temperate and equatorial zone during the Cadomian orogenesis in the Late Precambrian. The Caledonian folding of Gondwana in the early Paleozoic occurred in the southern hemisphere and only as late as in the Carboniferous period did the Bohemian Massif return to the equatorial zone, i.e. in the period of the Hercynian (Variscan) orogenesis. These mountain building processes formed the Bohemian Massif as a structurally complex unit, the central part of which is formed by collision-deformed and metamorphosed crystalline rocks of the Moldanubicum (Fig. 5.5). As early as the Carboniferous, rapid denudation led to the unroofing of deeper parts of the crust. The Central Bohemian granitoid pluton, separating the Barrandien (Horný and Turek 1999) from the Moldanubicum block, is represented in its northern part by granitoids and by their mantle of contact-metamorphosed Proterozoic and Paleozoic rocks.

Large granitoid intrusions occurred in extensional conditions in the mature stage of the Variscan orogeny, followed

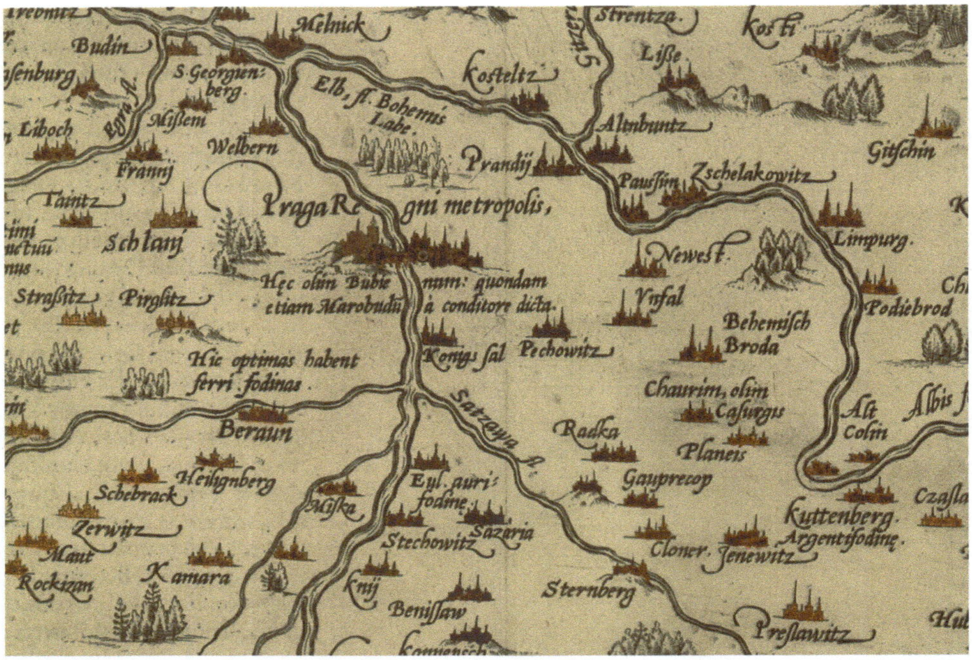

Fig. 5.3 Position of Prague (see locality "Praga regni metropolis") as seen on the map "Regni Bohemiae descriptio" (Abraham Ortelius, Antwerp 1570). Remarkable is a lifelike sketch of the drainage patterns in the middle Bohemia. However, they are conspicuous errors in drawing the confluence areas of the Sázava, Vltava and Berounka Rivers in the S of Prague. This historical map is a part of many editions of the "Teatrum orbis terrarum" by Abraham Ortelius and it is based on the Johann Criginger's map of Bohemia (1568). Dimensions of the sheet of map are 53 × 46 cm (Reproduction of the original map from the archives of the Map Collection of the Charles University in Prague.)

during its final stage by horizontal sliding movements. In the Late Paleozoic, some parts of the Central Bohemian Massif were deeply denuded and crystalline rocks from a depth of 15 km were exhumed exposing deep-seated granite massifs. The Prague region was dry land from the Late Permian to the Early Cretaceous and the Late Cretaceous transgression affected only the northern margin of this area. This period of tectonic stability saw the development of planation surfaces. The uplift of the Bohemian Massif at the end of the Santonian (some 65 million years ago) resulted from the ongoing Alpine and Carpathian orogenesis. These events marked the definitive retreat of the Late Cretaceous epicontinental sea which significantly receded leaving an erosional surface as a primary geomorphic surface for the region. The Bohemian Massif was also differentiated into a system of graben structures and tectono-volcanic zones.

At the beginning of the Tertiary, climate in the central part of the Bohemian Massif was humid and tropical, with a mean annual temperature of up to 26 °C and mean annual rainfall of 2 000–3 000 mm (Malkovský 1979). The occurrence of the pre-Oligocene planation surface is indicated by duricrust remnants in western and central Bohemia. In the Oligocene temperatures fell to 16 °C under savannah-type climate with dry winters, and a very dry climate prevailed also in the Middle Oligocene. The Late Oligocene was characterized by a permanently wet and warm climate, with subtropical rain forests remaining until the Middle Miocene (Malkovský 1975; Demek 2004). Up to the Paleogene, streams ran through shallow, wide vale-shaped low gradient valleys. However, at the end of the Oligocene, planation processes in the Bohemian Massif were interrupted by tectonic movements (e.g. Malkovský 1979; Chlupáč et al. 2002), accompanied by volcanic activities in its western part 35–17 million years ago.

The highest and oldest planation surfaces of Paleogene age are found westwards from Prague, at the present-day altitudes of 360–400 m, on Paleozoic and Cretaceous rocks. They are slightly inclined to the north. According to the geomorphic position of Miocene river sediments, it was originally an early Tertiary surface from which tropical regoliths were removed and the basal weathering surface was thus exposed during the Neogene. An example is the

Fig. 5.4 The location of Prague, dissected relief of its surroundings and a large incised meander of the Vltava River drawn by an anonymous artist as a part of *Ichnographia et orthographia metropolis Pragensis* (Reproduction of the cooper engraving (1740–1780) from the archives of the Map Collection of the Charles University in Prague.)

graded etchplain on Upper Cretaceous spongolites (argillites) at a locality west of Prague—at the Václav Havel Airport Prague.

During the Early Miocene, tropical humid climate with dry periods prevailed in the central part of the Bohemian Massif. This later changed to a subtropical wet climate in the Late Miocene. Periods of humid climate in the Neogene were characterized by very extensive erosion and denudation of the kaolinitic and lateritic weathering mantle, down to the basal weathering surface. The internal differentiation of this planation surface of Neogene age was dependent on rock resistance to weathering under tropical or subtropical climate. Moreover, the evolution of the relief of the Bohemian Massif was influenced by two stages of volcanic activities, in the Late Miocene between 9.0 and 6.4 Ma, and from the Late Pliocene to the Pleistocene, between 3.0 and 0.17 Ma ago (Ulrych et al. 2011). Morphostructural patterns of the Bohemian Massif, originating during the Miocene, determined the main elements of present-day river network.

The river valleys in the central part of the Bohemian Massif (Fig. 5.3), and thus also their terrace flights, are the product of processes of hydrographical capturing of several Miocene individual catchments with different drainage directions. For example, Neogene sediments near Jesenice, south of Prague,

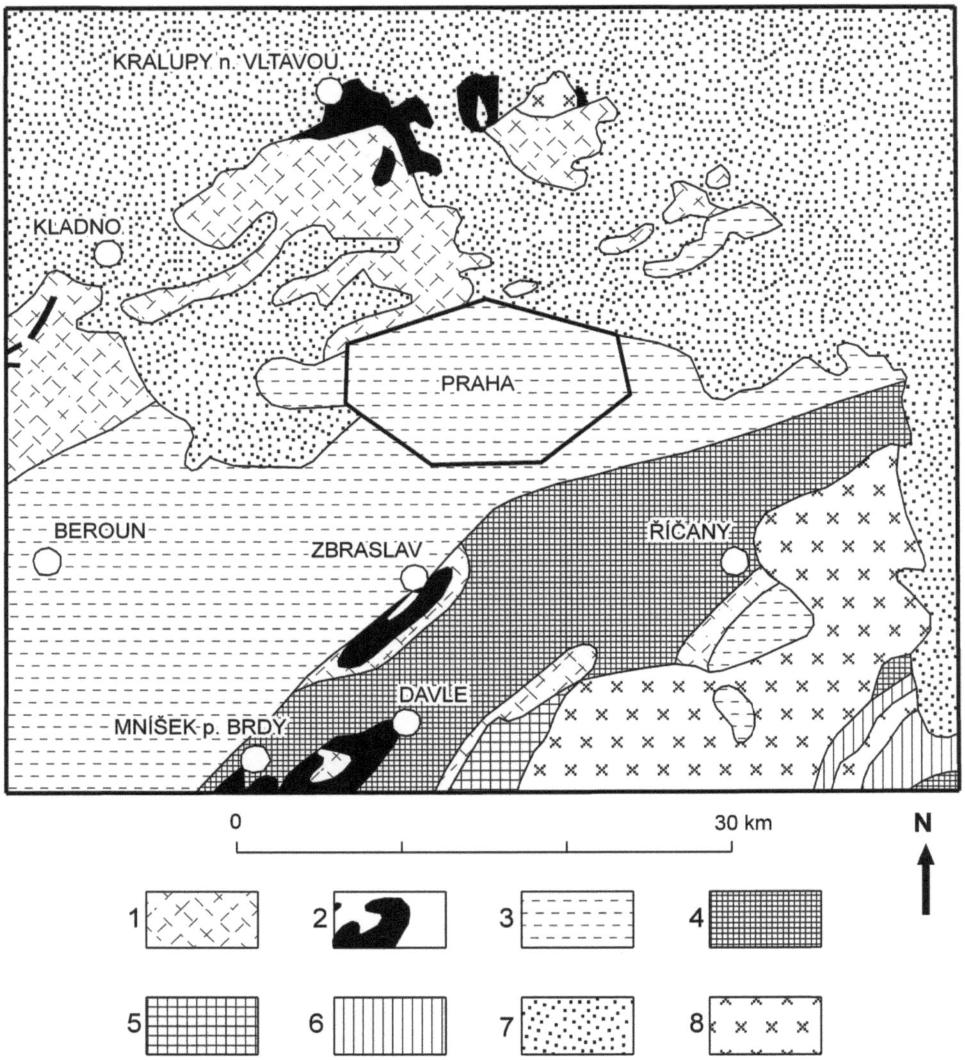

Fig. 5.5 Geological sketch of central Bohemia (Chlupáč et al. 2002). Explanations: *1* Proterozoic metamorphosed rocks of the Zbraslav and Kralupy units, *2* Proterozoic volcanites, *3* Cambrian to Devonian sediments and metamorphosed rocks of the Barrandian unit, *4* Proterozoic rocks of the Štěchovice unit, *5* metamorphosed subsilicic rocks, *6* metamorphosed rocks of an uncertain age, *7* Upper Carboniferous to Tertiary sediments, *8* Variscan (Hercynian) granitic rocks

fill deep channels near the Sázava—Vltava watershed (Kovanda et al. 2001). They indicate traces of drainage of the lower Sázava catchment to the north. In the Middle and Late Miocene, the substantial upper part of the Vltava catchment in the southern Bohemia was still drained towards the south (Tyráček 2001; Tyráček and Havlíček 2009). It is indicated by both relics of fluvial and lacustrine sediments and finds of river-transported moldavites (= specific rock types related to the meteorite impact) in the adjacent part of Austria. These tektites originated during the Ries Impact and are radiometrically dated at 14.3 million years.

The granular character of Pliocene river sediments is similar to those of Lower Pleistocene terrace deposits which indicate that the orographic situation of the central part of the Bohemian Massif was closely similar to one that occurs today (Balatka and Štěpančíková 2006; Kalvoda and Balatka 2006). The oldest and highest, mostly Early Pleistocene accumulation terraces survived only very sporadically and in small patches above the edges of the present-day valley incisions. Important changes in the fluvial network system occurred at that time with significant manifestations of epigenetic and antecedent evolution of river valleys through a rapid erosion.

5.3 River Terrace Evolution in the Prague Area During the Quaternary

The flight of fluvial deposits and related river terraces of the Sázava, Berounka, Vltava and Labe river valleys in the central part of the Bohemian Massif (Fig 5.3) has traditionally been used as a reference framework for the Quaternary stratigraphy of the region. It is also realised (e.g. Záruba et al. 1977; Tyráček et al. 2004; Balatka and Kalvoda 2008, 2010) that the terrace system, widespread along the Vltava and other major rivers in the central part of the Bohemian Massif (Fig. 5.6), developed as a result of regional neotectonic uplift.

As a part of geomorphological research in the central part of the Bohemian Massif, the longitudinal profiles of fluvial terrace accumulations and Neogene sediment localities, the

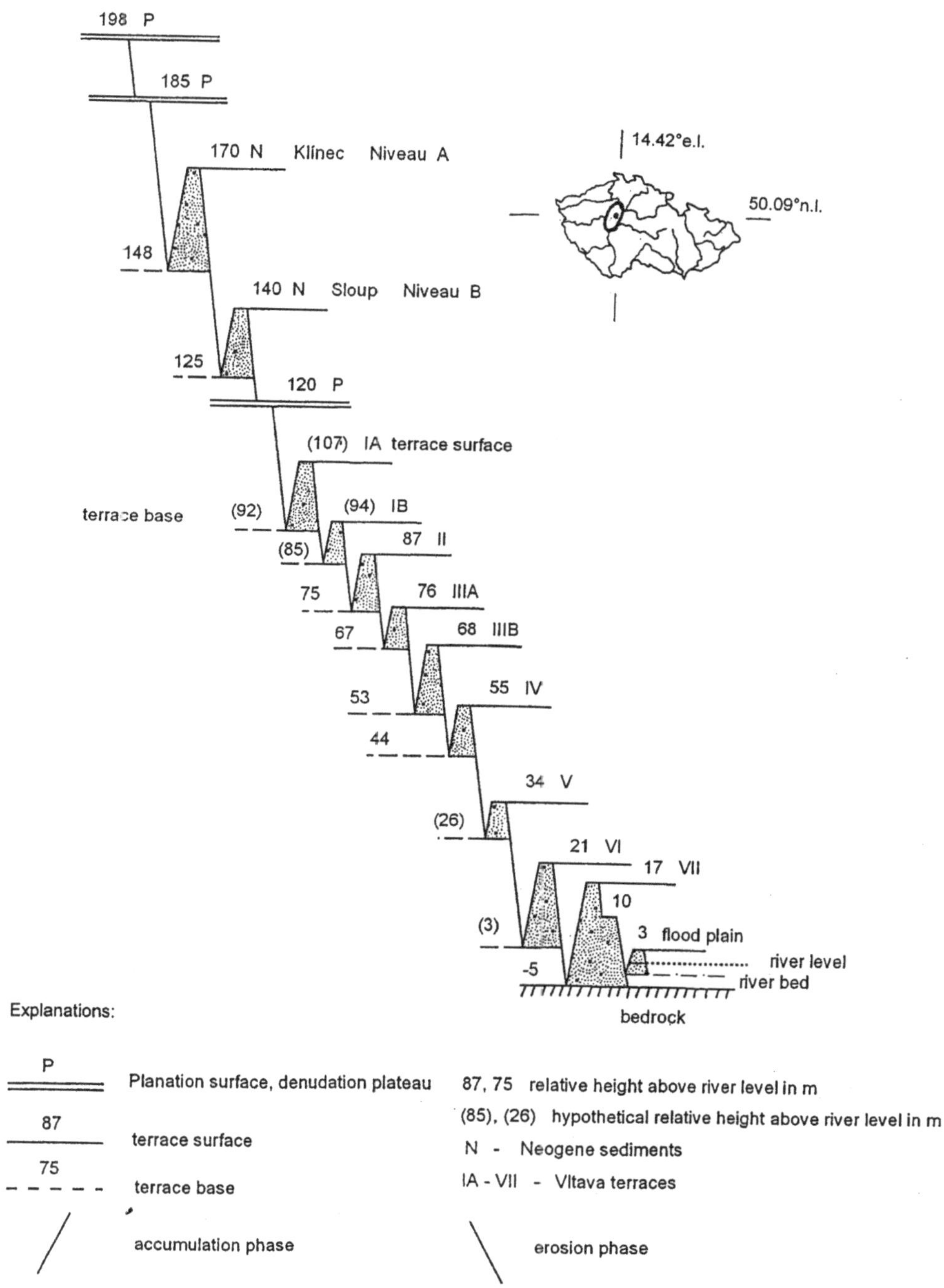

Fig. 5.6 Position of fluvial accumulation terraces in the Vltava valley between the mouths of the Sázava and Berounka Rivers south of Prague (Balatka and Štěpančíková 2006; Balatka and Kalvoda 2010). The chronostratigraphical correlation of accumulation terraces in the central part of the Bohemian Massif is demonstrated in Table 5.1

structure of valley cross-sections and the major occurrences of planation surfaces have been plotted (Balatka et al. 2010a, b; Balatka et al. 2015). This method of interpreting the valley evolution builds strongly on the assumption that the paleo-thalweg and the surfaces of each major terrace level maintained stable gradients that correspond to the contemporaneous longitudinal profiles. In this state, the discharge and transport capacity at each position along the river channel is in equilibrium with upstream sediment delivery and, averaged over millenia, the river thus neither erodes nor accumulates sediment but applies all its energy to the transfer of transported material. This state may be disturbed, either in the direction of net erosion or in that of net accumulation, as a consequence of differential tectonic movements and/or climate changes influencing discharge regime and sediment supply. In the Vltava canyon-like valley (Fig. 5.7) and other major valleys of the central part of the Bohemian Massif, increased water and sediment supply were associated with intensive cryogenic processes during the colder intervals in the Pleistocene. In these circumstances huge accumulation packages formed, altering the equilibrium profile to a new state over valley sub-reaches.

The sedimentary and morphological records of the Quaternary evolution of antecedent valleys and river accumulation terraces in the central part of the Bohemian Massif are

Fig. 5.7 Very steep rocky slopes of the antecedent Vltava River valley near its confluence with the Berounka River, consisted of Paleozoic calcareous sediments and metamorphosed rocks of the Barrandian unit, were modified by numerous anthropogenic changes as early as the nineteenth century. Historical drawing recorded the landscape of the southern periphery of Prague with an old railway and the St. John Church near the Chuchle area (Reproduction of the lithography (1887) from the archives of the Map Collection of the Charles University in Prague.)

Table 5.1 Chronostratigraphical correlation of river terraces in the central part of the Bohemian Massif related to global stratigraphical scheme for the Quaternary (Balatka et al. 2010a, b)

Regional stratigraphical stage/substage divisions of the Quaternary (Gibbard and Cohen 2008, Gibbard et al. 2009)	SÁZAVA Balatka and Štěpančíková (2006), Balatka (2007), Kalvoda (2007), Balatka and Kalvoda (2010)	BEROUNKA Balatka and Loučková (1992)	VLTAVA—LABE confluence area Balatka and Sládek (1962)	VLTAVA in the Prague region Záruba et al. (1977)	VLTAVA and LABE system Tyráček (2001), Tyráček et al. (2004)
Late Pleistocene Weichselian	Pikovice Terrace (VII)	Lipence Terrace (VIIa)	Hostín Terrace (VIIa, b, c, d)	Maniny Terrace (VII)	Maniny Terrace (Weichselian)
		Dobřichovice Terrace (VIIb)			Hostín 1 Terrace
Middle Pleistocene Saalian (Warthe)	Poříčí Terrace (VI)	Kazín Terrace (VI)	Mlčechvosty Terrace (VIa, b, c)	Veltrusy Terrace (VI)	Veltrusy Terrace (Warthe)
Middle Pleistocene Saalian (Drenthe)	Městečko Terrace (V)	Liblín Terrace (Va)	Cítov Terrace (Va, Vb)	Dejvice Terrace (V)	Dejvice 1 and 2 Terrace (Drenthe)
		Poučník Terrace (Vb)			
Middle Pleistocene Saalian (Fuhne)	Týnec Terrace (IV)	Zbraslav Terrace (IVa)	Hněvice Hill Terrace (IV)	Letná Terrace (IV)	Letná Terrace (Fuhne)
		Hýskov Terrace (IVb)			
Middle Pleistocene Elsterian	Buda Terrace (IIIb)	Srbsko Terrace (IIIb)	Straškov Terrace (IIIb)	Vinohrady Terrace (IIIB)	Vinohrady Terrace (Elster)
Middle Pleistocene Cromerian complex (Glacial c)	Chabeřice Terrace (IIIa)	Tetín Terrace (IIIa)	(IIIa)	Kralupy Terrace (IIIA)	Kralupy Terrace (Cromerian C)
Middle Pleistocene Cromerian complex (Glacial c)	Český Šternberk Terrace (II)	Pohořelec Terrace (IIa)	Ledčice Terrace (II)	Pankrác Terrace (II)	Pankrác Terrace (Cromerian C)
		Hlince Terrace (IIb)			
Middle Pleistocene Cromerian complex (Glacial b)	Hvězdonice Terrace (Ib)	Řevnice Terrace (Ib)		Suchdol Terrace (IB)	Suchdol Terrace (Cromerian B)
Middle Pleistocene Cromerian complex (Glacial a)	Střechov Terrace (Ia)	Skryje Terrace (Ia)	Krabčice Terrace (I)	Lysolaje Terrace (IA)	Lysolaje Terrace (Cromerian A)
Early Pleistocene Bavelian (Dorst) Menapian			Rovné Terrace		Rovné Terrace (Dorst)
					Vráž Terrace (Menapian)
Early Pleistocene Eburonian—Menapian	Niveau B Radvanice	Niveau B		Zdiby Stadium (Pliocene)	Zdiby Terrace (Eburonian—Menapian)
Early Pleistocene Tiglian					Stříbrníky Terrace (upper Tiglian)
Neogene	Niveau A Bojiště	Niveau A		Klínec Stadium	

correlated with the European chronostratigraphical scheme for the Quaternary (Table 5.1).

The oldest terrace accumulations in the Prague area are situated above the margins of the canyon-like valleys of Vltava, Berounka and Sázava rivers (e.g. Záruba-Pfeffermann 1941, 1942; Záruba et al. 1977; Kovanda et al. 2001; Tyráček et al. 2004; Balatka and Kalvoda 2008, 2010). Relics of Miocene gravels and sands at the Sulava lokality, near Radotín town have their surface lowered by erosion at 358 m a.s.l. and their base at 314 m a.s.l., i.e. 163 m or 119 m above the Berounka River. Other relics of sediments of Miocene and Pliocene age are recorded from the neighbourhood of Slivenec, near Suchomasty and on Bílá Hora (380 m a.s.l.). The surface of Early Pleistocene sands and gravels which are up to 40 m thick, between Kobylisy and Sedlec on the Zdibská plošina Plateau, is situated at 300–325 m a.s.l., i.e. 125–150 m above the Vltava level, and 35–60 m below the Ládví hill (359 m a.s.l.). Northwards from these Pliocene spreads on the Zdibská plošina Plateau are up to 20 m thick sediments with their surface 112 m above the Vltava River level. These sediments originated within the so-called Lysolaje group of terraces during the Middle Pleistocene (Table 5.1). In the Early Pleistocene, the Vltava and its tributaries were still freely meandering in shallow and wide valleys (Fig. 5.4) formed on Neogene planation surfaces.

Even as late as the beginning of the Middle Pleistocene, the basal boundary of which is the Matuyama/Brunhes paleomagnetic transition dated at 780 ka, new terrace steps in the valleys of the central part of the Bohemian Massif were progressively formed (70–100 m above the present water courses) together with a relatively rapid epigenetic and antecedent deepening of the river network. For example, the Suchdol Terrace (Fig. 5.8) in the Prague area is situated up to 2 km west of, and 96 m above the Vltava valley floor. The Straškov (IIIb) Terrace of Balatka and Sládek (1962) is now *ca.* 70 m above the Vltava River near Račiněves in the neighbourhood of the Říp mountain. It is described by Tyráček (2001) as the Straškov 2 Terrace and as an equivalent of the Vinohrady Terrace in Prague (Table 5.1). The fluvial deposits underlying the Straškov Terrace consist of a

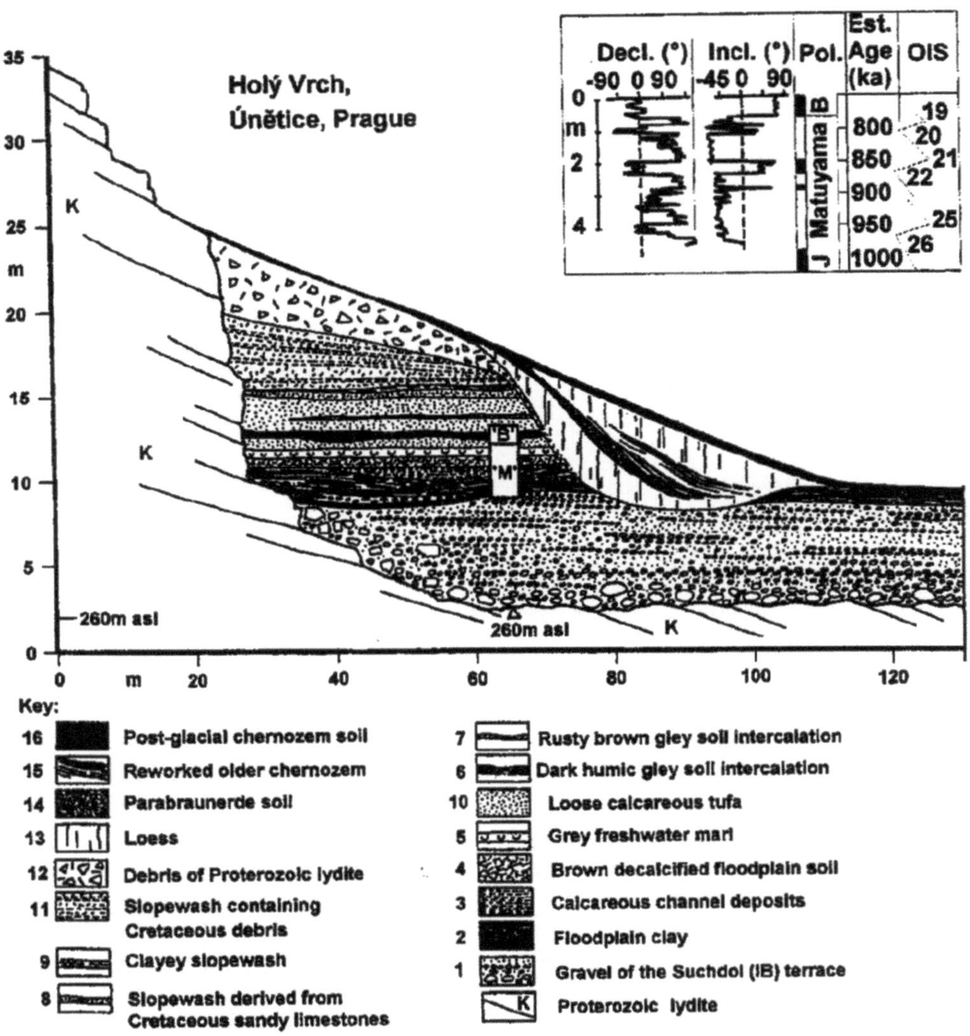

Fig. 5.8 Cross-section through the Suchdol Terrace of the Vltava River of the Middle Pleistocene age (Záruba et al. 1977; Tyráček et al. 2004). Oblong cut (B/M) in the middle part of the cross-section indicates a possible magnetostratigraphical boundary of Matuyama and Brunhes chrons

coarse lower and a finer upper unit (Tyráček et al. 2004). These sediments are overlain by loess and slope deposits that include paleosols probably representing two warm interglacial stages. The 12–14 m thick lower fluvial units with stratified sands and gravels indicate a cold-climate braided-channel environment. The 0.5–2 m thick upper fluvial unit is composed of sand and fine sandy gravel, disturbed by cryoturbation. It has yielded remnants of thermophilous mammals, interglacial molluscs and Paleolithic archaeological material.

After the end of the sediment accumulation of the IIIrd terrace in the Middle Pleistocene, the valley bends and meanders have been abandoned in several places, and a series of lower terraces developed as accumulation bodies in alluvial reaches of the valleys during a significant stage of long-term erosional valley deepening. River-bed dislocations and deep erosion (Figs. 5.7 and 5.9) were caused by the change in the local erosional base during highly variable erosional-denudational and accumulation processes. Changes in the intensity of these morphogenetic processes were related to neotectonic movements and non-uniform resistance of bedrock as well as to changes of climatic conditions in the late Cenozoic.

The values of the antecedent deepening of the Vltava River in the Prague area (stimulated by tectonic uplift), based on the position of remnants of river accumulation terraces (Fig. 5.6), are to be estimated with caution. Uncertainties include the possibility that terrace surfaces may have been irregularly lowered by erosion, and variability in the range and episodic rhythm of tectonic uplift. However, the results of the estimation provide data about the dynamics of fluvial bedrock erosion and transportation of weathered material in the region of central Bohemia during the late Cenozoic (Kalvoda and Balatka 2006; Kalvoda 2007) and are as follows: (a) Middle Miocene to Pliocene: rate of deepening about 2–4 cm ka^{-1}, (b) Early Pleistocene: 6–12 cm ka^{-1}, (c) the younger part of the Middle Pleistocene: 6–8 cm 10 a^{-1}, (d) a part of the Late Pleistocene (40–20 ka): 2–4 cm ka^{-1}. During the Holocene are mostly recycling of gravels and sands occurred and new slope accumulations in

Fig. 5.9 The Vyšehrad Castle was built on a large rocky platform and slopes (build by sandstones, slates and metamorphic rocks of the Paleozoic age) between two deeply incised erosion valleys of small right-side tributaries of the Vltava River. The painting records the Vyšehrad Castle with a new fortification (completed in the year 1668) as one of historical stages during permanent anthropogenic changes of relief in the Prague area (Reproduction of the lithography (1890–1910) archives of the Map Collection of the Charles University in Prague.)

the valley bottom originated. Besides of the system of river accumulation terraces, wind-blown sands, loess loams and loess (Fig. 5.8) also provide valuable sedimentary evidence of Quaternary landscape evolution (e.g. Šibrava 1972; Záruba et al. 1977; Tyráček 2001; Balatka et al. 2010a, b). These deposits are maintained in a significant thickness in depressions or on lower plateaux around Prague.

Geomorphological analysis of late Cenozoic fluvial sediments preserved in the central part of the Bohemian Massif confirm that seven main terrace accumulations in total, with several secondary levels, can be differentiated (Table 5.1). The relative height of the oldest fluvial terraces above the present-day bottom of river valleys in the Prague area exceeds 100 m (Fig. 5.6) which indicates the approximate depth of Quaternary erosion. An estimation of the values of the antecedent deepening of the Vltava in the late Cenozoic, based on the position of remnants of river accumulation terraces, suggests that the rate of downward erosion of the Vltava reached its maximum in the younger part of the Middle Pleistocene.

5.4 Geomorphic and Environmental Hazards

The dynamics of fluvial processes in the Prague area was deeply influenced by weathering, denudation and mass movements during the late Quaternary (Figs. 5.10 and 5.11). Apparently, main patterns and/or relics of pre-historic favourable natural and life conditions have been sustained in present-day Prague. For example, in the Prague area the mean year temperatures 8–11 °C with mild winters and precipitations between 400–800 mm, including relatively dry summer, are typical. Remarkable geo- and biodiversity, many kilometres of streams as well as patches of fertile soils on plateaux around the city still exist. However, the city and its surroundings are exposed to permanent serious threats from a variety of geomorphic and environmental hazards (compare Figs. 5.7, 5.11 and 5.12). Steep slopes built of hard Paleozoic rocks (e.g. lydite, quartzite and limestone) in deeply incised valleys of subsequent streams to the Vltava river emphasize picturesque landscape of the Prague area. At the same time, its dissected relief (e.g. Figs 5.9 and 5.10) is

Fig. 5.10 The Hradčany Castle is surrounded by dissected relief with ancient as well recent features of rapid erosion and variable slope movements including landslides. Extraordinary graphical work from the first-half of the nineteenth century presents the eastern area of the Hradčany Castle, including the Deer Ditch valley and rocky walls consisted of sandstones and greywacke of the Paleozoic age (Reproduction of the cooper engraving (1831) from the archives of the Map Collection of the Charles University in Prague.)

characteristic by different types of active slope processes, especially soil erosion and numerous mass movements, including rockfalls and very fast moving landslides.

The oldest reliable record of flooding in Prague is concerned with the disastrous flood in the year 1118 and more than 150 floods are mentioned in historical sources. Since the fifteenth century are some of them denoted by flood marks (e.g. near the Charles bridge) or recorded by flooding of buildings in the Old Town area of Prague. Main causes of the Vltava floods are extreme rainfall events a rapid thawing of snow in extensive watershed of the river. Before constructions of stone embankments and cascade of dams upstream was the area of Prague along the river also afflicted by floods with ice-jam effects. Substantial hazards are influences of extreme rainfall events to changes of groundwater flow systems, activity of landslides and flash-flood erosion and deposition. Actual research topics after two extreme events of 1997 and 2002 (Fig. 5.12) in the Czech Republic are concerned with quantifying feedbacks between climate variability and anthropogenic activities at various spacio-temporal scales. The challenge related to river management is the consideration of man-made floodplain modifications influencing the cross-section area and the hydraulic roughness significantly.

Prague is a city of great architectural and historic importance, but its ancient site and geomorphic position in deeply incised valleys and within dissected relief pose considerable problems in terms of environmental hazards and building foundation conditions. Many ancient buildings had been erected before ground conditions of valley side slopes and Quaternary deposits were understood (compare Figs. 5.10 and 5.12). Much of the valuable heritage of Prague is under threat, not only from changes of engineering-geological conditions (Píchal et al. 1979; Cílek 1995), but also from the present-day air pollution. There are many urban sources of particulates including traffic, combustion of fossil fuels and natural dust. Especially fine-grained particles pose health hazards and they contribute to soiling and damage to building, bridges, statues

Fig. 5.11 Conspicuous canyon-like valley of the Vltava River at the northern periphery of Prague developed by rapid incision and lateral fluvial erosion of Upper Proterozoic rocks during the Quaternary (Reproduction of the lithography (by A. Levý 1887) from the archives of the Map Collection of the Charles University in Prague.)

Fig. 5.12 Disastrous confirmation of an extent of floodplains in the Vltava valley passed in Prague during catastrophic summer floods in the year 2002. Aerial photography (14. 8. 2002, Raudenský and Dorazil 2002) displayed flooded areas around the National Theatre (down in the right) and a large part of the Lesser Town

and sculptures (Březinová et al. 1996). Profound changes in the urban development of Prague and a widespread industrial and agricultural activity throughout Central Europe have extremely severe impacts on historic and residential quarters of this beautiful city. Substantial reduction of risky co-existence of manifold environmental hazards in the Prague area is a topical challenge to heritage conservation endeavours supported by new applications of natural sciences.

5.5 Conclusion

Present-day landform patterns of the Prague area are determined by a long-term antecedent deepening of canyon-like valleys of the Vltava River and its tributaries to large planation surfaces of the central part of the Bohemian Massif during the Quaternary. The coupled occurrence of episodic tectonic uplift and variable climato-morphogenetic processes has led to the origin of stratigraphically significant river accumulation terraces as well as erosion and denudation slopes with weathered mantle of deposits. This extraordinary geodiversity and biodiversity of the landscape is, however, also associated with geomorphic hazards stimulated by human processes over the entire history of occupation of the Prague area, including devastating floods. During its centuries of history, the "golden and hundred-spire" Prague has become an architectural pearl of European and global significance. To effectively mitigate severe impact of modern urban development and related human activities on the architectural heritage of the city is the topical environmental issue in the Prague area.

Acknowledgements The paper was carried out under auspices of the project PRVOUK No. 43 "Geography" of the Charles University in Prague. The authors wish to thank Professor Dr. Philip Gibbard (University of Cambridge) for valuable comments on the manuscript and Dr. Eva Novotná for cooperation at evaluation of original works owned by the Map Collection of the Charles University in Prague, Faculty of Science.

References

Balatka B (1985) Geomorfologické členění pražského území. Staletá Praha, XV. Přírodovědecký význam Prahy. Praha, pp 454–456

Balatka B (2007) River terraces and the Sázava Valley evolution. In: Goudie AS, Kalvoda J (eds) Geomorphological variations, P3K Publ., Prague, pp 361–386

Balatka B, Kalvoda J (2008) Evolution of Quaternary river terraces related to the uplift of the central part of the Bohemian Massif. Geografie 113(3):205–222

Balatka B, Kalvoda J (2010) Vývoj údolí Sázavy v mladším kenozoiku. Evolution of the Sázava valley in the late Cenozoic. Nakl. ČGS, Praha 198 pp

Balatka B, Loučková J (1992) Terasový systém a vývoj údolí Berounky. Studia geographica 96:1–53

Balatka B, Sládek J (1962) Říční terasy v českých zemích. Geofond, NČSAV, Praha 580 pp

Balatka B, Štěpančíková P (2006) Terrace system of the middle and lower Sázava River. Geomorphologia Slovaca 6(1):69–81

Balatka B, Gibbard P, Kalvoda J (2010a) Evolution of the Sázava Valley in the Bohemian Massif. Geomorphologia Slovaca et Bohemica 10(1):55–76

Balatka B, Gibbard P, Kalvoda J (2010b) Morphostratigraphy of the Sázava river terraces in the Bohemian Massif. Acta Universitatis Carolinae, Geographica 45(1–2):3–34

Balatka B, Kalvoda J, Gibbard P (2015) Morphostratigraphical correlation of river terraces in the central part of the Bohemian Massif with the European stratigraphical classification of the Quaternary. Acta Universitatis Carolinae, Geographica 50(1):63–73

Březinová D, Bukovanská M, Dudková I, Rybařík V (1996) Praha kamenná. Přírodní kameny v pražských stavbách a uměleckých dílech, Národní museum, Praha 287 pp

Chlupáč L (1999) Vycházky za geologickou minulostí Prahy a okolí. Academia, Praha 279 pp

Chlupáč L et al (2002) Geologická minulost České republiky. Academia, Praha 436 pp

Cílek V (1995) Podzemní Praha. Soupis podzemních objektů hlavního města a vybraná bibliografie. Knihovna České speleologické společnosti 27, Praha 58 pp

Demek J (2004) Etchplain, rock pediments and morphostructural analysis of the Bohemian Massif (Czech Republic). In: Drbohlav D, Kalvoda J, Voženílek V (eds) Czech Geography at the Dawn of the Millenium. Czech Geographical Society, Palacky University in Olomouc, Olomouc, pp 69–81

Fridrich J (1997) Staropaleolitické osídlení Čech. Památky archeologie, Supplem., 10, Praha 235 pp

Fridrichová M, Fridrich J, Havel J, Kovařík J (1995) Praha v pravěku. Archaeologica Pragensia, Supplem., Muzeum hl. M. Prahy a Institut hl. M. Prahy, Praha 274 pp

Gibbard PL, Cohen K (2008) Global chronostratigraphical correlation table form the last 2.7 million years. Episodes 31:243–247

Gibbard PL, Head MJ, Walker MJC and the Subcomission on Quaternary Stratigraphy (2009) Formal ratification of the Quaternary System/Period and the Pleistocene Series/Epoch with a base at 2.58 Ma. J Quat Sci. doi: 10.1002/jqs.1338

Horný R, Turek V (1999) Joachim Barrande (1799–1883). Život, dílo a odkaz světové paleontologii. Národní museum, Praha 56 pp

Hrdlička L (1984) Nástin vývoje reliéfu historického jádra Prahy ve středověku. Archeologica Pragensia, 5, Praha, pp 197–209

Kalvoda J (2007) Morphostructural evolution of the relief in the locality of the Geodynamic Observatory at Pecný, the Ondřejovská vrchovina Highland, Czech Republic. In: Goudie AS, Kalvoda J. (eds) Geomorphological variations. P3K Publ., Prague, pp 387–407

Kalvoda J, Balatka B (2006) Morphostructural evolution of the relief of the Bohemian part of the Český masiv Massif. In: Balatka B, Kalvoda J (eds) Geomorphological regionalization of the relief of Bohemia. Kartografie a. s., Praha, 3 maps, 1 table, 68 pp

Kovanda J et al (2001) Neživá příroda Prahy a jejího okolí. Academia, Český geologický ústav, Praha 215 pp

Kříž J (1999) Geologické památky Prahy. Proterozoikum a starší prvohory, Český geologický ústav, Praha 278 pp

Kubíková J, Ložek V, Špryňař P et al (2005) Chráněná území ČR: Praha, vol XII. Agentura ochrany přírody a krajiny ČR, Ekocentrum, Brno 304 pp

Malkovský M (1975) Paleogeography of the Miocene of the Bohemian Massif. Věstník Ústředního Ústavu geologického 50(1):27–31

Malkovský M (1979) Tektogeneze platformního pokryvu Českého masivu. Academia, Ústřední ústav geologický, Praha 176 pp

Píchal Z et al (1979) Praha a inženýrská geologie. ČSVTS, PÚDIS, Praha 134 pp

Raudenský M, Dorazil I (2002) Povodně 2002. Letecké dokumenty, Atelier S-design, Brno 127 pp

Šibrava V (1972) Zur Stellung der Tschechoslowakei im Korrelierungssystem des Pleistozäns in Europa. Sborník geologických věd, Antropozoikum 8:5–218

Tyráček J (2001) Upper Cenozoic fluvial history in the Bohemian Massif. Quatern Int 79:37–53

Tyráček J, Havlíček P (2009) The fluvial record in the Czech Republic. A review in the context of IGCP 518. Global Planet Change. doi:10.1016/j.gloplacha.2009.03.007

Tyráček J, Westaway R, Gridgeland D (2004) River terraces of the Vltava and Labe (Elbe) system, Czech Republic, and their implications for the uplift history of the Bohemian Massif. Proceedings of the Geologists' Association, London, 115:101–124

Ulrych J, Dostal J, Adamovič J, Jelínek E, Špaček P, Hegner E, Balogh K (2011) Recurrent Cenozoic volcanic activity in the Bohemian Massif (Czech Republic). Lithos 123:133–144

Záruba Q, Bucha V, Ložek V (1977) Significance of the Vltava terrace system for Quaternary chronostratigraphy. Rozpravy ČSAV, Ř. mat.-přír. Věd, Academia, Praha, 87(4), 90 pp

Záruba-Pfeffermann Q (1941) Původ štěrků z terasy u Lysolaj a Suchdola. Zprávy Geologického Ústavu pro Čechy a Moravu, Praha, 17:298–308

Záruba-Pfeffermann Q (1942) Podélný profil vltavskými terasami mezi Kamýkem a Veltrusy. Rozpravy II. třídy České Akademie, Praha, 52(9), 39 pp

6 The Bohemian Karst: A Condensed Record of Landscape and Living Nature Evolution

Karel Žák, Pavel Bosák, and Jiří Bruthans

Abstract

Rich in historical heritage and natural beauties and located close to Prague, the Bohemian Karst has attracted visitors and scholars for centuries. Lower Paleozoic strata, folded and faulted during the Variscan Orogeny, have yielded thousands of fossil species. Well-exposed sedimentary rocks enabled definition of five international stratotype and/or parastratotype sections of geological boundaries. Most of the karst forms are inactive and largely sediment-filled (paleokarst). The area is poor in karren fields, dolines and underground active streams, but rich in complex maze caves, locally with small but deep cave lakes. Morphological evolution of the present-day landscape was initiated after Cretaceous and Paleogene planation, when the area started to be entrenched by low-gradient Oligocene and Miocene rivers. In the Middle and Late Pleistocene, the entrenchment of river valleys was accelerated, forming steep-walled rocky canyons combined with a system of river terraces. Following earlier thermal water karstification, river floodwater injection was the main speleogenetic process during the Neogene and Quaternary. Archaeological finds and human bones up to ca. 150,000 years old have been discovered in the area. Historical sites, like the former Slavic fortified settlement at Tetín, majestic Gothic Karlštejn Castle, the Baroque monastery and church at Svatý Jan pod Skalou, and the hiking and educational trails annually attract approximately half a million visitors.

Keywords

Bohemian Karst • Paleokarst • Hydrothermal karst • Maze caves • River terraces • Karlštejn Castle

6.1 Introduction

The Bohemian Karst (Český kras in Czech, sometimes translated into English as the Czech Karst) has been an important base of raw materials for the capital city of Prague since the Middle Ages. It can be reached from Prague within a single day—the castles and settlements located here are outside the capital, but close to it. The area has supplied the town with lime and stone for both construction and sculptures and, more recently, for cement. Limestone quarrying lasting hundreds of years dotted the Bohemian Karst with more than 100 quarries, some of which provide a unique record of the work of quarrymen in times past (e.g. the

K. Žák (✉) · P. Bosák
Institute of Geology of the Czech Academy of Sciences, Rozvojová 269, 165 00 Prague 6, Czech Republic
e-mail: zak@gli.cas.cz

P. Bosák
e-mail: bosak@gli.cas.cz

J. Bruthans
Institute of Hydrogeology, Engineering Geology and Applied Geophysics, Faculty of Science, Charles University in Prague, Albertov 6, 128 43 Prague 2, Czech Republic
e-mail: jiri.bruthans@natur.cuni.cz

T. Pánek and J. Hradecký (eds.), *Landscapes and Landforms of the Czech Republic*, World Geomorphological Landscapes, DOI 10.1007/978-3-319-27537-6_6

system of deep pit quarries and approximately 9 km of interconnecting underground galleries near Mořina, which were in operation from 1890 to the beginning of the 1960s).

Since the middle of the eighteenth century, the Bohemian Karst has been a pivotal study area for scholars and university students. It has had crucial importance in the development of Czech geology, botany, entomology and many other branches of science. The close relationship between Prague and the Bohemian Karst has also developed recently, when the suburbs of Prague increasingly spread into the karst and when parts of the area were converted into a maintained park-like landscape with educational nature trails and paths for cycling, in-line skating or hiking.

Because of the moderate heights, the area does not contain any spectacular world-class surface landforms. Its importance lies in the extremely long but traceable evolution of the current landscape since the Cretaceous and in the excellent record of former bioclimatic environments preserved in sediments trapped in caves and karst depressions. This presence of datable animal remains in sediments and of caves with datable speleothems was crucial for understanding the evolution of the landscape and river system during the Neogene and Quaternary. As the obtained knowledge can also be applied to adjacent non-karst areas, the Bohemian Karst acts as a key to the door to knowledge of landscapes in the whole Central Bohemian Region.

Because of the lack of typical surface karst features (such as blind valleys, poljes) and scarcity of others (sinkholes, karren fields), the Bohemian Karst was recognized as a coherent karst area relatively late, in the 1920s, in particular thanks to Jaroslav Petrbok, who excavated and archaeologically and biostratigraphically studied several dozen cave entrances. Efforts to protect nature, which were already initiated in the 1930s, culminated in declaration of a major part of the Bohemian Karst as a Protected Landscape Area in 1972.

6.2 Geology and Geography

The Bohemian Karst extends 35 km to the SW from the southern part of Prague. On the map, it has an oval shape that is up to 7 km wide (Fig. 6.1). The elevation and morphology were determined by the development of the whole karst region by river entrenchment below a surface planated during the Cretaceous and Paleogene. This former planation surface now has an altitude of about 450–500 m a.s.l. in the SW part of the karst area and is gently inclined towards the NE, where the preserved plains have an altitude of about 400 m a.s.l. In the SW part of the area, the planation surface was already more strongly dissected by valley entrenchment into a system of isolated hills during the Neogene. The shapes of these hills frequently follow their geological structure. The NE part was entrenched by the Vltava River and its tributaries, but large parts of the planation surface were preserved here. These plateaus are covered by thin blankets of Cretaceous or Neogene sediments. The highest geographic point of the whole karst area is Bacín Hill (499 m a.s.l.), located near the SW rim of the area. The lowest point is the water level of the Vltava River in Prague at 187 m a.s. l. The deepest known flooded cave section extends down to 142 m a.s.l. in the Podtraťová Cave.

Regional geological subdivision of the Bohemian Massif classifies the Bohemian Karst as a part of the unmetamorphosed or low-grade metamorphosed Teplá-Barrandian Unit. The unit consists of Neoproterozoic sedimentary and volcanic sequences followed, after Cadomian folding, by deposition of Cambrian continental and marine sediments. Cambrian evolution was terminated by volcanism. The sedimentary space then changed and sedimentation continued practically uninterrupted from the Ordovician until the late Middle Devonian. After the Variscan Orogeny and subsequent erosion, several structural segments of strongly folded and faulted Lower Paleozoic sediments remained preserved. The largest of them is a complex synclinorial structure called the Prague Basin (Chlupáč et al. 1998; Fig. 6.1). The Bohemian Karst occupies the central part of the Prague Basin and is developed in limestone of Silurian and mainly Devonian age. The area hosts several international stratotype and parastratotype sections, including the main Silurian/Devonian Global Boundary Stratotype Section at Suchomasty (Fig. 6.2).

As a result of the original sedimentary evolution and subsequent folding and faulting (Fig. 6.3), limestone layers of several lithological types alternate with non-karstic rocks, shales and submarine basaltic volcanic rocks. The whole district where limestone occurs either at the surface or at depth has an area of about 144 km^2. In a large part of this area, the limestone is covered by thin blankets of platform Cretaceous freshwater and marine sediments and by unconsolidated Neogene fluvial sediments. The area where the karst features can be directly observed is therefore much smaller.

Of the Ordovician to Middle Devonian sedimentary fill of the Prague Basin, the karst is partly developed in Silurian limestone types (containing about 2.3 % of the known cave length in Wenlock, 0.1 % in Ludlow and about 1.5 % in Pridoli Stage). The main karst rocks, and also the rocks containing the main karst aquifer, are the Lower Devonian Lochkovian and Pragian limestones, together hosting 90 %

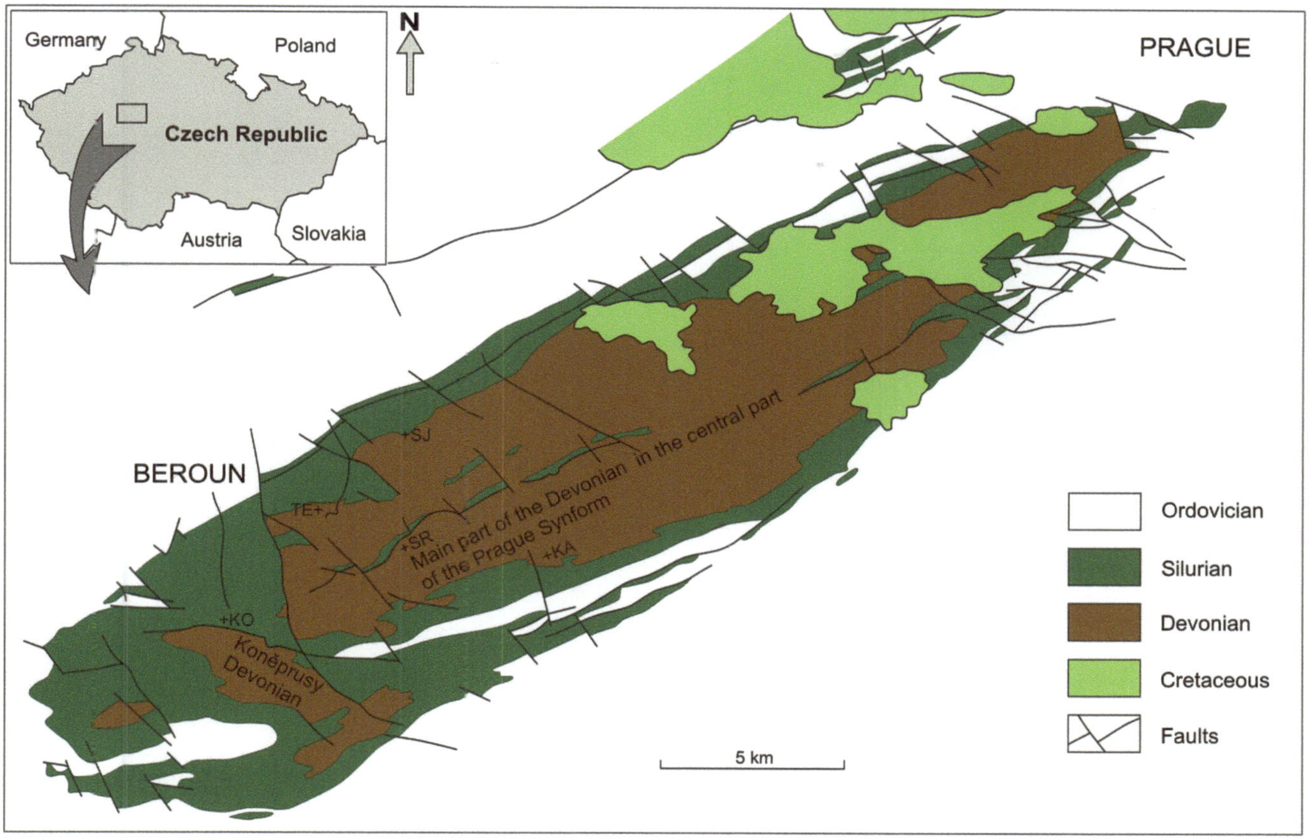

Fig. 6.1 Simplified geological map of the Prague Basin with the Bohemian Karst in its centre. Drawing by Š. Manda, based on geological maps published by the Czech Geological Survey. Location of the sites mentioned in the text: *KA* Karlštejn, *KO* Koněprusy, *SJ* Svatý Jan pod Skalou, *SR* Srbsko, *TE* Tetín

of the known caves of the area. Kotýs Limestone in Lochkovian, and Koněprusy Limestone, Slivenec Limestone, Loděnice Limestone in Pragian, belong to the lithostratigraphic units with the most widespread karst features. The overlying Emsian and Eifelian limestones of several types together host 6.1 % of known caves (most of them in the Suchomasty Limestone).

The primary porosity of all these limestone types is 1 % or less. Formation of karst cavities therefore followed the tectonic structures, faults in several directions, with the largest importance of the N–S to NW–SE-oriented, steeply dipping faults. Some cave sections are developed parallel to the bedding, along with zones of disharmonic folding of massive and layered limestone. The whole area was affected by thermal water circulation in several, so far poorly chronologically constrained phases, forming abundant calcite veins (Bosák 1998; Suchy et al. 2000). Large cave-sized crystal cavities developed locally. Some of the caves in the karst region are clearly of tectonic or of hydrothermal-corrosive origin.

6.3 Peculiarities of the Hydrogeology and Hydrology of the Bohemian Karst

An important feature which distinguishes the Bohemian Karst from a majority of other karst areas is its peculiar hydrogeology. The surface of the karst rocks is largely covered by unconsolidated sediments, which prevent concentrated infiltration. No permanent sinking stream or permanent underground stream is known in the whole area. Few caves contain lakes, and those that exist usually have a small lateral extent, but the depth can reach several tens of metres. Water in these cave lakes moves very slowly, but their water levels oscillate substantially, indicating low aquifer porosity and restricted outflow paths. After heavy rains, the water level in some cave lakes can rise by more than 30 m in a few days or weeks. It returns to its former level extremely slowly, over many months, reflecting changes of the water level in the karst aquifer. This rather uncommon feature for a karst environment indicates the predominance of paleokarst

Fig. 6.2 Silurian/Devonian Global Boundary Stratotype Section at Klonk near Suchomasty (*Photo* A. Komaško)

(Bosák 1997), the widespread blocking of cave corridors by cave sediments and the absence of cave systems with interconnected systems of permeable conduits extending to larger distances (cf. Bruthans and Zeman 2003).

Karst waters emerge in several tens of karst springs, which exhibit relatively small variability in their discharge. The discharge of the largest karst spring located at Svatý Jan pod Skalou (Fig. 6.4) varies between ca. 15 and 30 $l\ s^{-1}$. Thermal modelling shows that active groundwater circulation can penetrate down to 500 m below the surface. Tritium isotope data confirmed that the water underground residence time commonly extends to several dozen years.

Part of the karst drainage is directed towards the main allochthonous low-gradient streams (i.e. originating outside of the karst), especially the Berounka River, which divides the karst area into two parts. The average Berounka River discharge is 35.6 $m^3\ s^{-1}$; peak recorded flood discharge was above 2500 $m^3\ s^{-1}$ with observed flood water rise of more than 7 m. Average river longitudinal gradient of the river in the karst is 0.79 $m\ km^{-1}$. As will be shown below, the low-gradient major rivers crossing the karst for most of its evolution since the Oligocene have also been one of the major speleogenetic agents (Vysoká et al. 2012).

6.4 Evolution of the Karst Landscape and Caves During Pre-Quaternary Periods

After the Variscan folding and rapid erosion, parts of the area were covered by Carboniferous conglomerates, sandstones and claystones, locally with coal seams. The original extent of these sediments is unknown. The closest relics of Carboniferous sediments preserved in segments with tectonic subsidence are located at a distance of 5 km from the karst.

During the Permian, Triassic and Jurassic, the area was denuded with decreasing altitude differences and no surface sediments were preserved in the karst area or close to it. During the Late Cretaceous, the majority of the area was first covered by freshwater sediments and was later flooded by a shallow epicontinental sea, which deposited marine sands and marlstones. The Cretaceous sea probably left the area during the late Turonian. The flat character of the country was also preserved during most of the Paleogene. All these periods together were characterized by deep weathering of limestones and by the formation of a specific type of soft porous limestone affected by intergranular corrosion (called

Fig. 6.3 Overthrusted Silurian sequence (Pridoli, alternating limestone and shale, dark layer near the top of the quarry wall) on the massive Devonian Koněprusy Limestone (Pragian). Očkov Thrust, Kobyla Quarry near Koněprusy (*Photo* A. Komaško)

"white beds" in local terminology). Various sediment-filled karst depressions replacing the weathered limestone which formed during the Cretaceous, Paleogene and Neogene are abundant below all the surfaces with higher altitude.

Thermal water was an important speleogenetic agent participating in the formation of some of pre-Quaternary cavities (Cílek et al. 1994; Bosák 1998; Suchy et al. 2000). The regional thermal water flow is reflected in abundant calcite veins of several types and generations. The younger types of calcite veins typically contain crystal cavities sometimes reaching the size of a small cave, usually only several metres long. Fluid inclusion data indicate variable fluid temperatures from ca. 130 °C down to low ambient temperatures. The highly variable salinity data also agree with the existence of several thermal fluid circulation events. These events have not been exactly dated yet. Speculations cover periods from the Permian to the Neogene. The deep Únorová Chasm contains red carbonate sediments, with estimated Triassic age (Žák et al. 2007), indicating the existence of open cavities below the flat planation surface at that time.

Using cave morphology as evidence, the corrosive effects of low-temperature, deep circulation thermal water flow were also considered as a possible process in the formation of the most important cave in the area, the Koněpruské jeskyně Caves (Figs. 6.5 and 6.6; Bella and Bosák 2012). This cave hosts a unique speleothem type, corralites with the presence of opal, called "Koněprusy Rosettes" (Fig. 6.7). These up to 15 cm thick corralite layers grew independently of gravity, perpendicularly to the cavity walls. Opal (locally also α-quartz, at the speleothem base) forms irregular layers and isolated grains in the calcite mass of the speleothem. Its origin is usually interpreted as a thermal water precipitate, or a precipitate deposited near water level of a stagnant cave lake.

A more dynamic landscape was formed as a result of minor vertical movements of crustal blocks accompanied by river valley entrenchment during the latest part of the Paleogene and beginning of the Neogene. Formation of wide shallow valleys was accompanied by rapid erosion of Cretaceous sedimentary cover. Parts of these valleys were again filled with sediments. Preserved relics of this Neogene sedimentary fill locally exceed a thickness of 20 m.

Fig. 6.4 Tilted layers of Silurian (Lochkovian) and lowermost Devonian (Pragian) limestone in the NW flank of the Holyně-Hostim Syncline above Svatý Jan pod Skalou. The Baroque monastery located at the bottom of the valley was built at the site of emergence of the largest karst spring in the area. The spring water deposited an up to 17 m thick accumulation of Holocene calcareous tufa here, which can be seen immediately behind the church (*Photo* A. Komaško)

Paleogene, Neogene and Quaternary rivers were allochthonous, that is they had their large catchments outside of the karst area and they all had low longitudinal gradients. They became the main speleogenetic agent in the area. Regular floods resulted in cyclic floodwater injection into the karst cavities and their enlargement by corrosion (Bosák et al. 1992; Bruthans and Zeman 2001; Vysoká et al. 2012). The floodwater returned back to the river channel after the flood events. This specific process of cave formation resulted in the common presence of complicated maze caves, with large numbers of cave corridors concentrated in small areas. Caves with the largest cavities (Koněpruské jeskyně Caves, Martina Cave) are inactive today (i.e. without groundwater flow), isolated in apical parts of the hills and have very limited catchments in the present-day topography (Fig. 6.5). Their genesis was clearly related to pre-Quaternary periods (Bosák 1997).

The Bohemian Karst also provides evidence for Neogene life. The Červený Quarry, Plešivec Quarry and Zlatý kůň Hill in the Koněprusy area yielded the bones of animals of Late Miocene and Pliocene terrestrial and water-related animals at several sites (Fejfar 1990; Čermák et al. 2007). The bones of the latest Miocene/earliest Pliocene bats discovered in a chasmal cave in the Zlatý kůň Hill provide clear evidence that caves with open entrances already existed at that time.

6.5 Accelerated River Valley Incision and Cave Formation During the Middle and Late Pleistocene

There was no intensive river entrenchment during the Pliocene and Early Pleistocene. The rivers continued to flow in wide valleys, locally eroding earlier unconsolidated Miocene sediments. An abrupt change in the morphological evolution occurred near the boundary between the Early and Middle Pleistocene at ca. 780 ka before the present (BP). As a result of climatic changes, transition to longer and more severe glacial cycles, and probably also as a result of changes in longitudinal gradients in the whole river network, the Berounka River, Vltava River and their tributaries started to entrench narrow deep canyons with rocky walls. By the end of the last glacial (Weichselian), these canyons had reached,

Fig. 6.5 Aerial view of the Zlatý kůň Hill hosting the largest cave in the area, the Koněpruské jeskyně Caves. The cave is located below the highest part of the hill, which is strongly affected by limestone quarrying (*Photo* A. Komaško)

in several incision steps, the depth of about 70 m below the bottoms of the earlier pre-Quaternary and Early Pleistocene valley network. The Middle and Late Pleistocene canyons constitute the most prominent landscape feature of the area (Fig. 6.8). Cave formation was restricted mainly to rocky walls directly adjacent to these valleys, with floodwater injection continuing to be the main speleogenetic process to present day.

The alternation of Middle and Late Pleistocene bioclimatic events, glacials and interglacials, is demonstratively reflected in the system of river terraces (Tyráček et al. 2004). The cave environments also record these first-order climatic events quite well. Speleothem formation characteristic for interglacials was replaced by deposition of clastic cave sediments in the vadose and epiphreatic zones of the caves during glacials. Dating of vadose zone speleothems in caves along the Berounka River provides information on the precise limits on the ages of the individual terrace levels (Fig. 6.9). The Weichselian permafrost penetrated down to ca. 65 m below the surface (Žák et al. 2012) but, because of low precipitation, the permafrost had little effect on the landscape. Cryoplanation features developed only locally.

The succession of Pleistocene bioclimatic events is also recorded well in the karst sediments, as animal bones and terrestrial snail shells were either trapped in sediments in cave depressions or contained in surface aeolian sediments—loess blankets. The abundance of vertebrate bones increases from rather rare Early Pleistocene (Early Biharian) finds (Chlum near Srbsko, Sluj VI; Koněpruské jeskyně Caves) to the famous Middle Pleistocene and the most widespread Late Pleistocene (Weichselian) sites. Among the Middle Pleistocene localities, the bones of the latest Biharian karst depression C718 at the Zlatý kůň Hill, representing interglacial conditions, and from Chlum near Srbsko, recording several interglacial/glacial transitions, are of the greatest importance (Fejfar 1961; Horáček and Ložek 1988). Among Weichselian sites, the Komín (Chimney) of the Srbské jeskyně Caves at Chlum near Srbsko yielded the

Fig. 6.6 Morphology of the cavities with corrosion roof cupolas in the middle level of the Koněpruské jeskyně Caves (*Photo* M. Majer)

richest bone accumulation of animals of the last glacial, dominated by horses and interpreted as a hyena den (Diedrich and Žák 2006).

6.6 Record of the Holocene Climatic Evolution—Calcareous Tufa Deposits

The latest phases of Weichselian and the earliest part of the Holocene were characterized by deposition of fluvial river sediments and the formation of the youngest, 5–10 m thick terrace fills in the valley bottoms. The fine-grained floodplain sediments covering the sands and gravels of this terrace have been deposited here since the Middle Ages as a result of climatic changes, man-made deforestation and other changes in land use.

An increase in temperature and precipitation and gradual growth of forests since the beginning of the Holocene have led to elevated organic matter contents in the soils, causing an increase in the CO_2 content in the soil air. The dissolution rate of limestone was consequently accelerated. Discharged groundwater began to deposit abundant accumulations of freshwater tufa either near springs or in the subsequent sections of surface streams. The thickest spring-proximal Holocene tufa body has a thickness of 17 m (Žák et al. 2002). Sections with tufa cascades in the surface streams are several hundred metres long (Fig. 6.10). The tufa bodies provide quite a good record of Holocene climatic changes. The most massive, pure tufa deposited during the climatic optimum of the Holocene (between ca. 9 and 5 ka BP) was followed by alternation of tufa layers, limestone talus and interlayered fossil soils. The formation of the largest tufa accumulations ceased about 2,500–2,000 years BP, when more intensive erosion began. The tufa accumulations are rich in shells of terrestrial snails (up to more than 40 species in a single layer) and locally also in archaeological artefacts. They are also easily datable by several instrumental methods and Holocene climatic changes can be studied using stable isotope methods. The tufa bodies therefore provide a unique record of Holocene climatic changes and natural evolution of living species. The most important section is located at Svatý Jan pod Skalou near the Baroque church (Fig. 6.4; Žák et al. 2002).

6.7 Prehistoric Humans in the Bohemian Karst

The karst region with its rugged morphology, a variety of habitats and abundant rock and cave shelters attracted prehistoric humans since the Middle Paleolithic (Fridrich and Sklenář 1976). While proven Middle Paleolithic finds are rare, Late Paleolithic sites belonging to the Gravettian culture are somewhat more abundant. Finds related to the Magdalenian culture (17,000–12,000 years BP) are the most frequent.

Magdalenian sites have been excavated both in cave and surface settings (Vencl 1995). No cave wall paintings have been discovered to date but engravings of hunted animals on a shale plates have been found at two sites. The famous human remains collected on the surface of cave sediments in the largest Prošek's Chamber of the Koněpruské jeskyně Caves, earlier considered to be Gravettian or older, have also been dated to the Magdalenian culture (Svoboda et al. 2002). These bones are not interpreted to reflect any human activity directly in the Koněpruské jeskyně Caves. They were

Fig. 6.7 The "Koněprusy Rosettes" unique speleothem type of caves in the Koněprusy area (*Photo* A. Komaško)

probably buried into a karst chimney and redeposited in the cave by natural gravitational processes. The only evidence of the presence of prehistoric man deeper inside the caves is a find of large Mesolithic charcoal pieces below a flowstone layer deep inside the Martina Cave.

Neolithic, Bronze Age and Iron Age surface settlements are known from the valleys of the Bohemian Karst and adjacent areas. Fortified settlements on hills are typical of the late Bronze Age and early Iron Age period. Several fortified settlements were reused or constructed again after the arrival of the Slavic tribes in the sixth century AD. Interest in caves oscillated during these periods (Sklenář and Matoušek 1994). Several vertical chasms and caves were used again during the Bronze and Iron ages for deposition of human remains.

6.8 Historical Monuments and Tourism

The proximity of the Bohemian Karst to the country's capital Prague, together with the presence of a suitable hilly region led to the construction of several mediaeval castles here. By far the most important of them is Karlštejn Castle (Fig. 6.11), constructed by the Czech King and the Holy Roman Emperor Charles IVth between 1348 and 1365 on a rocky spur of Kněží Hora Hill in the heart of the karst area. Although the present-day appearance of the castle reflects its reconstruction in the neo-Gothic style by J. Mocker during the late nineteenth century, the castle belongs to the most important gothic castles of the country and is one of most frequently visited tourist destinations in the Czech Republic (annual attendance ~220,000 visitors).

The most valuable parts of the castle are its sacred parts, the Church of Our Lady, The Chapel of St. Catherine (former Sacristy), and especially the Holy Cross Chapel. The chapels are decorated with the original fourteenth century incrustation made from polished slabs of agate and chalcedony from the Krušné hory/Erzgebirge area. The set of panel paintings by Master Theodoricus decorating the Holy Cross Chapel represents one of milestones in the evolution of mediaeval painting in Europe.

From the geomorphological and karsological point of view, an interesting part of the Karlštejn Castle is its 76.5 m

Fig. 6.8 Aerial view of the Berounka River canyon near Srbsko. Wide Neogene to Early Pleistocene river terraces are cut by a narrow, steep-walled Middle and Upper Pleistocene canyon (*Photo* P. Pokorný)

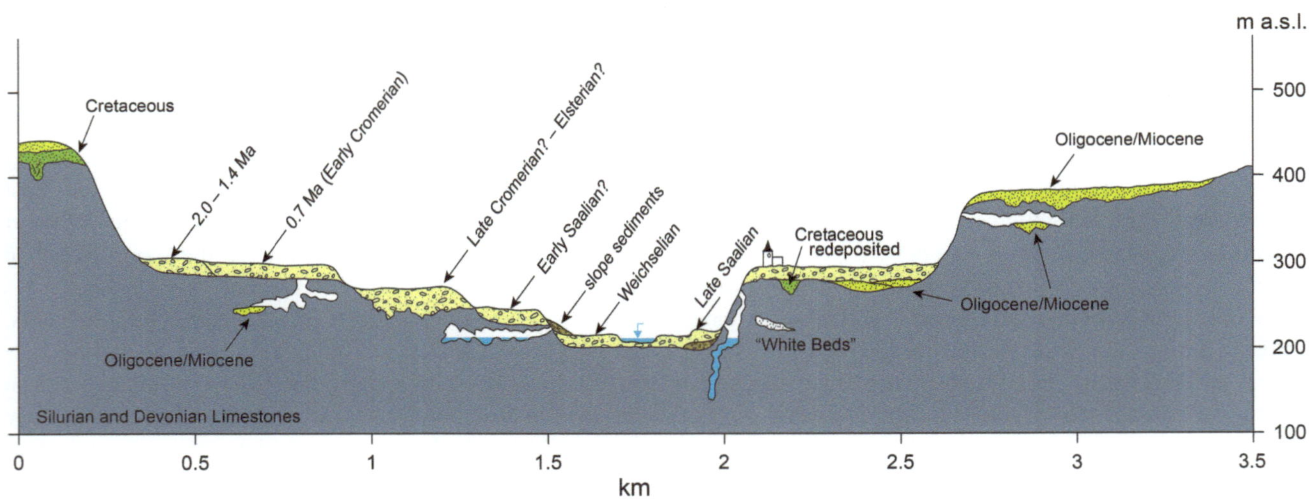

Fig. 6.9 Schematic cross section of the Berounka River valley, showing typical features of morphological and karsological evolution of the landscape

Fig. 6.10 Calcareous tufa cascade in the Císařská Gorge near Srbsko. The outcrops on both sides of the photograph were formed by massive tufa deposition mainly during the climatic optimum of the Holocene. This cascade was incised by erosion, most probably about 2000 years ago (Hlaváč et al. 2003). The new tufa cascade fills the entrenched central part (*Photo* K. Žák)

deep water well. Although the bottom of the well reached almost 28 m below the water level in the adjacent surface creek, because of the specific karst hydrology it was almost dry during the construction. The problem was resolved by digging an artificial gallery connecting the castle water well with the creek. This water supply was the weakest feature of the castle and many legends about the well were created during its more than 650-year history.

The settlement of Tetín, occupying a narrow rocky spur between the Berounka River valley and Tetín Gorge, is another place with extraordinary history and great importance in the early years of the country. The site was most certainly already populated in the Neolithic and cave finds go back to the Middle Paleolithic. The most famous part of Tetín history is connected with the Slavic fortified settlement of the ninth and tenth centuries, whose ramparts are still partly preserved. The Czech Princess Saint Ludmila, grandmother of the Czech Lord Saint Venceslas, was murdered here on 15 September 921 by killers hired by her daughter-in-law, Drahomíra. The attractions of Tetín include two small late Roman to early Gothic churches, a larger Baroque church and the ruins of a small late Gothic castle, from which a prominent geomorphological feature, the terrace system of the Berounka River, is clearly visible.

Fig. 6.11 Mediaeval Karlštejn Castle is the most frequently visited attraction in the Bohemian Karst (*Photo* A. Komaško)

The third site of high historical importance is the monastery and Baroque church at Svatý Jan pod Skalou, built close to accumulation of Holocene calcareous tufa and at the site of the largest karst spring in the area. The legendary history of the site goes back to Saint Ivan, a hermit said to live in a cave in the calcareous tufa in the nineth century. The site includes several historical Baroque monuments and magnificent views from the rocky cliffs surrounding the village (see Fig. 6.4).

The area of the Bohemian Karst started to become a tourist attraction during the late nineteenth century and the number of visitors has constantly increased since then. The discovery of the Koněpruské jeskyně Caves during the autumn of 1950 and its opening to the public in 1959 provided a further impetus for an increase in the number of visitors (ca. 90,000 people now visit the cave every year). The area is attractive not only because of the historical sites and the Koněpruské jeskyně Caves, but also for walks along educational nature trails or for outdoor activities such as hiking (Fig. 6.12), caving, rock climbing, bicycling, in-line skating, etc.

Fig. 6.12 One of the frequently visited hiking trails leads along the rims of deep open-pit limestone quarries near Mořina. This specific style of quarrying was necessitated by the vertical dip of the best quality limestone layers (*Photo* V. Konvička)

6.9 Conclusion

The Bohemian Karst can be viewed as an excellent example of the prolonged evolution of karst features under a planation surface and later in a landscape shaped by large, allochthonous low-gradient rivers. Earlier evolution of the karst was clearly influenced by a period of thermal water flow. Interactions between humans and the karst area have been well documented from prehistorical to historical periods, followed by an industrial period during which more than 100 limestone quarries were opened. At the present time, the areas of historical and modern limestone quarrying form a mosaic with nature reservations, historical monuments and areas for outdoor free-time activities.

The evolution and morphology of the caves of the Bohemian Karst display many similarities to some other karst regions of Europe, which were formed in similar geological units folded during the Variscan Orogeny, which then evolved under the influence of similar factors. Good examples are the karst of the Holy Cross Mountains in central Poland and the karst of the Rhenish Slate Mountains in the NW part of Germany.

Acknowledgments The geomorphological study of the Bohemian Karst and preparation of this chapter were supported by the programme RVO67985831 of the Institute of Geology of the Czech Academy of Sciences.

References

Bella P, Bosák P (2012) Speleogenesis along deep regional faults by ascending waters: case studies from Slovakia and Czech Republic. Acta Carsologica 41(2–3):169–192

Bosák P (1997) paleokarst of the Bohemian Massif in the Czech Republic: an overview and synthesis. Int J Speleol 24(1–4):3–39

Bosák P (1998) The evolution of karst and caves in the Koněprusy region (Bohemian Karst, Czech Republic), part II: hydrotermal paleokarst. Acta Carsologica, 27(2–3):41–61

Bosák P, Cílek V, Tipková J (1992) La Karst de Boheme au Tertiaire. In: Solomon JN, Maire R (eds) Karsts et évolutions climatiques. Presses Université de Bordeaux, Talence, pp 401–410

Bruthans J, Zeman O (2001) New data on character and evolution of underground karst forms in the Bohemian Karst and other areas with diffuse recharge mode in the Czech Republic. Český kras, XXVII:21–29. (In Czech with abstract in English)
Bruthans J, Zeman O (2003) Factors controlling exokarst morphology and sediment transport through caves: comparison of carbonate and salt karst. Acta Carsologica 32(1):83–99
Čermák S, Wagner J, Fejfar O, Horáček I (2007) New Pliocene localities with micro-mammals from the Czech Republic: a preliminary report. Fossil Rec 10(1):60–68
Cílek V, Dobeš P, Žák K (1994) Formation conditions of calcite veins in the quarry "V Kozle (Hostim I, Alkazar)" in the Bohemian Karst. J Czech Geol Soc 39(4):313–318
Chlupáč I, Havlíček V, Kříž J, Kukal Z, Štorch P (1998) Palaeozoic of the Barrandian (Cambrian to Devonian). Czech Geol Surv, Prague 183 p
Diedrich CG, Žák K (2006) Prey deposits and den sites of the upper Pleistocene hyena Crocuta crocuta spelaea (Goldfuss, 1823) in horizontal and vertical caves of the Bohemian Karst (Czech Republic). Bull Geosci 81(4):237–276
Fejfar O (1961) Review of Quaternary vertebrata in Czechoslovakia. Survey of Czechoslovak Quaternary. Institut Geologiczny Prace 34 (1):109–118
Fejfar O (1990) The Neogene vertegrate paleontology sites of Czechoslovakia: A contribution to the Neogene terrestric biostratigraphy of Europe based on rodents. In: Lindsay EH (ed). Proceedings of NATO advanced research workshop on European Neogene mammal chronology. Plenum Press, New York, pp 211–236
Fridrich J, Sklenář K (1976) Die paläolitische und mesolitische Höhlenbesiedlung des Böhmischen Karstes. Fontes Archaeologici Pragenses 16:1–122
Hlaváč J, Kadlec J, Žák K, Hercman H (2003) Deposition and destruction of Holocene calcareous tufa cascades in the Bohemian Karst (Czech Republic). Prace Geografyczne 189:225–253
Horáček I, Ložek V (1988) Palaeozoology and the Mid-European Qaternary past: scope of the approach and selected results. Rozpravy Československé akademie věd, řada matematických a přírodních věd 98(4):1–102
Sklenář K, Matoušek V (1994) Die Höhlenbesiedlung des Böhmischen Karstes vom Neolithikum bis zum Mittelalter. Fontes Archaeologici Pragenses 20:1–212
Suchy V, Heijlen W, Sykorova I, Muchez Ph, Dobes P, Hladikova J, Jackova I, Safanda J, Zeman A (2000) Geochemical study of calcite veins in the Silurian and Devonian of the Barrandian Basin (Czech Republic), Evidence for widespread post-Varsican fluid flow in the central part of the Bohemian Massif. Sed Geol 131:201–219
Svoboda JA, van der Plicht J, Kuželka V (2002) Upper Paleolithic and Mesolithic human fossils from Moravia and Bohemia (Czech Republic): some new ^{14}C dates. Antiquity 76:957–962
Tyráček J, Westaway R, Bridgland D (2004) River terraces of the Vltava and Labe (Elbe) system, Czech Republic, and their implications for the uplift history of the Bohemian Massif. Proc Geol Assoc 115:101–124
Vencl S (1995) Hostim. Magdalenian in Bohemia. Památky Archeologické, Supplementum, 4:1–264
Vysoká H, Bruthans J, Žák K, Mls J (2012) Response of karstic phreatic zone to flood events in a major river (Bohemian Karst, Czech Republic) and its implication for cave genesis. J Cave Karst Stud 74(1):65–81
Žák K, Ložek V, Hladíková J, Cílek V, Kadlec J (2002) Climate-induced changes in the Holocene calcareous tufa formations, Bohemian Karst, Czech Republic. Quatern Int 91:137–152
Žák K, Pruner P, Bosák P, Svobodová M, Šlechta S (2007) An unusual paleokarst sedimentary rock in the Bohemian Karst (Czech Republic), and its regional tectonic and geomorphologic relationship. Bull Geosci 82(3):275–290
Žák K, Richter DK, Filippi M, Živor R, Deininger M, Mangini A, Scholz D (2012) Coarsely crystalline cryogenic cave carbonate—a new archive to estimate the last glacial minimum permafrost depth in Central Europe. Clim Past 8:1821–1837

Brdy Highland: A Landscape Shaped in the Periglacial Zone of Quaternary Glacials

7

Karel Žák

Abstract

Brdy Highland represents the highest upland in the central part of the Czech Republic outside the mountain ranges distributed in the peripheral parts and along the borders of the country. Because of the existence of a military training ground in its highest part, Brdy Highland is the least known of all the Czech mountains and highlands. Termination of military use of the area by January 2016 opened the area to the public. Quartz-dominated Cambrian conglomerates and sandstones form the highest parts of the highland. The specific lithology of these sedimentary rocks produced extremely infertile, low pH soils. Because of the low fertility, the area was always almost unpopulated, forested and served as source of timber, charcoal and water for mining and ore processing activities in areas at the foot of the highland. Compared to higher mountains distributed along the borders of the Czech Republic, Brdy Highland receives less precipitation and fluvial processes have had a less pronounced effect on its morphology. The relics of pre-Quaternary planated surface forming the summit of the highland were dissected into structural-erosional ridges mainly by processes operating in the periglacial zone of Quaternary glacials. Solifluction on the gentle slopes and frost disintegration of the rock faces were the main processes forming abundant periglacial landforms. Fossil cryoplanation terraces and solifluction lobes, rocks of the tor type, frost-riven cliffs, block fields and block streams are abundant. The possible existence of other fossil features such as patterned ground, nivation hollows or possible rock glaciers requires further study.

Keywords

Brdy Highland • Cambrian • Periglacial landforms • Cryoplanation terraces • Tors • Frost-riven cliffs • Block fields

7.1 Introduction

There are few landscapes in the Czech Republic that are as little known and studied geomorphologically as the Brdy Highland. The reason is simple—an extensive area in the middle of the highland has been set aside as a military training ground, originally established in 1927 and designed mostly for artillery training, with strictly restricted access. After repeated enlargements and reductions, it reached an area of 260 km^2.

As a result of changes in the Armed Forces of the Czech Republic during the last 20 years, the area was returned back to normal civilian use and it was declared a *Protected Landscape Area* byJanuary 1, 2016. This fundamental change in the use and planned permitted access into majority of the area, together with its geomorphological and natural values and extraordinary beauty, were the main reasons for inclusion of the Brdy Highland in this book.

K. Žák (✉)
Institute of Geology of the Czech Academy of Sciences,
Rozvojová 269, 165 00 Prague 6, Czech Republic
e-mail: zak@gli.cas.cz

T. Pánek and J. Hradecký (eds.), *Landscapes and Landforms of the Czech Republic*,
World Geomorphological Landscapes, DOI 10.1007/978-3-319-27537-6_7

The Brdy Highland is an isolated upland reaching the highest elevation (865 m a.s.l., at Tok Hill) in the central part of the Czech Republic, exceeded only by the mountain ranges distributed in the peripheral parts and stretching along the borders of the country. The highest part of the Brdy Highland, characterized by plateau relics located at an elevation above 800 m a.s.l., is slightly higher than the Českomoravská vrchovina Highland (see Chap. 18 of this book), and is higher than the České středohoří Mts. (Chap. 12 of this book). The elevation difference of 400 m over a distance of 7 km from the valleys of the Litavka River to the highest point enable classifying the highest part of the Brdy Highland as a "mountainous environment" according to the definition of the UN Environmental Programme.

Even before 1927, the area was very sparsely populated and little known and studied, although it is located only 60 km SW of the country's capital, Prague. This was a land where it was easy to lose one's way and became entrapped in dense mysterious forests with abundant marshes, an area where robbers used to hide and wait for travellers. The Brdy Highland was known in the past for fierce battles between poachers and gamekeepers. It always served chiefly as a source of several main commodities—timber, charcoal, tar and potash produced from wood and water for ore mining and processing in the surrounding regions. Until recently, two of these commodities—water and timber—have remained the main products of the area. Several water reservoirs, acting as sources of drinking water for the towns in the surrounding areas, are located in the Brdy Highland.

The main reason for only extensive economic use of the area lies in its bedrock. The majority of the Brdy Highland consists of sedimentary rocks, sandstones and quartz conglomerates of Cambrian age, which are poor in all minerals other than quartz and therefore produce only extremely infertile low-pH soils. Forests grow only slowly in the highest parts of the highland and the spruce trees are partly damaged by air pollution in some areas, e.g. in the area between the hills called Praha and Malý Tok.

7.2 General Geographic and Geomorphological Settings

According to the geomorphological regionalization of the Czech Republic (e.g. Demek and Mackovčin 2006), a larger geomorphological unit, Brdská vrchovina, consists of two parts, the eastern Hřebeny Crests and its main and highest part, the Brdy Highland. Following these authors, the Brdy Highland covers an area of 566 km^2, has an average altitude of 601 m, and an average slope inclination of 5° 24′. While the deep valley of the Litavka River is usually considered in practice to be the dividing line between the Hřebeny Crests and the Brdy Highland, Demek and Mackovčin et al. (2006) located the boundary more to the east, in a shallow saddle east of the Litavka River valley.

For practical purposes, the Brdy Highland itself is usually subdivided into its highest central part (usually called the Central Brdy), coincident in area with the former military training ground, the southern part dominated by Třemšín Hill (827 m a.s.l.), and the lower elevation northwestern part dominated by Trhoň and Žďár hills (Fig. 7.1). This chapter focuses especially on the geomorphological features of the central part of the highland, within the borders of the newly declared *Protected Landscape Area Brdy*. Similar processes operated and similar geomorphological features can also be observed in the peripheral parts of the highland (area of Třemšín hill in the S and Žďár and Trhoň hills in the NW), which are fully open to the public and have a dense network of hiking trails.

The flat tops of the highest hills (Tok, 865 m a.s.l.; Praha, 862 m a.s.l.; and several other hills reaching above the 800 m contour line) form the highest part of the Brdy Highland. These flat summits of the highland are remnants of the former pre-Quaternary planation surface, an etchplain, resulting from deep weathering probably during the Cretaceous and Paleogene. The weathered material from its surface was removed during the Neogene and Quaternary. The slopes descending from these planation surfaces are usually gentle, except for those influenced by younger fault movements (e.g. the southern slopes of Praha Hill).

The shapes of the lower hills, below the 800 m contour line, are generally controlled by the lithology and inclination of the sedimentary rock layers (called *structural-erosional ridges* in local terminology). The most resistant rocks, lenses of Neoproterozoic chert, form isolated crags on the summits in the SW part of the area. The most common resistant rock type is Cambrian conglomerate, which forms either elongated outcrops with rock faces along the ridges (*frost-riven cliffs* in local terminology) or isolated cliffs and crags on the summits or on the slopes. The larger rocky outcrops of Cambrian conglomerates are typically accompanied by talus accumulations at their bases, commonly consisting of meter-sized blocks.

The Brdy Highland is drained by several creeks and small rivers. The Litavka River and its tributaries drain the western part of the highland, the Klabava River and their tributaries drain the north-western part and the tributaries of the Úslava and Skalice rivers drain the southern and south-western parts of the highland. Artificial, mainly nineteenth and twentieth century channels transfer part of the surface drainage from one creek catchment to another.

The majority of the area is now covered by coniferous spruce-dominated forests, which replaced the original broad-leaved forests cleared in the eighteenth and nineteenth centuries, mainly for timber and charcoal production for ore mining and processing. The original beech and oak forests,

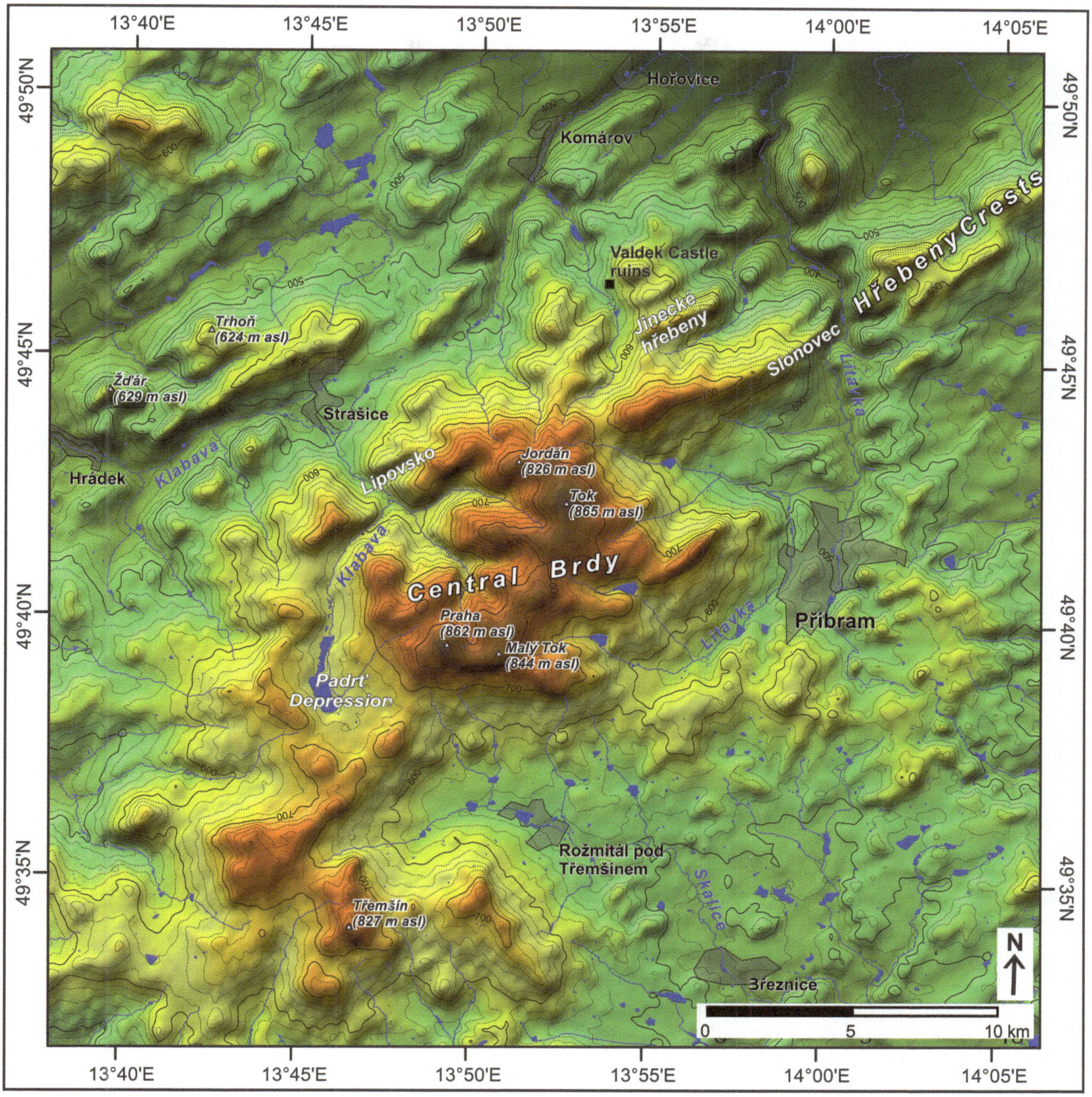

Fig. 7.1 Visualisation of the topography and morphology of the Brdy Highland (performed by V. Kapustová)

with pine on some rocky outcrops, survived only locally. They can be found only in rocky areas with difficult access for timber transportation. Several large areas in the middle of the territory were artificially deforested in the late 1920s and early 1930s to serve as artillery target areas. Specific ecosystems with numerous protected species of plants and animals developed on these frequently disturbed plains, which are kept permanently deforested. The future of these areas is uncertain. Some parts of them will probably be kept deforested, while other parts will be left for natural forest succession. The whole area is practically unpopulated, except for several forester's and gamekeeper's lodges along its periphery. All the villages and towns are located in the wide valleys on the periphery of the highland.

7.3 Geology and Paleontology

Regional geological subdivision of the Bohemian Massif classifies the Brdy Highland as a part of the unmetamorphosed or low-grade metamorphosed Teplá-Barrandian Unit (TBU). The unit consists of supposedly ~10 km thick Neoproterozoic sedimentary and volcanic sequences overlain by Paleozoic sedimentary basins (for current opinions on the structural position of the Neoproterozoic, see Hajná et al. 2010).

After deposition of the Neoproterozoic rocks and Cadomian folding, the development of TBU continued with deposition of Cambrian continental and marine sediments and volcanic rocks. Then the sedimentary environment changed again and the deposition continued in newly formed basins from the Ordovician until the late Middle Devonian, when it was terminated by Variscan folding (Chlupáč et al. 1998). Coal-bearing Carboniferous sediments in intramontane depressions of the rapidly eroded Variscan mountains represent the highest structural level (Fig. 7.2). Sedimentary and volcanic rocks of these four structural levels (Neoproterozoic, Cambrian, Ordovician–Middle Devonian, Carboniferous) are separated by angular unconformities, with the unconformity between the Cambrian and Ordovician being less pronounced than the other ones.

The Neoproterozoic sedimentary rocks (shale, greywacke, siliceous rock—chert) and basaltic submarine volcanic rocks form the basement of the whole area. In the Brdy Highland, the Neoproterozoic sequences crop out in the SW section of the area, around the depression of Padrť ponds and in the southern part of the highland, west of Třemšín Hill. The Neoproterozoic rocks are covered by thick Paleozoic sediments in the majority of the area.

Deposition of the Cambrian sedimentary succession (Fig. 7.2) began after the Cadomian folding and formation of new sedimentary basins. The oldest Cambrian sediments are coarse-grained conglomerates and breccia consisting of pebbles of variable rock types. These basal conglomerates were followed by a thick greywacke horizon. During the further Cambrian evolution, layers of conglomerates alternated with layers of sandstones. The proportion of quartz in these sediments increased during their evolution and a predominant proportion of the whole Cambrian sequence consists of quartz-rich rocks. The majority of this several km thick sequence of conglomerates and sandstones is thought to have been deposited in a fresh-water environment, including minor lenses of Paseky Shale. Paseky Shale, deposited in a fresh water or brackish environment, hosts the oldest macrofossils in the whole Czech Republic, unique endemic arthropod fauna with guide fossil *Kodymirus vagans* (Chlupáč et al. 1995). Proven marine Cambrian transgression occurred later and is equivalent to unnamed Cambrian Series 3 of the current international chronostratigraphic scale (Cohen et al. 2013).

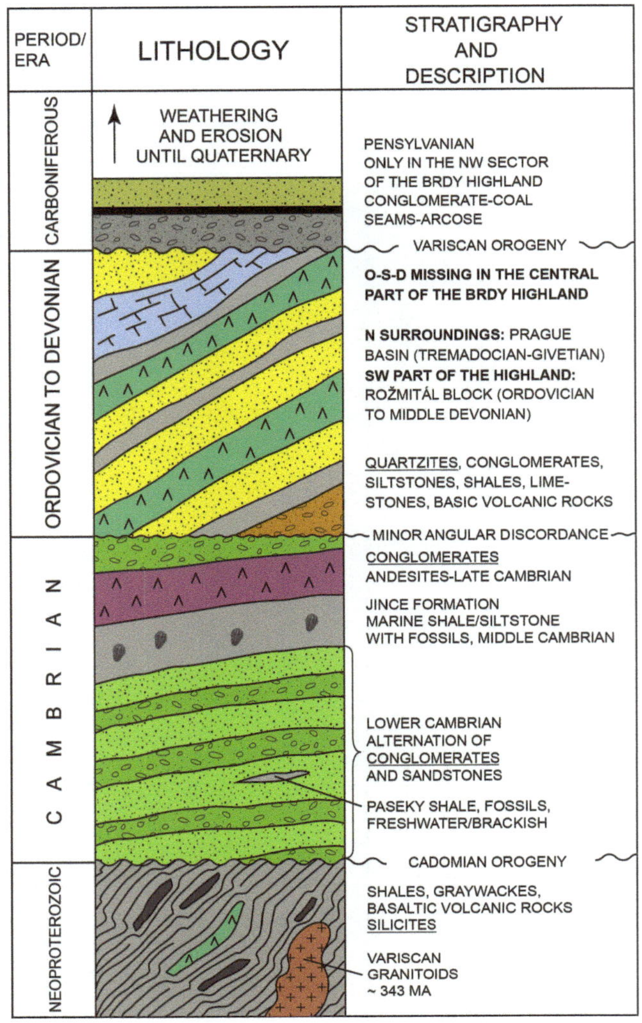

Fig. 7.2 Stratigraphic scheme of rocks forming the Brdy Highland. Modified after the stratigraphic schemes of Chlupáč et al. (1998). The rock types forming the prominent ridges and isolated rock formatios are underlined

During the marine transgression, the about 400 m thick world-famous shaly and silty Jince Formation was formed, known for its rich, nicely preserved fauna, dominated by trilobites (Fig. 7.3, Fatka et al. 2004; Fatka and Szabad 2014). Deposition of quartz conglomerates resumed after sedimentation of the shale and siltstone of the Jince Formation and the whole Cambrian evolution was completed by volcanism in the northern part of the area. Andesite predominates in the volcanic rocks of the Cambrian continental volcanism. The Cambrian sedimentary quartz-rich rocks, conglomerates and sandstones, are the basic material in the majority of the whole Brdy Highland and form all the highest hills.

Following reorganisation of the sedimentary basins at the Cambrian/Ordovician boundary, Ordovician deposition started in new sedimentary spaces with rapid bottom subsidence. The volcanism returned to the sea bottom and again

Fig. 7.3 Cambrian trilobites *Ellipsocephalus hoffi* (size of the individuals is 20–25 mm) from the Jince Formation (*Photo* P. Škácha)

produced rocks of basaltic composition. Sedimentation continued until the late Middle Devonian and was terminated by Variscan folding. Rock sequences of this structural level unconformably overlie the older Neoproterozoic or Cambrian rocks. As a result of later erosion, only in a few areas they are preserved (reviewed in Chlupáč et al. 1998). North of the highest part of the Brdy Highland, the extensive NE-SW elongated synclinorium of Prague Basin hosts the most complete sequence. The Lower Ordovician sediments of the rim of Prague Basin form the NW peripheral part of the Brdy Highland and are more widespread in the Hřebeny Crests more to the east. Ordovician to Devonian succession in the southern part of the Brdy Highland to the north and east of Třemšín hill contains is different and less thick. More details about the sedimentary environments and evolution of the whole area during the Lower Paleozoic can be found in Chlupáč et al. (1998), Fatka and Mergl (2009) or Patočka and Štorch (2004).

The Variscan mountain building processes were accompanied by granite intrusions. At ~343 Ma BP, a minor composite stock of granodiorite and granite intruded the area now occupied by the Padrť Ponds. The intrusion was accompanied by minor ore mineralization (Žák et al. 2014). Another, much larger intrusion strongly affected the area around Třemšín Hill, forming the Rožmitál Block, which is actually a sedimentary roof pendant within a large Blatná intrusion (~347 Ma BP). This southern part of the area experienced much greater uplift during the late Variscan events. The sedimentation of coal-bearing Carboniferous sediments occurred about 10 km to the north at approximately the same time. These Carboniferous sediments are already post-orogenic, i.e. unfolded, but commonly faulted and locally slightly inclined. They are preserved only in small, mostly fault-limited relics covering Neoproterozoic and Paleozoic sequences in the NW corner of the Brdy Highland near the town of Mirošov, and were exploited as a local source of coal.

7.4 Influence of the Geological Structure on the Evolution of the Landscape

The geological structure has a crucial influence on the formation of landforms. Geomorphic processes resulting in the present landforms operated in different ways in the area formed by Neoproterozoic rocks with local intrusions of

Variscan granitoids, and in the predominant part of the highland formed by Cambrian and locally also Ordovician sedimentary rocks. The Carboniferous rocks weather easily and do not support any distinctive morphology.

Among the Neoproterozoic rocks, shale and greywacke weather easily while the dark grey to black chert (silicite) lenses form prominent isolated crags on the summits. This rock, usually consisting of more than 95 % quartz, is very fine-grained and highly resistant to both chemical and physical weathering. The chert forms isolated lenses, or locally extended layers usually several tens of meters thick. Depending on the angle between the generally flat chert bodies and the weathering surface, these sediments form isolated outcrops of several morphological types. Up to 1 km long chains of up to 15 m high chert crags form *castle koppies* in the area surrounding Padrť Depression from west and south. Although the term *frost-riven cliff* is occasionally used for these rocks, some of them could have been exposed already in the pre-Quaternary period, without any participation of periglacial processes. The Neoproterozoic basaltic rocks are usually not morphologically productive. They form only several elevations to the west of the Padrť Depression (e.g. well-known Côte 718.8 m, planed in 2007–2008 to become a possible site for the US missile defence radar). Intrusion of the small ($\sim$5 km^2) Variscan Padrť Stock into the Neoproterozoic sequence produced a structural and contact-metamorphic aureole. Both the granitic rocks and the surrounding metamorphosed Neoproterozoic rocks weathered easily, forming the present-day Padrť Depression occupied by the large Padrť Ponds.

The most important rock for the morphology of the dominant and highest part of the Brdy Highland is resistant, well lithified Cambrian conglomerate, which alternates with less resistant sandstone and shale layers. Since the layers of conglomerate are usually inclined, they form elongated structural erosion ridges decorated with rocky formations, with abundant periglacial features on their slopes.

7.5 Periglacial Features

The present-day climatic conditions of the highest part of the Brdy Highland (average annual precipitation 800–900 mm; average annual temperature around 5 °C; January average temperature −3 to −4 °C; July average temperature 15–16 ° C; Quitt 2009) are less severe than those of the higher mountains along the country's borders. Especially, the amount of precipitation is lower which, together with the mostly gentle slopes and thick cover of weathered, water-permeable rock material on the slopes, resulted in less pronounced influence of fluvial processes on the morphology of the Brdy Highland.

The periglacial features shaped during the Quaternary glacials were formed in an unforested landscape. Possible patches of forests or isolated trees probably survived only along creeks and on the periphery of some marshes during glacials. The depth of the permafrost probably exceeded 100 m during the glacials, as it reached at least 65 m even in the surrounding lowlands (cf. Žák et al. 2012). During the Holocene, almost the whole area became covered with forests and features of the periglacial landscape were largely fossilized.

Vertical cliff faces on the outcrops of the hardest Cambrian conglomerate layers (frost-riven cliffs) are abundant along the ridges (Fig. 7.4). Block fields and occasionally also block streams exist below them. The blocks sometimes reach a size of several meters (Fig. 7.5). The frost-riven cliffs locally host underground cavities—crevice-type non-karst caves. The largest of them at Kloboućek Hill reaches a length of 20 m with 10 m vertical height. Several smaller cavities also exist within the block fields. The frost-riven cliffs typically display overhangs near their bases, resulting from higher water availability and thus more effective frost

Fig. 7.4 Rock face of Cambrian conglomerates (frost-riven cliff) with isolated rock tower in Kloboućek Hill (*Photo* K. Žák)

Fig. 7.5 Block field below the frost-riven cliff of Jindřichova skála (*Photo* K. Žák)

wedging. Rock formations of the tor type occur at several places, with the best developed examples on the crest and within the northern slopes of the Jinecké Hřebeny Ridge (Figs. 7.6 and 7.7).

The less resistant weathered sandstone and also the boulders released from the frost-riven cliffs were transported by solifluction. Up to four pronounced sequential morphological steps—cryoplanation terraces—were formed within the slopes, characterized by an alternation of conglomerate and sandstone layers. The best-developed examples of several successive cryoplanation terraces can be found on the SW slope of Lipovsko Hill near Strašice and in the SE slope of the Slonovec Crest near Čenkov (see Figs. 7.1 and 7.8). Nevertheless, the majority of the slopes in the whole area are gentle and quite regular, affected by solifluction transport of the weathered rock material over large distances. Material derived from the Cambrian rocks was commonly transported several hundred meters down the slopes and locally covered Ordovician rocks, thus providing clear evidence for the distance and extent of the solifluction transport. The pit-and-mound micro-topography of these slopes is related to tree uprooting (Pawlik 2013) and, during modern history, especially to the extensive windfall in November of 1941.

A number of morphological features require further study. Cílek and Ložek (2005) discussed the possible existence of fossil patterned ground on the northern slopes of

Fig. 7.6 Rock outcrop of the tor type on the crest of Jinecké hřebeny (*Photo* M. Majer)

Jordán Hill, within the Jordán artillery target area. The same authors also speculated about the existence of possible rock glaciers. The preservation of fossil patterned ground in an area located below 800 m a.s.l. in a region that was almost completely forested during the entire Holocene is rather improbable. Morphological changes related to tree uprooting (Pawlik 2013) would probably have destroyed any patterned ground. Moreover, the whole target area was strongly affected by explosions of ammunition. In contrast, larger structures like rock glaciers could have survived in the area. One possible place where a fossil rock glacier may exist is in the valley of Vološný Creek on the northern slopes of Praha Hill and another similar feature exists on the NE slope of Tok Hill. Confirmation or rejection of the existence of these features requires modern and detailed study.

7.6 Historical Land Use and Present-Day Tourist Sites

The highest part of the Brdy Highland is not suitable for agriculture and was therefore only occasionally used for hunting during most of prehistory. The ramparts of the Zavírka fortified site, which probably date back to the Iron Age, constitute the best known morphological evidence for permanent prehistoric human habitation. The site is close to the occurrences of the ores of the Příbram Ore Region(see below).

The area started to be used more intensely in the Middle Ages, when the whole highland was prospected for ores and when sites for production of tar from pine and birch wood (for production of traditional-type axle grease) were

Fig. 7.7 Rock outcrops of the tor type on the northern slope of the Jinecké hřebeny Ridge, where the layers of Cambrian conglomerates are inclined parallel to the slope (*Photo* M. Majer)

established. In the thirteenth century, in connection with a large colonization wave, several places along the foot of the highland were populated. The majority of these villages survived until the present day, although several ceased to exist a few hundred years later, leaving behind only the morphological traces of former buildings and ponds. Charcoal and potash (potassium carbonate, produced by leaching of wood ash) were the other traditional products of the area.

The ruins of Valdek Castle are the most important medieval historical monument in the area (Fig. 7.9). The castle was built by the Buzic aristocratic family on a rocky promontory in the northern part of the highland in the second half of the thirteenth century. The castle has been partly reconstructed but its main style is still Gothic. It was abandoned during the first part of the seventeenth century during the 30 Years' War. The ruin is partly repaired. It is located within the peripheral part of the military training ground. Therefore, it is accessible only during the weekends and is not yet suitable for full touristic use.

All the most important historical sights are located along the foot of the highland and are related to ore mining and

Fig. 7.8 Cryoplanation terrace in the slopes of Slonovec Ridge near Čenkov (*Photo* K. Žák)

processing. It should be mentioned that some areas at the foothill of the Brdy Highland belong to the earliest geologically mapped terrains of the Central Europe (e.g. Lipold and Krejčí 1860). The Příbram Ore Region located at the SE foot of the highland produced vein-type Ag-Pb-Zn ores for centuries. The Vojtěch Shaft at Březové Hory was the first one in the world that reached 1,000 m in vertical depth (on a single rope!) in 1875. This mining area hosts the Mining Museum at Březové Hory, with mineralogical and historical exhibits. Several abandoned mines with unique technical and mining equipment can be visited here (Fig. 7.10). After World War II, intensive mining focused on the more southerly located and newlydiscovered calcite-pitchblende (uraninite) veins (see Žák and Dobeš 1991 for a review of the ore district geology and ore genesis). The No.16 Háje shaft in the uranium mining area reached the depth of

Fig. 7.9 Inside the ruins of Valdek Castle (*Photo* K. Žák)

1,838 m in 1975. Of the 25 shafts in the uranium deposit, eight were more than 1,000 m deep. The total production of the whole district from the Middle Ages to 1985 (Ag, Pb, Zn ores) or 1991 (U ores) was about 500,000 metric tons of lead, 3,500 metric tons of silver, and 49,990 metric tons of uranium.

The northern surroundings of the Brdy Highland are sometimes called the *Blacksmith's Workshop of the Austro-Hungarian Empire*. This is because of the presence of sedimentary Ordovician iron ores, where the related iron works were the most important producer of iron in the whole empire in some parts of the eighteenth and nineteenth century. The area became famous for production of cast iron and artistic products made from this material. Both mining and ore processing at Příbram and in iron works based on Ordovician ores consumed enormous quantities of wood and charcoal and were the main driving force responsible for changes in the forest species and also anthropogenic changes

Fig. 7.10 The Ševčín Mine, which now hosts the Mining Museum (*Photo* Archive of the Mining Museum Příbram)

in the landscape of the Brdy Highland—construction of roads, platforms for charcoal production, etc.

Among several religious sites located along the foot of the highland, the Svatá Hora near Příbram pilgrimage site, consecrated in honour of the Blessed Virgin Mary, is one of the most important pilgrimage sites in Central Europe with excellent Baroque architecture, and is the most frequently visited place of the Brdy Highland area. The Augustinian monastery and Church in Svatá Dobrotivá (Komárov) and the churches in Mrtník, Strašice and Rožmitál pod Třemšínem are also interesting religious sites.

In addition to the above-mentioned Mining Museum in Příbram, several smaller museums are located around the Brdy Highland, e.g. in Strašice and Rožmitál pod Třemšínem. A unique historical mechanism, a nineteenth century water-driven hammer, is still in operation during the summer months in the village of Dobřív in the northwestern part of the Brdy Highland (Fig. 7.11).

Fig. 7.11 Historical water-driven hammer in the village of Dobřív (*Photo* K. Žák)

7.7 Conclusion

The Brdy Highland, especially its highest part, which was occupied by the military training ground, is currently an extensive unpopulated area covered by continuous forests (except for the former artillery target areas). Historically, land use was predetermined by the predominance of quartz-rich Cambrian sedimentary rocks producing extremely infertile and low-pH soils. After termination of military use and conversion to a Protected Landscape Area, the area will certainly become a popular tourist destination. Geomorphology of the Brdy Highland was shaped mainly by periglacial processes during the Quaternary glacials. These morphological features became forested and fossilized during the Holocene.

Acknowledgments The geomorphological study of the Brdy Highland and preparation of this chapter were supported by the program RVO67985831 of the Institute of Geology of the Czech Academy of Sciences. An early version of this chapter was read by V. Cílek and O. Fatka, their comments and additions are greatly appreciated.

References

Chlupáč I, Kraft J, Kraft P (1995) Geology of fossil sites with the oldest Bohemian fauna (Lower Cambrian, Barrandian area). J Czech Geol Soc 40(4):1–8

Chlupáč I, Havlíček V, Kříž J, Kukal Z, Štorch P (1998) Palaeozoic of the Barrandian (Cambrian to Devonian). Czech Geol Surv, Prague, p 183

Cílek V, Ložek V (2005) Reliéf a geomorfologie. In: Cílek V (ed) Střední Brdy. Czech Republic, Ministerstvo zemědělství ČR, Ministerstvo životního prostředí ČR, ČSOP Příbram a Kancelář pro otázky ochrany přírody a krajiny, Příbram, pp 59–69

Cohen KM, Finney SC, Gibbard PL, Fan JX (2013) The ICS International Chronostratigraphic Chart. Episodes 36:199–204

Demek J, Mackovčin P (eds) (2006) Hory a nížiny. AOPK, Brno, Czech Republic, Zeměpisný lexikon ČR, p 582

Fatka O, Mergl M (2009) The 'microcontinent' Perunica: status and story 15 years after conception. In: Basset MG (ed) Early

Palaeozoic peri-Gondwana terranes: New insights from tectonics and biogeography. Geol Soc London, Spec Publ, 325:65–101
Fatka O, Szabad M (2014) Biostratigraphy of Cambrian in the Příbram-Jince Basin (Barrandian area, Czech Republic). Bull Geosci 89(2):411–427
Fatka O, Kordule V, Szabad M (2004) Stratigraphical distribution of Cambrian fossils in the Příbram-Jince Basin (Barrandian area, Czech Republic). Senckenb Lethaea 84(1–2):367–381
Hajná J, Žák J, Kachlík V, Chadima M (2010) Subduction-driven shortening and differential exhumation in a Cadomian accretionary wedge: the Teplá-Barrandian unit. Bohemian Massif Precambr Res 176(1):27–45
Lipold MV, Krejčí J (1860) Verhandlungen der k. k. geologischen Reichsanstalt, Sitzungsbericht vom 24 April, Wien, pp 88-91
Patočka F, Štorch P (2004) Evolution of geochemistry and depositional settings of Early Palaeozoic siliclastics of the Barrandian (Teplá-Baranndian Unit, Bohemian Massif, Czech Republic). Int J Earth Sci 93:728–741
Pawlik Ł (2013) The role of trees in the geomorphic system of forested hillslopes-A review. Earth Sci Rev 126:250–265
Quitt E (2009) Typy topoklimatu. In: Hrnčiarová T, Mackovčin P, Zvara I (eds) Atlas krajiny České republiky. Ministerstvo životního prostředí ČR, Praha a Výzkumný ústav Silva Taroucy pro krajinu a okrasné zahradnictví, v. v. I., Průhonice, In Czech, 104
Žák K, Dobeš P (1991) Stable isotopes and fluid inclusions in hydrothermal deposits: the Příbram ore region. Rozpr. ČSAV, Ř. mat. přír. Věd, 1–109 (Academia Praha)
Žák K, Richter DK, Filippi M, Živor R, Deininger M, Mangini A, Scholz D (2012) Coarsely crystalline cryogenic cave carbonate-a new archive to estimate the Last Glacial minimum permafrost depth in Central Europe. Clim Past 8:1821–1837
Žák K, Svojtka M, Breiter K, Ackerman L, Zachariáš L, Pašava J, Veselovský F, Litochleb J, Ďurišová J, Haluzová E (2014) Padrť Stock (Teplá-Barrandian Unit, Bohemian Massif): Petrology, geochemistry, granodiorite U-Pb zircon dating, and Re-Os age and origin of related molybdenite mineralization. J Geosci 59:351–366

Bohemian Forest: Landscape and People on the Frontier

8

Pavel Mentlík

Abstract

The direct expression of variable rock resistance and fault lines of different ages is characteristic for the deeply denuded relief of the Bohemian Forest. Three types of relief are typical. First, an extensive plateau covered with peat bogs spreading out in the central part of the area. Second, canyon valleys incised in the edge of the plateaus with torrential flows in rocky and stony riverbeds and finally, a rugged glacial relief with glacial cirques and a few relicts of periglacial processes in the vicinity of the highest summits. Geomorphological mapping supported by ^{10}Be and radiocarbon dating showed that the culmination of the last glaciation in the Šumava range occurred during the Last Glacial Maximum, when valley glaciers appeared in some places. The cirque glaciers developed separately approximately 16,000 and 14,000 years ago. The recent findings suggest the presence of humans in the Šumava region since the Mesolithic period. The area was a source of raw material from the Middle Ages (gold and wood) for nearby populated regions and it was also a region of glass production. The most important recent change was connected with the end of the World War II when the predominant German population was expelled. The protection of natural heritage within the whole area is coordinated by the Šumava National Park, established in 1991.

Keywords

Bohemian Forest • Šumava • Planation surfaces • Glaciation • Periglacial

8.1 Introduction

Travellers to the Bohemian Basin do not have it easy. The hilly country and lowland areas are surrounded by a belt of forested uplands, which has always marked the border of the Bohemian Kingdom. In the south, this barrier consists of the Novohradské hory Mountains, Šumava and Český Les. The people of the Šumava respected the natural boundaries when determining the state borders, following the elongated and forested ridges. The Šumava is a significant natural boundary in terms of geology, topography and hydrology. The metamorphic rocks and granites of the Bohemian Massif here rise up over much younger rocks of the Tertiary Alpine Foredeep to the south. Even the present-day topography and river network create distinct units—the ridges of the Šumava form the watershed between the Black Sea and the North Sea. The name Bohemian Forest is usually used for the whole hilly and forested area including the Šumava, and Český Les on the Czech side and Bayerischer Wald on the German side of the border (Fig. 8.1).

8.2 Location and Long-Term Geomorphological Evolution

The Šumava is one of the central European Variscan mountain ranges, with a relatively high degree of metamorphism of rocks and the long-term history of denudation

P. Mentlík (✉)
Centre of Biology, Geoscience and Environmental Education, University of West Bohemia in Plzeň, Pilsen, Czech Republic
e-mail: pment@cbg.zcu.cz

T. Pánek and J. Hradecký (eds.), *Landscapes and Landforms of the Czech Republic*, World Geomorphological Landscapes, DOI 10.1007/978-3-319-27537-6_8

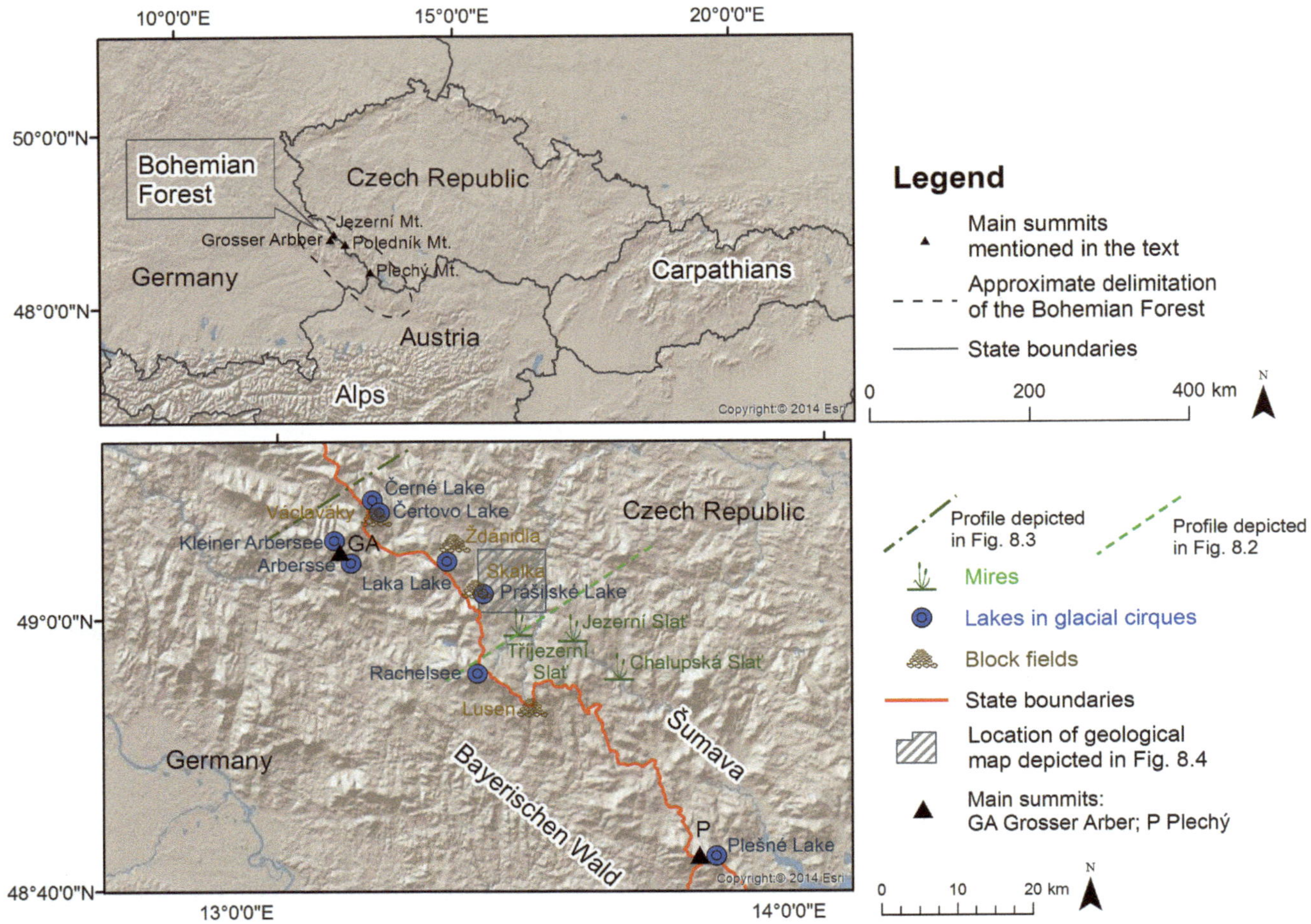

Fig. 8.1 Position of the Šumava Mountains in central Europe; Data: World Shaded Relief and State boundaries (ESRI 2014)

(together with the Vosges and the Black Forest). It is a mountain region about 110 km long and 30–40 km wide extending from the northwest to the southeast.

The relief of the Šumava today is a reflection of the long-term geological/geomorphological evolution that began with the origin of the Variscan Mountains and was followed by erosion to a flat planation surface. This surface, in turn, was subsequently uplifted and dissected by river erosion and small glaciers. The contemporary relatively flat relief in the watershed parts is the legacy of the original planation surface. In the central part of the Šumava we find the remains of the uplifted levelled surface in the form of large plateaus (the Šumava Plains) with an area of $\sim$670 km^2 (Fig. 8.2). In its peripheral part the incised remnants of this surface form peaks of similar altitude (the 'level peaks'). On the plateaus, at elevations of 1000–1200 m, numerous peat bogs of upland character have developed due to the flat topography and the relatively high rainfall and poor drainage system (the most extensive are Modravské Slatě mires, with an area of 3615 ha) (Janský et al. 2003). Their formation began at the end of the Pleistocene (about 11.5 thousand years ago), and have a peat thickness up to 7 m (Svobodová et al. 2002). Although the bogs are usually protected conservation areas, two typical raised bogs (Tříjezerní and Jezerní Slatě mires) and one valley/raised bog (Chalupská Slať mire) are accessible by wooden footpaths for tourists (see Fig. 8.1). The largest peat-originated lake occurs in Chalupská Slať mire with an area of 1.3 ha (Janský et al. 2003).

At the northwest and southeast edge of the Šumava the original flat surface is more incised by valleys of several rivers (Weisse Regen, Úhlava, Vltava) flowing to the northwest or southeast. Three prominent ridges run to the northwest of the central part (Fig. 8.3). The mountains here show the biggest difference in altitudes, up to 600 m relative relief between the bottoms of the valleys and the surrounding highest peaks The highest elevations are found in the southwest ridge (all in Germany), which starts with the highest mountain—Grosser Arber 1456 m.

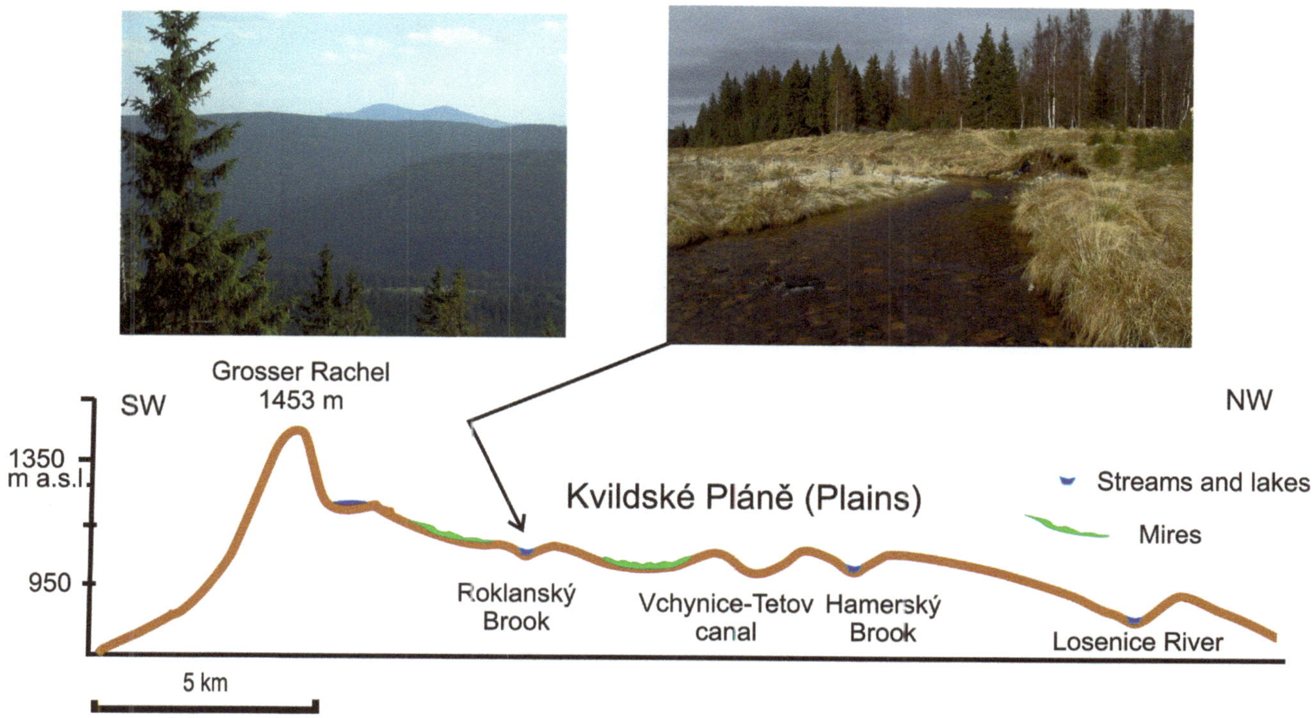

Fig. 8.2 Topographic profile across the central part of the Bohemian Forest (*Photo* P. Mentlík)

In the southeast, the depression of the Vltava river floodplain, with numerous peat bogs (filled from underneath by the Lipno Dam), is wedged between the high border ridge of Plechý peak (1378 m) and the raised granitic area of Boletice Highlands.

8.3 Geology the Background of the Landscape

8.3.1 Geological Setting and Development

The appearance of the contemporary Šumava, in addition to being controlled by the variability of rock resistance (see Sects 8.3.2 and 8.3.3), was mainly conditioned by two types of tectonic lines: older tectonic lines running north to south, which were deepened by fluvial and glacial erosion, and younger faults trending northwest southeast. The existence of younger faults was key to the formation of today's relief. These directions are the same as of the main Šumava ridges and along these tectonic lines the Šumava was uplifted to its current altitude. Earlier tectonic lines probably formed in connection with the Moravo Moldanubian Phase (345–330 Ma) and the younger can be remnants of the Bavarian Phase (330–315 Ma) (Finger et al. 2007), which had the greatest importance for the formation of the rocks which make up the Šumava today.

To the southwest, where the Šumava is tectonically separated by the peripheral Danubian watershed system from the Alpine Foredeep, two completely different geological worlds meet. Old (mainly Paleozoic) metamorphic rocks and granites of the Bohemian Massif are interwoven with adjacent units of the Tertiary Alpine Foredeep. The most significant unit, with a northwest southeast alignment, is the Pfahl (Bavarian quartz wall). It forms an unbroken line along the southwest foot of the highest Šumava, with a length of ~150 km. Most likely it served as a significant barrier to tectonic disturbance of segments of the Bohemian Forest, while it also prevented headward erosion into the core of the mountains from the Danube side. The Šumava therefore probably lies in a relatively compact tectonic segment, preserved by the Bavarian resistant quartz wall to form the largest central European relict of Late Tertiary plateaus (Mentlík and Šebesta 2003). To the northeast the range is limited by the Pošumavský Fault (Hartvich 2004; Hartvich and Valenta 2013) and by upfold systems, through which the Šumava passes to the Bohemian Massif (Fig. 8.4).

No marine sedimentary cover has ever been identified in the Šumava. Therefore, we assume that it is a long-term elevated and deeply denuded area where the resistance of

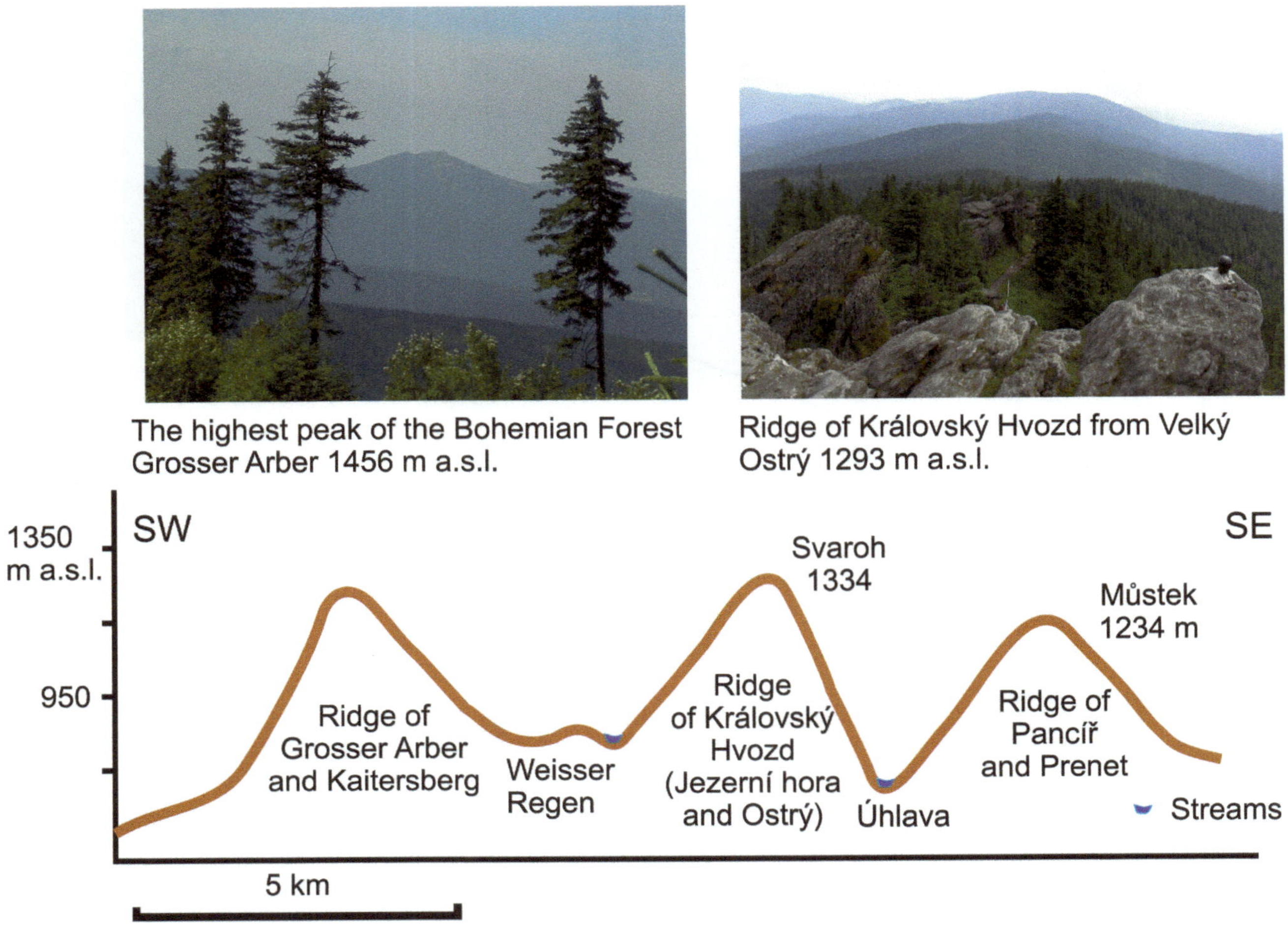

Fig. 8.3 Topographic profile across the northwest part of the Bohemian Forest (*Photo* P. Mentlík)

each type of rock, fracture systems and the presence of old faults are central to the shaping of the relief (Mentlík and Šebesta 2003).

The dominant rocks in the Bohemian Forest are metamorphic, Moldanubian, and large units of crystalline complex which form the majority of the bedrock in southwest Bohemia and partly in southern Moravia (Kodym 1961). The least metamorphic rocks (phyllites and mica schists) are found in the northwest Šumava in the Royal Forest/Královský Hvozd. The dominant rocks are then paragneisses, migmatites and quartzites, the so-called monotonous series of the Moldanubian forming the dominant part of the Šumava (Vejnar et al. 1991; Pelc and Šebesta 1994).

8.3.2 Relief on Metamorphic Rocks

The direction and inclination of the foliation, which mimics the layering of sedimentary rocks, is central for exposure of metamorphic rocks in the field and the shapes of rock outcrops. Thus, we can distinguish milder slope gradients controlled by the direction and tilt of the foliation, and in some cases, these areas partially stand above the surrounding terrain as rocky slabs. On the other hand, there are places where slope is perpendicular to foliation and here overhanging rock outcrops or extensive rocky defiles can be found (Fig. 8.5). Also, the presence of numerous quartz veins is of great importance, because the occurrence of crags is usually conditional on the existence of more resistive veins or quartziferous bedrock. Quartz was also a source of raw materials for glass production, widespread in the Šumava in the past. Šumava landmarks, where we can find such distinctive rock formations built of mica schist, are Velký and Malý Ostrý. The rocky ridge of Velký Ostrý (1293 m a.s.l.) is typical, extending along the border between the Czech Republic and Germany. Similar formations are also found on the nearby rocky peak Kleiner Osser (1200 m asl) in Germany (Fig. 8.6). The romantic landscape of these twin peaks and the stories that took place around it apparently inspired Carl Maria von Weber to compose the

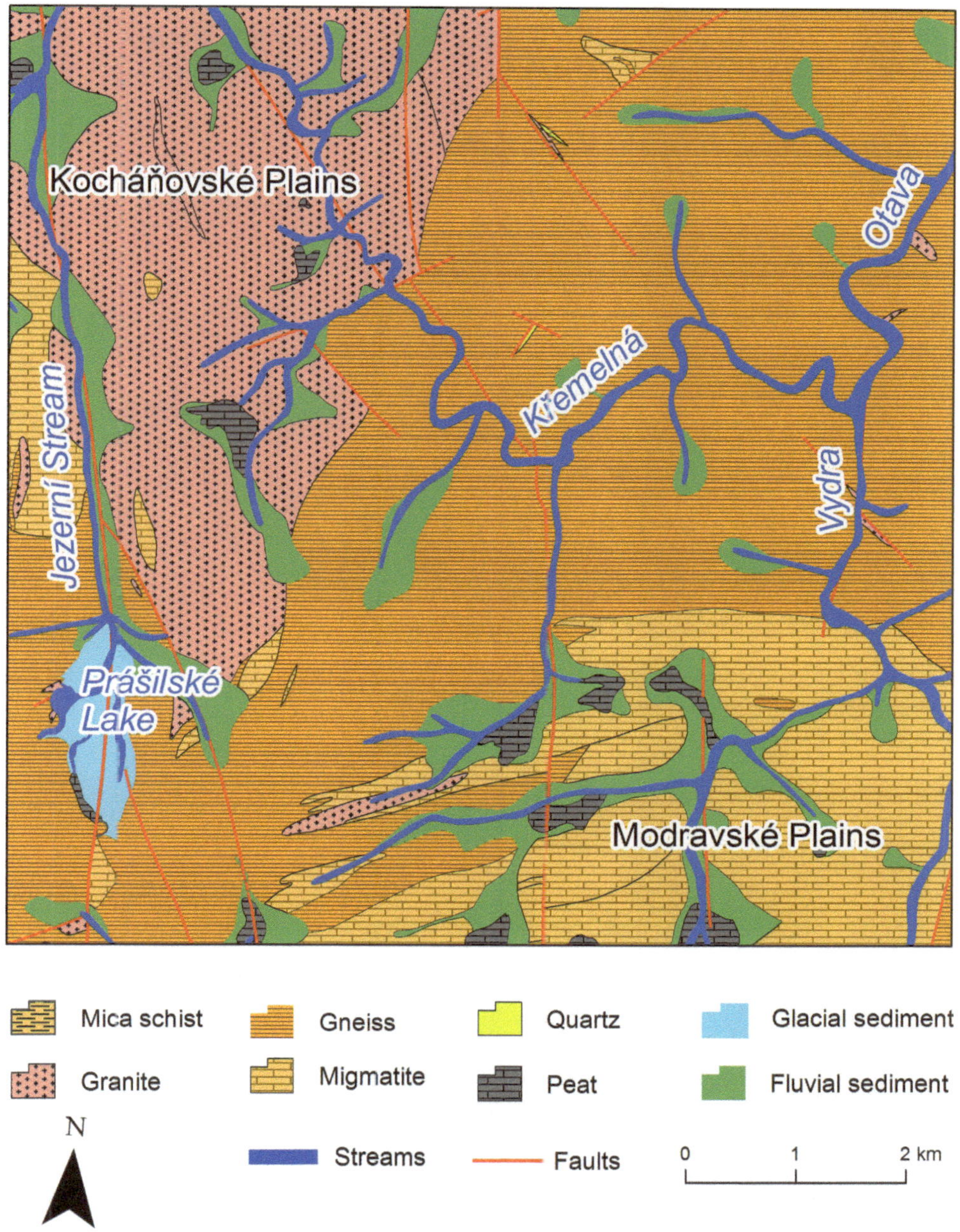

Fig. 8.4 Geological map of the central part of the Šumava; Note the systems of N–S older faults and NW–SE younger faults belonging to Pošumavský Fault zone (data Pelc and Šebesta 1994)

first German romantic opera *Der Freischütz*. It is said that the Bílá Strž creek with its powerful waterfall that flows over resistant quartzite is the "Wolf Ravine", where in the opera, in the presence of the devil, the legendary magical bullets were forged.

8.3.3 Relief on Granites

The geological structure of the Bohemian Forest is complemented by numerous granite plutons with specific geomorphic features which significantly differentiate this landscape from the areas of metamorphic rocks. Granites, in comparison with the metamorphic rocks in the Šumava, appear as less durable and in various locations they form topographic depressions. Granite rock outcrops are often distinctive landscape landmark and tourist destinations. What a granite rock will be like is decided deep below the earth's surface. Hot magma cools slowly there and solidifies. During this process, the whole mass of rock shrinks and changes its overall shape, which causes considerable stress. The rock fractures, often according to a regularly organized system of fissures. Perpendicular cracks are referred to as longitudinal and transverse. The generally horizontal cracks

Fig. 8.5 Rock outcrops on metamorphic rocks; from upright: **a** tor within a monotonous series of the Moldanubian gneiss; **b** foliation and quartz veins; **c** rock plate concordant with foliation (*Photo* P. Mentlík)

Fig. 8.6 The twin peaks of Velký and Malý Ostrý, named the "Breasts of the Mother of God" by old inhabitants of the Bayerischer Wald (*Photo* P. Ouředník)

are then pseudobedding fractures similar to the bedding fractures of sedimentary rocks. Longitudinal cracks are not conspicuous, they are often closed and without venous filling. On the contrary, the transverse cracks are usually wide and filled with quartz or ore veins. After many millions of years of the granite body lying beneath the earth's surface, the overlying rock is gradually eroded until the granite eventually reaches the surface. Here water and then plant roots penetrate along cracks created millions of years ago. Progressive weathering gradually separates whole rock blocks from each other, with their shapes and sizes corresponding with the arrangement of the fracture system.

Along the Bear Trail (Medvědí Stezka) in the granite area of Plechý one can find numerous interesting rock outcrops (tors) that result especially from the changes in the frequency of pseudobedding fractures (Fig. 8.7). These rocks have a typical "mushroom" shape. The "hat" of the mushroom consists of granite with a low frequency of pseudobedding fractures, whereas in the "leg" their spacing decreases. Along these numerous cracks weathering progressed rapidly, and so the "leg" is substantially narrower than the "hat", which consists of compact, unfractured granite.

The Bear Trail is also the site of characteristic relics of spheroidal weathering of granites. It rounds off the edges of joint planes and leads to isolation of corestones—rounded blocks within a matrix of weathering products (Selby 1985). This is some of the little remaining evidence of deep weathering under a humid tropical climate in the Tertiary, which apparently was accompanied by the formation of the planation surface in the contemporary Bohemian Forest. During the subsequent geological evolution they were then gradually brought to light by denudation.

Joint systems also determine the shape of larger surface forms. Their regular arrangement influenced the appearance of the cirque at Prášilské Lake. The hollow consists of two parts. The northern part is granitic and the glacier followed the fracture system (Fig. 8.8). The southern part is composed of gneiss and its shape reflects the directions of foliation. This cirque is a typical example showing the very close links between geological conditions and the shape of the earth's surface.

8.4 Present-Day Relief—Inheritance of the Pleistocene

8.4.1 Cold Periods in the Pleistocene

Essential for the appearance of today's relief, especially in the most elevated areas still not reached by retrogressive fluvial erosion, were geomorphological processes operating in cold periods of the Pleistocene. At the end of the last glacial period, the forest expanded and largely covered the now inactive glacial (cirques, moraines) and periglacial (block fields, frost-riven cliffs and scarps) forms. Prior to that, all slopes were modelled by periglacial processes such as gelifluction which left numerous angular scattered blocks and boulders (Fig. 8.9). Gelifluction sediments create a

Fig. 8.7 Rock outcrops on granites; from upright: **a** corestones—relics of the spheroidal weathering of granites; **b** tafone-like cavities—probable evidence of deep weathering under a humid tropical climate in the Tertiary; **c** rock mushroom (*Photo* P. Mentlík)

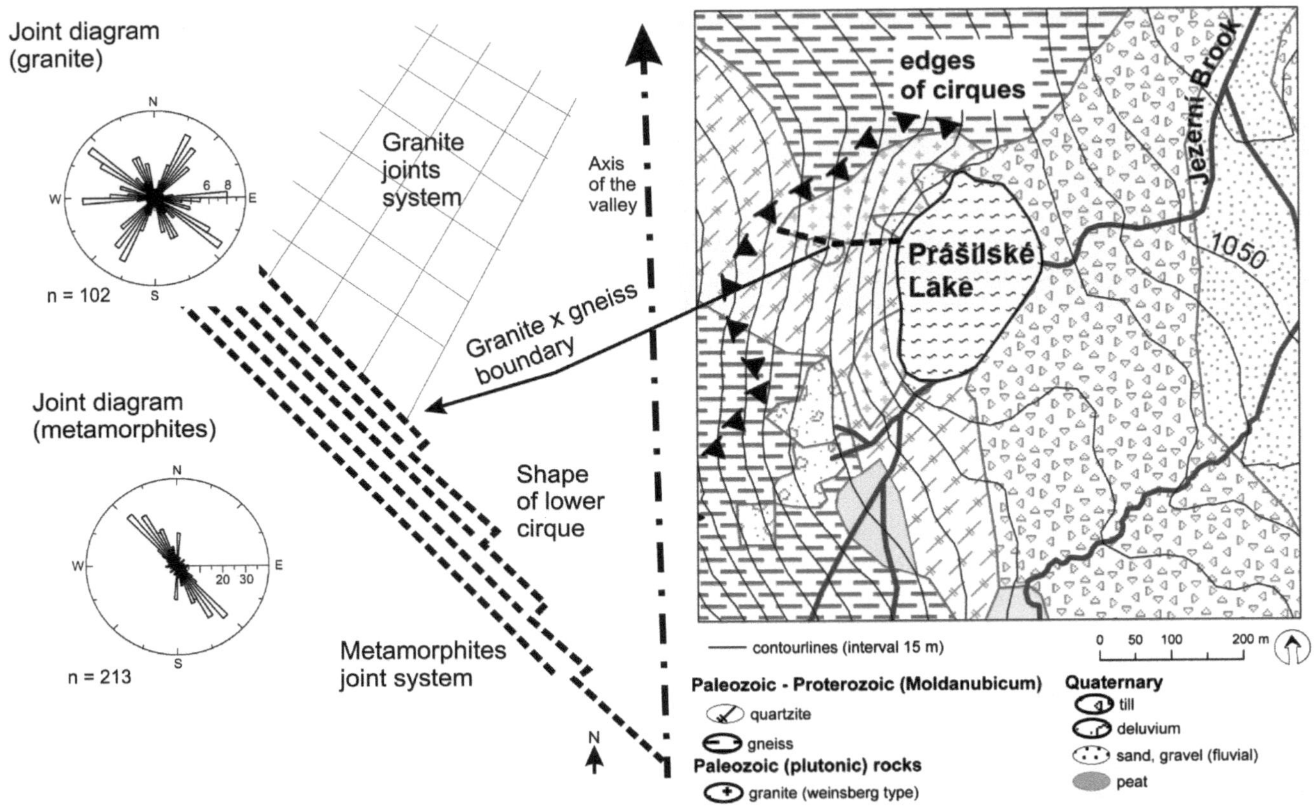

Fig. 8.8 Influence of joint systems in metamorphic rocks and granites on the form of the glacial cirque of Prášilské Lake (original data Mentlík et al. 2010)

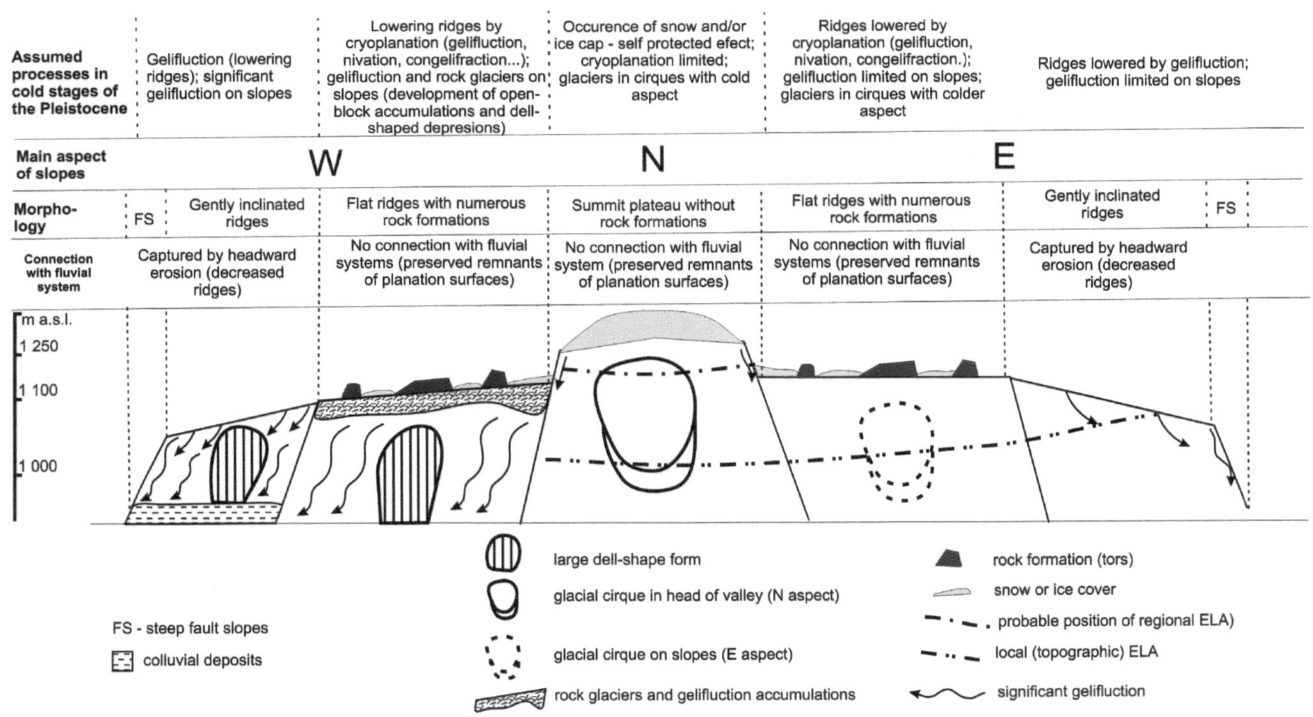

Fig. 8.9 Scheme of distribution of periglacial processes during cold periods of the Pleistocene in the most elevated parts of the Šumava Mts.

significant part of colluvial deposits, particularly thick at the foot of the slopes.

8.4.2 Between the North European Ice Sheet and the Alps

Due to its location, in the cold periods of the Pleistocene the Šumava was situated between the northern ice sheet and the heavily glaciated Alps. Glaciers in the Šumava left eight distinct glacial cirques, which are now filled by lakes (Table 8.1). Because these landforms are located adjacent to the highest parts of the Šumava, and in comparison with the predominantly smooth relief of the Šumava have very rugged topography, they are highly attractive for tourists. All glacial areas in the Šumava enjoy the highest degree of nature conservation. The shaded and inaccessible head walls maintain specific communities of plants and animals that are found nowhere else in the Šumava. The lakes themselves are habitats of rare plants (e.g. *Isoëtes*). The lakes were significantly affected by acidification, associated with the naturally acidic bedrock and the emissions that came from inconsiderate industrial production in the last century. This dramatically changed the flora and fauna of the lakes. Currently, the lake conditions are slowly returning to normal (Vrba et al. 2003).

8.4.3 Glacial Relief in the Šumava

Compared to the Variscan mountains lying farther west (e.g. Vosges) (Mercier 2014), the regional snowline of the Bohemian Forest in the cold periods of the Pleistocene was apparently at higher altitude. During the Last Glacial Maximum we assume its position at about 1300 m asl. This fact can be explained by the increasing continentality and generally lower humidity in the more easterly areas. Therefore, we do not assume the presence of more extensive glaciers on the higher plateaus of the Šumava plains, where the prevailing altitude is between 1000 and 1200 m. But it is likely that the highest peaks of the Šumava (with an altitude above 1300 m) were covered by more permanent ice caps (Fig. 8.9).

The following factors were the decisive local topographical conditions for the formation of glaciers:

- highest peaks exceeding 1300 m above sea level, ridge or summit plateaus at an altitude of 1200 m asl oriented against the west to northwest winds;
- north-facing valleys exploiting old faults;
- preglacial valley heads as well as older slope hollows and cirques that arose in previous cold periods in the valley bottoms at an altitude of about 1000 m.

The glacial forms dominant in the Šumava are cirques and enclosed valleys with their shoulders undercut by glacial activity. These are usually adjacent to several generations of moraines (Fig. 8.10). Depending on their size, one can distinguish two types of cirques in the Šumava. Cirques with a large area (such as Kleiner Arbersee cirque with an area of 353.8×10^6 m^3 are an order of magnitude larger than the small cirques, e.g. Prášilské lake cirque with an area of 11.0×10^6 m^3) (Mentlík et al. 2013).

Glaciers are important indicators of paleoclimate. Research on past glaciers in the Šumava provides information about local and regional climatic conditions in the key period at the end of the last glacial cycle, i.e. particularly since the last glacial maximum (about 24,000 years ago) until the end of the Late Glacial about 11,500 years ago (Fig. 8.11).

The research presented here is based on detailed geomorphological mapping (Mentlík et al. 2010; Mentlík and Novotná 2010). The mapped geomorphological forms (moraine ramparts) were dated by means of cosmogenic isotope (^{10}Be) concentrations on rock blocks located at the surface (Fig. 8.11). In the Czech part of the Bohemian Forest, drilling in the silted lakes was carried out at two sites (Stará jímka at Prášilské Lake and the depression between

Table 8.1 Basic characteristics of glacial lakes in the Bohemian Forest (data according to Janský et al. 2003, 2005; Raab and Völkel 2003; Reuther 2007)

Lake name	Water level altitude (m)	Area (ha)	Max. depth (m)	Approximate water volume (10^6 m^3)
Černé lake	1008	18.8	40	2.88
Čertovo lake	1028	10.7	36	1.85
Arbersse	934	7.7	16	0.45
Jezero laka	1096	2.6	4	0.05
Kleiner arbersee	918	8.6	10	0.25
Prášilské jezero	1079	4.2	17	0.35
Plešné jezero	1090	7.6	19	0.61
Rachelsee	1071	5.7	13	0.18

Fig. 8.10 Glacial cirque of Prášilské Lake (*Photo* P. Mentlík)

the moraine mounds at Černé Lake), where radiocarbon dating and optically stimulated luminescence (OSL) were used (Mentlík et al. 2010; Vočadlová 2011). Start of lake sedimentation confirmed the period for the end of glaciation (Mentlík et al. 2010). The results of radiocarbon dating proving the beginning of lake sedimentation and also the start of peat accumulation from the bottom of Kleiner Arbersee lake (German side) and the base layer of a peat bog to the south of this lake (Raab and Völkel 2003; Reuther 2007) also indicate the end of the last glacial period (Fig. 8.11).

Although the size of some cirques suggests that they were filled with glaciers repeatedly, no unequivocal evidence is available for any possible glaciations older than the last glacial cycle for the Šumava. The oldest moraines are dated from around Kleiner Arbersee (Reuther 2007). For this large cirque, glacial activity from the Last Glacial Maximum has been documented, which ended about 19,500 years ago. Glaciation left three generations of moraines dated at ~24,000, ~21,000 and ~19,000 years ago. At the time of the most extensive glaciation a glacier filled the valley with maximum length of 2600 m and maximal width of 800 m (Raab and Völkel 2003).

In the small cirques of Prášilské Lake and Laka Lake the end of the local most extensive glaciation (in the valley of Prášilské Lake the glacier reached a length of 2100 m and a width of 760 m) was dated for ~19,000 years ago. Here, in contrast to Kleiner Arbersee, smaller cirque glaciers responded to shorter climatic fluctuations in the Late Glacial. In the time interval from 19,500 to 14,000 years ago two generations of moraines in the surroundings of Prášilské and Laka lakes have been dated (Fig. 8.12) (Mentlík et al. 2013). Extensive glaciation probably ended ~16,000 years ago, leaving and outer, distinctive moraine in front of Prášilské Lake, and the youngest glaciation (at Prášilské Lake moraine dam lake) occurred ~14,000 years ago. Interestingly, no documentation of the presence of glaciers during the last very significant cold fluctuation of the Late Glacial (Younger Dryas ~12,500 to ~11,500 years ago) is provided by the Šumava moraines. This fact is confirmed by radiocarbon

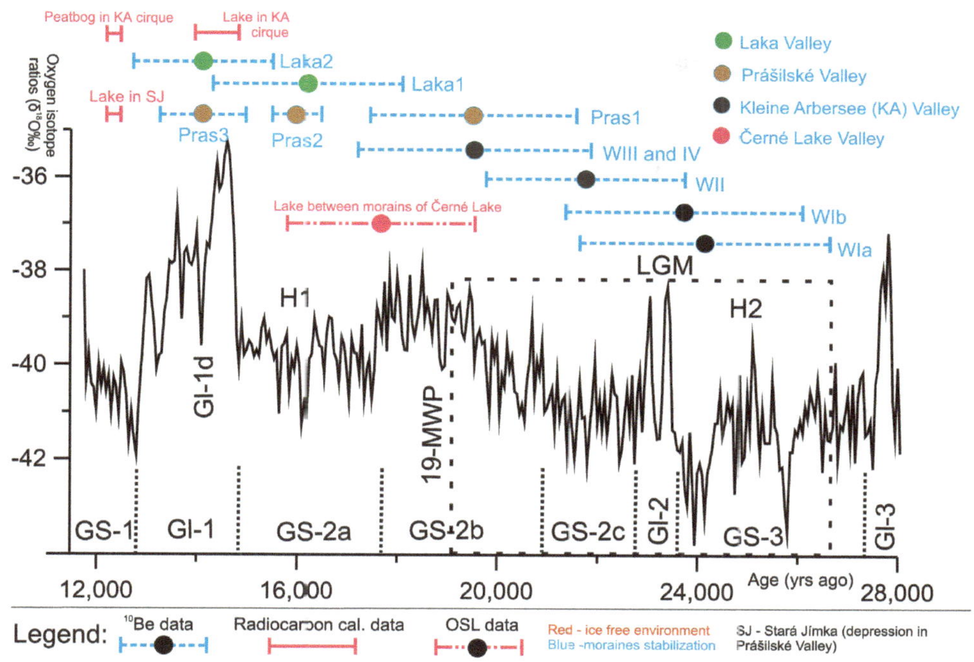

Fig. 8.11 Plot of exposure data (^{10}Be), OSL and calibrated radiocarbon data from the Prášilské, Laka, Černé Lakes and Kleiner Arbersee valleys against the GRIP ice-core record (Blockley et al. 2012). The dashed rectangle marks the extension of the LGM and 19-MWP (Clark et al. 2009), as well as the timing of H1 and H2 (Peck et al. 2007). Pras1, 2, 3 and Laka1, 2 (Mentlík et al. 2013); WIa, WIb, WII and WIII based on Reuther (2007). Calibrated radiocarbon data: South part of Kleiner Arbersee (13.8–14.9 ka) (Raab and Völkel 2003; Reuther 2007); peat bog south of Kleiner Arbersee (12.4–12.8 ka) (Raab and Völkel 2003; Reuther 2007); Stará Jímka (12.2–12.6 ka) (southern of Prášilské Lake—Fig. 8.2) (Mentlík et al. 2010); OSL data silted lake between two moraine walls in the glacial cirque of Černé Lake (Vočadlová 2011)—see text for additional details

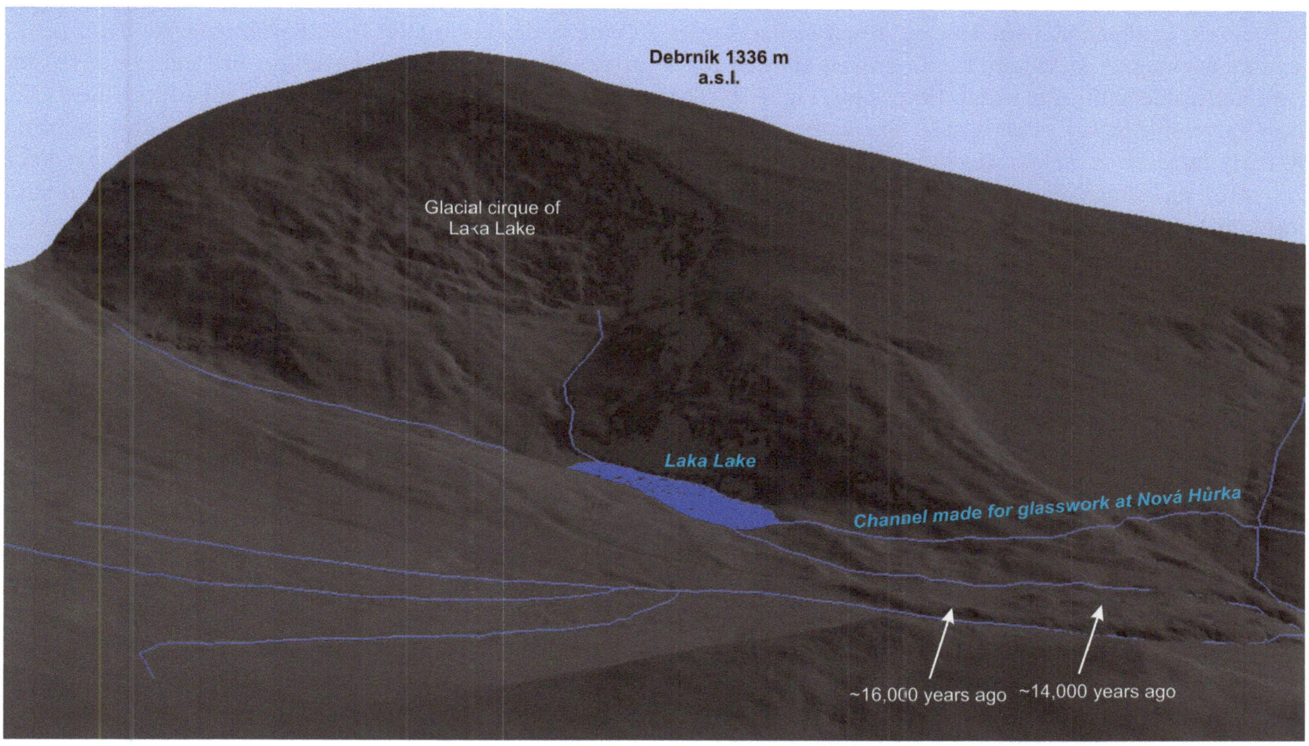

Fig. 8.12 Geomorphology of the surroundings of Laka Lake; moraine dating according to Mentlík et al. (2013)

dating of lacustrine sediments from Stará Jímka and Kleiner Arbersee, indicating that during the Younger Dryas these cirque depressions were no longer glaciated, but were filled by lakes (Fig. 8.11). On the contrary, the dating of the youngest moraines shows that the glaciers in the Šumava were present during a shorter, earlier cooling period at ~14,000 years ago (Older Dryas). This view is primarily supported by pollen analyses of lake sediments from Switzerland and Austria (Lotter et al. 1992; Ammann et al. 1994), but dating of moraines from the Alpine region is missing; it is likely that they were covered by younger and more extensive glaciation in the Younger Dryas (Ivy-Ochs et al. 2008). It seems that preserved relics of glaciation in the Older Dryas are specific to the Variscan mountain ranges of central Europe, as moraines associated with this period have been also documented from the Krkonoše, at the opposite north-eastern rim of the Bohemian Massif (Engel et al. 2014).

8.4.4 Inheritance of Periglacial Processes

In the ridge and summit areas of the Bohemian Forest and on adjacent slopes the evidence of periglacial processes is ubiquitous. In cold periods of the Pleistocene, there was widespread permafrost throughout the Bohemian Massif (Czudek 1997; Czudek 2005). In the mountainous regions slopes with cold orientation seem to have remained frozen even during the summer period. These conditions limited the efficacy of gelifluction, whereas on south-facing slopes summer thawing of seasonally frozen ground allowed for a more noticeable shift of material downslope (Fig. 8.9). The most impressive legacy of periglacial conditions is provided by fairly large block fields, which are assumed to partially consist of rock fragments disintegrated in situ, with additional material transported there from the top plateau by means of talus creep within rock glaciers. The most extensive (~5 ha), continuous and accessible block field surrounds the summit of Lusen Mt. (1373 m asl). Other block fields occur in the surroundings of the previously glaciated summits. Unlike cold-oriented cirques, block fields cover upper parts of slopes with warmer (south to west) aspects. Typical block fields with an area between 2 and 5 ha can be found in the surroundings of Jezerní hora Mt. (1343 m asl), Ždánidla Mt. (1309 m asl) and Poledník Mt. (1315 m asl) (see Fig. 8.1 for location). Broad dells usually covered by sparse block slope sediments sometimes cover the middle and lower parts of slopes with warmer aspects positioned below the block field accumulation. On the summit and ridge plateaus rock outcrops include elongated (up to 80 m long and 20 m wide) walls and isolated tors (up to 11 m high). Typical outcrop formations can be visited on the ridge between Špičák Mt. (1202 m asl) and Rozvodí Mt. (1189 m asl) (Mentlík 2001).

8.5 People on the Border

8.5.1 Early Settlement and Mountain Colonization

Because of its location in Central Europe the Šumava has been, since time immemorial, associated with the presence of humans and human activity. The latest archaeological research shows that the flat topography of the Šumava Plains with its numerous rivers and streams attracted hunter-gatherers from the Mesolithic period, despite the altitude of over 1,000 m (Fig. 8.2). Findings include a chipped stone industry, which was located on the edge of the river terrace at the confluence of Javoří and Roklanský streams near Modrava (Čuláková et al. 2012). Čuláková et al. (2012) reported that people could come here for food, namely maturing blueberries and cranberries, hazelnuts and migrating salmon. Rather than being a border, the Šumava at the time was a place of visits and seasonal migrations. The findings from the chipped stone industry indicate sources of raw material from sites 40 km away in Bavaria, and also in Central Bohemia and western Moravia (Čuláková et al. 2012).

In historical times the forests of the Šumava were probably used more by the Celtic Boii. A site with a probable Celtic settlement is Obří Hrad towering above the Losenice river valley. It is a very interesting site with preserved fort ramparts and an adjacent, relatively large block field. It is likely that the abandonment of the fort followed partial destruction of the site by a landslide, initiated by deforestation, slope destabilization and erosion by the Losenice river. Even today, it is a place with distinctive slope process dynamics (Hartvich and Mentlík 2010).

With the development of settlements in Central Europe and the growing importance of the Bohemian Kingdom it was necessary to ensure stability even in border areas. The Šumava was crossed by land routes, which (along with gold mines) were necessary to protect. Therefore, in the twelfth century Czech King Přemysl II Ottakar began greater colonization of the region. The settlers who came here were mostly of German origin. They received numerous privileges from the King for their services (the right to hunt, brewing, clearing trees). These settlers were not subordinate to any of the nobility, but directly to the king. Therefore, they began to be called 'Králováci' ('Kingsmen'). Their settlements were scattered as solitary homesteads, which stood in their own grounds. They were relatively large, with a characteristic bell tower.

Subsequently, probably from the fourteenth century, the Šumava was a source of raw materials, first gold and later wood. Gold in the Šumava in particular was panned for. Soil heaps representing the mining residues are found at relatively high elevations along streams and rivers (such as the Horská Kvilda village or Javoří Pila over 1000 m above sea level). Widespread logging was mainly associated with glass production. The glassworks in the area were responsible for large-scale forest clearance. The rapid development of the glass industry began in the sixteenth century and continued until the eighteenth century (Neuhäuslová et al. 2001). The greatest natural wealth of the Šumava therefore was wood. For owners of Šumava estates it was a problem getting wood from the remote border region to the cities. Wood was at that time rafted along larger rivers, but impassable canyon sections were an obstacle. Therefore, in the eighteenth and nineteenth centuries canals were built (Schwarzenberg in the southeast and later Vchynicko-Tetovský in the northwest). The construction of these canals initiated extensive logging that caused the destruction of the native forests of the Šumava (Hubený 2011).

8.5.2 Šumava Mts. as a Part of Iron Curtain

Very significant changes occurred in the Šumava in the mid-twentieth century. After World War II the German population was expelled and consequently the whole border area was closed—as a border zone and military area (Šumava lost approximately 2/3 of its inhabitants). In the 1950s the communist regime erected the Iron Curtain and Šumava happened to be an external boundary of the Soviet Union Empire. The Iron Curtain was formed by a continuous border zone with a system of fences and barriers, interwoven with signalling devices. It prevented the movement of the population of the Eastern bloc into western Europe. Attempts to cross this border cost the lives of many people. That was the last time when the "smugglers" cooperated in the Šumava. They guided people from the communist Czechoslovakia to Bavaria by taking advantage of local knowledge. Settlements that had stood here were largely destroyed, utterly changing the land use. A unique, relatively large area with a low density of settlements and population was thus created in the middle of Central Europe. The only economic activity was forestry which was managed under military control. The Iron Curtain fell and was dismantled (except for small parts, which now serve as museums) just after 1989.

8.5.3 Nature Conservation

After 1989 people started to return to the Šumava, occupying anew the previously depopulated areas. The villages began to be renovated, expanded and tourism developed. Nowadays, tourism is probably a major source of income for the whole area as logging is limited because of increased protection of nature. In 1991 the Šumava National Park was established, which is responsible for specialist and practical management of the natural environment and landscape throughout the region. The romantically deserted spaces encourage maintaining these conditions and creating a piece of naturally developing area in Central Europe. Nonetheless, even today the Šumava is facing problems. Perhaps, the most prominent is the bark beetle calamity (following extensive wind calamities) and the felling of affected trees, which arouses considerable emotion among conservationists.

8.6 Conclusion

In common with other Variscan mountains, the Šumava is a long-term raised and deeply denuded area where the resistance of each type of rock (mainly metamorphites and granites), fracture systems and the presence of old faults is central to the shaping of the relief. In comparison with other mountainous areas of the Bohemian Massif two geomorphic phenomena are typical for the Šumava: large plateaus (plains) occurring ~1000 m asl and glacial cirques with tarns—remnants of the Pleistocene glaciations.

The plateaus are covered with extensive peat bogs further reducing the already mild relief of the central part of the mountainous region. Theses mires at high altitude, positioned at the main European watershed, create a water accumulation of European importance.

In contrast to the mild relief of the plains, the rugged glacial cirques and adjacent periglacial landforms occur in the vicinity of the highest peaks. Eight fairly well-developed glacial cirques with tarns present the highest concentration of Pleistocene glacial landforms in the Bohemian Massif.

The most important current change to the landscape was connected with the end of World War II when the predominantly German population was expelled and several years later the communist regime erected the Iron Curtain. At that time the whole border area was closed—as a border zone and military area and it was depopulated. Many villages and hamlets were destroyed and land use was changed. People started to return to the vacated region after 1989. At present the protection of natural heritage is organized by the Šumava National Park which was established in 1991.

References

Ammann B, Lotter AF, Eicher U, Gaillard MJ, Wohlfarth B, Haeberli W, Lister G, Maisch M, Niessen F, Schlechter C (1994) The Würmian Late-Glacial in lowland Switzerland. J Quat Sci 9:119–125

Blockley SPE, Lane CS, Hardiman M, Rasmussen SO, Seierstad IK, Steffensen JP, Svensson A, Lotter AF, Turney CSM, Ramsey CB, INTIMATE members (2012) Synchronisation of palaeoevironmental records over the last 60,000 years, and a next ended INTIMATE event stratigraphy to 48,000 b2 k. Quat Sci Rev 36:2–10

Clark PU, Dyke AS, Shakum JD, Carlson AE, Clark J, Wohlfarth B, Mitrovica JX, Hostetler SW, McCabe AM (2009) The Last Glacial Maximum. Science 325:710–714

Čuláková K, Eigner J, Metlička M, Přichystal A, Řezáč M (2012) Horské mezolitické osídlení u Javoří pily, obec Modrava, okr. Klatovy. Archeologie ve středních Čechách. 16:19–28

Czudek T (1997) Reliéf Moravy a Slezska v Kvartéru. Tišnov, Sursum 213 pp

Czudek T (2005) Vývoj reliéfu krajiny České republiky v Kvartéru. Moravské zemské muzeum, Brno 238 pp

Engel Z, Braucher R, Traczyk A, Laetitia L, and ASTER team (2014) ^{10}Be exposure age chronology of the last glaciation in the Krkonoše Mountains, Central Europe. Geomorphology 206:107–121

Finger F, Gerdes A, Janoušek V, René M, Riegler G (2007) Resolving the Variscan evolution of the Moldanubian sector of the Bohemian Massif: the significance of the Bavarian and the Moravo-Moldanubian tectonometamorphic phases. J Geosci 52:9–28

Hartvich F (2004) Morfostrukturní analýza SV okraje Šumavy v okolí Pošumavského zlomu. Miscellanea Geographica 10:115–128

Hartvich F, Mentlík P (2010) Slope development reconstruction at two sites in the Bohemian Forest Mountains. Earth Surf Process 35:373–389

Hartvich F, Valenta J (2013) Tracing an Intra-montane Fault: An Interdisciplinary Approach. Surv Geophys 34:317–347

Hubený P (2011) Jaké vlastně jsou ty šumavské lesy? Veronica 5:20–21

Ivy-Osch S, Kerschner Reuther A, Preusser F, Heine K, Maisch M, Kubik P-W, Schlüchter C (2008) Chronology of the last glacial cycle in the European Alps. J Quat Sci 23:559–573

Janský B, Šobr M, Hrdinka T, Zbořil A, Vránek T, Pošta P, Oulehle F, Šnajdr M, Klouček O, Chalupová D (2003) Jezera České republiky. Praha, Katedra fyzické geografie a geoekologie na PřFUK v Praze 216 pp

Janský B, Šobr M, Kocum J, Česák J (2005) Nová batymetrická mapování glaciálních jezer na české straně Šumavy. Geografie—Sborník ČGS, 110, 3:176–187

Kodym O (ed) (1961) Vysvětlivky k přehledné geologické mapě ČSSR 1: 200 000, M-33-XXVI—Strakonice. Academia, Praha 149 pp

Lotter AF, Eicher U, Seigethaler U, Birks HJB (1992) Late-glacial climatic oscillation as recorded in Swiss lake sediments. J Quat Sci 7(3):187–204

Mentlík P, Šebesta J (2003) Geologie a geomorfologie Šumavy. Šumava 8. Podzim 2003:28–29

Mentlík P (2001) Zarovnané povrchy ve vrcholových partiích Špičáku a Rozvodí (Královský hvozd). Silva Gabreta 6:7–18

Mentlík P, Minár J, Břízová E, Lisá L, Tábořík P, Stacke V (2010) Glaciation in the surroundings of Prášilské Lake (Bohemian Forest, Czech Republic). Geomorphology 117:181–194

Mentlík P, Novotná M (2010) Elementary forms and "scientific reliability" as an innovative approach to geomorphological mapping. J Maps:564–583

Mentlík P, Engel Z, Braucher R, Léanni L, Team Aster (2013) Chronology oftheLateWeichselianglaciationintheBohemianForestinCentralEurope. Quat Sci Rev 65:120–128

Mercier J-L (2014) Glacial imprint on the Main Ridge of Vosges Mountains. In: Fort M, André M-F (eds) Landscape and Landforms of France. World Geomorphological Landscapes, Springer, Dordrecht, pp 161–169

Neuhäuslová Z, Buryová B, Ložek V, Majer J, Petruš J, Prach K, Procházka F, Sádlo J, Sofron J, Soukupová L, Svobodová H, Štěch M, Vokoun J, Vorel J, Wild J, Zatloukal V (2001) The map of potential natural vegetation of the Šumava National Park, Explanatory text, Silva Gabreta Suppl:175–129

Peck VL, Hall IR, Zahn R, Grousset F, Hemming SR, Scourse JD (2007) The relationship of Henrich events and their European precursors over the past 60 ka BP: a multi-proxy ice-rafted debris provenance study in the North East Atlantic. Quatern Sci Rev 26:862–875

Pelc Z, Šebesta J (1994) Geologická mapa ČR. List 22–33 Kašperské Hory, 1:50 000. Český geologický ústav, Praha

Raab T, Völkel J (2003) Late Pleistocene glaciation of the Kleiner Arbersee area in the Bavarian Forest, south Germany. Quatern Sci Rev 22:581–593

Reuther A (2007) Surface exposure dating of glacial deposits from the last glacial cycle. Evidence from Eastern Alps, the Bavarian Forest, the Southern Carpathians and the Altai Mountains, Vol. 21, Gebr. Borntraeger, Berlin

Selby MJ (1985) Earth´s Changing Surface. Clarendon Press, Oxford 607 pp

Svobodová H, Soukupová L, Reille M (2002) Diversified development of mountain mires, Bohemian Forest, Central Europe, in the 13,000 years. Quatern Int 91:123–125

Vejnar Z, Kopecký A, Růžička M (1991) Geologická apa ČR, List 21-44 Železná Ruda. Český geologický ústav, Praha

Vočadlová K (2011) Pleistocenní zalednění Šumavy (Případová studie —Černé a Čertovo jezero). Dissertation theses. UK Praha, Katedra fyzické geografie a geoekologie

Vrba J, Kopáček J, Fott J, Kohout L, Nedbalová L, Pražáková M, Soldán T, Schaumburg J (2003) Long-term studies (1871–2000) on acidification and recovery of lakes in the Bohemian Forest (central Europe). The Science of the Total Environment 310:73–85

Morphology of the Youngest Little Volcanoes in Western Bohemian Massif

9

Jan Mrlina

Abstract

The two little volcanoes, Komorní hůrka and Železná hůrka, are the youngest volcanoes in the western part of the Bohemian Massif, the principal regional geological unit of the Central Europe. The two volcanoes do not impress by their height, or complex petrological composition, but rather by their age uniqueness, as the dating results vary between 100,000 and 450,000 years. There are numerous other Cenozoic volcanoes in the western part of the Bohemian Massif, but they range between 31 and 8 million years. From this perspective, the two youngest ones prove that after certain break in volcanic activity since 8 million years ago, a new phase of "recent" volcanism has occured. Nowadays, we can still observe some ongoing signs of magmatic activity, e.g. many sites with intensive CO_2 degassing, mofettes, hot and cold mineral water springs including a small hot geysir, and relatively significant earthquake swarms with frequent repetition and magnitudes up to M4.5 in the nearby region of the Cheb Basin and surroundings. The little volcanoes were, anyway, a subject of investigation for centuries. The most famous is the research performed by J.W. Goethe, a worldwide known poet and scientist. We have, moreover, recently discovered the only Quaternary maar in the Bohemian Massif by geophysical surveying in the near vicinity of Železná hůrka.

Keywords

Volcanism • Morphology • West Bohemia • Geodynamics

9.1 Introduction

West Bohemia is a geologically complex region where many different geological units are present. The principal neotectonic structures are the so-called Ohře/Eger Rift of NEE–SWW direction, the Cheb–Domažlice Graben trending NNW–SSE and the N–S-oriented Regensburg–Leipzig–Rostock zone (Mrlina et al. 2009). At the junction of these tectonic zones, the Cheb Basin (CB) originated in the Cenozoic (Fig. 9.1). This tectonic setting is explained by the residual deformation and stress from the Alpine collision with the European platform. At the same time, there are many elements indicating active geodynamic and magmatic processes in the region, e.g. earthquake swarms (Fischer et al. 2014), mineral springs, gas emissions (Bräuer et al. 2011), surface movements (Mrlina and Seidl 2008), etc. Among important geological features of the region are the Quaternary volcanoes that are expressed by small scoria/lava hills (Komorní hůrka, Železná hůrka), but also by an explosive maar structure (Mytina Maar).

Volcanism in West Bohemia was studied among others by Ulrych et al. (2003), who distinguished three phases of Cenozoic alkaline volcanic activity in W Bohemia:

i. Early Oligocene–Early Miocene (31–26 Ma) predominantly basaltic volcanism of the Ohře/Eger Rift (Graben) wider zone and its continuation as far as to the Franconian Line.

J. Mrlina (✉)
Institute of Geophysics, v.v.i, Academy of Sciences of the Czech Republic, Boční II/1401, Praha 4, Czech Republic
e-mail: jan@ig.cas.cz

T. Pánek and J. Hradecký (eds.), *Landscapes and Landforms of the Czech Republic*, World Geomorphological Landscapes, DOI 10.1007/978-3-319-27537-6_9

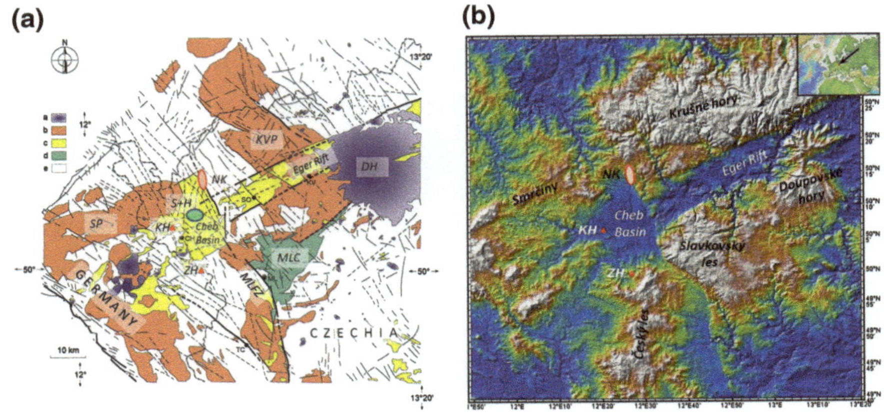

Fig. 9.1 **a** Geological scheme of West Bohemia: *a* Tertiary volcanics, *b* granites, *c* Tertiary sediments, *d* mafic and ultramafic rocks, *e* Paleozoic and Proterozoic formations. Geological/geomorphological structures: *DH* Doupovské hory Mountains, *KVP* Karlovy Vary pluton, *MLC* Mariánské Lázně Complex, *MLFZ* Mariánské Lázně Fault zone, *SP* Smrčiny Massif, *NK* Nový Kostel epicentral zone, *KH* Komorní hůrka volcano, *ZH* Železná hůrka volcano and Mytina Maar. Cities: *CH* Cheb, *KV* Karlovy Vary, *ML* Mariánské Lázně, *SO* Sokolov (and Sokolov Basin), *TC* Tachov, S + H—SOOS and Hartoušov degassing sites (adopted after V. Cajz, unpublished). **b** Digital Elevation Model (DEM) of West Bohemia, identical scale and symbols with geological scheme; *white letters*—basins and grabens, *black letters*—mountains (map from GeoMapApp 3.3.9., Columbia University)

ii. Middle to Late Miocene (16.5–8.3 Ma) differentiated polymodal volcanism associated with the NE flank of the Cheb–Domažlice Graben.
iii. Pleistocene (1.0–0.26 Ma) volcanism of the Ohře Rift and the Cheb–Domažlice Graben junction in the area of the Cheb Basin with melillite-bearing olivine nephelinite.

There is also evidence for ascending upper mantle-derived magma intrusions into the lower crust beneath the Cheb Basin, based on new $^3He/^4He$ data of CO_2-rich gas exhalations from the Cheb Basin (Bräuer et al. 2011).

Thus, the western Ohře/Eger Rift area is a potential natural laboratory that allows direct observation of fluid/magma distribution from upper mantle depths to the Earth´s surface, as well as further study of morphological and geological expression of existing young volcanoes. The importance of the young volcanoes was recognized by establishing natural reserve sites.

9.2 Geomorphological Units in the Cheb Region in West Bohemia

The geomorphological structure of the region consists of numerous highlands/mountains that meet around the depression of the Cheb Basin (Fig. 9.1). Despite they geologically represent Variscan (Paleozoic) epoch, the current topography is due to the Tertiary–Quaternary tectonic evolution (Peterek et al. 2011).

The **Cheb Basin** (CB; English and German: Eger Basin) is a Tertiary sedimentary basin originated among numerous mountainous units at the crossing of regional tectonic zones (Fig. 9.1). The elevation ranges between 400 and 480 m a.s.l. The basin was *formed* between the late Oligocene and Pliocene by reactivation of basement fracture systems, and its fill includes coal seam, but mainly a thick clay formation. The eastern part is the deepest, with the maximum depth of about 300 m. The central part of basin is characterized by a flight of five terraces between 1 and 20 m high above the recent floodplain of the Ohře river that runs through the basin in W to E direction. Minor displacements of the terrace surfaces are related to tectonic activity. The age of the uppermost terrace was determined as minimum 145 ka (Peterek et al. 2011). Slightly undulating relief of the outer parts of the basin is due to subsidence/uplift processes. As an example, the local segment of the N–S trending *Regensburg–Leipzig–Rostock* zone caused a relative uplift of the eastern block of CB by up to 10–20 m with morphologically very clear fault scarp.

The most striking limit of CB, however, is the NNW–SSE Mariánské Lázně Fault zone (MLFZ) at the eastern side of the basin. The zone separates CB from the *Krušné hory Mountains* (English: Ore Mountains, German: *Erzgebirge*) which is a NE–SW prolonged mountain range of about 130 km length forming a natural Czech–German border (Fig. 9.1b). On the SE side they decline to the Eger Rift valley by steep slopes (with total vertical difference of up to 700 m), while to NE the terrain gradually descends into lowlands. The highest mountain is Klínovec with elevation of 1244 m a.s.l. Thanks to intensive magmatic/volcanic

Fig. 9.2 The hanging wall of the Mariánské Lázně Fault forms a steep slope of Slavkovský les Mountains; Cheb Basin on the left horizon (*Photo* J. Mrlina)

activity, the Krušné hory Mountains are rich in metal ores and curative springs, which gave rise to numerous spa resorts. In addition, the most impressive negative gravity anomaly in the Czech Republic is located here, related to the enormous Karlovy Vary granite pluton.

At the eastern corner the Cheb Basin is limited by the *Slavkovský les* (English: Slavkov Forest). It is distinguished by its unique landscape rising steeply above the Cheb–Domažlice Graben limited by MLFZ. The whole area has a character of a planation surface. The highest points are Lesný 983 m a.s.l. and Lysina 982 m a.s.l. It is formed partly by granitic massifs and partly by the ultrabasic Mariánské Lázně Complex.

To the south, CB has a morphologically smooth boundary with the *Český les* (English: Bohemian Forest, German: *Oberpfälzer Wald*), a highland range elongated N–S along the Czech–German border, with the highest peak Čerchov (1042 m a.s.l.). Geologically, the area is not much varied, with the rocks of the so-called Moldanubicum Unit of Paleozoic to Proterozoic age being widespread (granites, gneiss, schists, slates).

At the western side, CB ends at the foothill of the *Smrčiny* (English: Fichtel Mountains, German: *Fichtelgebirge*), which can be considered as a prolongation of the Krušné hory Mountains to SW to Bavaria (Germany). The highest peak is located in Bavaria (Schneeberg, 1051 m a.s.l.), while the highest one on the Czech territory is Háj (758 m a.s.l.). The range is built mainly of Variscan granites plunging into its root under the Cheb Basin.

Further to NEE from the Cheb Basin are the *Doupovské hory Mts*. (English: Doupov Mountains, German: Duppauer Gebirge)—the biggest stratovolcano in the Bohemian Massif. The highest point is Hradiště, a forested hill rising to 934 m a. s.l. The volcanic hilly country of the Doupov Mountains is a uniform mountain mass formed by the separation of a mighty Tertiary stratovolcano with a diameter of about 30 km and covering an area of approximately 650 km^2.

The Quaternary volcanoes are located at the western and southern edges of the Cheb Basin; they erupted through Paleozoic Saxothuringian basement formed mainly by phyllite and mica schist. The geological map (Fig. 9.1) documents a high number of faults in this tectonically exposed region, despite it represents only one of the numerous tectonic concepts. Figure 9.2 shows one of the most important tectonic zones, with its clear morphological expression.

9.3 Komorní Hůrka Volcano

Komorní hůrka Hill, whose name can be translated as "Chamber Hill", is actually only a small hill with the elevation of 503 m a.s.l. (50° 06′ 04″ N, 12° 20′ 14″ E), not more than 30 m of relative height to the surrounding plain of the Tertiary Cheb Basin (Fig. 9.3). It forms an elliptical cone prolonged in the E–W direction (about 450 m E–W and 250 m N–S). It is located near the Františkovy Lázně Spa famous for numerous mineral water springs used as a spa treatment medium.

Komorní hůrka is one of the two youngest (Quaternary) volcanoes in the Bohemian Massif, and its age is estimated for 250,000–400,000 years. The volcano erupted into an already drying salt lake extending over the territory of today's Cheb and Sokolov Basins. Its origin is nicely presented in the old sketches reproduced as Fig. 9.4.

Fig. 9.3 Komorní hůrka in the Cheb Basin—view from west. Small volcano still represents a notable morphological element (*Photo* J. Mrlina)

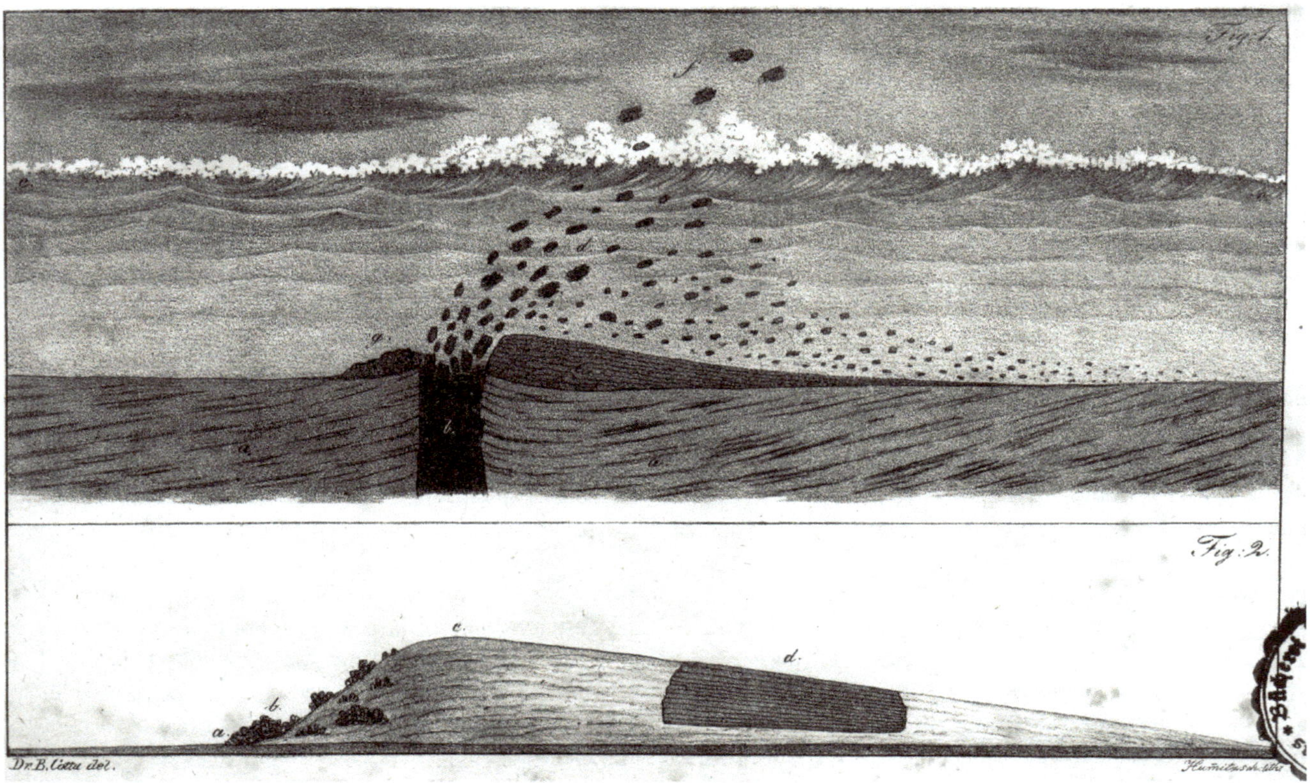

Fig. 9.4 Komorní hůrka—old painting shows the assumed origin of the volcano in the lake (*top*) and its final shape as seen in the nineteenth century (*bottom*) (*Source* NKC/New Kozák's Collection in Prague)

As for geological composition, the hill of a stratovolcano type is composed of olivine nephelinite scoria cone formed by intrusions through basement consisting of phyllite and quartzite (Hradecký 1994). The exsolution of the gas phase and its rapid ascent due to low viscosity of nephelinitic magma triggered fountain-fed eruptions. Rapid cooling

Fig. 9.5 Komorní hůrka. *Left* painting of J.W. Goethe from nineteenth century (“View at Komorní hůrka from SW”). Legend: *a* opening to old gallery, *b* basalt wall, *c* basalt boulders and bombs, *d* small quarry opening, *e* rock material from gallery. *Right* meeting of J.W. Goethe with the Swedish scientist Berzelius and others, discussing the neptunist and plutonist versions of the volcano origin. Subsequently, a gallery cut into the hill proved its volcanic origin by finding the volcano chimney (*Source* NKC/New Kozák’s Collection in Prague)

prevented substantial crystallisation, so that there is rather high volume of rock mass with glassy texture (Gottsmann 1999). The original profile of the volcano is broken up by former extensive quarrying of rock (basaltic lavas and scoria) and powdery volcanic effusive material (volcanic ash and pyroclastics ejected by Strombolian type of explosion) in the eastern section of the location. The quarry used to display loose tuffs with basaltic bombs up to 0.5 m in length and xenoliths of granite and metamorphic rocks. In the exposed area we can still find abundant volcanic material (ash and smaller pieces of volcanic pumice), as well as burnt fragments of rocks from the basement of the volcano. If we compare the current situation with the old paintings, it is clear that significant volume of pyroclastic material has been excavated. The residua of lava flows are forming the central hill, but mainly about 90-m-long lava flow with maximum height of 7 m at the south-western slope of Komorní hůrka. The blocky olivine nephelinite lava was probably erupted from a flank fissure and not from the central vent (Gottsmann 1999). The morphology of the volcano’s central part is nowadays covered by a forest, contrary to an open view in the nineteenth century (compare Figs. 9.5, 9.6 and 9.7).

The extraordinary cultural and historical significance of Komorní hůrka is mostly due to the fact that it may be the best explored volcano in Europe, and possibly in the world. First, reports of a targeted research date back to around 1760–1770, when the excavation of a 100-m-long adit in hope of finding coal beds was ordered. The volcano was also investigated by Berzelius, and J.W. Goethe (Fig. 9.5) who was not only a famous poet, but also a passionate naturalist. He visited the volcano for the first time in summer of 1808 and later proposed to dig another gallery to prove, whether it really was a volcano. The work started in 1826 and about 300 m of galleries were dug. The volcanic origin was confirmed when crossing a conduit in the so-called “Dwarfs Hole”. The last reminders of the period of exploration are the decorated entry to the gallery (Fig. 9.6) and historical photographs (Fig. 9.7).

9.4 Železná Hůrka Volcano

Železná hůrka (Iron Hill, Eisenbühl) scoria cone is situated approximately 11 km SSE of Cheb and 1 km S of Mytina, just about 200 m from the Czech–German border—this was the reason the site was very difficult to visit during the old times due to the military protection of the border. In the nearby valley there was a village of Boden that was abandoned after the WWII.

The hill elevation is 622 m a.s.l., with relative height of not more than 15 m (49° 59′ 30″ N, 12° 26′ 39″ E). It forms a small morphological elevation visible from up the slope to the Mytina village (Fig. 9.8). The diameter of the volcano scoria cone is approximately 200 m. It was again J.W. Goethe who called attention to the Železná hůrka Hill in the nineteenth century.

From the southern side the volcano was opened by a quarry for the scoria material. This fact actually made the volcano very interesting with almost vertical cuts through the central part of its body. The outcrops are both picturesque, as well as scientifically remarkable (Fig. 9.9).

This Quaternary volcano pierced phyllites of the Cheb–Dyleň Crystalline Complex. The rock has a character of simple mantle basalts. According to Hradecký (1994), the

Fig. 9.6 Komorní hůrka—current situation with natural reserve information panels, and the entry (*left*) to the famous gallery. Outcropping basalt lava flow (at the *back*) erupted from a fissure and form substantial part of the current volcano morphology (*Photo* J. Mrlina)

Fig. 9.7 Besides other investigations, the volcano was also surveyed and photographed in detail (*photo* of 1880, NKC/New Kozák's Collection in Prague)

lower part of the whole scoria cone corresponds to Strombolian activity with pyroclastics deposition. With the extension of the conduit, the Strombolian activity passed into that of Hawaiian volcanic style with phreatomagmatic or hydroclastic events involving ground water and less gases. The two eruption types need not to be separated by a longer hiatus, and the unconformity can be explained by a relatively short shift of the original vent or by eruption from the newly formed fissure. The presence of both types of pyroclastic rocks is a frequent phenomenon of cinder cones. The idea of the extent of the effusive volcanic products was presented by Lochman (1961) with tuff and tephra deposits as far as 2 km to north (Fig. 9.10). This extensive deposit was a subject of questioning and inspired a trench investigation at the northern side of Mytina (Fig. 9.11) (Mrlina and Kämpf 2012). As the results indicated possibly another source of tuff and tephra deposits than the Železná hůrka itself, a reconnaissance geophysical survey was performed in the morphological depression (Mrlina et al. 2007).

The age of the volcano was estimated to late Pleistocene, with wide range of age from 200,000 to 500,000 years, according to radiometric dating. However, some volcanologists think that it is not older than 100,000 years. The latest age dating by Ar–Ar method on phlogopite, and olivine nephelinite rock matrix by Mrlina et al. (2007) from samples taken out of the tephra layer about 1.5 km N of Železná hůrka, provided the age of 288 ± 17 ka. Either way, it is the youngest volcano in the western part of the Bohemian Massif.

Fig. 9.8 Železná hůrka—view from NW; the little volcano scoria cone may even be difficult to notice, see the *red dashed line* showing a morphological profile over the highest point. The opposite side is undercut by a quarry, see Fig. 9.9 (*Photo* J. Mrlina)

Fig. 9.9 Železná hůrka—*Left* view from the quarry to W, with the detail of the contact of different volcanic products/phases. The black inclined Hawaiian phase 3 effusive scoria overlays discordantly the phase 1 Strombolian tuffs deposited in semi-horizontal layers of various colours from *yellow* to *orange* and *brown–grey*. A man with GPS in the middle of the slope as a scale. *Right* View from the quarry to N–NE, at the contact of the phases 1 (*stratified orange tuffs*), 2 (*dark grey* to *black scoria*) and 3 (*black scoria*) (*Photo* J. Mrlina)

9.5 Mytina Maar

Based on results of previous investigations of tephra-tuff volcaniclastic deposits and a geophysical survey in the surroundings of the Železná hůrka (Mrlina et al. 2007), a detailed geophysical survey using gravimetry and magnetometry was performed. Striking anomalies in a morphological depression near Železná hůrka were observed as a strong evidence of an assumed, and to-date unknown maar-diatreme structure. Magnetic survey also showed pronounced local anomalies outside the depression that can reflect relicts of the tephra rim of the maar.

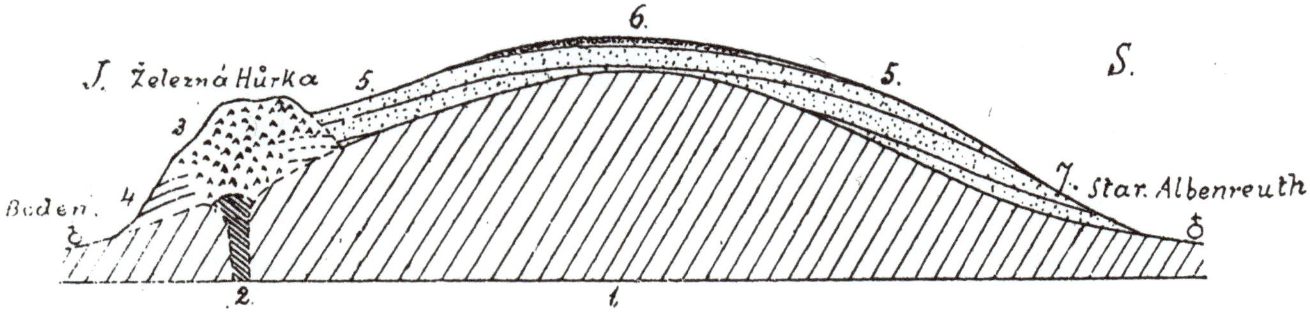

Fig. 9.10 Železná hůrka—old schematic painting on the assumed structure of the volcano originated in the southern part of a phyllite–mica schist hill. Legend: *1* mica schist, *2* assumed volcano conduit, *3* scoria cone, *4* lapilli, *5* tuff and lapilli, *6* tuff with schist and quartz breccias, *J* South, *S* North. (after Lochman 1961)

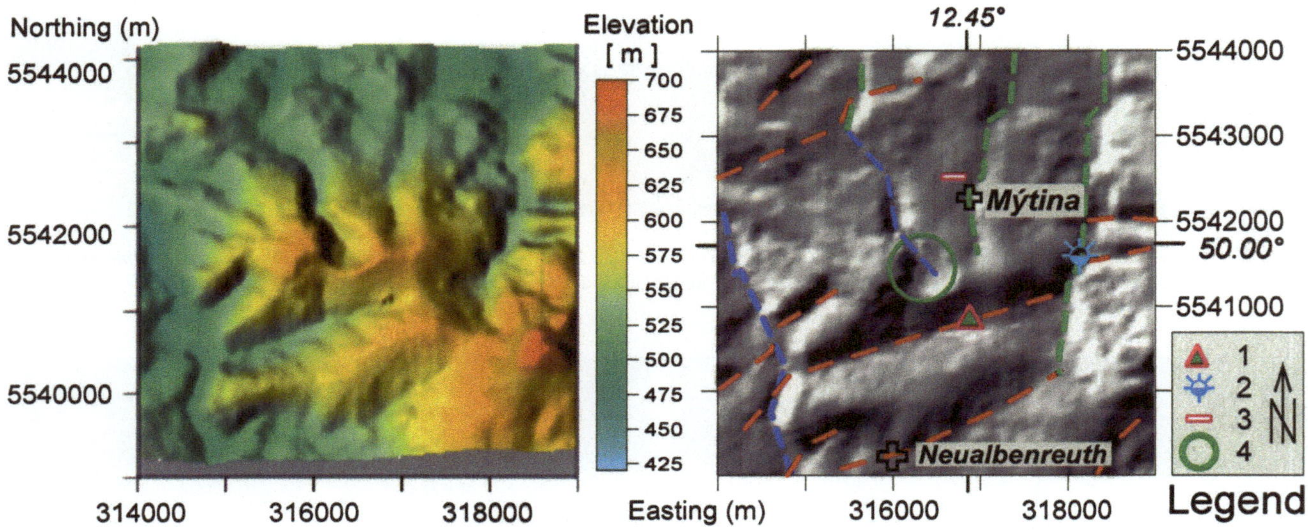

Fig. 9.11 Mytina Maar and Železná hůrka surroundings: *left* coloured shaded topographic model (SRTM90 + ZABAGED); *right* identical *greyscale* topography with important features: *1* Železná hůrka; *2* mineral water spring with CO_2 emissions; *3* exploration trench; *4* Mytina Maar. Drainage systems (faults?) are presented as *coloured dashed lines* (Mrlina et al. 2009)

This geophysical evidence was then proven by an exploratory drilling near the centre of the gravity anomaly. Macroscopic on-site evaluation of the core, and more detailed sedimentological, petrochemical, palynological and microbiological laboratory analyses further, confirmed the existence of a maar structure filled by 84 m of Quaternary lacustrine sediments reflecting a succession of several warm and cold climatic periods. At the bottom of the MY-1 borehole (84–85.5 m), country rock breccia was found, containing also volcanic bombs and lapilli (Mrlina et al. 2009).

The discovered volcanic structure is considered to be the first known Quaternary maar-diatreme volcano on the territory of the Bohemian Massif. It means that at this time we already know three Quaternary volcanic bodies in the western Bohemian Massif.

The isometric morphological depression of the Mytina Maar is located on top of a hill, contrary to the nearby Železná hůrka scoria cone situated almost at the valley bottom (Fig. 9.11). The distance between the two volcanic structures is less than 700 m and we may expect that they represent two different edifices of the same magmatic source. Their mutual temporal relationship has not yet been determined.

It is evident that the maar is a much larger structure than the little volcano (Fig. 9.11). The maar depression/crater has a diameter of 500 m and relative depth of 50 m. This crater is open to NW, which is the trend of the Cheb–Domažlice

Fig. 9.12 Mytina Maar—view of the maar crater depression from the eastern rim; Cheb Basin behind the far right horizon of the snap

graben and its western principal Tachov fault in particular. There are three drainage systems recognized at the area shown in Fig. 9.11 that can be associated with principal tectonic zones orientation:

ENE–WSW	Ohře/Eger Rift (red lines);
NNW–SSE	principal Variscan tectonic direction (blue lines);
N–S	Regensburg—Leipzig—Rostock zone (green lines).

The crater could not be recognized from field observations during the geological mapping, as the whole structure is covered by various types of forest (Fig. 9.12). Except the eastern edge, no site for a good view and photograph can be found. There is also only one outcrop at the edge of the crater that is showing the country rock—phyllite, and no trace of volcanics. However, a detailed view at the digital elevation model (DEM) in Fig. 9.11 proves that the circular morphological depression is definitely an exotic topographic feature within the wider area. That is why it was decided to perform geophysical surveys to identify whether the

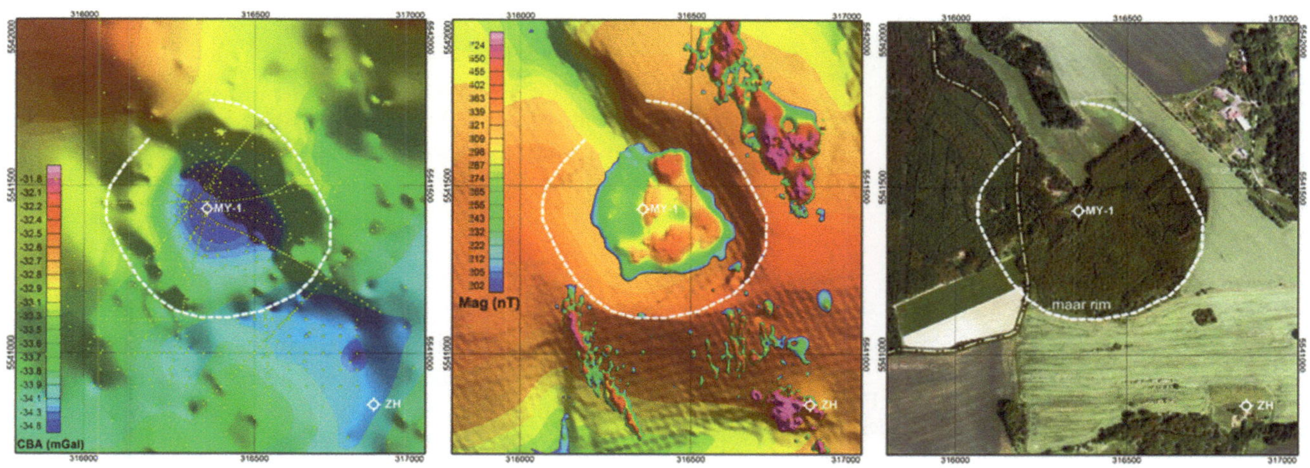

Fig. 9.13 Mytina Maar—images of gravity (*left*) and magnetic (*centre*) anomalies indicating a volcanic maar structure; the negative gravity "*blue eye*" is caused by the low density of the diatreme (*chimney*) disintegrated rocks including volcanics in the deeper section, and by very low-density sediments that filled a lake originated inside the crater. Positive extracted magnetic indications shown on top of DEM image reflect the presence of volcanic rocks with high content of magnetic minerals inside the diatreme, as well as in the so-called tephra forming a cover of the maar surroundings outside the morphological rim. Intensive magnetic anomaly is also related to Železná hůrka in the SE corner of the map (adopted after Mrlina et al. 2009). Google map (*right*) showing forested maar, surrounding fields and Železná hůrka; topographic contour (*edge*) of the maar shown by *white-dashed line*; relative depth of the maar depression is about 50 m, it is opened to NNW by a valley; MY-1—exploration borehole

Fig. 9.14 National Nature Reserve SOOS near Františkovy Lázně, NE of Cheb. This protected area is a small post-glacial moorland of about 2 km^2 with hydrochemical types of mineral and surface waters (**a**), exceptional gas emission zones, e.g. Devil's Face (**b**), mofettes, and organogenic sediments like diatomite shield that is known as European rarity; it was formed from the shells of diatom algae living in the lake (at the back of **c**). The bizarre terrain is churned up by erosion and covered with yellow and white efflorescence of mineral salts (front of **c**). The topography of the site is flat, but degassing and diatomites are causing small changes in surface, like dry gas holes (**b**), or small hummocks. The reserve is accessible by a 1.2-km-long planked nature trail (*Photo* J. Mrlina)

depression can indeed be a volcanic structure. Very intensive gravity low (Fig. 9.13) corresponds well with the idea of explosive maar/diatreme structure filled by low-density rocks. Positive magnetic indications suggest the presence of volcanic rocks containing minerals with high magnetization (magnetite, etc.). The contours of the rim and geophysical anomalies are presented in Fig. 9.13.

9.6 Exhibition of Current Magmatic Activity

As mentioned above, the region of western Bohemia exhibits numerous phenomena related to the assumed magmatic and tectonic activity. Besides the presented young volcanoes, there are ongoing semi-regular earthquake swarms with magnitudes up to M4.5 with the principal epicentral zone near Novy Kostel (Fig. 9.1). Despite the swarms are well described (e.g. Fischer et al. 2014), the exact source of stress is not yet defined. The magmatic process and related active fluids are, however, considered as the most probable.

The fluids represent an important geological phenomenon. Besides numerous mineral water springs, accumulated mainly in the famous spa areas of Karlovy Vary, Mariánské Lázně and Františkovy Lázně, the emissions of mantle CO_2 and Helium are monitored at a few locations (Bräuer et al. 2011). An extraordinary strong degassing process can be observed in the central part of the Cheb Basin near Hartoušov, about 10 km east of Komorní hůrka. Besides active moffetes with dry CO_2 emissions, the water spring Bublák with very intensive CO_2 flux serves as the best example. Moffetes and Bublák are protected natural reserves in the valley of the small river Plesná.

A few kilometres closer to Cheb, another gas flux structure forms a unique natural reserve SOOS with impressive gas emissions. CO_2 forms interesting bubbling features in a flat muddy and swampy landform, with special fauna and flora (Fig. 9.14).

9.7 Conclusion

The chapter documents three young Quaternary volcanic structures in western Bohemia with the age determination smaller than 500,000 years. Two of them are small hills (scoria cones with lava flows), while the third one forms a significant morphological depression of a maar-diatreme structure. Considering other geological and geodynamical elements and features existing in the region, we may conclude that western Bohemia is by far the most geodynamically active part of the Bohemian Massif, which is also documented by many different landscape units and complex topography. Active magmatic chamber at depth may still be a principal reason of earthquake activity, as well as numerous CO_2 degassing localities, hot mineral water springs and small surface movements. We also cannot exclude the possibility of another volcanic eruption in the future (in geological time) that may add some new features to the current topography.

References

Bräuer K, Kämpf H, Koch U, Strauch G (2011) Monthly monitoring of gas and isotope compositions in the free gas phase at degassing locations close to the Nový Kostel focal zone in the western Eger Rift, Czech Republic. Chem Geol 290:163–176

Fischer T, Horálek J, Hrubcová P, Vavryčuk V, Bräuer K, Kämpf H (2014) Intra-continental earthquake swarms in West Bohemia and Vogtland: A review. Tectonophysics 611:1–27

Gottsmann J (1999) Tephra characteristics and eruption mechanics of the Komorni hurka Hill scoria cone, Cheb Basin. Czech Repub Geolines 9(1999):35–40

Hradecký P (1994) Volcanology of Železná and Komorní hůrka in Western Bohemia. Věst. Čes. geol. Úst. 69(2):89–92

Mrlina J, Kämpf H (2012) Quaternary magmatic activity—Železná hůrka scoria cone and Mytina maar, West Bohemia. State of geomorphological research in 2012. Field trip guide (Štěpančíková-Peterek-Marek, Edits.), Sokolov 18.-20.4.2012, p. 13–15

Mrlina J, Kämpf H, Geissler WH, van den Boogart P (2007) Assumed Quaternary maar structure at the Czech/German boundary between Mýtina and Neualbenreuth (western Eger Rift, Central Europe): geophysical, petrochemical and geochronological indications. Z Geol Wiss 35(4–5) 213–230

Mrlina J, Kämpf H, Kroner C, Mingram J, Stebich M, Brauer A, Geissler WH, Kallmeyer J, Matthes H, Seidl M (2009) Discovery of the first Quaternary maar in the Bohemian Massif, Central Europe, based on combined geophysical and geological surveys. J Volc Geoth Res 182:97–112

Mrlina J, Seidl M (2008) Relation of surface movements in West Bohemia to earthquake swarms. Stud Geoph Geod 52(4):549–566

Peterek A, Reuther CD, Schunk R (2011) Neotectonic evolution of the Cheb Basin (Northwestern Bohemia, Czech Republic) and its implications for the late Pliocene to Recent crustal deformation in the western part of the Eger Rift. Z. geol. Wiss., 39 (2011) 5/6, 335–365, Berlin

Ulrych J, Lloyd FE, Balogh K (2003) Age relations and geochemical constraints of Cenozoic alkaline volcanic series in W Bohemia: a review. Geolines 15:168–180

The Krušné Hory Mts.—The Longest Mountain Range of the Czech Republic

10

Vít Vilímek and Pavel Raška

Abstract

Located in the borderland between the NW part of the Czech Republic and Saxony, the Krušné hory Mts. (highest peak Klínovec, 1244 m a.s.l.) represents the longest mountain range in the Czech Republic. Its current geomorphologic character with a steep SE fault scarp and extensive planation surfaces at the top has been influenced by long-term tectonic activity, which has uplifted the original denudational surface underlain by Paleozoic magmatic and metamorphic rocks. The neotectonic uplift has taken place since the Neogene and has resulted in vertical exaggeration of the mountain range above the neighbouring basins reaching more than 500 m. It is mainly the SE fault scarp of the mountain range with its extreme gradient, which is frequently affected by flash floods and different kinds of mass movement. The natural environment has been fundamentally reshaped during hundreds of years of human activity. First, this activity was focused on mining of rich mineral deposits (silver, tin, lead, copper and other ores). Subsequently, the inhabitants of the mountain range changed their focus to manufacturing and agriculture, which is illustrated by technical monuments and agricultural landforms. The extraordinary interrelations between natural conditions and cultural heritage are currently the main argument behind the proposal to enlist the Krušné hory Mts. among the World Heritage sites.

Keywords

Krušné hory Mts. • Fault scarp • Planation surface • Anthropogenic transformation • Ore mining

10.1 Introduction

Stretching along the historical boundary between NW Czech Republic and Saxony in Germany, the Krušné hory Mts. (*Ore Mountains* in English, *Erzgebirge* in German) was once the most densely populated mountain range in Europe. The basic character of the mountain range is given by gently inclined NW slopes, culminating in the planation surface at an altitude of approximately 900 m a.s.l. This flat to gently sloping surface is suddenly interrupted by a steep fault-generated escarpment inclined towards the SE, resulting in a local gradient of more than 500 m within a distance of only a few kilometres (Fig. 10.1). Despite the severe climate, the planation surface at the top of the longest mountain range in the Czech Republic (approximately 130 km) has always been suitable for settlers, who hoped to profit from the rich mineral deposits of silver, tin, lead, copper and other ores present there. The character of this unique mining landscape was later completed by a

V. Vilímek (✉)
Department of Physical Geography and Geoecology, Faculty of Science, Charles University in Prague, Prague, Czech Republic
e-mail: vilimek@natur.cuni.cz

P. Raška
Department of Geography, Faculty of Science, J. E Purkyně University in Ústí nad Labem, Ústí nad Labem, Czech Republic
e-mail: pavel.raska@ujep.cz

T. Pánek and J. Hradecký (eds.), *Landscapes and Landforms of the Czech Republic*, World Geomorphological Landscapes, DOI 10.1007/978-3-319-27537-6_10

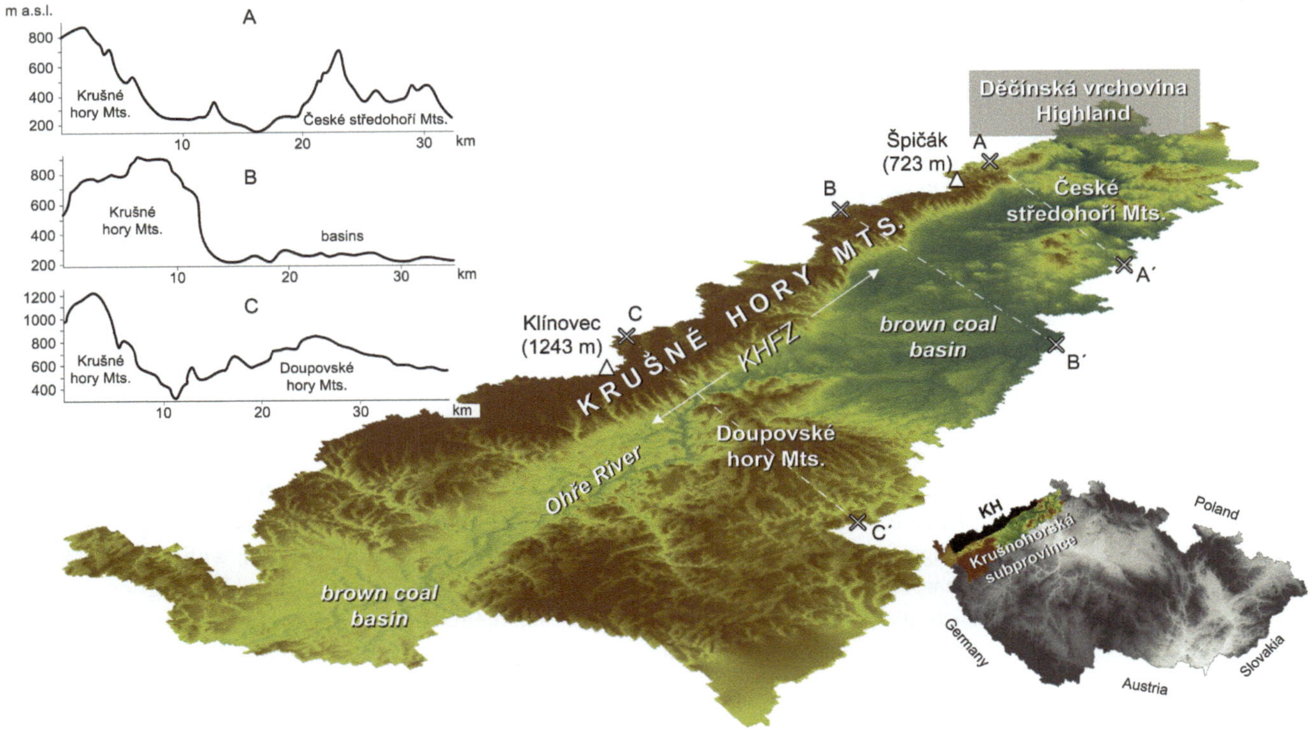

Fig. 10.1 Geographical setting of the Krušné hory Mts. in the NW part of the Czech Republic and cross-sections showing the SE-facing fault scarp of the mountain range

settlement structure of small towns and hundreds of villages (some of them currently abandoned) with fields, meadows and pastures interlaced with agrarian walls and levees. Thus, the Krušné hory Mts. provide an excellent example of tectonic, erosional, and anthropogenic relief, all in close interaction with one another.

10.2 Geographical Setting

The Krušné hory Mts. (Fig. 10.2) represent a mountain range stretching from SW to NE, with the highest mountain Klínovec (1244 m a.s.l.) in its western part. They have represented a natural barrier and a boundary for human interaction since prehistoric times. In fact, the mountain range reaches far into Saxony in Germany, where it is called Erzgebirge, with the highest peak Fichtelberg (1214 m a.s.l.), thus only one half of the area is located in the territory of the Czech Republic. However, it was not the flat surface in the upper part of the mountain range, but rather the steep fault scarps in the SE that were difficult to cross for such a long time. This "protective wall", representing a historical frontier of the Bohemian kingdom, also forms a barrier in a natural sense. Annual precipitation reaching more than 1100 mm at the top of the mountains falls to about 400 mm in the basins in the SE lee side of the mountain range because of the rain shadow, while differences also exist in the temperature regime. Both the precipitation and average temperature influence vegetation

Fig. 10.2 Southern view to the Krušné hory Mts. (mountain range in the background) from the České středohoří Mts. (*Photo* P. Raška)

patterns and conditions for agriculture. Most of the slopes in the Krušné hory Mts. were covered by forests for a long time, influencing erosion rates as well as hydrological systems (water storage capacity of soils, drainage patterns, etc.). This has changed with the emergence of medieval colonisation, resulting in an increase in settlement density, deforestation and cultivation, which started hundreds of years of intercultural relations between Germans and Czechs.

10.3 Structural Setting and Geological Evolution

10.3.1 Pre-Cenozoic Evolution

The Krušné hory Mts. represents the part of Saxothuringian zone, a northwest part of the Bohemian Massif. The Bohemian Massif originated during the Variscan orogeny and amalgamation of pre-existing terranes, but the current mountainous character of the Krušné hory Mts. is a result of Tertiary tectonic uplift. The documented pre-Variscan evolution of the area of the current Krušné hory Mts. included a process of planation in late Neoproterozoic after the Assynthian orogeny, whereas the Caledonian orogeny did not touch this area. A geosyncline appeared there in the Lower Paleozoic. Sediments of the geosyncline were metamorphosed regionally during the Variscan orogeny. By the end of the Carboniferous, the Central Bohemian sedimentation area may have been connected to this area (Malkovský et al. 1985). The Upper Paleozoic sediments were subject to weathering and a denudation, which continued during the whole of the Triassic, Jurassic and Early Cretaceous (Pekárek 1960). The eastern part of the current area of the Krušné hory Mts. suffered from the transgression of the Late Cretaceous sea. The current boundary between the surface built by Cretaceous sediments in the east and the older crystalline rocks in the west is located nearby the Tisá village (approximately profile A in Fig. 10.1) and is not distinct in topography due to protracted planation. The original Cretaceous sediments could have reached further to the west, however, as implied from their relics preserved at the Špičák hill nearby the Krásný Les village.

10.3.2 Pre-Quaternary Evolution

During the Tertiary, in general, the Bohemian Massif went through an uplift followed by the creation of fault zones and grabens (Chlupáč et al. 2002). Late Cretaceous sedimentation and sea regression were followed by denudation at the beginning of the Paleogene. Therefore, the Paleogene system is represented by a stratigraphic hiatus. During the Eocene the Krušné hory Mts. did not exist as a mountain unit, and NW Bohemia appeared as a peneplain at approximately 150 m a.s.l. Kopecký (1972) gives the altitude of the peneplain between 100 and 200 m and denudation is considered as the main process by which the topographic surface evolved during that period. The development of the peneplain was stopped by interference of the second phase of Saxonian radial tectonics in the Early Oligocene. During the Middle Oligocene the sea entered Leipzig bay in the N, which raised the niveau of the lower erosional base and forced increasing sedimentation by rivers flowing persistently from Central Bohemia to NW. The presence of a tropical climate in the Bohemian Massif during the Paleogene (more exactly from the Late Cretaceous to the Early Oligocene) is considered to be due to the continental drift of Variscan Europe to the S. The area of the Krušné hory Mts. was therefore under action of a more humid and warmer climate than today. On the contrary, during the Miocene the climatic conditions changed to subtropical, or even temperate, with favourable conditions for red soil weathering (Malkovský et al. 1985).

The subsequent evolution is characterized by intensive volcanic activity in the so called Ohře/Eger Rift structure, which was temporarily connected with the Oligocene/Miocene transition, and by the formation of thick sedimentary layers in the adjacent basinal region in the Miocene. The presence of basalt extrusions of Kamenný Vrch (892 m a.s.l.), Špičák (723 m a.s.l.), and Jedlová and Mračný Vrch (both 853 m a.s.l.) in the Krušné hory Mts. is the result of this volcanic activity. During the Early Miocene, the Eocene planation surface suffered from definite destruction. After the volcanic period, the newly formed volcanic rocks were subject to kaolinitic and lateritic weathering. During the Pliocene uplift of the Krušné hory Mts. an important vertical differentiation between the mountain area and the adjacent Mostecká pánev Basin has emerged. This differentiation resulted in the formation of a new watershed. The hydrographic link with Saxony was therefore disconnected. Flexure bending of the Krušné hory Mts. developed into fault movements; however, not all faults are post-Miocene. Older Miocene faults were smaller and vertical differentiation was often levelled by sedimentation.

10.3.3 Quaternary Evolution

During the Quaternary, further differentiation between the mountains and the adjacent basin occurred. The intensity and timing of the uplift is difficult to determine because of a lack of dated proluvial and deluvial deposits (Vilímek 1995). Malkovský et al. (1985) concluded that no volcanism accompanied the Quaternary uplift. Kopecký (1972) described an acceleration of movement exaggerating the difference between the basin and the mountains and dated them to the end of the Pliocene. Nevertheless, he ascribed the main uplift to the Pleistocene (without closer specification) due to the occurrence of coarse, rapidly laid-down gravels. Hurník (1982) dates the main uplift to the Mindel and Lower Riss glacial periods. However, tectonics had a greater impact on landform development in this area than climatic oscillations.

10.4 Typical Landforms and Processes

10.4.1 Planation Surfaces

The vast majority of the area of the Krušné hory Mts. is represented by planation surface(s) at the top of the mountain range (Král 1968). This surface developed during the long-term denudation that exposed the Paleozoic crystalline complex of the fundament and individual magmatic bodies of the Karlovy Vary granite pluton (West) and the rhyolite rim of the Altenberg–Teplice caldera (East) (Fig. 10.3). The geometry of this caldera was reconstructed using geophysical methods by Mlčoch and Skácelová (2010), because it is currently present only as a geological structure of the Paleozoic age, without significant surface morphological features.

The original denudation surface located at low altitudes of approximately 100–200 m (see Sect. 10.3) was later modified and uplifted during the Neogene to its current position as a result of tectonic movements within the volcano-tectonic zone of NW Bohemia (the so-called Eger Rift zone). As a result, the areas on the top of the mountain range also include some bodies of Cenozoic volcanism from the Eger Rift zone that have been uplifted (see Sect. 10.4 and Fig. 10.3). The contemporary planation surface at the top of the Krušné hory Mts. is characterized by average slope inclination varying between 3–5° at an altitude of 700–1000 m a.s.l. (Figs. 10.4 and 10.5a). Viewed in more detail, however, the distinct variations caused by differential tectonic movements are apparent. In general, the planation surface declines from its highest positions in the W (>900 m a.s.l.) to lower altitudes in the E (700–800 m a.s.l.). The illustrative examples of planation surface at the Czech side of the mountain range may be seen in its central part (near the Přísečnice and Fláje water dams) and eastern part (between the Horní Blatná and Boží Dar towns).

During the Quaternary period the mountain range was not subject to glaciation, but the area was located in a periglacial climate that accelerated frost weathering processes, resulting in the formation of many rock forms. These are best developed in granites in the western part of the mountain range (Fig. 10.5b) and represent rock walls, rock slopes and individual tor-like forms, frequently exceeding 10 m in height and 30 m in length. Many representative tor-like forms are concentrated in the N and W surrounding of the Nejdek town, for instance. In contrast, around10 large peat bogs evolved in flat areas in the highest segments of the planation surface. According to palynological and stratigraphical analyses, the most prominent peat bog, Boží dar, has continuously evolved since the Late Glacial period (Břízová 1995). The current planation surface is further diversified by dells, which developed under periglacial climate conditions of the Pleistocene, and mainly by many wide open valleys of mountain streams. These streams, which continue to the slightly NW-inclined part of the Krušné hory Mts., preserve the wide and shallow character of their valleys, whereas the streams continuing to the tectonic slope in the SE suddenly increase their gradient and form deep V-shaped erosional valleys.

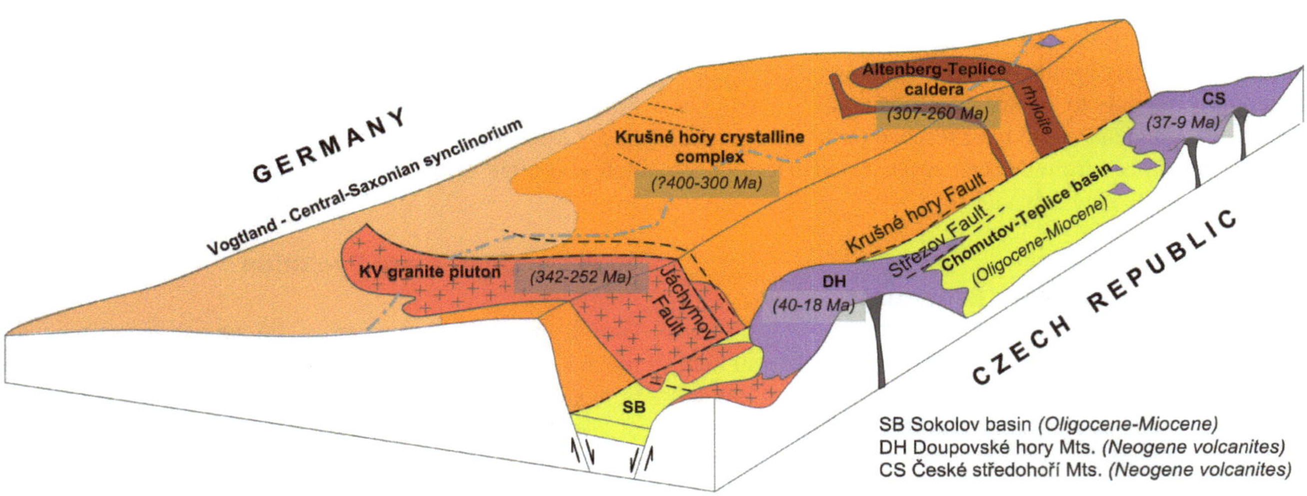

Fig. 10.3 Schematic geological cross-section of the NW part of the Bohemian Massif. K-Ar data compiled from a summary in Klomínský (1994) and Cajz (2000)

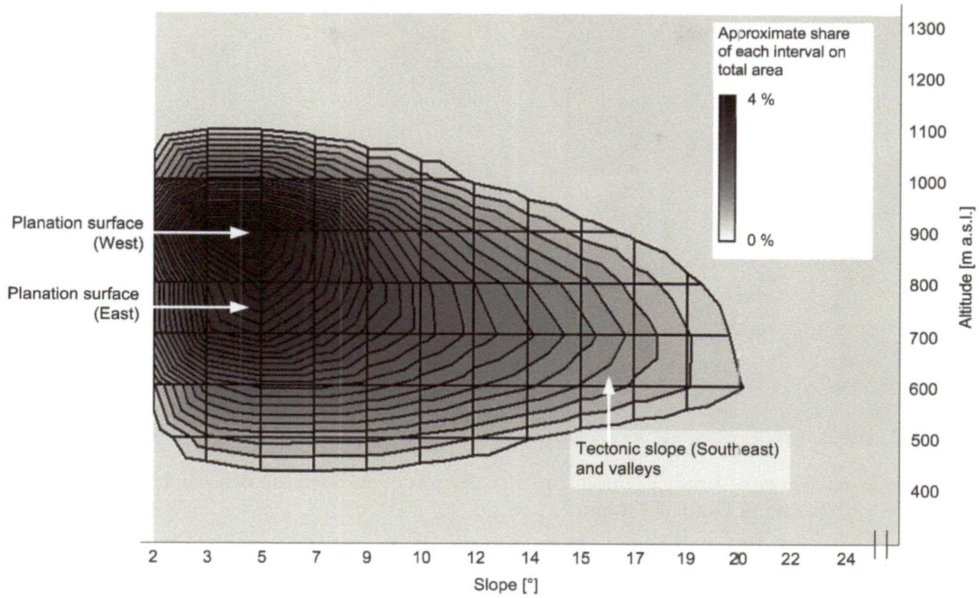

Fig. 10.4 Morphometric characteristics of the Krušné hory Mts. indicating the position of the planation surface. Note: only the part of the graph with the highest frequency of slope categories is displayed. Maximum slope inclination in the area exceeds 30°, but is less frequent

Fig. 10.5 **a** Planation surface at an altitude of approximately 900 m a.s.l. in the central part of the mountain range. **b** Granite tor on the planation surface in the W part of the mountain range. (*Photo* P. Raška)

10.4.2 Fault-Generated South-Eastern Escarpment

The SE slope of the Krušné hory Mts. evolved along the Krušné hory fault line during the tectonic uplift of the mountain range, mainly during the Neogene. Its height above the foothill ranges from approximately 300 m to more than 500 m and reflects the vertical differences between the W and E parts of the planation surface at the top of the mountains. The highest gradients are reached in the central and western parts of the range, e.g. 550 m between Medvědí skála (924 m) and Jezeří castle (350 m) at a distance of 3.4 km. Landslides on the fault scarp of the Krušné hory Mts. were important components of landscape evolution in the Quaternary (e.g. a rockfall from Jezeří hill—see Rybář 1982). Nevertheless, due to the recent open pit mining activity at the foothill, landslides in the slope sediments have to be carefully investigated and monitored (e.g. Burda et al. 2013). Slope instability has also been studied by dendrogeomorphologic techniques, which showed that landforms like landslides on dumps, which have not been mapped yet, could be dated by Silver Birch—*Betula pendula* (Tumayer and Burda 2013),

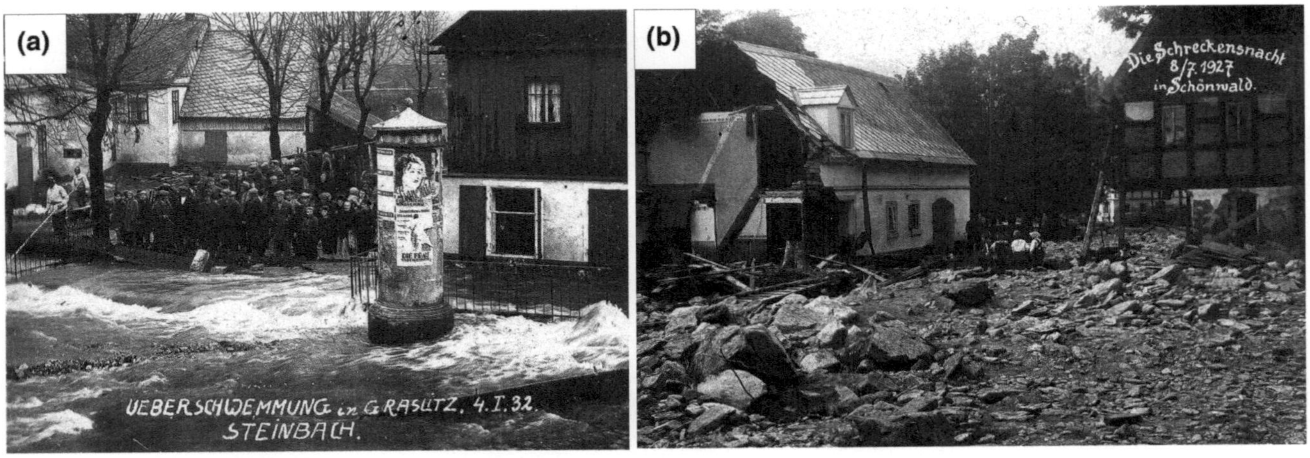

Fig. 10.6 Flash floods in 1932 in the town of Kraslice (Graslitz; **a**) in the W part of the mountain range and in 1927 in the village of Krásný Les (Schönwald; **b**) in the E (*source* Museum of Ústí nad Labem)

and which also enabled to exclude the effect of underground mining on the slope deformation on Salesius Hill (Vilímek et al. 2002).

The fault scarp is diversified by many SE-oriented watercourses coming from the planation surface. While some of these streams are rather ephemeral, many of them exhibit high discharges depending on the volume of anomalous precipitation in the most elevated part of the mountain range. Extreme discharges frequently result in disastrous flash floods in gorges and valleys within the fault scarp (Fig. 10.6). In the area where open pit mines in the Most basin reached the foothill of the Krušné hory Mts., several streams had to be relocated into pipes.

The western part of the Krušné hory Mts. is famous for earthquake swarms (around the town of Kraslice) which have been documented since the sixteenth century (e.g. 1552) and monitored since the turn of the twentieth century. The highest magnitude (M) was 4.8 during the swarm of 1985/86; the most recent event happened on May 31st 2014, showing magnitude of 4.5. Recent geodynamic activity within the Krušné hory Mts. has been recorded for several years. Košťák et al. (2011) identified a significant pressure phenomenon during 2003, which they named a "pressure pulse". The pressure pulse initiated a series of tectonic deformations (earthquakes occurred during the later stages of the deformation process).

10.4.3 Human Impact on Landforms

According to Jankovská and Pokorný (2013) the first clear evidence of a human effect on vegetation in promoting the expansion of secondary grasslands is dated in pollen diagrams to around 4,000 uncal. yr BP (i.e. the Subboreal period). The mineral resources of the Krušné hory Mts. have been exploited at least since the Bronze Age (3rd–1st Millennium BC), which facilitated the rise of the well-known Únětice culture in Central Europe. At that time, copper was abundant in many deposits across Europe, but the tin deposits that could be extracted directly from natural outcrops and by placer mining were only present in the Krušné hory Mts., in Cornwall (SE England), and the W Iberian peninsula, thus making the Krušné hory Mts. the prominent source of tin in Central Europe (Muhly 1985). The real boom of mining activities, however, came in the medieval period. The colonisation of the mountain range on the border between Saxony and Bohemia (historical part of the Czech Republic) was predisposed by findings of ore deposits that made the area a centre of innovation in metallurgy. The culmination of mining dates back to the sixteenth century, when the famous scientist Georgius Agricola summarised his knowledge on mining and metallurgy in the most influential work in the field, *De re metallica libri XII* (Fig. 10.7), published in Basel a year after his death. The inspiration for his book also came from time spent in the mining town of Jáchymov in the Krušné hory Mts. (Figure 10.8).

Mining accelerated development of many towns and villages in the area and as a result, their names frequently include the name of the extracted ore, such as Stříbrná (Silvertown) or Cínovec (Tintown) (Fig. 10.8). Anthropogenic transformation connected with mining is apparent in many areas (Fig. 10.9), mostly in the form of entrances to complex shaft systems as well as water ditches and mills (Table 10.1 and Fig. 10.7). The ore deposits were still being mined during the seventeenth century, however, only a few remained into the twentieth century, such as uranium ore in Jáchymov. The decline of mining forced people to re-orientate themselves to other economic activities, such as manufacturing and agriculture, which left agrarian levees

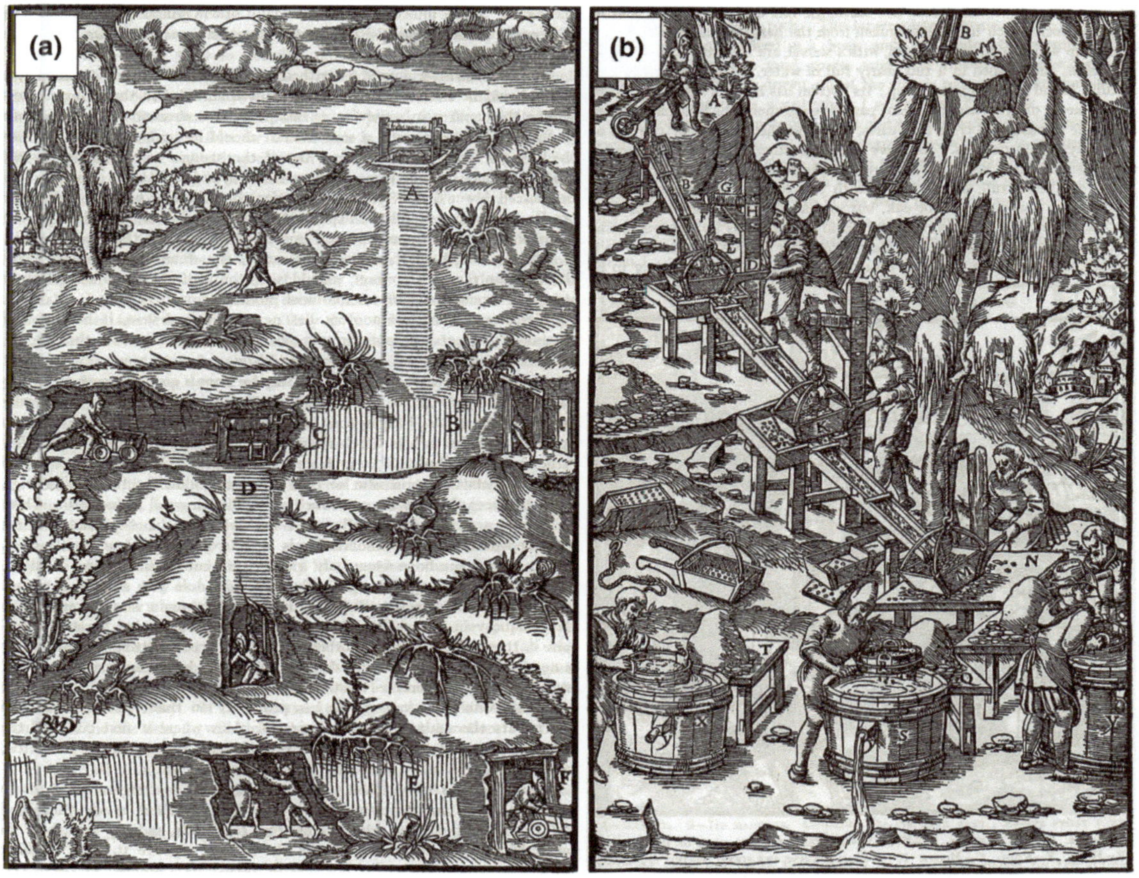

Fig. 10.7 **a** Depiction of subsurface ore mining and **b** of placer mining in the famous work *De re metallica libri XII* (twelve books on mining and metallurgy) by Georgius Agricola, published in 1556

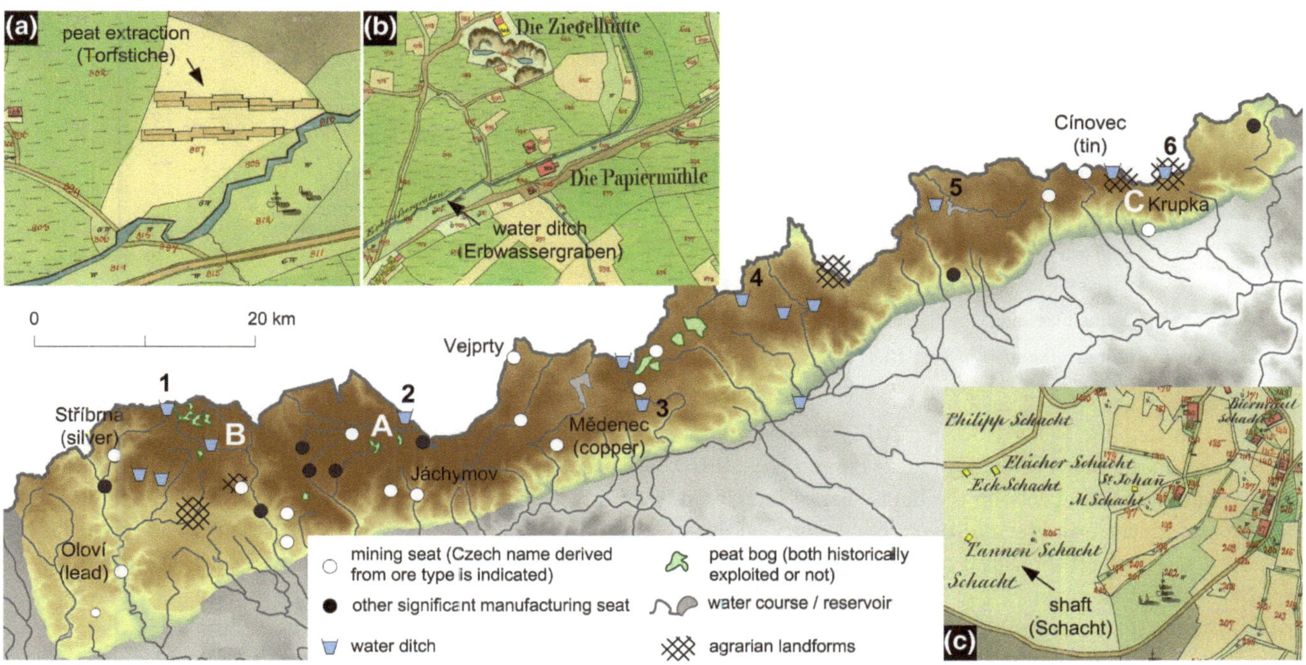

Fig. 10.8 Major anthropogenic transformations in the Krušné hory Mts.: ore mining, water ditches, agricultural landforms and bog extraction. **a**, **b** and **c** show these transformations on cadastral maps from the 1840s (*source* ČÚZK Praha). The numbers of the water ditches refer to Table 10.1

Fig. 10.9 Historical shaft in the Sv. Martin mining area. (*Photo* L. Bobr)

Table 10.1 List of major water ditches in the Krušné hory Mts. with their basic characteristics (after Mohelník 2013)

No. in map	Name	Length (km)	Average width (m)	Built in	Technical monument
1	Přebuz ditch	4.7	2.7	Sixteenth century (?)	no
2	Blatenský ditch	13.0	1.0	Sixteenth century	yes
3	Výsluní ditch	2.8	3.2	Seventeenth century	no
4	Kalek ditch	1.6	1.8	Nineteenth century	no
5	Flájský channel	3.7	4.9	Seventeenth century	no
6	Mohelnice ditch	1.8	2.8	Fifteenth century(?)	no

Note numbers according to the map in Fig. 10.8

and walls still visible today (Riezner 2011), or peat extraction on the summit areas of the mountain range (Fig. 10.10).

The main human impacts on the landscape of the Krušné hory Mts. are dated from the mid-twentieth century and were caused by mining and industrial activities in the basin area SE of the mountain range. Open pit brown coal mining in basins directly next to the Krušné hory Mts. raised the necessity to study the stability conditions of the adjacent slope. This is why slope movements were investigated by repeated levelling measurements (Kalvoda 1995; Burda and Vilímek 2010).

Furthermore, emissions from power plants using brown coal both from the NW part of the Czech Republic and SE Germany were considered to be the major source of air pollution in the Krušné hory Mts. in the 1980s and 1990s. The most affected forests of spruce monocultures (*Picea sp.*)

Fig. 10.10 **a** Agrarian wall in Fojtovice in the W part of the mountain range (*Photo* J. Riezner). **b** Fláje water reservoir built in 1951–1964. (*Photo* P. Raška)

had to be cut down and removed, leading to a risk of intensive sheet and rill erosion. Fortunately, the slopes were quickly covered by brush and new trees were artificially grown. The changes in emissions and pollution concentrations in the area during most of the past decade, and their relation with meteorology were reviewed by Bridgman et al. (2002), where, for instance sulphur dioxide (SO_2) is used as the example pollutant. The results show a decrease in pollution concentrations since 1996, as air pollution control and management strategies for important point sources took effect.

10.5 Geoheritage and Landscape Conservation

With its extraordinary nature as well as long-lasting history of settlement and human activities, the Krušné hory Mts. could have easily become a centre of geoscientific research. The specific location on the country borders, which was most distinctly proclaimed by displacement (expulsion) of Germans after the World War II (Mikšíček 2005) and by controlled (and partly limited) access to frontiers during communism, has unfortunately limited research activities since the beginning of the twentieth century. It was also probably due to these facts that the Krušné hory Mts. represent the only mountain range along the Czech border without the large-scale landscape or nature conservation. Moreover, nature conservation has been limited by the environmental impacts of brown coal mining and the related industrial activities beneath the mountain range. In addition to the above-mentioned effects, such as mass movement and emissions, these impacts also include a threat to cultural monuments. One example is Jezeří castle, which is located directly in a landslide prone area on the footslope of the Krušné hory Mts., near to a brown coal mine.

During the last two decades, efforts on both the Czech and German side have brought about significant results in environmental protection and landscape conservation in the Krušné hory Mts. Firstly, the environmental quality of the area has improved thanks to significant investment. Secondly, over the past few years, cooperation between Czech and German scientists as well as the public and politicians has led to significant emphasis being placed on the appreciation of the historical landscape of the area. Mapping of natural ecosystems and inventories of mining monuments was focused on the establishment of the Euroregion of Krušnohoří/Ergebirge, proposed for inclusion on the UNESCO World Heritage List (e.g. Urban 2011). Currently, the area is included in the tentative list of UNESCO World Heritage properties (http://whc.unesco.org/en/tentativelists).

10.6 Conclusions

The Krušné hory Mts. represent a historical cross-border landscape which has been predisposed by long-term tectonic evolution and shaped by hundreds of years of human activity. The culmination segment of the mountain range at an altitude of between 700 and 1000 m is represented by planation surfaces, developed by tectonic uplift of the original denudational surface during the Neogene to its current position. The geological record in this part of the mountain area is now apparent as sporadic rock forms, such as tor-like formations on Paleozoic granites in the west, or hills in the central and eastern part, built on uplifted Neogene basalts. Toward the SE, the surface continues in a fault scarp dropping by more 500 m to the sedimentary basins, thus forming a distinct natural barrier. The tectonic and geomorphic peculiarities of the mountains result in various kinds of natural hazards, such as earthquake swarms and mass movements.

The most elevated part of the mountain range displays various ways in which its inhabitants adapted to difficult natural conditions across centuries and exploited resources of different kinds, from ore deposits to peat bogs. The natural diversity was accompanied by cultural contrasts, resulting in a specific course of the region's history. Paradoxically, this caused the geoscientific value of the Krušné hory Mts. to be overlooked for a long time. Fortunately, we are currently in a time when the beauty of the mountain range is newly appreciated and hopefully will result in the development of sustainable human activities and the establishment of the mountain range as a protected area.

Acknowledgement Authors are indebted to Jiří Riezner and Ladislav Bobr for providing photos of anthropogenic objects in the area.

References

Bridgman HA, Davies TD, Jickells T et al (2002) Air pollution in the Krušné hory region, Czech Republic during the 1990s. Atmos Environ 36(21):3375–3389

Břízová E (1995) Reconstruction of the vegetational evolution of the Boží Dar peat bog during Late Glacial and Holocene. Geolines 2:10

Burda J, Hartvich F, Valenta J, Smitka V, Rybář J (2013) Climate-induced landslide reactivation at the edge of the Most Basin (Czech Republic)—progress towards better landslide prediction. Nat Hazards Earth Syst Sci 13(2):361–374

Burda J, Vilímek V (2010) The influence of climate effects and fluctuations in groundwater level on the stability of anthropogenic foothill slopes in the Krusne hory Mountains. Czechia. Geografie 115(4):377–392

Cajz V (2000) Proposal of lithostratigraphy for the České středohoří Mts. volcanics. Bull Czech Geol Survey 75:7–16

Chlupáč I, Brzobohatý R, Kovanda J, Stráník Z (2002) Geologická minulost České republiky, Academia, p 436

Hurník S (1982) Dnešní představy o vzniku severočeské hnědouhelné pánve. Zprávy VÚHU 1–2:47–59

Jankovská V, Pokorný P (2013) Reevaluation of the palaeoenvironmental record of the former Komoranske jezero lake: late-glacial and Holocene palaeolimnology and vegetation development in north-western Bohemia. Czech Republic. Preslia 85(3):265–287

Kalvoda J (1995) Geomorphological analysis of levelling measurements between Mikulovice village and Jezeri Castle in the Krusne Hory Mountains. Acta Universistatis Carolinae, Geographica, Supplement 30:139–160

Klomínský J (1994) Geological Atlas of the Czech Republic—Stratigraphy. Česká geologická služba, Praha

Kopecký A (1972) Hlavní rysy neotektoniky Československa. Sbor. Geol Věd, řada A 6:77–155

Košťák B, Mrlina J, Stemberk J, Chan B (2011) Tectonic movements monitored in the Bohemian Massif. J Geodyn 52(1):34–44

Král V (1968) Geomorfologie vrcholové oblasti Krušných hor a problém paroviny. Rozpravy ČSAV 77, 9, 66, Academia, Praha

Malkovský M et al (1985) Geologie severočeské hnědouhelné pánve a jejího okolí. Academia, Ústřední ústav geologický 424 pp

Mikšíček P (2005) Das wiederentdeckte Erzgebirge. Obec Boží Dar, Boží Dar

Mlčoch B, Skácelová Z (2010) Geometry of the Altenberg-Teplice Caldera revealed by the borehole and seismic data in its Czech part. J Geosci 55(3), Special Issue: SI, 217–229

Mohelník P (2013) Vodní kanály Krušných hor. Thesis, UJEP, Ústí nad Labem

Muhly JD (1985) Sources of tin and the beginnings of Bronze Metallurgy. Am J Archaeol 89(2):275–291

Pekárek O (1960) Nová geologická mapa a přehled geologických poměrů chomutovsko—mostecko—teplické pánve. Sbor. 1. geol. konference o chomutovsko—mostecko—teplické pánvi a blíže přilehlých oblastech, 23–44, Osek, Teplice

Riezner J (2011) Krajinný ráz území typických agrárními valy a mezemi a jejich vegetací na vybraných příkladech ze severozápadních Čech. Studia Oecologica 5(2):65–79

Rybář J (1982) Fosilní svahové deformace na jižním svahu Krušných hor. Zpr. Geotechn. Symp, Prague, pp 56–62

Tumajer J, Burda J (2013) Landslide-induced changes of vessel shape in Betula pendula Roth.—A preliminary study. Acta Universistatis Carolinae. Geographica 48(1):59–68

Urban M (ed) (2011) Tři studie k hornické a kulturní krajině českého Krušnohoří. Oblastní muzeum v Mostě, Most

Vilímek V (1995) Quaternary development of Kateřinohorská Vault relief in the Krušné hory Mountains. Acta Universitatis Carolinae, Geographica 30 Supplem., 115–137

Vilímek V, Fantucci R, Stemberk J (2002) Dendrogeomorphological investigations of slope deformation on Salesius Hill in the Krušné Hory Mts. In: Rybář J, Stemberk J, Wagner P (eds.) Landslides Proceedings of the 1st Europe Conference on Landslides. Prague, Balkema, Brookfield, Rotterdam, pp 321–326, 24–26 June 2002

Elbe Sandstones

11

Zuzana Vařilová

Abstract

The Elbe Sandstones is a sandstone area of the Czech Republic constituting the north-western part of the Bohemian Cretaceous Basin. Characteristic landscape with plateaus, deep canyons, rock cities and a wide range of small landforms developed on massive quartzose sandstone deposits mostly of Middle Turonian age. The region has special status compared to other similar areas by its large areal extent, its vertical diversity causing the high-energy relief, the presence of volcanic bodies forming significant elevations and relative wilderness of inaccessible rock massifs and extensive forests. The dominant feature and also the connecting element is the huge Elbe River Canyon, meandering through sandstones between Děčín and Pirna in Saxony. The most famous object of the whole area is the Pravčická brána Arch—the largest sandstone arch in Europe. The table mountain of Vysoký Sněžník Hill and the typical rock city called the Tiské Stěny Cliffs belong to other interesting places here. The landscape is unique not only because of its geomorphic diversity but also because of the close relationship between the living and inanimate nature and the presence of relicts after ancient settlement. A higher occurrence of natural geohazards, especially the threat of rockfall together with the degradation of sandstone cliffs due to weathering processes belong among the most pressing challenges of present time.

Keywords

Elbe Sandstones • Bohemian Cretaceous Basin • Turonian sandstones • Tertiary volcanics • Rock city • Table mountain • Pravčická brána Arch

11.1 Introduction

The Elbe Sandstones area (=Labské pískovce in Czech) constitutes an extensive unique geological–morphological unit and significantly exceeds the state border towards Germany; the entire territory covers an area of around 700 km^2 (the Bohemian part about 288 km^2). This erosional landscape surrounds the course of the Elbe (Labe in Czech) River between the town of Děčín and the town of Pirna in Saxony. German literature has an orographic apposite name *Elbsandsteingebirge* for this whole territory, however, a better known name is just the romantic Saxon–Bohemian Switzerland (*Sächsisch-Böhmische Schweiz*) (Mikuláš et al. 2007).

Altitudes within the Elbe Sandstones area range from 115 to 726 m a.s.l. This sandstone area is considered one of the most rugged in Europe. The landscape is mainly characterized by the contrast between the plateaus and deep gorges lined with steep rock walls (Fig. 11.1), by the huge Elbe River Canyon and the numerous distinctive table mountains (situated, however, on the Saxonian side, with only one exception) that give the landscape an unmistakable character. The wide range of different rocky landforms and their dimensions is unparalleled elsewhere in the Czech Republic.

Z. Vařilová (✉)
Museum of the City of Ústí nad Labem, Ústí nad Labem, Czech Republic
e-mail: varilova@muzeumusti.cz

T. Pánek and J. Hradecký (eds.), *Landscapes and Landforms of the Czech Republic*, World Geomorphological Landscapes, DOI 10.1007/978-3-319-27537-6_11

Fig. 11.1 Rocky slopes to the northeast of the Pravčická brána Arch (Geognostische Skizzen aus Sächsische Schweiz—1858, Gutbier, Leipzig)

Now, the Elbe Sandstones are primarily known as a significant European tourist area. This tradition dates back to the Romanticism period at the end of the eighteenth century, when the region has been "discovered" to a wider public by the Swiss and German landscape painters, the most famous among them being Caspar David Friedrich (Richter 2007). Since the end of the nineteenth century, the entire area has been made accessible by a traditional network of marked trails. And it should be remembered that the Elbe Sandstones is not only a hiking region but also the cradle of mountain climbing in sandstone rocks. The first peaks were climbed in the latter half of the nineteenth century in Saxony (Stein 2007). In Bohemia, the development for climbers started in the Pravčická brána Arch area and later also in Elbe River Canyon and in the Tisá and Ostrov districts. Local climbers made history not only because of their sport performance but mainly by setting the rules for sandstone climbing.

11.2 Geology and Tectonic Setting

11.2.1 Geological History

The Elbe Sandstones are built of shallow marine sediments of Mesozoic age, lying on the Early Paleozoic and (Late) Proterozoic crystalline bedrock. The latter is composed mostly of granites and granodiorites in the east (the granitic Lusatian massif) together with phyllites, greywackes, schists (Elbe Slate Mountains) and also by gneisses in the west (Erzgebirge Crystalline Complex)—Fig. 11.2a.

The Elbe Sandstones belong to the Lusatian Development of the Bohemian Cretaceous Basin (BCB), to geographical subregion of the Děčínská vrchovina Highland (e.g. Glöckner 1995). The outcrops within this territory are mostly built of relatively massive, quartzose sandstone of Upper Cretaceous age (Fig. 11.2b). For a period of 10–12

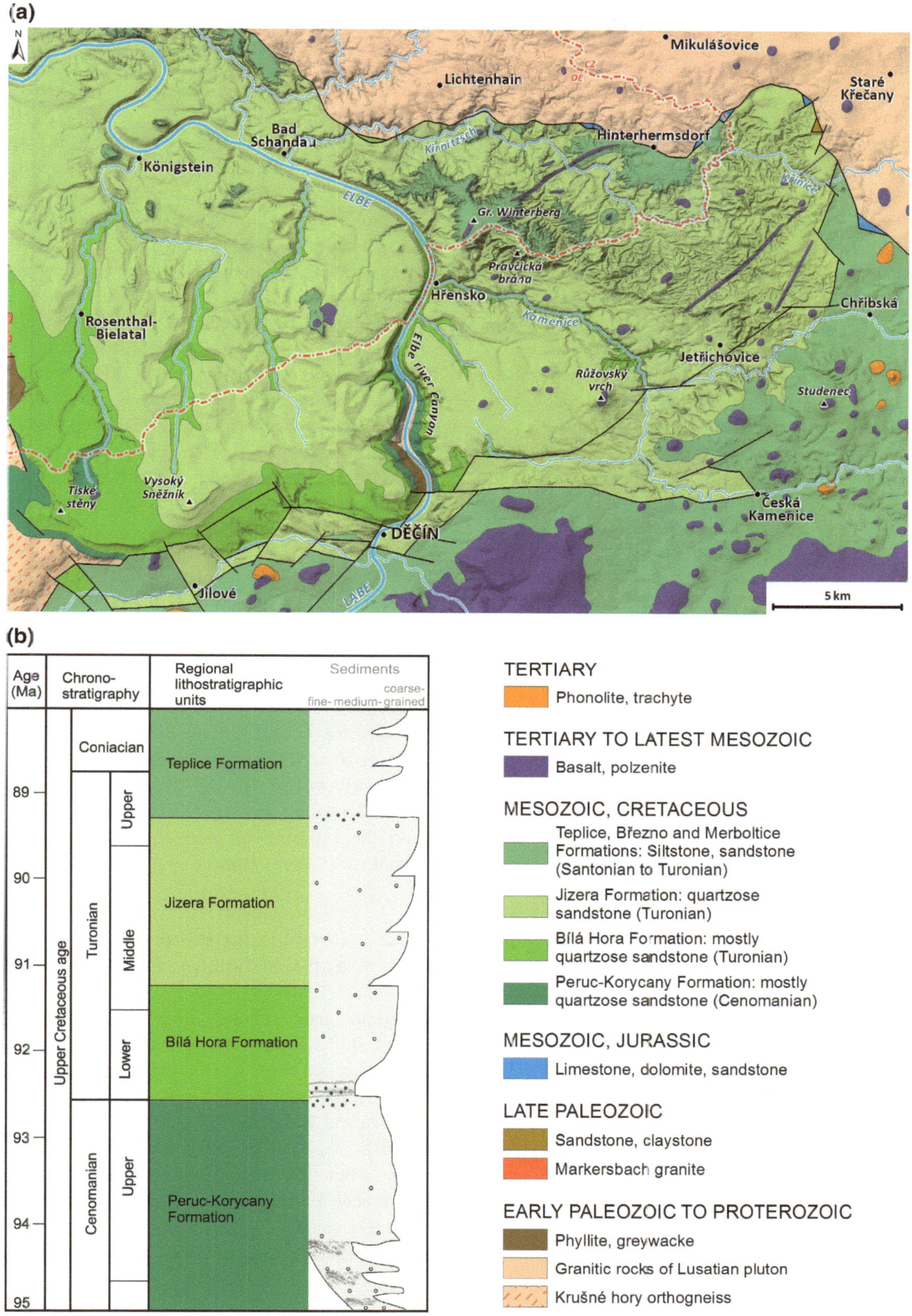

Fig. 11.2 **a** A simplified geological map (O. Holešinský, modified after Cílek et al. 2010) and **b** Table of Cretaceous stratigraphy of the Elbe Sandstones: Local sandstones represent the interval from Cenomanian to Coniacian age, but the majority of outcrops are of Middle to Upper Turonian age, called the Jizera Formation (e.g. Čech et al. 1980). The rocks are sub-horizontally stratified with upward-coarsening (progradational) cycles which start with fine-grained sandstones and terminate with medium- or coarse-grained sandstones, locally with conglomerates with sharp or even erosive boundaries (e.g. Valečka 1989). Older formations of Lower Turonian age and younger formations of Upper Turonian age contain scarce intervals of calcareous mudstone or alternating claystone/siltstone beds (Valečka 1997)

million years, the permanent subsidence of sea floor led to the deposition of sub-horizontally stratified sequences up to 1000 m in thickness, which is the highest thickness within the BCB. In the Elbe Sandstones area, the uppermost strata were removed by erosion and the thickness of sandstone-dominated packages does not exceed 650 m here. According to a paleographical analysis, granitoids of the Western Sudetic Island were a major source of clastic material for building the marine sandy facies (e.g. Valečka 1989). There are two different theories explaining the origin of the sandstone bodies: (1) sandstones originated either in a shallow sea, where sand bodies formed during the migration of large submarine sand dunes (environment corresponding to prograding sandy shoreline—Skoček and Valečka 1983) or (2) sandstones are remnant/sediments of offshore coarse-grained deltas (Uličný et al. 2009).

Geological evolution was influenced by intensive volcanic activity in the Tertiary (Oligocene–Miocene). Intrusions of volcanic rocks form isolated, notable rounded hills (necks) and hummocks in the sandstone landscape. The highest and most massive one is Růžovský vrch Hill (*Rosenberg*, elev. 619 m). The slopes of volcanic necks are frequently covered with a thick layer of blockfields and slope debris, which represent the Quaternary accumulations caused by frost weathering during the Pleistocene. Other types of colluvial sediments, fluvial sediments, loess loam accumulation and organic deposits also belong to the youngest Holocene sediments in this region. (e.g. Valečka 1997)

11.2.2 Tectonic Framework

General morphostructure has been strongly affected by tectonic activity (Engel and Kalvoda 2002). The area lies at the intersection of two long-active fault zones: (1) the Lusatian Fault zone (a part of the Elbe Lineament, NW–SE) limiting the extent of Cretaceous sediments in the northeast, forming the boundary against the uplifted block of the Krkonoše–Jizera Crystalline Complex; (2) the continuation of the Erzgebirge Fault (NE–SW), which separates the block of Elbe Sandstones from the subsided block of the Eger Graben in the south. These two major fault zones are crosscut by younger faults of different strikes (most important strike being E–W). Of significance is also fracture zones related to these faults, either parallel to them or forming splay structures. Zones of dense jointing striking NE–SW can be observed, together with other prevailing strikes NW–SE, NNE–SSW, NNW–SSE and ENE–WSW (e.g. Figs. 11.2a and 11.3).

11.3 Formation of Sandstone Landscape and Its Changes Over Time

11.3.1 Quaternary Morphological Evolution

Typical sandstone relief was formed after the retreat of the Cretaceous sea, when the massive accumulation of marine sediments became gradually eroded and denuded. Tertiary denudation, taking place mainly in subtropical climate conditions, removed most of the first product of volcanic activity and then the younger Cretaceous sediments from the surface. A generally flatter landscape relief with small altitude differences developed. Dramatic changes and significant reshaping of the landscape occurred at the end of the Tertiary and during the Quaternary. The whole area was uplifted, and the originally flat-topped sandstone massif was broken to a number of separate uplifted and subsided blocks during the peak phase of the Saxonian/Alpine Orogeny. There was also an asymmetrical uplift of massive blocks of Cretaceous sediments, today sloping at an angle of 1–3° towards the NW to N. Blocky disintegration of the sandstone proceeds along a system of rectangular joints, which became the access paths for a subsequent action of external factors—especially water, frost, wind and living organisms. Since the beginning of the Quaternary, the local area became a foreland of continental glaciers. The alternation of glacial and interglacial periods led to intensive and rapid erosion along tectonically weakened zones in sandstones. Tectonic uplift of the area was offset primarily by accelerated incision of the stream network (in this area particularly of the Elbe River and its tributaries) and by the formation of a typical canyon-like valley (Fig. 11.3). Less intensive lateral erosion led to the exposure of hard volcanic bodies, while less resistant rocks were removed (e.g. Valečka 1997).

11.3.2 Rock Slopes Development and Geodynamic Proceses

The specific geomorphology of the Elbe Sandstones area has evolved in long term through complex processes taking place under particular morphogenetic regimes, which were dominated by rockfall together with erosion and weathering (Vařilová and Zvelebil 2007). The Pleistocene is also considered as a period of maximum activity of slope movements, when extensive follow-up frost loosening and gravitational deterioration of slopes (including rockfalls and landslides) came into effect, especially during glacial and early interglacial periods (Kalvoda and Balatka 1995).

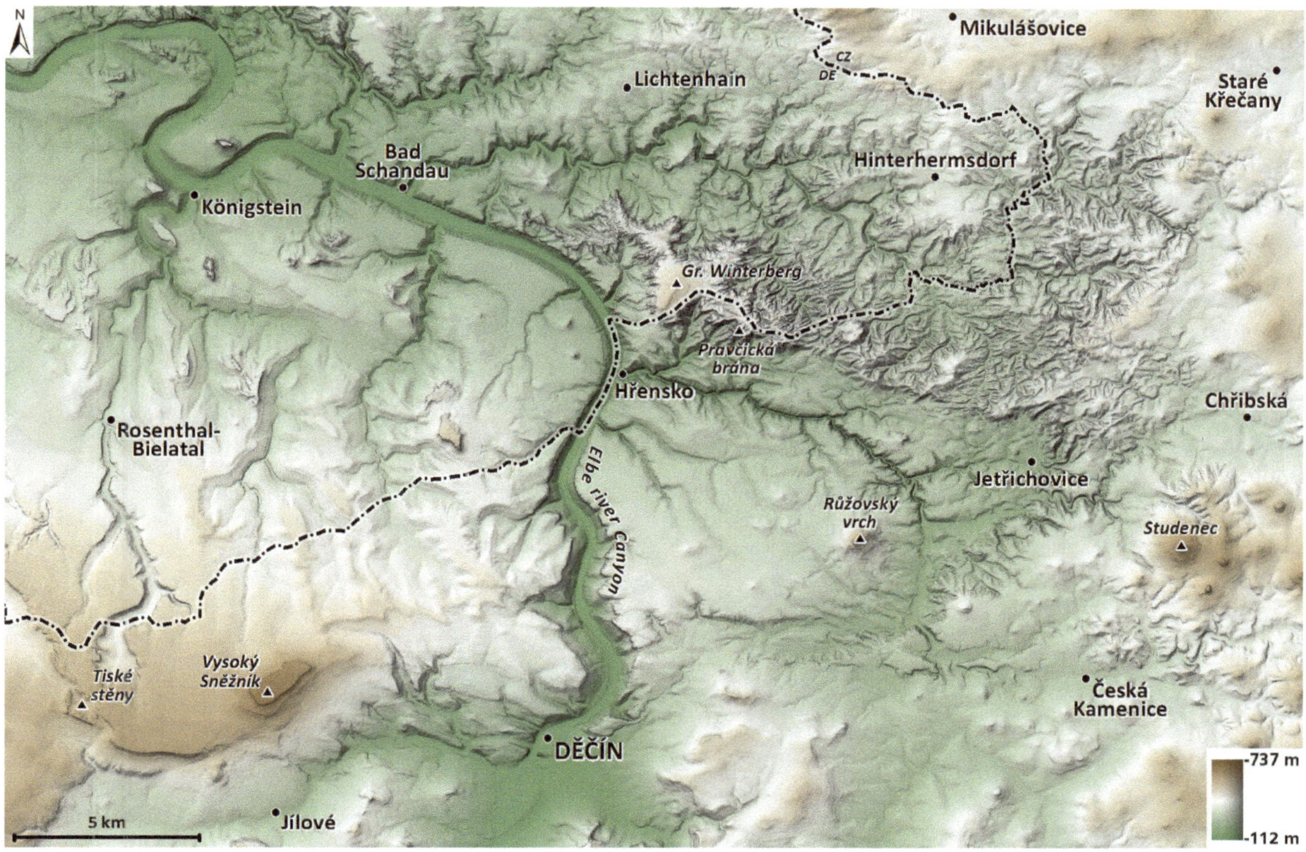

Fig. 11.3 Colour shaded relief map of the Elbe Sandstones area (O. Holešinský)

The type of slope process then also decided on the type of emerging forms and individual developmental stages of the shape of the rock slope (Kalvoda and Zvelebil 1983; Vařilová and Zvelebil 2007). Excellent permeability, coupled with intensive tectonic fracturing of sandstone rocks, allowed for a relatively rapid infiltration of rainwater. A significant role in the younger geomorphic history of this area was played by surface hardening and salt weathering in combination with mass movements.

11.3.3 Main Landforms

The shape of relief can be classified according to various criteria. Classification of rock landforms according to their relative proportions into three major size classes is frequently used. Currently, the landscape of the Elbe Sandstones is divided into three height levels, which represent large landforms (*macroforms*) of sandstone relief (Fig. 11.4). The lowermost level is formed by the Elbe River Canyon with its small tributaries and narrow deep valleys of the Kamenice River (Edmundova, Divoká and Ferdinandova soutěska Gorge) and Křinice (*Kirnitzsch*) River. The middle level is a structural plateau with an altitude of around 350 m. The upper level consists of rugged rock cities and table mountains at an altitude of around 450 m and more. The origin and development of large sandstone landforms (such as deep valleys and table mountain) is linked to the retreat of rock slopes and proceeds in several stages (development stages of the shape of the rock slope, described e.g. by Vařilová and Zvelebil 2007).

The group of medium-sized landforms (*mesoforms*) is represented by rock ledges, shelters, rock towers and rock pillars, mushroom rocks, beaker forms on top chimney rocks, along with a variety of rock perforations—from rock windows to arches and tunnels (Fig. 11.5). Most of the mesoforms, their orientation and the actual shape reflect the jointing pattern in sandstones. Jointing in massive rocks usually follows two perpendicular systems, which—together with horizontal jointing along bedding planes—leads to

Fig. 11.4 **a** A view from the rock massif of the Křídelní stěny Cliffs across the Elbe River to the Saxonian part of the Elbe Sandstones with a monumental table mountain called Grosser Zschirnstein which represents the upper level of sandstone landscape, **b** the Kamenice River Gorge which represents the lowest level of sandstone landscape (*Photo* V. Sojka)

Fig. 11.5 Examples of typical mesoforms of sandstone relief: **a** view from a rock shelter to the Pravčický Důl Valley (*Photo* V. Sojka), **b** rock pillars within the Tetřeví stěny Cliffs with an unusual view of the Pravčický kužel Pillar (*Photo* V. Sojka), **c** an open crevasse with a hiking trail called Úzké schody (*Narrow Stairs*) or Myší díry (*Mousehole*) (*Photo* V. Sojka), **d** a rock labyrinth with tunnels in the Tiské stěny Cliffs (*Photo* J. Bálek)

disintegration into blocks several metres in size. This phenomenon is characteristic especially for the Elbe Sandstones and therefore the name of "block of sandstone" is usually used there for local sandstone rocks.

The smallest relief forms (*microforms*) come into existence through a series of micro-scale processes (especially salt weathering) which operate on the surface and in subsurface parts of the sandstone massif (e.g. Mikuláš 2007). Most typical are abundant honeycombs, weathering pits, tafoni and minute rock cavities, and also pseudokarst karren which cover the rims of the plateaus and cliffs (Fig. 11.6).

A specific phenomenon, spanning the meso- and microforms in size, is secondary Fe mineralization in sandstones whose formation is closely related to Tertiary volcanic activity. This ferruginization is one of the processes of surface hardening that control geomorphology of the Elbe Sandstones and occur in many different forms: subvertical tabular bodies, irregular undulating parallel crusts (circles or tubes) or subhorizontal stratabound bodies (more in Vařilová 2007; Adamovič et al. 2010).

11.4 Remarkable and Typical Landforms

11.4.1 Elbe River Caynon

The monumental Elbe River Canyon is the largest river canyon in the Czech Republic. The valley shape is much tighter in the Elbe Sandstones region, and its both sides are lined with high cliffs (Fig. 11.7a). The stream pattern in Bohemia became drained to the north across the Elbe Sandstones region in the Early Miocene. The Elbe River Canyon also functions as the current and historical base level of erosion for the majority of the Bohemian Highlands area. The current shape of the canyon was formed due to deep, stepwise erosion in the Early and Middle Pleistocene, being

Fig. 11.6 Examples of typical microforms of sandstone relief: **a** hour-glass columns and a rock window constituting a landform called the "Rock chapel", the surface is affected by salt weathering **b** pinnacles/solution runnels on the edge of the Křídelní stěny Cliffs, **c** a high cliff face lining the Kamenice River, covered with honeycomb pits, **d** typical concentric ferruginous crusts near the Zadní Doubice border crossing (*Photo* V. Sojka)

Fig. 11.7 The Elbe River Canyon: **a** Perspective representation of digital terrain model (O. Holešinský) and **b** view from the upper edge to the canyon with the known rock tower "Strážce Dolního Žlebu", the gradual river erosion formed the characteristic cross section of the valley (*Photo* V. Sojka)

controlled by climatic (glacial and interglacial period alternation) and tectonic factors (especially repeated tectonic uplift) (Kalvoda and Zvelebil 1983).

The Elbe River cuts through the entire thickness of Cretaceous sediments down to the crystalline bedrock. The total difference in altitude between the river and the upper edge of the canyon locally reaches 200–300 m, and almost 500 m near Grosser Winterberg at the Bohemian/Saxonian border. The lower part of the slopes are milder as the river cuts into softer Cenomanian and Lower Turonian sandstones, which are easily subject to erosion and weathering. In contrast, upper level rocks (with high cliff faces in the upper part of the slope) are Middle to Upper Turonian sandstones (Jizera Formation), harder and more resistant. The height of cliff faces between Děčín and Hřensko gradually decreases (from 80 m to a mere 30 m) towards the north, and so does their height position within the slope. The sporadic rock towers reach the heights of a few tens of metres here (Kalvoda and Balatka 1995).

The shapes of side valleys also show a prominent block structure associated with the existing network of joints. This is evidenced not only by the conspicuous ridges (e.g. the uplifted and tilted tectonic blocks of the Pastýřská stěna Cliff, Zámecký vrch Hill or Kvádrový vrch Hill) and depressions, but also by the shape and course of the side valleys of several smaller streams and erosive gorges. Confluence of the Elbe River with its main tributary, the Kamenice River, is located in the Hřensko village (the lowest point of the Czech Republic, 115 m above sea level).

The progress in river erosion can be usually reconstructed from the main features of river terraces of various ages. They are, however, relatively rare in this area because incision prevailed over accumulation for most of the time. Nevertheless, remains of the erosion–denudation zones were found by the geomorphological analysis in the Elbe canyon, which can help to identify the partial levels of the stream network incision (Kalvoda and Balatka 1995). Morphostructures and surface shapes of this antecedent valley also contain a relatively detailed record of neotectonic and paleogeographic history of whole sandstone area. It is possible to study the different stages of geomorphic evolution of rocky slopes in sandstone blocks, including all phases of catastrophic rockfalls (e.g. Vařilová and Zvelebil 2007). Especially the steep rocky slopes on both sides of the canyon, due to its altitude exposure and tectonic deformation of the sandstone, provide favourable conditions for regular geodynamic processes (Fig. 11.7b). A characteristic feature is also the existence of more than 50 pseudokarst caves (crevasse caves), which were formed especially on the right bank of the river. One of the longest cave systems reaches 160 m in length and 40 m in depth (Winkelhöfer 1997; Kukla in Adamovič et al. 2013).

11.4.2 Vysoký Sněžník Hill

Vysoký Sněžník Hill (*Hoher Schneeberg*) impresses with its unusual shape compared to other peaks in the Elbe Sandstones on the Bohemian side. It is the only table mountain in the Czech part of the region and is situated on the uplifted tectonic block northwest of the Krušné hory (*Erzgebirge*) Fault (Děčín Fault zone), to the west of the town of Děčín. Its peak, reaching the altitude of 726 m a.s.l., is also the highest point of the whole sandstone area. The plateau surface is approximately 1.7 by 0.6 km in size, thus representing the largest table mountain in the Saxon–Bohemian Switzerland and also in the whole Czech Republic (Fig. 11.8a, b). The top plateau constitutes the remnants of the original platform—i.e. the top surface of the local Cretaceous sandstones. The edge of the table mountain forms an escarpment with numerous partially developed rock towers.

Fig. 11.8 Vysoký Sněžník Hill: **a** Perspective representation of a digital terrain model (O. Holešinský), **b** the southern edge of the table mountain (*Photo* J. Preclík), **c**, **d** fluorite mineralization covering open crevasess on the southern slopes (*Photo* by V. Sojka)

Most of the Vysoký Sněžník Hill slopes are covered by sandstone boulders to block debris accumulation. Their origin is connected with periglacial climate during the Pleistocene. Several balanced boulders on the top plateau are also sites of geomorphological interest.

Nearly the whole stratal succession of the Jizera Formation is preserved here (with a total thickness of over 140 m). All sandstone outcrops are inclined NE at an angle of about 1.5°, consistently with the dip of the whole tectonic block (Adamovič et al. 2010). The sandstones are characterized by intensive silicification related to tectonic movements and circulation of mineralized solutions.

On the southern slopes of Vysoký Sněžník Hill, a fluorite deposit was industrially mined until 1994. Thick coatings of banded fluorite, locally also with crystallized fluorite (CaF_2) occur on the surfaces of open fissures striking E–W or NE–SW (Fig. 11.8c). These crevasses were formed mainly in sandstones of the Bílá Hora Formation and are related to the Tertiary gravitational sliding of blocks by gravity. The connection of the discovered pseudocavities in sandstones with the hydrothermal mineralization is one of the unique phenomena of the region (e.g. Veselý 1999).

11.4.3 Pravčická Brána Arch

The Pravčická brána Arch (*Prebischtor*) is located in the Jetřichovické stěny Cliffs, about 3 km northeast of Hřensko and approximately 2 km north-west of Mezní Louka, near the border with Germany. The arch is developed in the top part of a narrow rock rib more than 100 m in length, limited by a system of parallel joints striking NE–SW (elev. 442 m). This arch, longest of all European sandstone arches, has a span of 26.5 m and reaches 16 m in height (Fig. 11.9). It is a

Fig. 11.9 Pravčická brána Arch: **a** A typical view of the arch from the west, with Růžovský vrch Hill in the background (*Photo* by K. Brož) and **b** perspective representation of points from terrestrial laser scanning draped over the digital terrain model—aerial view from the south (O. Holešinský)

symbol of the Elbe Sandstones region and is considered a nature monument of great significance.

The arch was formed in the highest part of the Jizera Formation, at the base of the original cliff face, by the process of progressive deepening of rock shelters on the two flanks (at the level of the lower conglomerate bed). These perforations were then further widened by the fall of blocks and scale delamination along arcuate exfoliation planes, with the contemporaneous action of erosional and weathering processes according to the distribution of internal stress (Cílek et al. 2010; Bruthans et al. 2014). The current status of the arch can be described as a mature form. The width of its delicate ceiling is 7.5 m at the narrowest point, and its vertical thickness is a mere 2.5 m. Due to its geometry and exposure, it is threatened not only by the stress posed by its own weight, but also by extreme microclimatic factors (seasonal and daily temperature changes, insolation, etc.). Geological and tectonic patterns of the rock body together with the external effects include dynamic effects of volumetric changes and irreversible movements (Zvelebil et al. 2002; Vařilová et al. 2014) and the driving force of physical and chemical weathering processes. These result in gradual deterioration of physicomechanical parameters of the sandstone and stability threatening (see Vařilová et al. 2011a, b, 2015). The Pravčická brána Arch therefore became a subject to interest of professionals as early as in the 1990s and today belongs to the well-explored and long-term monitored rock objects. The aim is to propose the most suitable protective measures and evaluate its lifetime.

Entrance to the arch itself has not been permitted since 1980 due to excessive mechanical erosion of the surface, which was caused by the large attendance of tourists. However, the accessible rocky outcrop in the vicinity of the arch provides wonderful views of the landscape of the Bohemian and Saxonian Switzerland.

11.4.4 Tiské Stěny Cliffs

One of the most beautiful rock cities, which forms a noticeable rock wall towering 70 m above the Tisá village buildings, lies on the left bank of the Elbe River, at the north-western end of the Elbe Sandstones. This rock city has been for a long time a wild and uncharted territory, where people rarely ventured; they could enter the rock labyrinth only with a local guide starting from 1918. Today, the Tiské stěny Cliffs (*Tyssaer Wände*) belong to the most popular tourist sites, also frequently visited by climbers. Its area (about 1000 m^2) is relatively small, but it is a perfectly developed rock city in blocky-jointed sandstone. It is divided into two main separate parts—Velké stěny (Large Cliffs) in the east and the Malé stěny (Small Cliffs) in the west, separated by a "rock square" (Fig. 11.10a).

The Tiské stěny Cliffs are built of sandstones of the Bílá Hora Formation, which have been preserved here in an uplifted tectonic block very gently dipping to the NE, situated northwest of the Krušné hory (*Erzgebirge*) Fault zone. The most frequent sedimentary structures include horizontal stratification and tabular cross-bedding (decimetres to metres in thickness). Sandstone is less resistant to erosion at the levels of densely spaced bedding plans or tabular cross-bedding, giving rise to elongate notches, ledges, rock niches and windows at such levels on the cliff faces (Adamovič et al. 2010).

Local sandstones are significantly vertically divided by a dense network of grikes and narrow ravines along dilated joints (Fig. 11.10b). Approximately a hundred perfectly developed rock towers are present. Due to the relatively high altitude of the sandstone plateau (613 m), the rock towers reach a height of only twenty to thirty metres. Common forms also include rock tunnels, smaller arches and rock mushrooms. Tens of bizarrely shaped rock blocks were given specific names by human imagination: the "Mushroom" and "Turtle" (Fig. 11.10c), "Napoleon's shoe", "Janu's Head", "Thick mayor", "Thin doctor" or a "Mummy".

11.5 Traces of Human Activity in the Sandstone Landscape

11.5.1 Rock Shelter Settlement, Mining and Other Artificial Remains

Although the Elbe Sandstones area was once considered barren, covered by impenetrable forests, much evidence of land use by our ancient ancestors can be found here. The Elbe River Canyon itself was probably a key route for prehistoric and historic cultures in the distant past.

Remains of the oldest documented settlement come from the Mesolithic period, from the time when a small group of prehistoric hunters searched for rock shelters for seasonal dwelling. More than 7 sites with high density of lithic artefacts have been discovered and subjected to a detailed exploration (Svoboda 2003). The finds from the rock shelters belonging to the Agricultural prehistoric period (Eneolithic Age, period from Early Bronze Age to Late Bronze Age and the Iron Age) are scattered throughout the area. The turning point in the history of the settlement then came with the colonization of in the High Middle Ages (ongoing until the end of the Fourteenth century). Several small castles built directly inside the rock massif—for example the Šaunštejn Rock Castle or Falkenštejn Rock Castle—and one of the finds of glassworks are also dated to this period. Numerous archival reports from the seventeenth century have been preserved with information about the use of local natural resources. Since then, natural sandstone relief has been

Fig. 11.10 Tiské stěny Cliffs: **a** Colour shaded relief map (O. Holešinský), **b** a view of the vertically articulated rock city (*Photo* by J. Bálek), **c** the most famous mushroom rock in the Tiské stěny Cliffs, its cap is 6 × 3.5 m in size and its maximum height is 2.5 m; its stipe developed in cross-bedded sandstone less resistant to erosion and weathering (*Photo* by Z. Vařilová)

significantly modified by man. The degree of human activities cannot be fully reconstructed, although they are still noticeable in the form of large and small quarries, remains after water supply activities, tunnels, railway tracks and common paths through the rocks (scarped roads and footpaths), rock dwellings and stables, abundant small rock chapels and sculptures, and a number of rock engravings (e.g. Belisová 2007; Jenč and Peša 2007).

The most important quarries were located in the Elbe River Canyon (Fig. 11.11). Sandstone quarrying on the Bohemian side of the Elbe Sandstones never reached such expansion as on the German side. Nevertheless, a number of known historical objects were built from local sandstone: e.g. the Terezín Citadel, the Prague–Dresden railway or the completion of the St. Vitus Cathedral in Prague. Quarrying activities became subdued at the turn of the nineteenth and twentieth centuries, and sandstone was last quarried in years 1938—1942 during the period of Děčín–Hřensko road construction. In addition, stone and sand were not only quarried at the surface but also extracted by underground

Fig. 11.11 A rockfall during sandstone quarrying in the Elbe River Canyon (Preyss J.: Quarry in Goldnen Ranzen near Hřensko, 1832)

mining (e.g. the "Sandhöhlen" underground tunnel or the "Labyrinth" underground sand pit at Ludvíkovice, over 110 m long (Belisová 2013)). Other examples of human activities linked to sandstone relief include traces of iron ore mining (e.g. Vařilová 2007; Adamovič et al. 2010) or remnants of tar ovens (Belisová 2007).

11.5.2 Present-Day Conservation and Hazards

The main purpose for the declaration of officially protected areas within the Elbe Sandstones area (i.e. the *Bohemian Switzerland National Park*, the *Elbe Sandstones Protected Landscape Area* and several small protected areas—AOPK ČR 1999) is to protect the unique geomorphology of sandstones and the variety of ecosystems. The vertical diversity of the sandstone landscape is expressed in the very specific environmental characteristics that depend particularly on micro- to mesoclimatic conditions (e.g. Riebe et al. 1999). Obvious is also the very close relationship between the rock relief and characteristic biodiversity. The rugged relief provides numerous habitats difficult to access, where the remains of indigenous forest ecosystems and endangered species of fauna and flora have survived. Even the occurrence of mountain to subalpine plant species in deep inversion gorges has been reported from extremely low altitudes, around 200 m a.s.l. (e.g. Härtel et al. 2007). The Elbe Sandstones are, among others, also an area of recharge and circulation of groundwater for the Bohemian Cretaceous Basin. The Turonian aquifer, which communicates with sandstone surface (surface infiltration and many springs), represents the main regional source of potable water. And it should be neither forgotten that the Elbe Sandstones are not only a vast complex of unpopulated forest and rock massifs, but also a country landscape that was gradually shaped to produce the present-day harmonious cultural landscape with agricultural land and a number of human settlements. Preservation of remains of past land use, local folk architecture and the overall landscape character also belongs among important tasks of nature conservation authorities.

Although the whole area is under the long-term conservation management, there are several challenges that threaten the natural environment and the people themselves. The attractive environment of the area is the target of a great number of tourists who have an adverse impact on nature. From the perspective of inanimate nature the most common problems include gradual mechanical erosion of rock surface, building constructions in the rock massifs, degradation of sandstone due to burnout at camps or by wildfire, etc.

The cliff faces are relatively high (150 m, locally even 250 m) because the base level of the Elbe River Canyon is deeply incised below the level of the surrounding plateaux. This is related to the high dynamics of relief, hence favourable conditions for the slope movement activity.

Block movements linked to the steep edge of the rock plateaux (toppling) are characteristic for this type of relief (e.g. Vařilová and Zvelebil 2007). The frequent rockfalls are a disastrous phenomenon in this area, but they also constitute an integral part of the contemporary development of sandstone landscape, hence they are a part of the conservation of natural environment. At the same time, there is also a strong need to ensure the safety and protect the lives of tourists as well as the property of local inhabitants. To meet both these tasks, an integrated system of effective management of rockfall risks has been launched, applied especially in the Hřensko village, Elbe River Canyon and in other key areas during the last decades (e.g. Kalvoda and Zvelebil 1983; Vařilová and Zvelebil 2005).

The Elbe Sandstones territory lies within the atmospheric pollution-impacted area of the Black Triangle, and is a well-known example of salt weathering. Wet and dry atmospheric deposition has caused salt accumulation within bedrock outcrops and accelerated sandstone chemical weathering. Sulphur isotopic data proved that the source of secondary sulphate efflorescences was the emissions from combustion of fossil fuels (e.g. Schweigstillová et al. 2009). The deposition of sulphur and nitrogen in the area of the Bohemian Switzerland NP has decreased significantly but still remains higher than in the rest of the Czech Republic (Vařilová et al. 2011a), and no further significant improvements in deposition are expected. Consequences of accelerated weathering are mostly manifested by the process of disintegration with a direct impact on the dynamics of changes of the micro-relief and the related accelerated loss of near-surface parts of the cliff faces. The redistribution of stress due to progressive disintegration and weakening of individual portions of the rock massif through weathering processes is an important finding from the perspective of stability of rock massifs, added to the processes of near-surface weathering that operate mostly in the form of flaking, spalling and granular disintegration (Vařilová et al. 2014).

11.6 Conclusion

The landscape development within the Elbe Sandstones territory has been controlled by a number of processes in the last 2 million years. These include exposure to climatic/microclimatic changes, vertical and lateral erosion, geodynamic activity and selective weathering. Properties of Cretaceous sandstones themselves have been one of the factors determining the formation of landforms and their evolution since the earliest Quaternary. A wide range of characteristic landforms have been created. The very unique geomorphology of the sandstone relief has become the main subject of nature protection. Due to the singularity of this sandstone area, there are efforts towards its inclusion in the World Natural Heritage List of UNESCO. However, a controversial discussion is conducted at various scientific and political levels. Especially the capacity of the sandstone landscape and its potential still remains an unresolved question as the considerable increase in attendance would pose a direct impact on the natural environment.

The ruggedness of the Elbe Sandstones relief with its unique natural monuments has been in the focus of tourists as well as conservationists and researchers for a long time. In many cases the Elbe Sandstones represent a model area for the description of various geoscientific phenomena (understanding sedimentary basin sequences, the development of rock slopes, trial testing of a new control methods of rock massif displacements, documentation of specific landforms, study of salt weathering processes, etc.). In spite of this, it provides a yet greater potential for further study and better understanding of the functioning of the sandstone phenomenon.

Acknowledgments Many thanks to Jiří Adamovič for a linguistic revision. Figures 11.2a, 11.3, 11.7a, 11.8a, 11.9b and 11.10a were created on basis of geodata provided by the Bohemian Switzerland National Park Administration. We acknowledge to Chair of Remote Sensing, Institute for Photogrammetry and Remote Sensing, Dresden University of Technology, for providing the digital terrain model created from airborne laser scanning data in the frame of the EU INTERREG IIIA project "Geoinformation Networks for the cross-border National Park Region of Saxon-Bohemian Switzerland". Digital elevation models SRTM DEM (Courtesy NGA and NASA) and ASTER GDEM (Courtesy NASA and METI) were used to visualize terrain of the area outside the Elbe Sandstones Region.

References

AOPK ČR (1999) CHKO Labské pískovce. In: Mackovčin P (ed) Edice Chráněná území ČR, Ústecko, sv. I, Praha, pp 282–315

Adamovič J, Mikuláš R, Cílek V (2010) Atlas pískovcových skalních měst České a slovenské republiky. Academia, Praha, 460 pp

Adamovič J, Migoń P, Golab Z, Kopecký J, Jenka O, Mertlík J, Peša V, Havránek P, Kukla J, Komaško A (2013) Sandstone caves and rock cities of Bohemia. Czech Speleological Society, Praha, 56 pp

Belisová N (2007) Humans and the landscape: how landscape character influences architectonical and artistic works in the sandstone areas of central Europe. In: Härtel H, Cílek V, Herben T, Jackson A, Williams R (eds) Sandstone Landscapes. Academia, Praha, pp 286–291

Belisová N (2013) Lomy v kaňonu Labe. Kámen revue, 3/2013

Bruthans J, Soukup J, Vaculíková J, Filippi M, Schweigstillová J, Mayo AL, Mašín D, Kletetschka G, Rihošek J (2014) Sandstone landforms shaped by negative feedback between stress and erosion. Nat Geosci 7:597–601

Cílek V, Adamovič J, Vařilová Z (2010) Das Prebischtor: Aus Sand geboren. In: Vařilová Z, Belisová N (eds) Das Prebischtor. Das grosse Buch über das grosse Tor. Academia Praha, České Švýcarsko o. p. s, Krásná Lípa, pp 39–58

Čech S, Klein V, Kříž J, Valečka J (1980) Revision of the Upper Cretaceous stratigraphy of the Bohemian Cretaceous Basin. Věstník ÚÚG 55:277–296

Engel Z, Kalvoda J (2002) Morphostructural development of the sandstone relief in the Bohemian Cretaceous Basin. In: Prikryl R, Viles HA (eds) Understanding and managing of stone decay (SWAPNET 2001). The Karolinum Press, Prague, pp 225–231

Glöckner P (1995) Fyzickogeografické a geologické poměry okresu Děčín. Vlastivěda okresu děčínského, ř. Příroda Děčín, 194 pp

Härtel H, Adamovič J, Mikuláš R (2007) General overview of European sandstone landscapes. In: Härtel H, Cílek V, Herben T, Jackson A, Williams R (eds) sandstone landscapes. Academia, Praha, pp 321–324

Jenč P, Peša V (2007) Sandstone landscapes of the Bohemian Cretaceous Basin - prehistory, history and present (Czech Republic). In: Härtel H, Cílek V, Herben T, Jackson A, Williams R (eds) Sandstone landscapes. Academia, Praha, pp 275–285

Kalvoda J, Zvelebil J (1983) Dynamika a typy porušování svahů při vývoji údolí Labe v Děčínské vrchovině. Acta Montana, 63, Praha: 5–73

Kalvoda J, Balatka B (1995) Chronodynamics of the Labe River Antecedence in the Děčínská vrchovina Highland, Acta Montana. Ser. A 8:43–60

Mikuláš R (2007) Microforms of the sandstone relief. In: Härte H, Cílek V, Herben T, Jackson A, Williams R (eds) Sandstone Landscapes. Academia, Praha, pp 66–75

Mikuláš R, Adamovič J, Härtel H, Benda P, Trýzna M, Kučerová L (2007) Elbe Sandstones (Czech Republic/Germany). In: Härtel H, Cílek V, Herben T, Jackson A, Williams R (eds) Sandstone Landscapes. Academia, Praha, pp 326–328

Riebe H, Härtel H, Bauer P, Benda P (1999) Die Naturausstattung der Sächsisch-Böhmischen Schweiz. Nationalpark Sächsische Schweiz, Bad Schandau 3:20–57

Richter F (2007) Artists discover Saxon-Bohemian Switzerland (Germany/Czech Republic). In: Härtel H, Cílek V, Herben T, Jackson A, Williams R (eds) Sandstone Landscapes. Academia, Praha, pp 292–295

Schweigstillová J, Přikryl R, Novotná M (2009) Isotopic composition of salt efflorescence from the sandstone castellated rocks of the Bohemian Cretaceous Basin (Czech Republic). Environ Geol 58:217–225

Skoček V, Valečka J (1983) Paleogeography of the Late Cretaceous Quadersandstein of Central Europe. Paleogeography, Paleoclimatology, Paleoecol. 44:71–92

Stein K (2007) History of tourism and mountaineering in Bohemian Switzerland (Czech Republic). In: Härtel H, Cílek V, Herben T, Jackson A, Williams R (eds) Sandstone Landscapes. Academia, Praha, pp 295–299

Svoboda JA (ed) (2003) Mesolithic period in Northern Bohemia. Complex excavation of rockshelters in the Česká Lípa and Děčín areas, 1978–2003. The Dolní Věstonice Studies, Brno 9, 328 pp

Uličný D, Laurin J, Čech P (2009) Controls on clastic sequence geometries in a shallow-marine, transtensional basin: the Bohemian Cretaceous Basin, Czech Republic. Sedimentology 56(4):1077–1114

Valečka J (1989) Sedimentology, stratigraphy and cyclicity of the Jizera Formation (Middle-Upper Turonian) in the Děčín area (N Bohemia), Praha. Věst Ústř Úst geol 64(2):77–90

Valečka J (ed) (1997) Die Böhmische Schweiz. Geologische Wanderkarte 1:25 000. ČGÚ, Praha

Vařilová Z (2007) Occurrences of Fe-mineralization in sandstones of the Bohemian Switzerland National Park (Czech Republic). In: Härtel H, Cílek V, Herben T, Jackson A, Williams R (eds) Sandstone Landscapes. Academia, Praha, pp 25–33

Vařilová Z, Zvelebil J (2005) Sandstone relief geohazards and their mitigation: Rock fall risk management in the Bohemian Switzerland NP, Czech Republic. Ferrantia 44:51–56

Vařilová Z, Zvelebil J (2007) Catastrophic and episodic events in sandstone landscapes: slope movements and weathering. In: Härtel H, Cílek V, Herben T, Jackson A, Williams R (eds) sandstone landscapes. Academia, Praha, pp 115–128

Vařilová Z, Navrátil T, Dobešová I (2011a) Recent atmospheric deposition and its effects on sandstone cliffs in Bohemian Switzerland National Park, Czech Republic. Water Air Soil Pollut 220:117–130

Vařilová Z, Přikryl R, Cílek V (2011b) Pravčice Rock Arch (Bohemian Switzerland National Park, Czech Republic) deterioration due to natural and anthropogenic weathering. Environ Earth Sci 63:1861–1878

Vařilová Z, Zvelebil J, Hubatka F, Beneš V, Frolka J (2014) The application of non-destructive methods to assess the stability of the national nature monument of the Pravčická Brána Rock Arch, Czech Republic. AUC Geographica, Charles University of Prague, pp 49–59

Vařilová Z, Přikryl R, Zvelebil J (2015) Factors and processes in deterioration of a sandstone rock form (Pravčická brána Arch, Bohemian Switzerland NP, Czech Republic), Zeitschrift für Geomorphologie, Supplementary Issues, 59(1):81–101

Veselý M (1999) Jeskyně v areálu fluoritových dolů na Děčínském Sněžníku. Děčínské vlastivědné zprávy 3/IX: pp 29–35

Winkelhöfer R (1997) Durch Höhlen der Böhmischen Schweiz: Höhlenführer und Katasterdokumentation. Der Höhlenforscher, B.m

Zvelebil J, Cílek V, Stemberk J (2002) Partial results of monitoring of stability deterioration on Pravčice Rock Arch, NW Bohemia. In: Přikryl R, Viles HA (eds) Understanding and managing stone decay, SWAPNET 2001. Karolinum, Praha, pp 243–261

Neovolcanic Terrain of the České Středohoří Mountains

12

Pavel Raška and Vladimír Cajz

Abstract

The České středohoří Mts. represent the largest volcanic terrain in the Czech Republic. Volcanic evolution dates from 37 to 9 Ma ago, but the largest volumes of volcanic rocks were produced in the early phases of volcanic activity. Current geomorphological characteristics of the area are the result of long-lasting erosional history, being predominantly influenced by the Labe/Elbe River and its tributaries in the central part of the mountain range. The original surface has been locally eroded by hundreds of metres, preserving peculiar conic hills, domes and rock faces. The geomorphological diversity is also enhanced by effects of mass wasting processes from the youngest era of its geological history and by anthropogenic transformations that started in prehistoric times, together forming a landscape that is protected for its geodiversity and biological significance and that helped the world community to understand certain basic volcanic processes and phenomena.

Keywords

České středohoří Mts. • Volcanism • Denudation • Mass movements • Labe/Elbe River Valley

12.1 Introduction

Formerly known as the Czech Paradise, the České středohoří Mts. (CS; *Böhmisches Mittelgebirge* in German; Fig. 12.1) represent the largest volcanic terrain in the Czech Republic, unique in Europe for its landforms, as well as flora assemblages that developed on suitable lithologies. Thanks to a warm and dry climate, the southern region of the area also favoured early Neolithic settlement. The Labe (Elbe) River runs through the centre of the mountain range in a deep erosional valley, which does not only expose the long record of geological history but also represents the primary route for all of those coming from the north. For this reason, the mountain range gained a metaphoric name of Porta Bohemica—Gate to the Bohemia. The diversity of natural systems in the area resulted in the foundation of a protected landscape area in 1976. In turn, the area always attracted the attention of famous scientists, such as Alexander von Humboldt, who used the České středohoří Mts. as an example of typical volcanic rocks and landforms in his *Personal Narrative of Travels to the Equinoctial Regions of the New Continent* (1814) and *Geognostical Essay on the Superposition of Rocks, in both Hemispheres* (1825).

P. Raška (✉)
Department of Geography, Faculty of Science, J. E. Purkyně University, Usti nad Labem, Czech Republic
e-mail: pavel.raska@ujep.cz

V. Cajz
Institute of Geophysics, v.v.i, Academy of Sciences of the Czech Republic, Prague, Czech Republic
e-mail: v.cajz@ig.cas.cz

T. Pánek and J. Hradecký (eds.), *Landscapes and Landforms of the Czech Republic*,
World Geomorphological Landscapes, DOI 10.1007/978-3-319-27537-6_12

Fig. 12.1 The general view of the W part of the České středohoří Mts. with its highest peak Milešovka (837 m) in the background (*Photo* P. Daško)

12.2 Geographical and Structural Setting

The České středohoří Mts. is a 60-km-long, NW–SE-oriented mountain range stretching from the town of Kamenický Šenov in the N Czech Republic to the historical town of Louny, which is approximately 50 km NW from Prague (Fig. 12.2). The range is divided by the Labe River, Czech largest river, running here from south to north, and representing a boundary between the two very different geomorphological units of the area (Král 1966). The western part is typically represented by slightly undulating terrain on eroded Cenozoic sediments (less than 300 to ca 500 m a.s.l.), which is locally interrupted by distinct volcanic domes and cones (up to 837 m a.s.l.) (Figs. 12.2c, 12.3). The area east of the river is characterised by planation surfaces (ca 550–650 m a.s.l.; Fig. 12.2d) incised by many erosional valleys of small water courses. The differences between the W and E part of the range are also apparent in distinct gradients in climate and land cover patterns.

Geologically, the České středohoří (CS) Mts. represent an eroded Cenozoic volcanic range, situated in the Ohře/Eger Rift, which has developed inside the stabilised block of the Bohemian Massif. The volcanic evolution of the CS started approximately 37 Ma ago and the youngest volcanic activity is dated to 9 Ma. Pre-volcanic rocks are represented by uplifted crystalline rocks of the Saxothuringian basement, exposed mainly in the valley of the Labe River (Porta Bohemica as a part of the Oparno-Břvany Elevation—OBH, see Fig. 12.2), and by Late Cretaceous marine sediments and Lower Cenozoic fluvial sediments. Continental freshwater sediments of Miocene age, containing lignite deposits, reach the W and NW margins of the volcanic range. The landforms which we can now see are results of erosional history following the last volcanic event. The largest transportation medium of eroded material, the Labe River, laid gravels and sands in several levels and wind blowing from glaciated areas deposited loess on some of the leeward slopes.

Tectonic activity in the region has followed the tectonic evolution of the entire rift structure. The stress field changed several times over time and the periods of volcanism were most likely associated with relative extension. Grabens and horsts were detected within the volcanic range and faults can be dated in the context of volcanic activity (Cajz and Valečka 2010). Movements on faults were mostly vertical, but a strike-slip component is present as well. The current knowledge of the tectonic setting emphasises the occurrence of rhomboidal blocks between the margins of the rift structure in the area of the CS (Cajz et al. 2004).

12.3 Geological Evolution

12.3.1 Pre-volcanic Evolution

During the Late Cretaceous the area of the future volcanic range was a part of the intracontinental sea, making interconnection between northern and southern oceans. This sea deposited sediments of the Bohemian Cretaceous Basin. Sedimentation in the relatively shallow sea started in the Cenomanian and lasted, at least, up to the Santonian. Sediments vary from conglomerates and coarse grained sandstones to calcareous mudstones (marlstones) and limestones, depending on the depth of the sedimentary basin, distance from source areas of clastic material and sedimentary process itself. Lithofacial changes are relatively rapid within the stratigraphical units. The youngest Upper Cretaceous sediments of the Bohemian Cretaceous Basin—those that are Coniacian to Santonian (Merboltice Fm. sandstones)—are

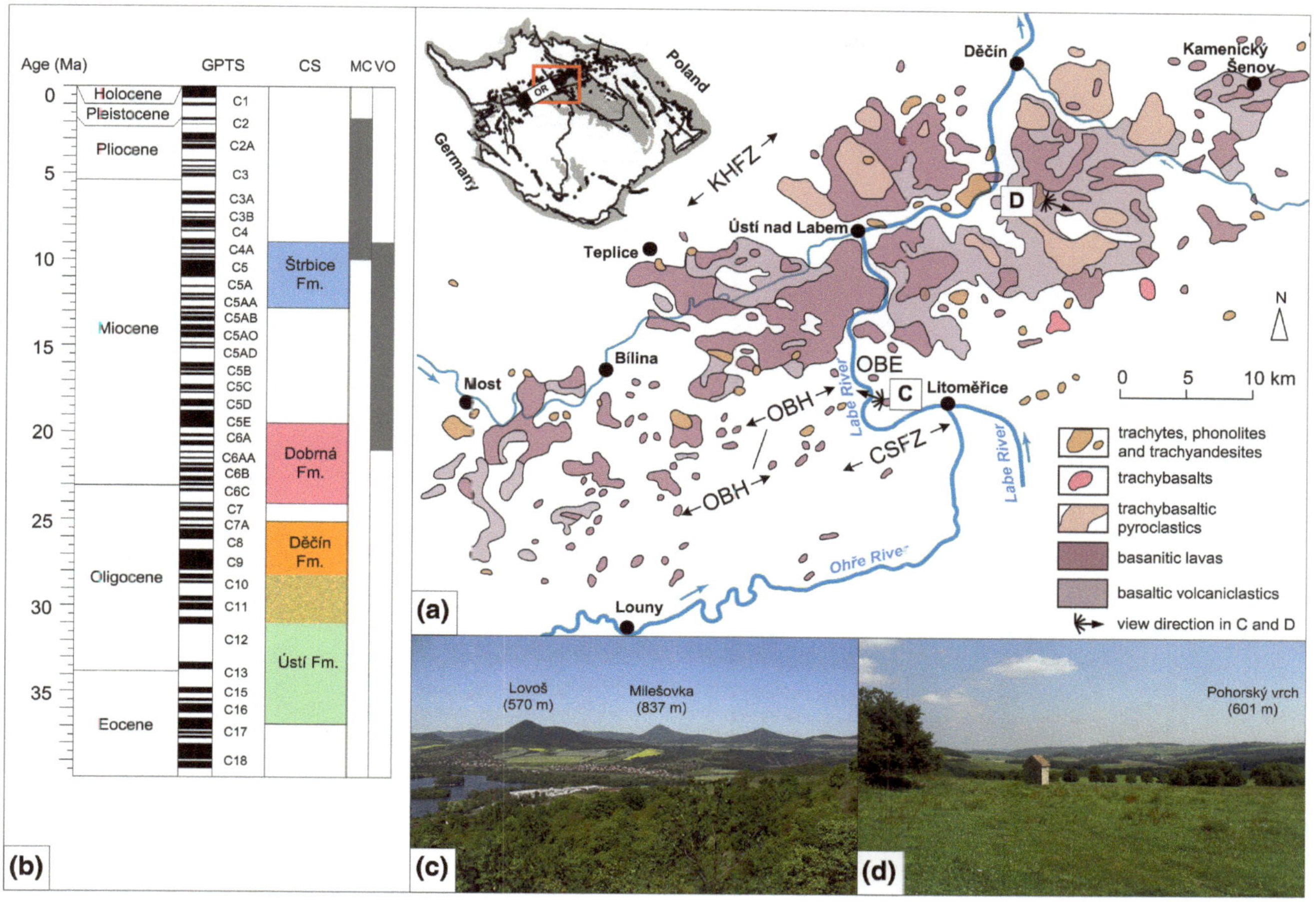

Fig. 12.2 **a** Location and a simplified geological sketch of volcanic rocks. *KHFZ* Krušné hory Fault Zone, *CSFZ* České středohoří Fault Zone, *OBH* Opárno-Břvany Horst, *OBE* Opárno-Břvany Elevation. **b** Stratigraphy of the České středohoří Mts. volcanic activity (Cajz 2000). *GPTS* Geomagnetic Polarity Timescale (http://www.stratigraphy.org), the *grey stripes* indicate volcanic activity in Massif Central (MC) and Vogelsberg (VO) for comparison. **c** View on conical hills in the W part, and **d** view on the post-volcanic denudation level in the E part of the mountain range (*Photo* P. Raška)

preserved only inside the CS area, due to the combination of tectonic activity and covering of volcanic material.

The erosion of Cretaceous marine sediments was caused by rivers in Eocene/Oligocene times, which were arranged in a network significantly different from the current one. These fluvial sediments are preserved in certain parts of the CS, beneath the volcanic rocks.

12.3.2 Cenozoic Volcanism

The rocks of the CS volcanic complex are of alkaline chemistry, with basaltic types prevailing above phonolitic types. The beginning of volcanic activity in the CS was related to the large tectonic remobilisation that formed grabens and/or sunken and water-filled areas. Olivine-bearing basaltic magma (olivine basalts, basanites, olivine foidites) ascended along fault and fracture zones (see Sect. 12.4.1.1). Contact with water in the vents caused phreatic and phreatomagmatic explosions, resulting in the origin of maars (see Sect. 12.4.1.2). Today, they are eroded down to the level of explosive vents—diatremes (Kopecký 1971). Because these structures are mostly concentrated along the suture of two crystalline terranes, the diatremes can contain the famous Bohemian Garnet or pyrope, which is present at lens-like peridotitic bodies of the Teplá-Barrandian Crystalline Complex. Much more frequent water contact with the basaltic magma occurred at the surface. Lava flows were decelerated, and their parts destroyed by hyaloclastic processes, forming totally argillised clastics in some places. This effect was accompanied by mass movement transport into the grabens (see Sect. 12.4.1.4) and re-washing of sandy to fine-grained particles that sedimented together with non-volcanic material, such as re-transported Cretaceous sandstones, marlstones, and newly originated organic matter. Large volumes of volcano-sedimentary (epiclastic) rocks are known from this period. They contain limestones, diatomites and small-sized seams of brown coal. Paleontological finds

Fig. 12.3 The view of the Labe River valley and the southwest part of the České středohoří Mts. with conic hills of Lovoš (570 and 489 m a.s.l.) (*Photo* P. Raška)

document species such as frogs and fishes. The contemporaneous relief, still formed by tectonic activity, was concurrently flattened. This is the reason why volcanic rocks overlay either Santonian sandstones or Coniacian marlstones in different blocks. The period of dominantly effusive volcanism lasted from 37 to 28 Ma (Eocene to Oligocene—the Ústí Fm.).

Assumed change in the stress regime highly limited the ascent of, or closed the magma in the crust and magmatic evolution produced more explosive volcanism of non-olivine basaltic type (basalts, tephrites, foidites to trachybasalts) from a magmatic chamber. A large composite volcano grew, considerably overlapping the recent extent of the CS Volcanic Complex. A central, complicated, multiphase vent was completed by a number of smaller parasitic vents, situated over a wider area. The Volcanic Centre of the CS, now exhumed by the Labe River, was located between the towns of Ústí nad Labem and Děčín. It uncovers the deepest and the largest Cenozoic subvolcanic intrusions of the Bohemian Massif and a large, highly explosive vent (Mrlina and Cajz 2006). During the formation of the composite volcano, lahars (volcanic mudflows) originated on its slopes and reached distances of tens of kilometres from the volcanic centre. They destroyed vegetation—holes after trunks have been preserved. The lahar accumulations were rewashed in distal facies. Leaves were found in fine-grained sediment of this origin, documenting climatic conditions during this volcanic period, lasting from 31 to 25 Ma (Oligocene—the Děčín Fm.). Continuing tectonic activity placed individual blocks at different height levels and subsequent erosion destroyed a large volume of volcanic rocks, especially those of the composite volcano. Thus, this volcanic unit is presently preserved in erosional remnants only (Cajz ed. 1996). It is estimated that approximately 300–400 m of volcanic rock is missing at the central part of the CS.

Differentiated phonolitic and trachytic magmas reached mostly sub-horizontal lithological discontinuities at the basis of the Volcanic Complex or inside the volcanic rocks, and formed laccoliths and sills. Sub-vertical discontinuities contributed to stock formation. Linear vertical discontinuities (faults and fracture zones) allowed for the formation of dykes, especially in the vicinity of the Volcanic Centre. This is still the situation in the central part of the volcanic range. At the region near the cities of Most and Bílina (W to SW part of the CS), the phreatic structures of now eroded maars are present with intrusions of differentiated magmas (Hněvín and Bořeň Hills). In these regions, bodies of phonolite and trachyte are situated closer to the surface of that time and may have reached the surface itself. Possible lavas of this chemistry have not yet been proved.

The same alkaline volcanic rocks (of basaltic and phonolitic groups) occur in the NE continuation of the rift structure, outside the CS Volcanic Complex, and form the morphology of the Lužické hory Mts. Towards the Lusatian Overthrust, they are exposed at different levels of erosion, mostly in the subvolcanic position. This seems to be the main reason for the initial difference between these two mountain ranges—the prevalence of basalts (superficial volcanics) in the České středohoří Mts. and the prevalence of phonolites (intrusives) in the Lužické hory Mts.

After the intermission in volcanic activity, accompanied by erosion and flattening of relief, another tectonic remobilisation caused repeated ascent of primitive olivine-bearing basaltic magma to the surface. But, the volume of magmatic production was significantly smaller. Isolated Strombolian volcanoes were formed, very often being completed by lava effusions. This occurred in the period between 24 and 19 Ma (Upper Oligocene to Miocene—the Dobrná Fm.). This was also the time interval during which similar volcanic styles widened the interior of the larger part of the rift structure (western Bohemia, the Doupovské hory Mts.) and the Krušné hory/Erzgebirge Mts.—the future rift shoulder.

Afterwards, no long-term volcanic activity occurred. The formation of the Krušné hory/Erzgebirge Mts. in the north started and the river network changed. Lacustrine sedimentation of lignite continued in the basins in the E and SE part of CS, initially with re-transported material from the Volcanic Complex. Later, deposition of the lignite seam reached the margins of the Volcanic Complex and overlaid some of the eroded, volcanic rocks.

Another remobilisation allowed the ascent of primitive basaltic magmas for the last time in the span 13–9 Ma (Late Miocene—the Štrbice Fm.). The volumes of volcanic rocks are the smallest, as only a few Strombolian volcanoes were formed. Some of them were close to the Hawaiian volcanic style (Cajz et al. 2009). Several vents or dykes penetrated basinal sediments and caused caustic metamorphoses of lignite into coke and clays into porcelanites. However, much larger volumes of porcelanites were constituted by lignite seam burning than by magmatic effect (see Sect. 12.4.1.3).

12.3.3 Post-volcanic Evolution

Four volcanic episodes (Cajz 2000) produced a large amount of volcanic rocks, but they were eroded continuously since the time of their origin. The volcanic activity finished, but tectonic activity has continued. At that time, the paleo-Labe River started to erode a canyon across the volcanic range, and selective erosion occurred. River gravels were laid down to form several river terraces, and loess, as thick as 10 m, and blown-up sands were deposited along the volcanic range. Creep movement (solifluction) during interglacial periods left colluvial sediments of mixed material, and later, landslides and rockfalls have become significant phenomena of the recent relief formation.

12.4 Typical Landforms and Processes

12.4.1 Landforms on Volcanic Rocks

12.4.1.1 Vrkoč—Columnar Jointing of the Basanitic Dyke

Vrkoč is a basanitic dyke (Fig. 12.4a) with an inverted, fan-shaped, columnar jointing (charcoal kiln). Aged at 32.5 Ma, this possible fissure vent is classified as one of the oldest signs of volcanic activity (the Ústí Fm.). It is developed along the Vrkoč Fault (Raška et al. 2014), having displaced the base of volcanic rocks by more than 100 m vertically. Three cooling planes are supposed to be responsible for the shape of the columns. Two of them are represented by the sides of the dyke and the third occurred at the top, formed by rapidly cooled lava produced from this fissure as it moved to the water-saturated environment. A touristic trail leads to the Humboldt Viewpoint at the top of the dyke, which passes the highest waterfall in the mountain range, the 12 m high Vaňov waterfall, which originated on a morphological step formed by lava flow older than the dyke. Morphology of the active Čertovka landslide and a retaining wall are visible, as well.

12.4.1.2 Hnojnice—the Maar Vent

This is the location where the phreatic activity and maar formation was interpreted for the first time on the territory of the Bohemian Massif. In the early 1970s, this outcrop was interpreted as a maar rim and the uppermost part of the explosive vent of a maar (diatreme—Kopecký 1971). Morphology of the volcanic body could have lead to this idea. The erosion of surrounding rocks of that time (down to Cretaceous marlstones) places the recent outcrop into the deeper parts of the vent, however. Ascending basaltic magma along one of the faults of the Ohře Fault Field met water from sedimentary collectors and a phreatic to phreatomagmatic explosion took place. The thermal shock caused alteration of magma and an explosion, via an overheated steam, destroyed the wet marlstones. Their larger clasts had effects on the cooling of the diatreme as chilling centres (Fig. 12.4c). The radiating joints running from sedimentary clasts gave the name "Stone Suns" to this famous and unique outcrop.

Fig. 12.4 **a** Basanitic dyke of Vrkoč; **b** Porcelanites near the Dobrčice village; **c** "Stone Suns" near the Hnojnice village; **d** the Divoká rokle Gorge; **e** landslide at the D8 highway; **f** view on the western part of the České středohoří Mts. from the Deblík Hill (the Libochovany quarry) (*Photo* V. Cajz (**a**), P. Raška (**b**, **c**, **d**, **f**), R. Čmelík (**e**)

12.4.1.3 Dobrčice—the Porcelanite Quarry

Porcelanites, produced by caustic metamorphosis of Miocene basinal fill, are not a result of volcanic activity at this location. The covering of such sediment by a younger, hot, lava flow is not able to cause such large metamorphic changes, due to its rapid cooling and the thermal isolation properties of the sediment itself, as well as the cooled lava bottom. The former clastic sediments with organic material were baked by spontaneous combustion. The lignite seam with FeS_2 matter (marcasite) was exposed, and weathering (=oxidation together with water presence) produced sulphuric acid (H_2SO_4), which attacked the organic material up to the ignition of the fire. Burning caused varicoloured changes to former relatively grey, clayey sediment, as well as volume changes that resulted in folding of the formerly horizontal layering (Fig. 12.4b). The weathering of the lignite seam at this location can be partly associated with tectonic activity younger than 16 Ma (Cajz and Valečka 2010), but only the continuing erosion was responsible for the lignite seam exposure and initiation of the process.

12.4.1.4 Volcanic Sequence at the Divoká Rokle Gorge

This locality (Fig. 12.4d) illustrates the mechanism of gravitational mass movement (i.e. sliding) in the midst of the Volcanic Complex rocks over geologic history. During the oldest volcanic activity, a narrow tectonic graben was constituted and is now filled with more than 100 m of volcaniclastic material. The repeated mechanism of subaquatic slides of incoherent material coming from hyaloclastic brecciation was employed, together with re-sedimentation of Cretaceous rocks from the margins of the graben. The depression was filled, possibly up to the water level, but mass movement of destroyed volcanic material continued from the surrounding area. It is documented by the presence of well-shaped basaltic columns in the upper part of the outcrop, constituting the largest clasts. Afterward, during the evolution of the adjacent Volcanic Centre, dykes and sills were emplaced. The dykes are known by interesting mineralogical finds (e.g., large crystals of augite, tabular crystals of Ti-phlogopite, analcite, natrolite and calcite in vesicules after magmatic gases), and, now being exhumed from volcaniclastic material, they constitute steep walls at the outcrop.

Significantly after last volcanic activity, landslides associated with the recent relief evolution occurred several times. The upper part of the outcrop itself represents a headscarp of a prehistoric landslide. The lowermost part of the gorge is eroded in the accumulation of a younger landslide. As the gorge is 20–30 m deep in this part, we can observe the dynamics of the erosion in time, probably not longer than several hundreds of years.

12.4.2 Labe River Valley

The valley of the largest Czech river in the České středohoří Mts. represents a deep lineament induced by erosion into the tectonically raised terrain of the mountain range during the late Cenozoic. Reaching a depth of up to 500 m from the nearest peaks and width that usually does not exceed a few hundred metres, the valley bottom forms a narrow corridor that has concentrated major transportation routes and settlement. The geological evolution of the Labe River in the České středohoří Mts. dates back to the Pliocene/Pleistocene, when it started to cut through the so-called post-volcanic denudation level (ca 550–650 m a.s.l.), developed after the decay of volcanic activity. The Pretegelen/Biber(?) river terrace, preserved in the central part of the mountain range at the altitude of approximately 250 m (Balatka and Kalvoda 1995), would suggest immediate post-volcanic valley deepening by at least 300 m, with further deepening of 120 m to the current level of the valley bottom at approximately 130 m (cf. Fig. 12.5a, b). The real value may differ due to secular tectonic movements in the area. The fluvial terraces of the Labe River cannot be identified easily due to intensive Late Cenozoic erosion and anthropogenic transformation along the river channel. sMost

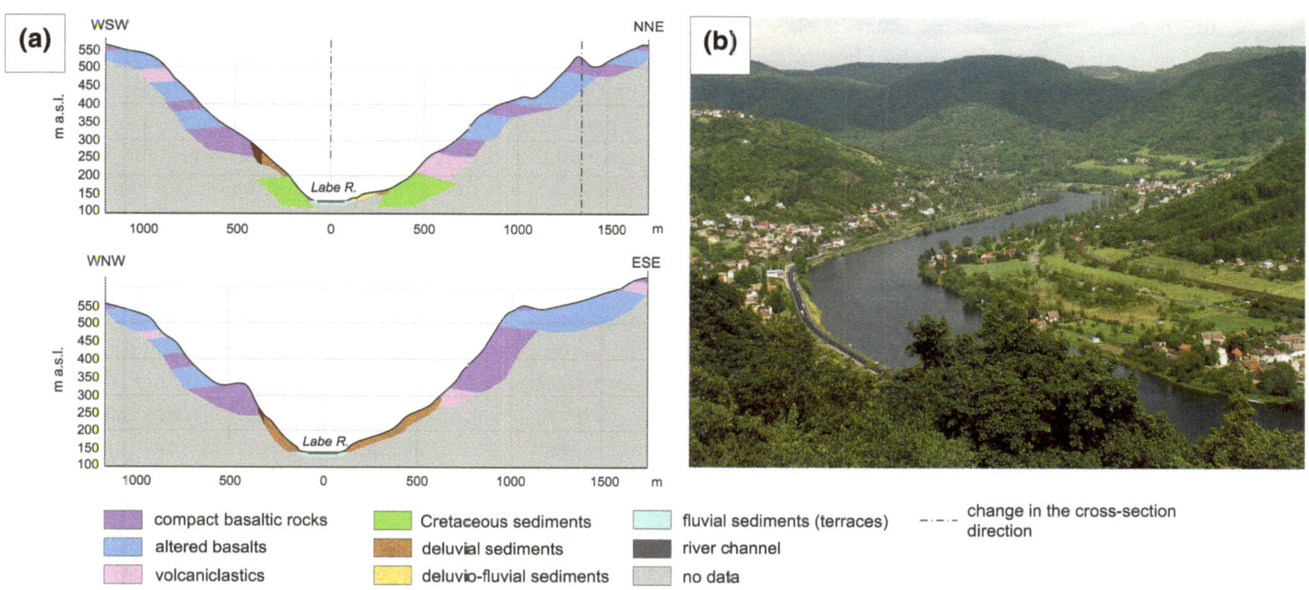

Fig. 12.5 **a** Typical geological cross-sections of the Labe River valley in the *central* part of the České středohoří Mts. **b** Labe River Valley approximately corresponding to the lower geological cross-section on *left* (*Photo* P. Raška)

of the modern built-up areas are located on the flat terrain of the Riss/Saalian terrace, preserved along the river channel and in depositional segments of river meanders. Older terraces are preserved sparsely due to tectonic activity and rejuvenation of steep slopes by sheet and gully erosion and mass movements.

Current slope morphology of the Labe River valley displays distinct steps predisposed by lithological change and variable resistance to weathering. Most typically, the generally steep slopes between the post-volcanic denudation level and a flat terrain of river terraces are diversified by rockwalls on compact basaltic rocks, whereas the altered basalts and volcaniclastics are represented by gently inclined slope segments. Waterfalls, up to 12 m high, are frequent where the small tributaries of the Labe River cut the slopes within the above-mentioned lithologies.

The current channel morphology is influenced by both human-induced changes, such as regulation (e.g., lock chambers), and by frequent floods. As the largest river in the Czech Republic, the Labe has a high average annual discharge and suffers from frequent high-magnitude floods (Table 12.1; Fig. 12.6). The most severe flood occurred in 2002, but six more floods with Q_{10} and higher were recorded in the area over the last 170 years. In addition to the impacts on built-up structures and human activities, the floods also cause geomorphologic changes to the river floodplain and lowest river terraces, mainly through significant deposition of sands and clays.

12.4.3 Mass Movements

12.4.3.1 Stony Debris Accumulations

Stony debris accumulations, built mainly by rockfall-derived clasts of volcanic rocks, are the important building blocks of landform diversity and biodiversity of the mountain range. Currently, they represent only 165 identified patches within the forested slopes (Fig. 12.7a) of the Labe River valley and individual hills, but they were much more frequent in the periglacial environment of the Pleistocene period, covering almost all slopes. The origin of present-day stony debris accumulations is linked with Dansgaard-Oeschger cycles at the end of the last glacial period in central Europe when the frequent freeze-thaw cycles accelerated rock cliff disintegration (Cílek 2000). In subsequent years, the accumulations have further developed with minor intensity and under increasing influence of other factors, such as bioturbations (Raška 2011). The distribution of stony debris accumulations is limited to an elevation interval between ca 200 and 600 m a.s.l. and slope inclination of 5–40°, with only a few locations exceeding these limits (Fig. 12.7b).

As a result of their geomorphic characteristics, including thickness frequently exceeding 10 m and the presence of an open void system between individual clasts, the stony debris accumulations also represent unique ecosystems in low- to mid-altitude mountains of Central Europe (Raška 2011). In many accumulations, the air circulation pattern within the open void systems is typical of negative thermal anomalies during the year, allowing survival of stenoecious species, including plant communities (Kubát 1971) or glacial relic mites (Zacharda et al. 2005).

Table 12.1 The flood events at the Labe River in Ústí nad Labem since the 1840s Century, based on the database of Povodí Ohře and Povodí Labe companies

Flood magnitude [Q recurrence—approximate water level] (cm)	Year
Q_{100}—1150	2002
Q_{50}—1080	1845
Q_{20}—980	1862, 1890, 2013
Q_{10}—910	1920, 1940
Q_5—820	1923, 1941, 2006
Q_2—690	1926, 1947, 1954, 1981, 1988, 2003

12.4.3.2 Landslides and Rockfalls

Together with the Outer Western Carpathians (see Chap. 27), the CS ranks among the areas of the Czech Republic most prone to different types of mass movements. Most typically, the predispositions for landsliding and rockfall are present in the Labe River valley and its tributary valleys, within their steep slopes frequently exceeding the angle of 30°. The lithology of these slopes, with alternating bodies of basaltic rocks, volcanic breccias and volcaniclastic material, as well as underlying Cretaceous sediments, results in significant mass strength anisotropies. The localities suffer from rockfalls, where the rivers and small streams cut their way through the solid rocks of the volcanic basement. In contrast, landslides are frequently triggered by heavy rainfalls (e.g., period of 1890–1900, mid-1920s, 1940s, 2002) that increase instability in the underlying lithologies. In this case, the sliding surface is usually located at the boundary between volcanic rocks and Cretaceous sediments (Rybář et al. 2000), but sliding inside the volcanic complex was documented as well (Raška et al. 2014).

The revision of historical records from the interval starting with the Enlightenment period and ending in mid-twentieth Century has revealed a total frequency of 1.39 landslide and rockfall events per year in this and adjacent areas (Table 12.2). Mass movements are well documented until the present day and have caused significant impacts to settlement and infrastructure. The Čertovka landslide

Fig. 12.6 Painted shooting target depicting disastrous flood on the Labe River in the Ústí nad Labem town in 1845 (see Table 12.1 for flood magnitude). *Source* Museum of the Ústí nad Labem town

occurred between 1994 and 1995 at the southern margin of the Ústí nad Labem town. The landslide needed to be stabilised with a retaining wall because the landslide body propagated toward the residential quarter and railway and main road corridor. Most recently, the deep-seated landslide has affected the highway (Fig. 12.4e), which is under construction, even though it was well known before its construction that the planned highway route crosses the dangerous, landslide prone area.

12.4.4 Human Impact on Landforms

While the higher altitude terrains, with mild climate in the north and east of the mountain range, evince the long-term evolution of forest covers and sparse settlement, the corridors of the Labe River and its tributaries and the south and west parts of the mountain range are characteristic of historical cultivation dating to as early as the Neolithic settlement, although agricultural cultivation has significantly

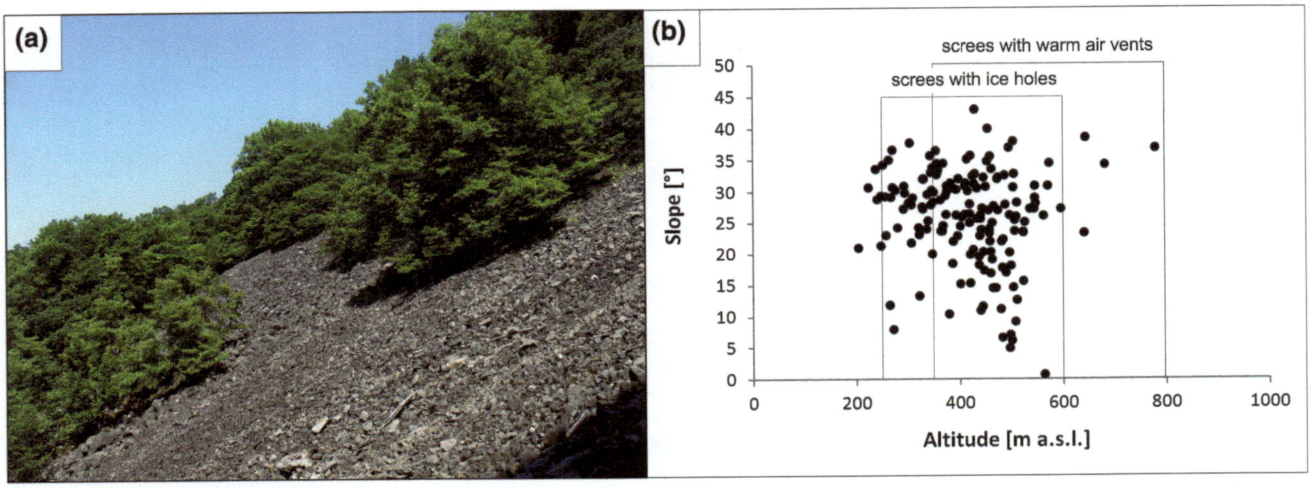

Fig. 12.7 **a** A patch of stony debris accumulation in the broad-leaved forest in the central part of the České středohoří Mts. **b** Distribution of 165 stony debris accumulations according to their average altitude and slope inclination (*Photo* P. Raška)

Table 12.2 Historical retrospective of landslide and rockfall events in the České středohoří Mts. and surrounding areas between 1770–1945

	Landslides	Rockfalls	Both events
Number of events	103	23	126
Recurrence (in years)	1.71	7.65	1.39
Number of localities	68	14	55

increased since the early Medieval period. Since at least the nineteenth Century, the human impact on landforms has gained new forms. The growing population raised demand for space (increase in built-up areas) as well as natural resources (quarrying), increasing the vulnerability of growing communities to various natural hazards (floods, mass movements). This resulted in the adoption of different prevention measures. Both rivers (Labe and Bílina) and small streams were regulated after the disastrous floods in 1845 and 1920, and flash floods in 1897 and the mid-1920s (Fig. 12.8a, b). For instance, the channel of the Labe River has undergone significant changes due to construction of lock chambers S of the town and river ports of Ústí nad Labem (Fig. 12.8c).

Perhaps the most significant human impact caused to specific volcanic landforms was quarrying. A multi-temporal assessment of quarrying resulted in the identification of 80 visible abandoned quarries, but the historical records mention approximately 160 quarries present in the area. While the historical quarries were of small extent and located regularly throughout the region, the number of quarries gradually decreased through time (Table 12.3), and only eight quarries are operating today. The present-day quarries have unprecedented area and massive annual volumes of extracted stone and are more concentrated along transportation routes (Fig. 12.4f). Furthermore, the present-day quarries have immense direct impact on landform and habitat destruction, as well as on the visual quality of the landscape when compared to historical, rather local quarrying (Beranová and Raška 2012).

12.5 Geoheritage

The scientific significance of the České středohoří Mts. was discovered by German geologist and mineralogist F.A. Reuss in the late eighteenth Century. Since then, the area has been subject to geological studies contributing to the development of many geological disciplines. In addition to the regional protoscientific and scientific reports, the area has also been studied by internationally renowned scientists, such as Alexander von Humboldt. The most significant contribution to the knowledge of the geological evolution of the mountain range, however, was made by Professor J.E. Hibsch (1850–1940) of the Agricultural Academy in Liberda near the town of Děčín. Between 1896 and 1932, he conducted detailed field geological mapping of the area, resulting in 19 maps at a scale of 1:25000 with detailed explanatory texts, which still represent a significant source of our current knowledge (see Fig. 12.9). These early works

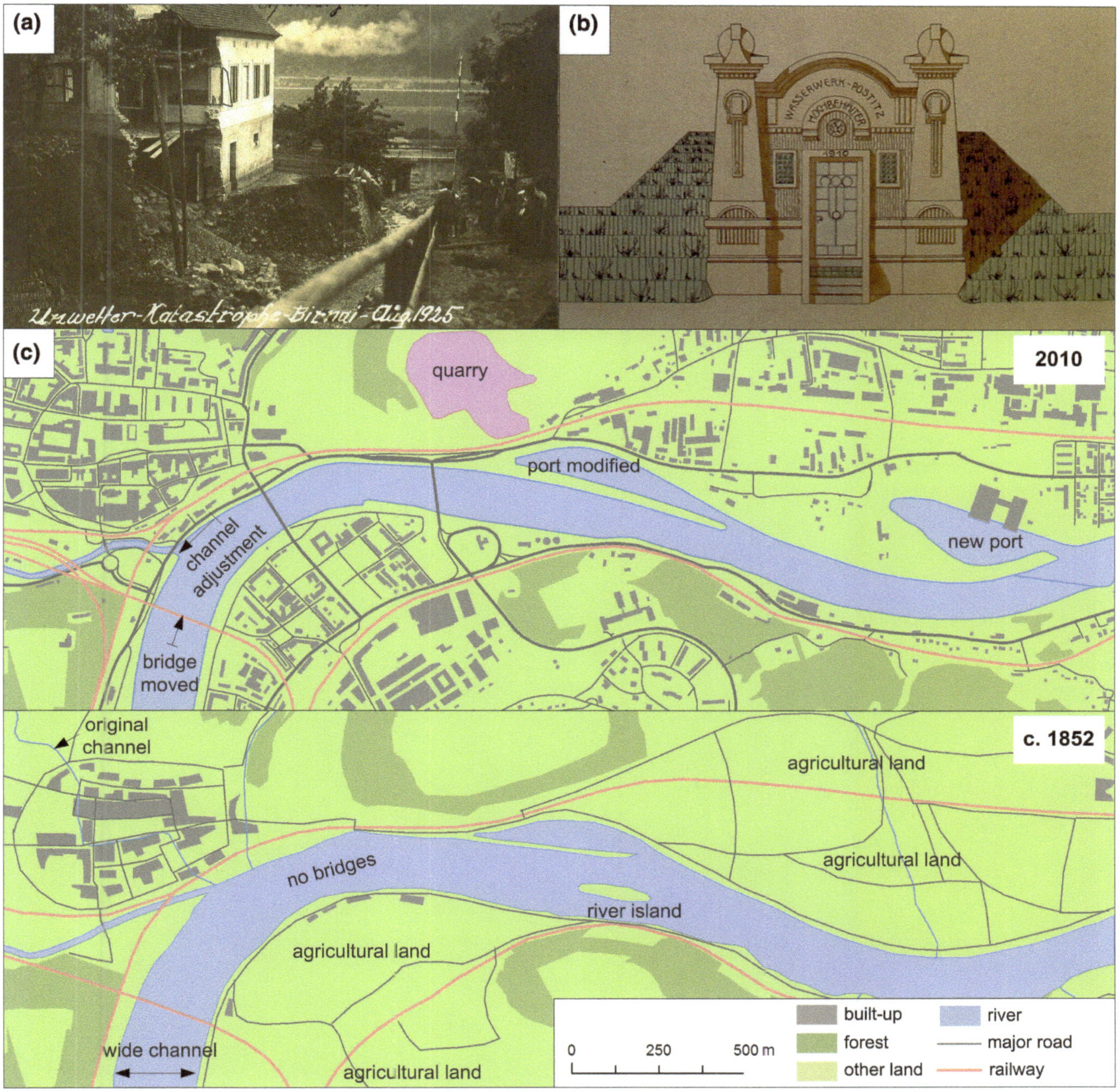

Fig. 12.8 **a** Impacts of 1925 flash flood at the Labe River tributary in Brná nad Labem. **b** Project of piping of the Klíšský potok Brook in the Ústí nad Labem town from 1923. *Source* Archives and Museum of the Ústí nad Labem town. **c** Anthropogenic transformation of the Labe River and its surrounding in the Ústí nad Labem town between c. 1852 and 2010

also helped to establish the (Central) European Volcanic Province (Wimmenauer 1974) as a parallel to other world regions.

Following the geological attention, the vegetation assemblages have gained recognition since the nineteenth Century. Volcanic landforms such as solitary cones, scree slopes and rockwalls, which, together with microclimatic factors, allowed the unique species to inhabit the area, were firstly described in a classic work of botanist K. Domin in 1904. In the early 1970s, when the mountain range was considered for nature and landscape conservation, the most unique habitats were classified into four groups: (a) meadow assemblages including orchids (*Orchidaceae Juss.*), (b) scree forests and scree slopes with xerothermic species,

Table 12.3 Number and area of abandoned quarries and their visual impact to landscape (according to Beranová and Raška 2012)

Year	Number of quarries[a]	AAQ (m^2)	TAQ (km^2)	VIS (%)	VIS in PA (%)
1840s	80	2999	0.2	55.32	72.69
1953	55	17919	1.0	55.17	65.89
2007	37	59350	2.2	48.54	64.28

Explanation [a]numbers refer to those quarries that were visible in a certain year; *AAQ* average area of a quarry; *TAQ* total area of quarries; *VIS* share of the České středohoří Mts. area visually affected by quarries; VIS in *PA* share of the protected areas in the České středohoří Mts. visually affected by quarries. The geodatabase of quarries was based on old maps and aerial image analysis. The visual impact was assessed using the *viewshed* analysis in ArcGIS 10.1 upon the DEM, resulting in an extent of areas, from which quarries are/are not visible

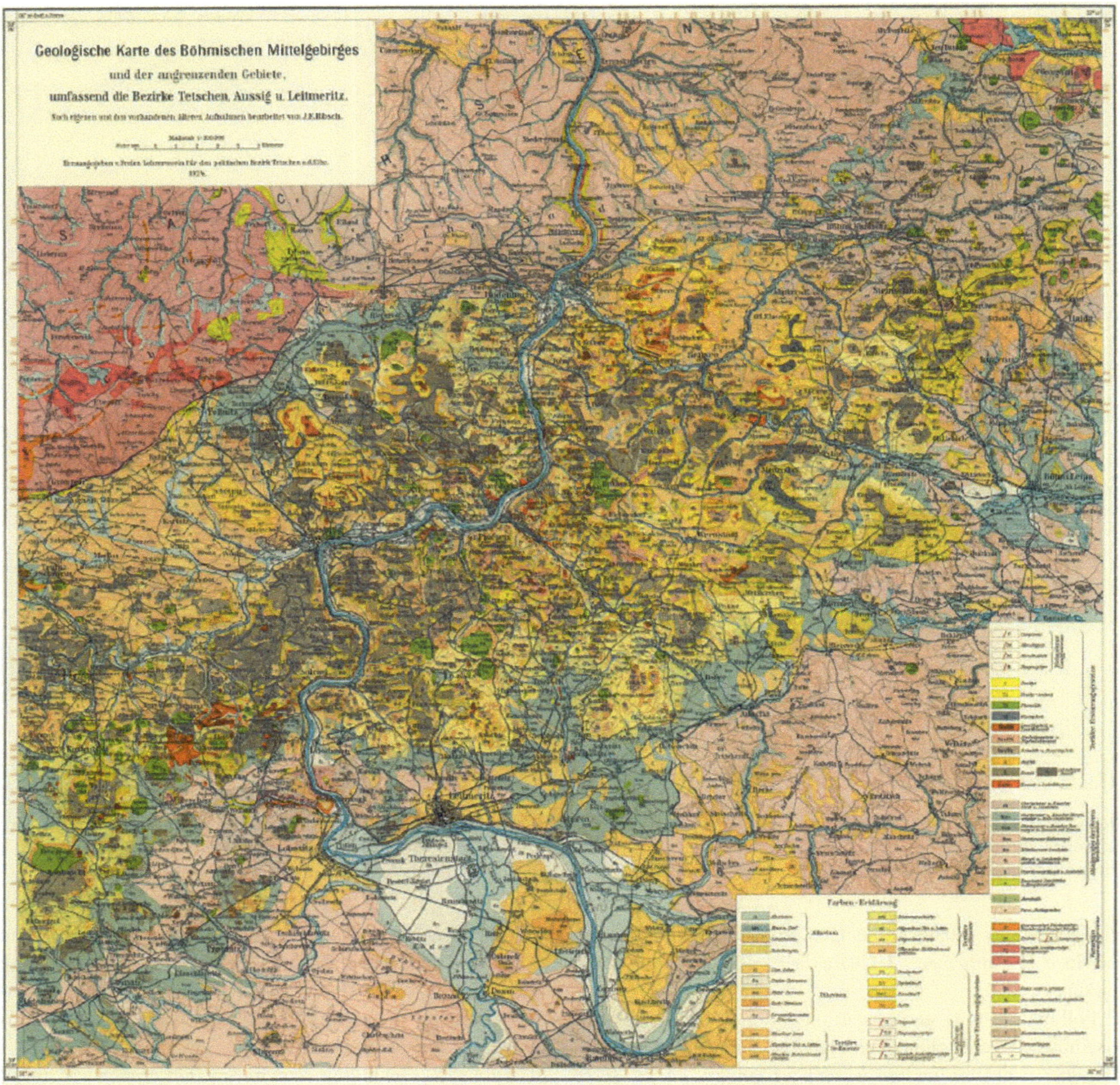

Fig. 12.9 Geological map of the České středohoří Mts. by J. E. Hibsch (1924) in scale of 1:100 000, based on more detailed maps 1:25 000 of the same author

Fig. 12.10 Painting "Pilgrimage church near the Ziegenberg close by the Ústí nad Labem town" by Ernst Gustav Doerell (approx. 1873). The Kozí vrch Hill (*Ziegenberg*) on the right is an exhumed trachytic laccolith. The Divoká rokle Gorge is visible just above the tower of the church

(c) ice holes and undercooled scree slopes with psychrophilic and cryophilic species, and (d) south-facing rock-mantled slopes with xerothermic species.

The unrivalled blend of lithologies, landforms and organisms was further strengthened by the cultural value of the landscape with prehistoric and medieval archaeological relicts, traditional architecture and literary and painterly reflection of natural beauties (Fig. 12.10).

Recently, appreciation has been given to features that originated due to human activities as well. Tens of abandoned quarries from the nineteenth and early twentieth Century represent displays of geological, mineralogical and geotechnical peculiarities and are preserved as geo-heritage sites and endangered species habitats. With its extraordinary example of basalt disintegration, Radobýl hill (399 m a.s.l.), at the southern margin of the mountain range, has been used as a comparative reference for Giants′ Causeway in Ireland, when it was proposed to the UNESCO World heritage list (UNESCO 1985).

12.6 Conclusions

The České středohoří Mts. is an extraordinary area with geomorphologic features and a research history that significantly contributes to the complexity of our knowledge of volcanic terrains. Ranging from basaltic lava flows, lahars and eroded subsurface volcanic bodies to recent fluvial and anthropogenic landforms, the mountain range exposes a complexity of features valuable for geo-heritage and geo-education. The ecological characteristics of these abiotic features enabled the evolution of unique communities and assemblages, resulting in legal protection of the area since 1976. As an outstanding mountain range, it also draws attention of many tourists from the Czech Republic and abroad, studying volcanism and volcanic terrains.

Acknowledgments Parts of the chapter result from the research projects GAČR No. 13-02080P, GAAV No. IAA300130612 and IAA3013102. Roman Čmelík is greatly acknowledged for providing the aerial photo of a study site. We are greatly indebted also to Peter Daško for the general photo of the area and to the staff of regional museum and archives for their help during the search in archival sources.

References

Balatka B, Kalvoda J (1995) Evolution of the Elbe River valley in the Děčínská vrchovina Mts. (in Czech). Sborník České geografické společnosti 100:173–192

Beranová L, Raška P (2012) Multitemporal assessment of visual impacts of abandoned quarries in the České středohoří Mts. using GIS tools. In: Brtnický M, Brtnická H, Foukalová J, Kynický J (eds) Regenerace - revitalizace - rekultivace krajiny. Mendel University, Brno, pp 106–113 (in Czech)

Cajz V (1996 ed.) České středohoří - geologická a přírodovědná mapa 1: 100 000 [The České středohoří Mts. - Geology and Nature Features; Das Böhmisches Mittelgebirge - Geologische Wanderkarte]. – Český geologický ústav, Praha

Cajz V (2000) Proposal of lithostratigraphy for the České středohoří Mts. volcanics. Bull Czech Geol Survey 75:7–16

Cajz V, Adamovič J, Rapprich V, Valigurský L (2004) Newly identified faults inside the volcanic complex of the České středohoří Mts., Ohře/Eger Graben, North Bohemia. Acta Geodyn. Geomater., 1, 2(134):213–222

Cajz V, Rapprich V, Erban V, Pécskay Z, Radoň M (2009) Late Miocene volcanic activity in the České středohoří mountains (Ohře/Eger Graben, northern Bohemia). Geol Carpath 60(6):519–533

Cajz V, Valečka J (2010) Tectonic setting of the Ohře/Eger Graben between the central part of the České středohoří Mts. and the Most Basin, a regional study. J Geosci 55:201–215

Cílek V (2000) Scree Slopes and Boulder Fields of Northern Bohemia: Origin, Processes and Dating. In: Kubát K et al.: Stony Debris Ecosystems. Acta Universitatis Purkinianae 52, Studia Biologica IV, UJEP, Ústí nad Labem, 5–18

Kopecký L (1971) Excursion guide into the maar-area in the western periphery of the České středohoří Mts. Československá společnost pro mineralogii a geologii, Praha. (in Czech)

Král V (1966) Geomorphology of central part of the České středohoří Mts. Rozpravy Československé akademie věd, Řada matematických a přírodních věd 78:1–65 (in Czech)

Kubát K (1971) Ice holes and ventarols in the České středohoří Mts. II. Vlast. Sbor. Litoměřicko 8:67–89 (in Czech)

Mrlina J, Cajz V (2006) Subsurface structure of the volcanic centre of the České středohoří Mts., North Bohemia, determinated by geophysical surveys. Studia Geophys. Geod. 50:75–88

Raška P (2011) Paleogeomorphic significance and environmental change of scree slopes in the České středohoří Mts. Ph.D. thesis, Masaryk University, Brno. (in Czech)

Raška P, Klimeš J, Dubišar J (2015): Using local archive sources to reconstruct historical landslide occurrence in selected urban regions of the Czech Republic: examples from regions with different historical development. Land Degradation and Development 26 (2):142–157, doi:10.1002/ldr.2192

Raška P, Hartvich F, Cajz V, Adamovič J (2014) Structural setting of the Čertovka landslide (Ústí nad Labem, Czech Republic): morphostructural analysis and electrical resistivity tomography. Geol Q 58(1):85–98

Rybář J, Vilímek V, Cílek V (2000) Process analysis of deep slope failures in České středohoří neovolcanites. Acta Montana Ser AB 8 (115):39–46

UNESCO (1985) Nomination to the World heritage list - Giants′ Causeway. ICOMOS UNESCO, Paris

Wimmenauer W (1974) The alkaline province of central Europe and France. In: Sorensen H (ed) The Alkaline Rocks. Wiley, London, pp 286–291

Zacharda M, Gude M, Krause S, Hauck Ch, Molenda R, Růžička V (2005) The Relict Mite Rhagidia *(Acari, Rhagidiidae)* as a Biological Cryoindicator of Periglacial Microclimate in European Highland Screes. Arct Antarct Alp Res 37:402–408

The Kokořín Area: Sandstone Landforms Controlled by Hydrothermal Ferruginization

13

Jiří Adamovič

Abstract

The area between the Jizera and Elbe rivers in north-central Bohemia features a southerly dipping package of sedimentary rocks of Cretaceous age, subjected to modest uplift throughout the Quaternary. The Kokořín sandstone (Middle to Upper Turonian) comprises five superimposed bodies with generally high permeability and low-to-medium resistance to weathering. Most valleys are dry, shaped by occasional flash floods and gravitational processes. Disintegration of vertical cliff faces is dominated by salt weathering. Specific landforms develop on sandstones cemented by iron oxyhydroxides of hydrothermal origin. These form thin (centimetres to metres), sheet-like bodies, either subvertical or bedding-parallel, and give rise to structural plateaus and mesas, steep erosional ridges and mushroom rocks. The variety of small-scale relief forms on ferruginous sandstone is unique at a global scale. The highest elevations in the landscape are formed by exhumed subvolcanic bodies. The Kokořín Area is a perfect example of a sandstone-dominated erosional landscape whose high relief complexity is largely due to the contrasting resistance of rocks to weathering.

Keywords

Kokořín area • Cretaceous sandstones • Landforms • Hydrothermal activity • Ferruginization

13.1 Introduction

Kokořínsko (Kokořín area) is the area of excellent sandstone exposure between the towns of Mělník and Česká Lípa to the north of Prague. Its name is derived from the name of the Kokořín Castle, a historical landmark founded above the stream of Pšovka in the thirteenth century. The extent of the Kokořín area tends to be synonymized with the limits of the Kokořínsko Protected Landscape Area (PLA), established in 1976 with the aim of conservation of unique geomorphological features, flora and fauna. The Kokořín area covered by this contribution is best characterized as the Kokořínsko PLA area before its 2014 expansion to the northeast. South of Dubá, it includes the basins of the southerly flowing streams of Pšovka and Liběchovka. In the north, areas of vast sandstone exposure form a divide between the Obrtka flowing to the Elbe River (=Labe in Czech) and the Robečský potok flowing north, to the river of Ploučnice. Most of the valleys are dry, lying well above the present groundwater table. In the geomorphological subdivision of Bohemia, the whole Kokořín area belongs to the unit of the Ralská pahorkatina, subunit of the Dokeská pahorkatina and the district of the Polomené hory (Balatka and Kalvoda 2006). In the southeast, the rugged relief passes to the flat, gently sloping unit of the Jizerská tabule Table. The altitudes range between 160 m in the Liběchovka valley and 614 m at Vlhošť Hill (Balatka and Sládek 1981), the average annual temperatures are 7–8.5 °C, and the annual precipitations are 480–680 mm (Mikuláš et al. 2007).

J. Adamovič (✉)
Institute of Geology of the Czech Academy of Sciences, v.v.i, Prague, Czech Republic
e-mail: adamovic@gli.cas.cz

T. Pánek and J. Hradecký (eds.), *Landscapes and Landforms of the Czech Republic*, World Geomorphological Landscapes, DOI 10.1007/978-3-319-27537-6_13

The Kokořín area lies in the central part of the Bohemian Cretaceous Basin, the sedimentary fill of which has been subjected to several episodes of tectonic deformation, emplacement of volcanic rocks, and hydrothermal activity. In quartzose sandstones, circulation of hydrothermal fluids resulted in a localized dissolution of quartz grains and re-precipitation of silica in the inter-granular space, known as *silicification*. Even more notable are the effects of *ferruginization*: introduction of ferrous iron by fluids from deep-reaching convective cells and its precipitation in the form of iron oxyhydroxides. The importance of ferruginization and silicification of Cretaceous sandstones for the landscape evolution in the Kokořín area has been recognized already by Bruno Müller who published a classical paper on this subject almost 90 years ago (Müller 1928).

13.2 Geological Background

13.2.1 Cretaceous Sandstones

The floor of the Bohemian Cretaceous Basin actively subsided in the Cenomanian to Santonian (ca. 97–84 Ma) to allow deposition of a sediment pile >1000 m in thickness, mostly in a shallow marine setting (e.g., Uličný et al. 2009). Coarse detrital material was supplied to the basin from the north, where crustal blocks dominated by granitic plutons were constantly rising along faults of the Elbe Zone. This paleotectonic situation produced a series of sand-dominated wedges reaching from the northern basin margin to its axial part. Geometries of these wedges were controlled primarily by sea-level fluctuations of either eustatic or tectonic origin.

The Kokořín sandstone, outcropping in the wide region between the Jizera and Elbe rivers in north-central Bohemia, occupies a distal part of a clastic wedge protruding towards the southeast along the axis between Blíževedly and Mšeno (Fig. 13.1). This wedge is ranked within the basin-wide Jizera Formation sensu Čech et al. (1980) whose age is Middle to Upper Turonian based on fossil fauna (ca. 92–89 Ma). The Kokořín sandstone consists of five superimposed sandstone bodies generally coarsening from base to top, each of them 30–80 m thick (Fig. 13.2). The bases of the bodies consist of fine-grained silty sandstone, mostly bioturbated, deposited in lower shoreface conditions (depths 30–40 m). Middle and upper parts of the bodies consist of medium- to coarse-grained sandstones with trough cross-bedding and other sedimentary structures indicative of deposition in upper shoreface conditions (depths 5–30 m). Beds of gravelly sandstone and conglomerate, deposited on gravelly beaches, form the very tops of the bodies but may equally occur on tops of lower-order upward-coarsening cycles within each body (Adamovič 1994). Over 90–95 % of the sandstone is composed of quartz, 1–3 % is feldspar, <1 % is glauconite. The sandstone also contains a variable amount of kaolinite. No pervasive mineral cement is usually present, and the strength of the sandstone is rendered by the packing of quartz grains due to sediment compaction, and by poor quartz cement at points of mutual grain contact. The overall thickness of the Kokořín sandstone package is 230 m in its axial part.

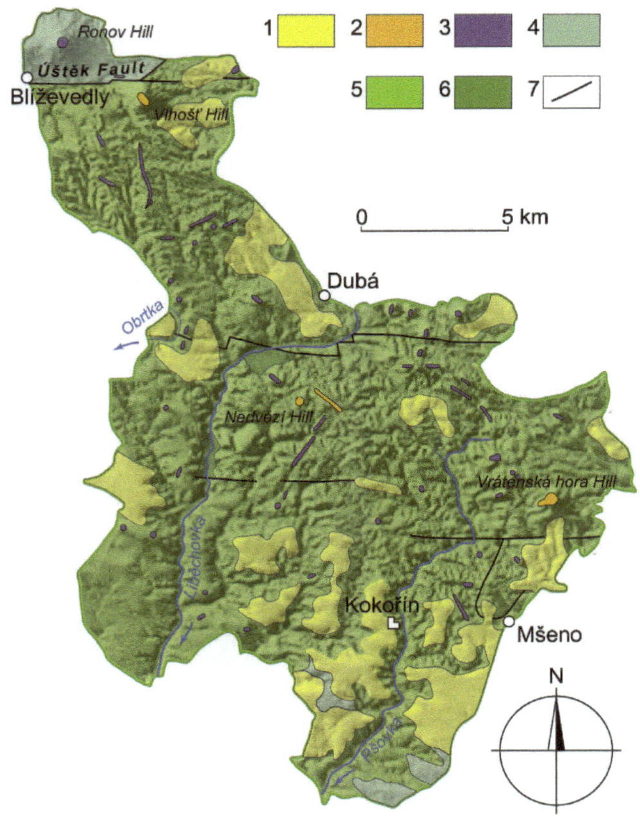

Fig. 13.1 Geology of the Kokořín area on a shaded relief map. Quaternary: *1* loess and loess loam; Tertiary: *2* trachyte and phonolite, *3* basalt; Mesozoic, Upper Cretaceous: *4* marlstone (Upper Turonian to Coniacian), *5* sandstone (Middle to Upper Turonian), *6* calcareous siltstone (Lower to Middle Turonian); *7* prominent faults. Shaded relief: Czech Institute of Geodesy and Cadastre, www.cuzk.cz

13.2.2 Tertiary Volcanics

Magmatic activity in the Kokořín area was related with crustal relaxation and formation of the Ohře Rift graben in the Late Eocene to Early Miocene (ca. 35–20 Ma; Adamovič and Coubal 1999; Ulrych et al. 2011). Phonolitic and basaltic magmas were emplaced almost simultaneously, as suggested by sparse radiometric data. Phonolites typically form laccoliths (Vrátenská hora) or plugs (Nedvězí and Vlhošť hills), surrounded by smaller basaltic dykes. Two basaltic intrusive centres with a number of plugs, dykes and

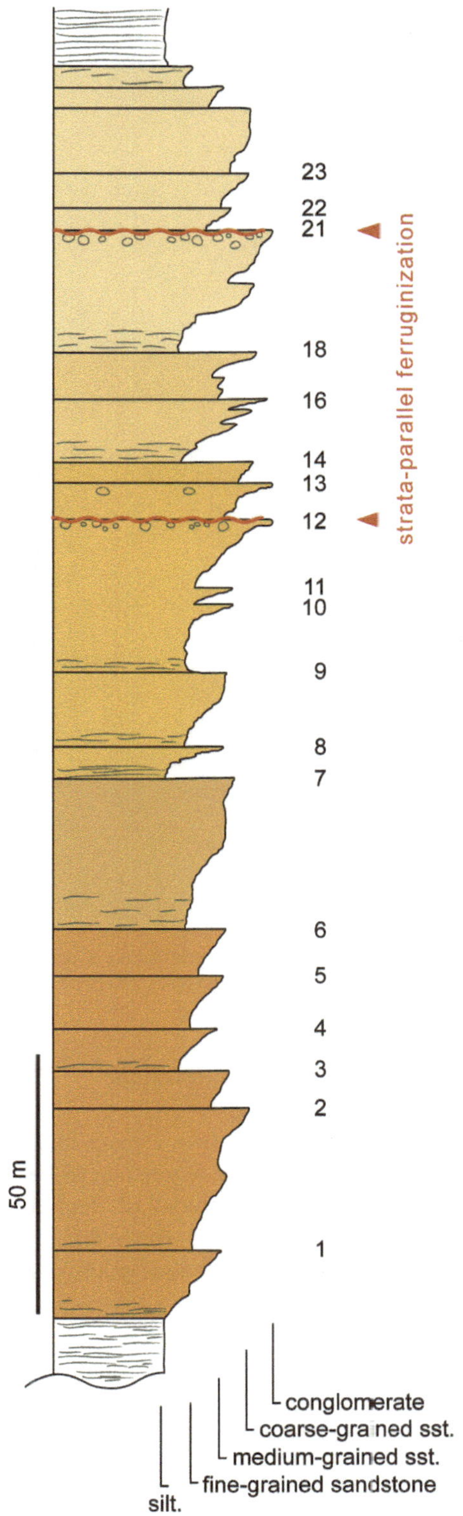

Fig. 13.2 Vertical section of the Jizera Formation in the axial part of the Kokořín sandstone body NW of Kokořín, showing the most complete succession of superimposed sandstone units. Numbers 1–23 denote the upward-coarsening cycles, each of the five sandstone units (higher-order cycles) are marked in different colours. Positions of beds usually cemented by iron oxyhydroxides are marked by arrows

bodies of intrusive breccia lie to the SE of Dubá (Houska and Beškov) but thinner (<1 m) basaltic dykes are scattered throughout the whole area. No superficial volcanic products have been reported from the Kokořín area yet.

Smaller intrusive bodies became subjected to hydrothermal alteration (kaolinization) immediately following their emplacement and produce no prominent landforms. On the other hand, large bodies are responsible for the origin of basaltic necks with more-or-less circular bases, rising 30–120 m above the surrounding sandstone plateaus. The most important of these hills are Lipový kopec (471 m), Kuželík (482 m) and Zámecký vrch (432 m) SE of Blatce, Velký Beškovský kopec (474 m) and Korecký vrch (465 m) SE of Dubá, and Ronov (552 m). Basaltic dyke swarms typically give rise to elongated hills and ridges like Supí hora (434 m), Dubová hora (397 m) and Kostelec (433 m). Phonolitic bodies tend to produce large, symmetrical cupola-form hills with radial valley patterns independent of the local tectonic setting: Vrátenská hora (508 m), Nedvězí (458 m), and Vlhošť (614 m)—the highest peak in the area. For the locations of the hills, see Figs. 13.1, 13.7 and 13.8.

The interaction between alkaline magma and water-saturated porous host rock is generally associated with a number of phenomena, most of which are well documented in outcrops in the Kokořín area. These include various thermal effects associated with the origin of columnar jointing in host sandstone (Fig. 13.3), deformation bands due to shear stress produced by penetrating magma, chemical corrosion of quartz grains and subsequent silica re-precipitation in inter-granular spaces. Especially, the redistribution of silica has a great impact on the development of sandstone macro- and microrelief at a local scale. Even more important in this respect are the processes of syn-magmatic and early post-magmatic ferruginization, covered in Sect. 13.3.

13.2.3 Tectonic Setting

Since its deposition, sedimentary fill of the Bohemian Cretaceous Basin has been affected by several paleostress fields of compressive or tensional character (Coubal et al. 2015). By effect of these stresses, the body of the Kokořín sandstone became segmented into separate tectonic blocks, and magma ascent was permitted along tensional fractures. Tectonic dips of strata of 2–3° towards the south are observed in the northern part of the area; these are interpreted as a result of the oldest stage of ductile deformation which took place in the latest Cretaceous times (Coubal and Klein 1992). Analogous dips of ca. 1° towards the south-southwest are common in the southern part of the area, although these may locally reach as much as 4° due to subsequent strike-slip movement along faults NNE–SSW and NW–SE and the related rotation of

Fig. 13.3 Columnar-jointed sandstone in the proximity of a basaltic dyke at Čiř hill SW of Dřevčice (*Photo* J. Adamovič)

Fig. 13.4 A dense fracture zone accompanying a bedding-plane slip fault in Černý důl Valley W of Vrátenská hora (*Photo* J. Adamovič)

smaller (ca. 1.5 by 1.5 km) tectonic blocks. Steeply and moderately dipping dense fracture zones are very typical for sandstones in the Kokořín area (Fig. 13.4); they accompany minor thrust faults and bedding-plane slip faults (Adamovič and Coubal 2012).

The most intensive faulting in the Kokořín area took place during the period of crustal relaxation and rifting in the Late Eocene to Early Miocene. The southern marginal fault of the Ohře Rift graben, locally called the Úštěk Fault, transects the northernmost part of the Kokořín area. The subsided northern block with Ronov Hill (a throw of ca. 400 m) exposes Upper Turonian to Coniacian siltstones with thin sandstone intercalations, building a flat landscape completely different from that in most of the Kokořín area. Minor shear faults of the Úštěk Fault zone cut the Kokořín sandstone in the slopes of Vlhošť Hill and display silicified, polished and striated fault planes.

13.3 Ferruginization Process

In bodies of ferruginous sandstone, cement fills all voids in the rock. It is composed mostly of goethite (α-FeOOH), less commonly with some proportion of lepidocrocite (γ-FeOOH) or hematite (α-Fe_2O_3). Three different morphological types of bodies of ferruginous sandstone are distinguished in the Kokořín area (Fig. 13.5):

- Subvertical planar bodies, mostly developed as fillings of open joints and faults. As revealed by geological survey, such fractures almost invariably intersect a body of volcanic rock further along their strike in the Kokořín area (Adamovič et al. 2001). Where developed at the very contact with a basaltic dyke, the layer of ferruginous sandstone passes directly to a layer of hard, ferruginized basaltic rock.

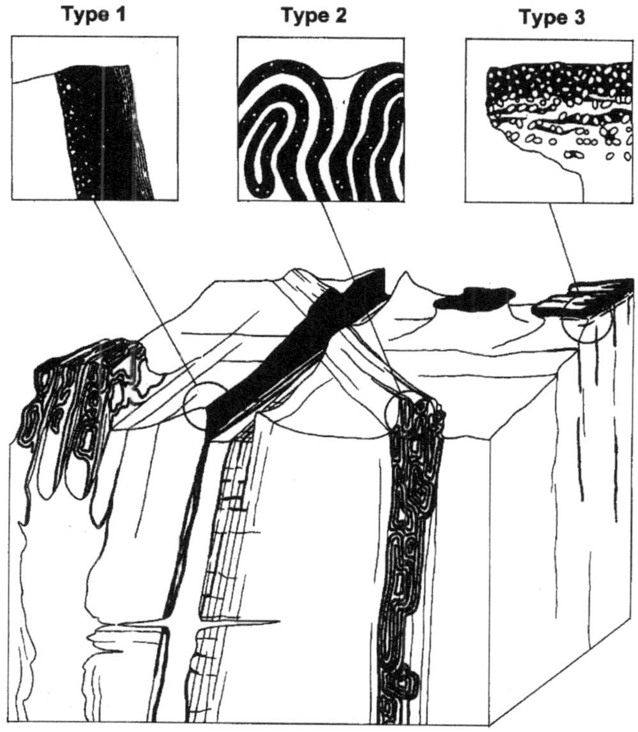

Fig. 13.5 A schematic block-diagram showing the different macro-morphological types of ferruginization in relation to dykes and pipes of volcanic rocks and fracturing of the host sediment in the Kokořín area. *Type 1* subvertical planar bodies lining contacts of basaltic dykes or filling joints and faults; *Type 2* undulating ferruginous crusts; *Type 3* strata-bound subhorizontal bodies. After Adamovič et al. (2001)

- Thin, undulating crusts, often tube- and sheath-like in shape, clustered within subvertical zones several metres or first tens of metres wide in sandstone. Characteristic morphotypes are single-layered tubes and their bundles, or multi-layered tubes with low plunge angles, trending parallel or somewhat oblique to the strike of the zone. This pattern occasionally becomes rather distorted, with locally developed crusts of vermiform, rose-like or cauliflower-like appearance with variable trends of axes of tubes. Some of these zones stretch at a distance of several kilometres. Their genesis is explained by the Liesegang phenomenon describing periodical precipitation structures formed by diffusion (e.g. Ortoleva 1994). Its application to the precipitation of iron compounds in the Kokořín sandstone was first pointed out by Graber (1904). The Liesegang phenomenon explains the formation of ferruginous rings and spherical concretions by diffusion in static-fluid conditions, i.e. in open joints and in water-saturated sandstone, and the formation of tube-like crusts by diffusion in sandstone with a steady fluid flow.
- Strata-bound subhorizontal bodies with ferruginous cement, which follow subhorizontal zones of higher permeability in sandstones, such as conglomerate beds or major unconformities. Some of these bodies reach several km^2 in area (Močidla Gorge) and a thickness of 1 m or more (max. 6 m thick). Ferruginous cement is massive or composed of a number of undulating laminae.

The origin of ferruginous cement has been interpreted in different ways in the past (see review in Adamovič 2002b). Watzel (1862) suggested that iron was produced by the heat effect of magma emplacement. His "contact theory" has been soon abandoned and the role of groundwater circulation along intrusions of basaltic rocks has been adopted for the origin of morphological types 1 and 2 (Müller 1914, 1928; Zima 1950). Strata-bound bodies of ferruginous sandstone have been, however, long considered products of precipitation of iron from groundwater, reflecting its successive levels during a progressively lowering base level during the Quaternary (e.g. Müller 1928; Zima 1950). Iron was believed to have been released from weathered volcanic rocks or from iron-bearing detrital and authigenic minerals contained in the sandstone (glauconite, pyrite). Later studies (Adamovič 1988; Adamovič et al. 2001) found a close spatial relationship between the distribution of ferruginous cement in type-3 ferruginization and basaltic bodies, suggesting a hydrothermal origin of iron-bearing fluids and their ascent along dyke contacts simultaneous with, or slightly post-dating, the dyke emplacement. In this process, types 1 and 2 functioned as feeders for type-3 strata-bound bodies. The source of iron for all types of ferruginization is provided by basaltic rocks, with a contribution from granitic rocks in the basement, as suggested by trace element contents in the goethite cement.

Association of massive type-3 ferruginization with high-permeability sediments is well illustrated by the Močidla Gorge NW of Mšeno. Here, ferruginization is best developed below the top of a conglomerate bed overlain by less permeable fine-grained sandstones, in the axial part of a flat anticlinal structure (Adamovič 1988). Large bodies of intrusive breccia to the east of the gorge are interconnected with the body of ferruginous conglomerate by joints striking NW–SE and NE–SW, some of which contain thin basaltic dykes. Many of them show linings of iron oxyhydroxides and are interpreted as additional ascent paths for iron-bearing fluids (Fig. 13.6).

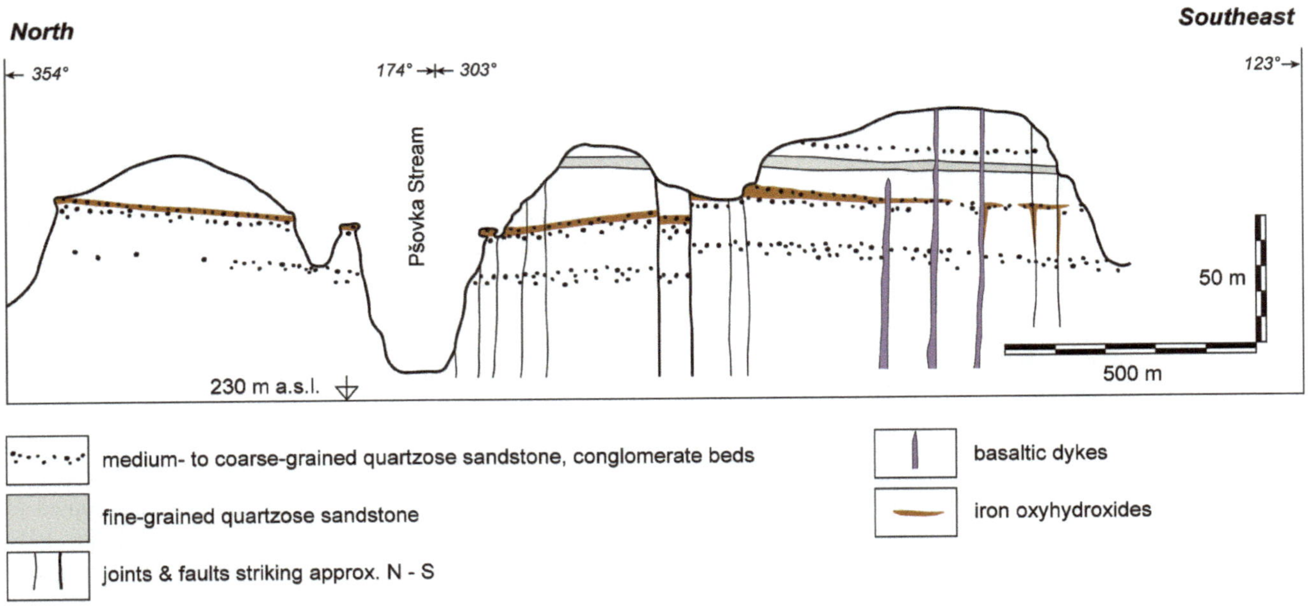

Fig. 13.6 A longitudinal section of the Močidla Gorge NW of Mšeno. Iron oxyhydroxides precipitated in the form of pervasive cement in a flat-lying conglomerate bed. Ferruginous conglomerate is cut by two N–S-striking faults, clearly post-dating the hydrothermal mineralization. Iron-bearing fluids were supplied from a large body of intrusive basaltic breccia at the eastern end of the Močidla Gorge

13.4 Landforms Due to Ferruginization

13.4.1 Structural Plateaus and Ridges

Ferruginous sandstones and conglomerates show a higher strength and a higher resistance to weathering compared to uncemented sandstone (Figs. 13.7 and 13.8). Almost horizontal strata-bound bodies of ferruginous sandstone (type 3) therefore tend to produce flat surfaces. Two principal beds of ferruginous conglomerate can be distinguished in the Kokořín sandstone between Mšeno and Dubá. Although showing an uneven areal distribution of ferruginous cement depending on the location of the sources of iron-rich fluids, each of these two beds was responsible for the origin of flat-topped hills—structural plateaus (Fig. 13.7). Tops of hills developed on the stratigraphically lower bed can be traced from the Močidla Gorge area in the SE (300 m a.s.l.) in northwesterly direction as far as to Supí hora (ca. 370 m a.s.l.) and northerly-lying Rač hills (380–390 m a.s.l., Fig. 13.9a). Those on the stratigraphically upper bed include flat surfaces at Zadní Žluč (408 m a.s.l.) and Nedvězí hills (ca. 400 m a.s.l.). Tops of the hills formed by ferruginous conglomerate are inclined very gently (1–3°) south to south-southwest, following the regional tectonic dip.

Structural ridges controlled by the courses of vertical zones of ferruginization are very common throughout the Kokořín area. They rise ca. 10–50 m above their surroundings, being topped by a straight line of cliffs, isolated pillars and walls of ferruginous sandstone up to 15 m high (e.g. Houska SE of Blatce, Čap Hill—Fig. 13.9c). Slopes of some of the ridges are covered with blocky talus (Kamenný vrch). The role of structural ridges is most notable in the northern part of the Kokořín area between Dubá and Vlhošť Hill (Fig. 13.8).

13.4.2 Pillars and Mushroom Rocks

Ridges dissected by a regular network of vertical joints are sites favourable for the origin of “rock cities” in which groups of pillars separated by grikes were formed by erosion due to the lowering of base level. Although smaller in area (max. 200 by 200 m) than those in other sandstone regions in Bohemia, some of the “rock cities” are perfect labyrinths providing shelters and hideouts (e.g. Kobylka and Pastuší gorges NW of Mšeno, several sites in the valley of Planý důl, Rač plateau). Free-standing pillars, separated from the main rock mass by deeply eroded fracture zones, are common. Where the tops of the ridges and hills are flat, formed by a resistant ferruginous layer, pillars on their rims tend to produce mushroom-shaped forms. Their stipes of gravelly quartzose sandstone become thinner due to the combined effect of salt and frost weathering, while their caps of ferruginous conglomerate show a higher resistance to weathering. Occasionally, perforations beneath the caprock have the character of rock arches and rock windows.

The best example of mushrooms rocks is present at the mouth of the Močidla gorge (Fig. 13.9d) where the pillars

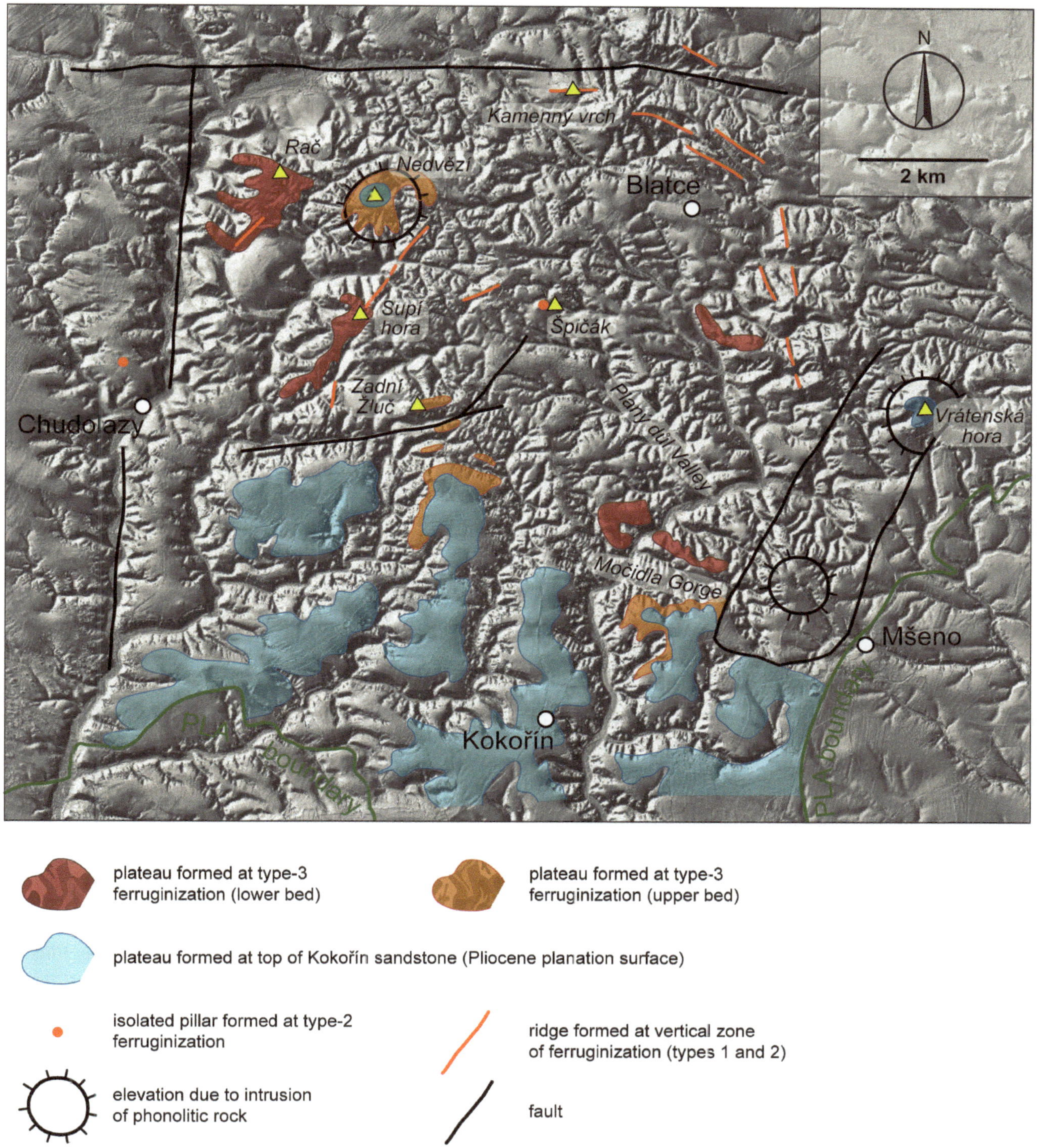

Fig. 13.7 A shaded relief map of the southern part of the Kokořín area between Kokořín and Dubá showing the main structural plateaus and ridges due to ferruginization (*red* and *orange red*) and relicts of the Pliocene planation surface (*blue*). Shaded relief: Czech Institute of Geodesy and Cadastre, www.cuzk.cz

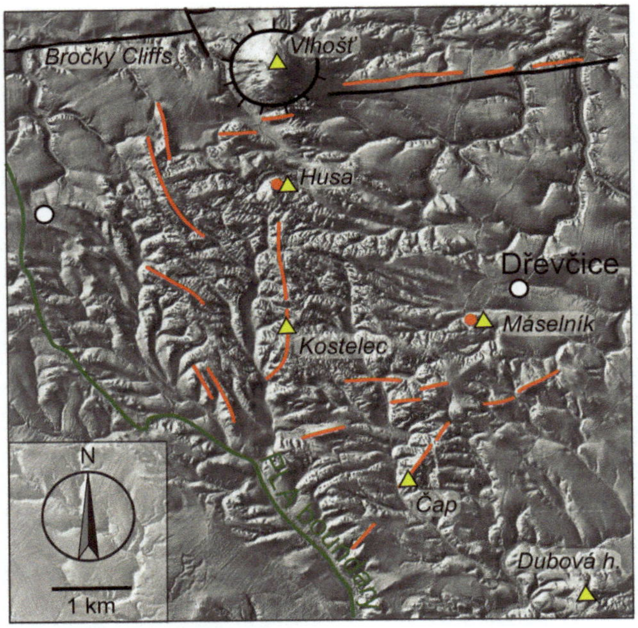

Fig. 13.8 A shaded relief map of the northern part of the Kokořín area between Dubá and Vlhošť hill. Some of the ridges are clearly connected with the courses of vertical zones of ferruginization. Shaded relief: Czech Institute of Geodesy and Cadastre, www.cuzk.cz

are 12–15 m high and their tops lie 60 m above the bottom of the Pšovka valley (Balatka and Sládek 1981). Mushroom rocks at the same level of ferruginous conglomerate lie only 1 km farther north, in the Vojtěšský důl. Mushroom rocks of the same type commonly appear along the circumference of relict plateaus topped by the upper layer of ferruginous conglomerate, such as Zadní Žluč, Supí hora and Rač hills.

Prominent rocky crests and hilltops, occasionally dissected into separate pillars, form at occurrences of ferruginous sandstone of morphological type 2. The most notable examples in the southern part of the Kokořín area are the hills of Špičák near Střezivojice and Kamenný vrch near Dubá (Figs. 13.7 and 13.9; Adamovič et al. 2001; Adamovič 2002a). Mushroom rocks and pinnacles at sites of type-2 ferruginization represent prominent landmarks in the northern part of the Kokořín area (Figs. 13.8 and 13.9), e.g. at Čap, Husa, and Vlhošť hills, and in the Bročky Cliffs east of Blíževedly.

13.4.3 Microforms

The variety of small-scale relief forms specific to ferruginous sandstone observed in the Kokořín area can be hardly paralleled elsewhere. This is largely due to the wide variety of forms of type-1 and type-2 ferruginization in sandstones now exposed at the surface (Fig. 13.9). These structurally controlled microforms include:

- systems of concentric rings of positive relief on steep or vertical surfaces of ferruginized joint planes;
- mammillary surfaces on sandstones with scattered iron-rich enclaves, such as spherical ferruginous concretions or ferruginized bivalve shells;
- "wrinkled" surfaces consisting of parallel ridges formed by mm- to cm-thick laminae of ferruginous sandstone: wavy, subspherical or contorted, very complex patterns. Irregular concentric patterns have been previously often referred to as "rock roses" (Balatka and Sládek 1981), e.g. Špičák and Kamenný vrch hills (Fig. 13.9g). Bowl-shaped depressions max. 2 m in diameter, resembling weathering pits, may form on horizontal surfaces in the centres of the concentric structures (Husa Hill, Fig. 13.9e);
- tubes and their bundles with straight, parallel axes. Tube walls are formed by mm- to cm-thick laminae of ferruginous sandstone. The tubes are max. 0.5 m across and 5 m in length, producing positive relief with their outer surfaces, and negative relief with their interiors (east of Vlhošť, Fig. 13.9i).

13.5 Geomorphic Evolution in the Quaternary

The highest structural plateaus in the Kokořín area follow the top of the Kokořín sandstone and are probably Pliocene in age (Balatka and Sládek 1981). Their distribution (Fig. 13.7) suggests that they were originally interconnected to form a more-or-less continuous planation surface extending from the Elbe River in the south to the Úštěk Fault in the north, now completely destructed in the northern part

Fig. 13.9 Landforms on ferruginous sandstone/conglomerate in the Kokořín area. **a** The Rač plateau topped by a bed of ferruginous conglomerate; **b** An uneven layer of ferruginous sandstone lining a basaltic dyke–Laka valley N of Mšeno; **c** An isolated pillar of ferruginous sandstone near a basaltic dyke–Čap Hill; **d** Mushroom rocks topped by a bed of ferruginous conglomerate —"Pokličky" at the mouth of Močidla Gorge; **e** Bowl-shaped depressions at vertical columns of ferruginous sandstone (also called "geyser stalagmites" in literature) —Husa Hill; **f** "Wrinkled" surface due to parallel laminae of ferruginous sandstone—Strážný vrch NW of Mšeno; **g** Parallel laminae of ferruginous sandstone constituting large concretionary forms—Kamenný vrch; **h** Subtle elevations on top rock surface due to thin bedding-parallel laminae of ferruginous sandstone; **i** Tube-shaped forms with parallel axes at extremely elongated single-layered concretions of ferruginous sandstone—Švábský důl east of Vlhošť Hill (*Photo* J. Adamovič)

of the area. The only elevations reaching above this surface were those formed by exhumed Oligocene to Miocene subvolcanic bodies.

In the Pleistocene, a network of valleys was incised in the Kokořín sandstone as a result of base level lowering. During the last 2.6 My, the valley of the Vltava River around Prague deepened by 125 m with the highest estimated incision rate in the Mid Pleistocene (3–10 cm/100 years; Balatka and Kalvoda 2006). Further downstream, the valley of the Elbe River near Děčín deepened by 180–200 m during the same time interval (Kalvoda and Balatka 1995). This suggests that the incision of major river valleys in northern Bohemia was at least partly driven by tectonic uplift whose intensity increased towards the north; this also explains the higher degree of destruction of the Pliocene surface in the northern Kokořín area. In the Kokořín area, incision of major valleys by 120–200 m can be observed, with two to four levels of cliffs, each representing one of the superimposed bodies of the Kokořín sandstone (see Sect. 13.2.1).

The general southerly inclination (SSW to SE) of the Pliocene planation surface was responsible for the early south-directed drainage of the Kokořín area, mediated by the Pšovka, Liběchovka and Obrtka streams. Apparent offsets of the courses of the Liběchovka and Obrtka streams along the E–W line south of Dubá were explained by piracy inflicted on the hypothetical westerly flowing "Dubá River" in Mid Pleistocene by Müller (1939) but may equally result from the progressive destruction of a Pliocene fault scarp along the E–W-striking fault (see Fig. 13.1 for a general view). Alluvial plains of the Pšovka and Liběchovka streams lie at the level of the groundwater table within the Kokořín sandstone, being fed by valley springs. In contrast, floors of their tributary valleys lie well above the groundwater table and host no permanent streams. They deepen during periodical flash floods while their slopes are modelled by a steady gravitational redeposition of weathering products.

Initial valleys, broad and shallow, were possibly formed by stream erosion in the Early Pleistocene, at a higher groundwater table and a slow uplift rate. The main valley incision occurred in the Mid Pleistocene, based on parallelism with the Elbe River valley (Balatka and Sládek 1964). The tributary valleys could not keep pace with the increasing rate of stream erosion in the main valleys and their broad transverse profiles changed into narrow ones. Slot canyons tens of metres deep were locally formed, following fractures and fracture zones and often showing zig-zag courses depending on tectonic offsets of the master fracture. In the Late Pleistocene, loess was deposited on the plateaus and on right-hand slopes of the south-directed valleys, thereby pushing their courses somewhat to the east and giving them asymmetrical transverse profiles (Balatka and Sládek 1964; 1981). Higher erosion rates in the main valleys at the Pleistocene/Holocene transition can be inferred from the presence of hanging valleys with almost 10 m drops in the valley floor at their mouths.

Directions of the uppermost reaches of valleys and gorges conform to the strikes of fractures. In contrast, courses of middle and lower reaches of the valleys are largely controlled by the orientation of tectonic dips. This is well demonstrated by radial patterns of valleys around phonolitic intrusions associated with periclinal dips of strata in the host sandstone (Na Rovinách elevation NW of Mšeno, Vrátenská hora, Fig. 13.7).

Deepening of valleys within the Kokořín sandstone was uneven due to the presence of harder, cemented layers several metres thick in the sedimentary succession. This is demonstrated by stepped longitudinal profiles of upper and middle reaches of valleys (Balatka and Sládek 1964) where waterfalls up to 6 m high develop after torrential rains. Hydrothermal hardening of sandstone, i.e. ferruginization and silicification related to volcanic processes in the Tertiary, had a profound inhibitory effect on the progress of headward erosion. Structural plateaus were formed at levels of areally extensive beds of ferruginous sandstone (type-3 ferruginization), and ridges and necks similar to those on sub-volcanic bodies were formed at type-1 and type-2 bodies of ferruginous sandstone. In the northern part of the Kokořín area, the distribution of volcanic rocks and ferruginous cement seems to be the most effective factor controlling the valley network pattern (Fig. 13.8).

13.6 Geological Heritage and Human Impacts

Throughout the history, the dense network of deep, cliff-lined valleys in the Kokořín area provided shelters for prehistoric hunters, places of solitude for hermits, hideouts for outlaws but—most of all—refuges for local villagers during wartime. The cliff dwellers left behind a number of inscriptions and carvings on vertical rock faces. Sandstone clifftops and tops of volcanic necks were perfect sites for the construction of fortifications.

Elevations of ferruginous sandstone were used as observation points in the Mesolithic (Strážník Cliff near Dřevčice) and sites for timber-made castles (Čap and Rač hills) in the Middle Ages (Gabriel and Panáček 2000). In the northern Kokořín area, fragments of ferruginous sandstone were used by Mesolithic hunters and foragers in their dwellings under rock shelters: as boiling stones in fireplaces and as heat-accumulation stones (Svoboda 2003, V. Cílek, pers.-comm.). Silicified sandstone was extensively used for the production of millstones in the eighteenth–nineteenth centuries (e.g. Kostelec and Supí hora hills).

In the mid-nineteenth century, the Kokořín area became renowned as a Romantic landscape—an attitude supported

Fig. 13.10 The incised valley of the Pšovka stream with three levels of sandstone cliffs has been mentioned in the writings of the famous poet Karel Hynek Mácha. The Kokořín Castle (*left*) was reconstructed in the early twentieth century in a style enhancing the Romantic spirit of the valley (*Photo* J. Adamovič)

especially by the journeys and writings of the famous Romantic poet Karel Hynek Mácha (1810–1836). Two of his masterpieces were situated in this area: the poem *Máj* (May) and the novel *Cikáni* (Gypsies). The rugged, densely forested country started to be visited by pilgrims of all sorts, and legends appeared on mysterious knights, hidden treasures, robbers and magic creatures. The ruin of the Kokořín Castle—the central point of Mácha's experience—was purchased by the Špaček family and reconstructed in a Romantic style in 1911–1918 (Fig. 13.10).

Tourist activities in the southern part of the Kokořín area flourished after World War I, being supported by the Czech Tourist Club. Sandstone country around Dubá, then known under its Romantic name "Dubá Switzerland" (*Daubaer Schweiz*) became increasingly visited by members of the German tourist club *Nordböhmisches Excursions-Club* after its foundation in Česká Lípa in 1877. Much like in the south, a boom of tourism around Dubá is linked with the sign-posting of tourist paths after World War I. Rock climbing activities in the Kokořín area started in the same period: the earliest ascents date to the year 1909. They culminated in the 1980s, and over 1000 pillars are registered by climbers today.

13.7 Conclusion

Among the many sandstone areas in the Bohemian Cretaceous Basin, the area between Kokořín and Česká Lípa displays the widest variety of products of hydrothermal ferruginization. It equally provides a number of good outcrops documenting the close link between the distribution of ferruginous cement and the origin of landforms of all different sizes. Such a textbook of "hydrothermal geomorphology" can be hardly paralleled at a global scale. Geosites related to hydrothermal sandstone alteration therefore deserve due attention. Conservation of geomorphic features and living nature has been rendered since the establishment of the Kokořínsko Protected Landscape Area in 1976: the most valuable landforms are protected within the status of nature monuments; others lie in areas protected within the category of nature reserves.

Acknowledgments This contribution was prepared with the support of Institutional Research Plan RVO 67985831 of the Institute of Geology CAS, v.v.i. and by project 16-19459S from the Czech Science Foundation. The colour graphics were kindly drawn by Jana Rajlichová.

References

Adamovič J (1988) Distribution of iron oxides in the Kokořín sandstones. Student thesis, Faculty of Science, Charles University, Praha, 37 pp

Adamovič J (1994) Paleogeography of the Jizera Formation (Late Cretaceous sandstones), Kokořín area, central Bohemia. Sbor geol Věd, Geol 46:103–123 (in Czech)

Adamovič J (2002a) Kokořín area. In: Adamovič J, Cílek V (eds) Ironstones of the Bohemian Cretaceous Basin. Knih Čes speleol spol 38, Praha, pp 7–36 (in Czech) http://adamovic.euweb.cz/katalog.pdf

Adamovič J (2002b) Occurrences of Fe-oxyhydroxides in sandstones of the Bohemian Cretaceous Basin. In: Adamovič J, Cílek V (eds) Pseudokarst proceedings 2. Knih Čes speleol spol 37, Praha, pp 7–40

Adamovič J, Coubal M (1999) Intrusive geometries and Cenozoic stress history of the northern part of the Bohemian Massif. Geolines 9:5–14

Adamovič J, Coubal M (2012) Bedding-plane slip movements in Cretaceous sandstones in the Bezděz area. Zpr geol výzk v roce 2011:9–12 (in Czech, English abstract)

Adamovič J, Ulrych J, Peroutka J (2001) Geology of occurrences of ferruginous sandstones in N Bohemia: famous localities revisited. Geol Saxonica – Abh Mus Miner Geol Dresden 46/47:105–123

Balatka B, Kalvoda J (2006) Geomorphological regionalization of the relief of Bohemia. Kartografie Praha, Praha 80 pp

Balatka B, Sládek J (1964) Development of valleys in pseudokarst rocks in south-eastern part of Polomené Hills, northern Bohemia. Československ Kras 15(1963):37–50 (in Czech, English abstract)

Balatka B, Sládek J (1981) Geomorphology of the Kokořínsko Protected Landscape Area and vicinity. Bohemia centralis 10:7–53 (in Czech)
Čech S, Klein V, Kříž J, Valečka J (1980) Revision of the Upper Cretaceous stratigraphy of the Bohemian Cretaceous Basin. Věst Ústř úst geol 55(5):277–296
Coubal M, Klein V (1992) Development of the Saxonian tectonics in the Česká Lípa region. Věst Čes geol Úst 67(1):25–45
Coubal M, Málek J, Adamovič J, Štěpančíková P (2015) Late Cretaceous and Cenozoic dynamics of the Bohemian Massif inferred from the paleostress history of the Lusatian Fault. J Geodyn 87:26–49
Gabriel F, Panáček J (2000) Castles of the Česká Lípa district. Argo, Praha, 204 pp (in Czech, German abstract)
Graber HV (1904) Geologisch-petrographische Mitteilungen aus dem Gebiete des Kartenblattes Böhm.-Leipa und Dauba, Zone 3, Col. XI der österr. Spezialkarte. Jb k-k geol Reichsanstalt 54, 3–4:431–460
Kalvoda J, Balatka B (1995) Chronodynamics of the Labe River antecedence in the Děčínská vrchovina Highland. Acta Montana, A 8(97):43–60
Mikuláš R, Adamovič J, Hoffmann A, Beran L, Honců M (2007) Kokořín and Doksy area (Czech Republic). In: Härtel H, Cílek V, Herben T, Jackson A, Williams R (eds) Sandstone Landscapes. Academia, Praha, pp 339–343
Müller B (1914) Der Grossteich bei Hirschberg in Nord-Böhmen. Der Geologische Aufbau des Hirschberger Teichgebietes. Monogr Abh Int Revue ges Hydrobiol Hydrogr 5/III:1–81
Müller B (1928) Der Einfluss der Vererzungen und Verkieselungen auf die Sandsteinlandschaft. Firgenwald 1(4):145–155
Müller B (1939) Erdgeschichte und Bau des Sudetenlandes. Deutscher Boden 9:1–150
Ortoleva PJ (1994) Geochemical self-organization. Oxford Monogr Geol Geophys 23:1–411
Svoboda JA ed (2003) Mesolithic of northern Bohemia. Dolnověstonické studie 9:1–305 (in Czech, English abstract)
Uličný D, Laurin J, Čech S (2009) Controls on clastic sequence geometries in a shallow-marine, transtensional basin: the Bohemian Cretaceous Basin, Czech Republic. Sedimentology 56:1077–1114
Ulrych J, Dostal J, Adamovič J, Jelínek E, Špaček P, Hegner E, Balogh K (2011) Recurrent Cenozoic volcanic activity in the Bohemian Massif (Czech Republic). Lithos 123:133–144
Watzel C (1862) Beschreibung der im Horizonte vom Böhmisch = Leipa vorkommenden Gesteine und Mineralien. Progr k-k Ober-Gymn B.-Leipa Schl Schulj 1862:3–28
Zima K (1950) The geological conditions of the southwestern part of the Polomené hory. Sbor Stát geol Úst Českosl Rep, Geol 17:289–339 (in Czech, English abstract)

Jizerské Hory—an Interplay of Rock Control, Faulting and Inland Glaciation in the Evolution of a Granite Terrain

14

Piotr Migoń

Abstract

The granite massif of Jizerské hory in the northern part of the Czech Republic is an excellent area to examine different controls on the geomorphic evolution, acting over a long time scale. A protracted period of denudation in the early Cenozoic was followed by uplift and tilting of the plateau to the south in the Neogene. In consequence, the subdued upland topography is truncated by a steep, fault-generated escarpment from the north, incised by valleys with bedrock and boulder-filled channels. Characteristic medium-size landforms include tors and crags of multiple shapes, boulder blankets, waterfalls, whereas weathering pits and other forms of microrelief typify many exposed granite surfaces. In the Pleistocene the Scandinavian ice sheet reached the footslopes of the mountains and remodelled bedrock elevations into roches moutonnées. Very small glaciers may have existed on lee-slopes of upland elevations, fed by snow blown-in by westerly winds. Torrential floods and occasional debris slides and flows are main geomorphic processes acting nowadays.

Keywords

Jizerské hory Mts. • Planation surfaces • Neotectonics • Granite landforms • Ice sheets • Debris flows

14.1 Introduction

Granite massifs abound in the territory of the Czech Republic, occupying approximately 10 % of the country. They are distinctive units within the geological structure of the Bohemian Massif and occur in diverse topographic settings, from lowlands (e.g. Žulovská pahorkatina, see Chap. 21) to the most elevated parts of the Krkonoše—the highest mountain terrain in the country. It has long been recognized that landform inventories within granite massifs are not only very specific and even endemic to this type of rock, but they also record various phases of geomorphic evolution of the region, going back to at least mid-Cenozoic (Czudek et al. 1964; Demek 1964). Among these numerous granite massifs, the one that builds most of the Jizerské hory has a particularly rich assemblage of medium-size and minor rock-controlled landforms of different origins, providing a nearly complete catalogue of granite landforms identified worldwide (Twidale 1982; Twidale and Vidal Romaní 2005; Migoń 2006). Looking from a broader perspective, the present-day morphology of the Jizerské hory bears evidence of various geomorphic processes, environments, and controls, hence escapes simple landscape classification by origin. The principal aims of this chapter are thus to reveal the complexity of granite morphology of the area and to outline its long-term geomorphic history, including the contemporary morphodynamics.

14.2 Setting and Location

Jizerské hory (=mountains) are a transboundary region within the Sudetes, which themselves are an extended tract of mountainous and upland terrains in the north-eastern part

P. Migoń (✉)
Institute of Geography and Regional Development, University of Wrocław, pl. Uniwersytecki, 1 50 137 Wrocław, Poland
e-mail: piotr.migon@uwr.edu.pl

T. Pánek and J. Hradecký (eds.), *Landscapes and Landforms of the Czech Republic*, World Geomorphological Landscapes, DOI 10.1007/978-3-319-27537-6_14

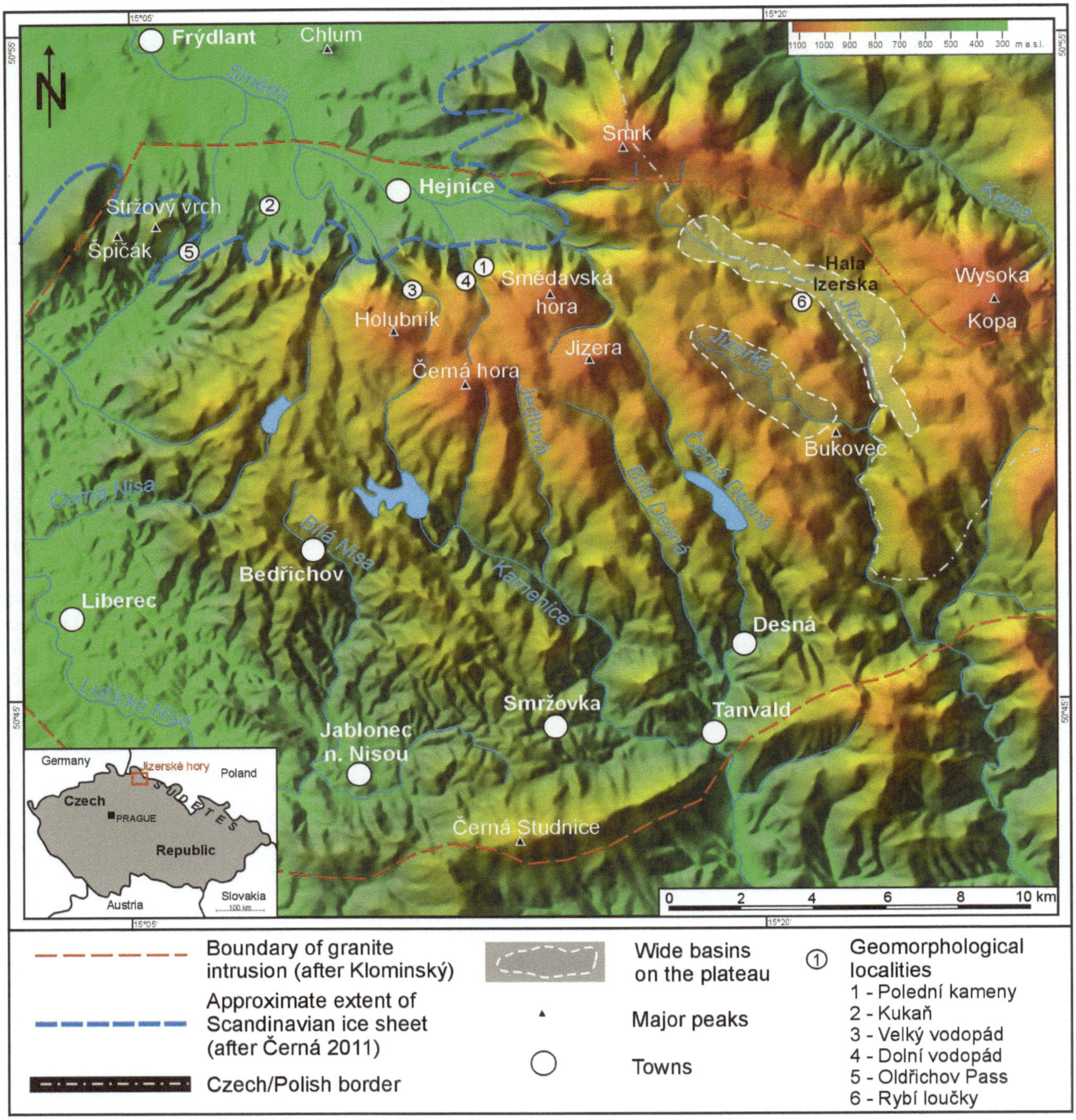

Fig. 14.1 General relief of the Jizerské hory and location of sites and features mentioned in the text

of the Bohemian Massif (Fig. 14.1). Altitudes in the Sudetes reach 1603 m a.s.l. (Mt Śnieżka in the nearby Krkonoše Mountains) and a few other massifs exceed 1000 m a.s.l., including the Jizerské hory. From a morphotectonic point of view, the Sudetes is a large fault-bounded block, uplifted in the late Cenozoic and broken along numerous fault lines into a spatially complicated pattern of minor horsts, tilted blocks, subdued uplands, and intramontane basins (Migoń 2008). However, these effects of differential tectonics have been superimposed onto a pre-existing relief which developed through long-term deep weathering, saprolite removal, and differential erosion, and continued to be shaped by rock-controlled denudation processes concurrently with uplift and subsidence (Demek 1975; Migoń 2011).

Therefore, the contemporary morphology of the Sudetes is a palimpsest of tectonic landforms and geomorphic features primarily moulded by exogenous processes in changing climatic conditions of the past.

Jizerské hory are a part of the West Sudetes and the second highest massif within their limits. The most elevated spot is located in Poland (Wysoka Kopa, 1126 m), but the highest summits in the Czech Republic are only a few metres lower (Smrk 1124 m, Jizera 1122 m). The general appearance is that of a subdued but considerably elevated upland (Fig. 14.1). To the east of the Jizerské hory the terrain rises to the summit plateaus of the Krkonoše, whilst from both the south/south-west as well as the north the massif is surrounded by low-lying grounds, with altitudes dropping to 300–400 m a.s.l. The drop is particularly pronounced on the northern side. These height relationships suggest that the Jizerské hory are one of those uplifted blocks within the Sudetes and as a whole may be considered as a large horst (Balatka and Pilous 2009).

14.3 Geology

14.3.1 Granite Intrusion and Metamorphic Cover

Geologically, the Jizerské hory in the Czech part are a very homogeneous area, underlain mostly by a large granitoid intrusion of late Carboniferous age (330–310 Ma). Only 5 % of the area in the north-east, with the highest Mt. Smrk, and north-west is built of rocks making the metamorphic cover—gneiss, mica schist, and leptinite (Fig. 14.1).

The granite intrusion is actually a part of a much larger body, the Krkonoše-Jizera pluton, which extends further east, beyond the state border and into Poland. The structure of the intrusion is complex and several lithological variants are distinguished, of different compositions and textures (Klominský 1969; Knotek 2009). In the Czech part, the so-called Jizera Granite is most widespread. It is medium grained, usually porphyritic, with individual feldspars even up to 7 cm long. The Liberec Granite is similar in terms of crystal size, but shows more pinkish or reddish colour. The third more common type is the Tanvald Granite, present in the south, where it occurs as an elongated body c. 20 km long and up to 4 km wide. It supports the Černostudniční hřbet (ridge) (see Sect. 14.4.3). It contains both biotite and muscovite and is relatively even grained. Within each variant one can find dark, fine-grained basic enclaves, usually less than 0.5 m long and often protruding from the rock faces. The granite mass is further cut by numerous aplite veins.

For geomorphology, perhaps the most important characteristics of granite bedrock is its jointing. This is fairly regular in the Jizerské hory and consists of three sets, perpendicular to one another. Two are nearly vertical and strike NW–SE and NE–SW, although towards the east the strikes change to more N–S and W–E. The third set is close to horizontal or gently inclined. This jointing system is responsible for evident fracturing of granite into more or less cubic blocks of different dimensions, depending on joint spacing. On this system younger partings are superimposed, usually parallel to the topographic surface and likely related to erosional unloading. On the plateau they are close to horizontal and give rise to pseudobedding, whilst on steep northern slopes and under valley sides they may be considerably inclined, up to 45–50°. The characteristics of jointing have a decisive influence on the shape of many medium-size granite landforms: rock slopes, tors, boulders, and bedrock channel features (see Sect. 14.5).

14.3.2 *Cenozoic Volcanism*

In addition to the dominant granites, in a few places in the Jizerské hory, one can find much younger volcanic rocks which record crustal extension in the Cenozoic and associated volcanic activity. Whilst most of the volcanism occurred to the north-west and west of the area, within the major tectonic structure known as the Ohře/Eger Rift, several outcrops of basalt and phonolite break the geological monotony of the mountains and their northern foreland. The most evident among them is the cone of Bukovec (1005 m), which rises sharply above the elevated upland surface near the village of Jizerka. A line of volcanic outcrops in the northern foreland, including Chlum (495 m), runs parallel to the northern margin of the Jizerské hory, a few kilometres north of it. The ages of volcanic rocks are likely to be Oligocene/early Miocene, similar to those in Lusatia to the north-west (Tietz et al. 2013), but are yet to be precisely determined.

14.4 Main Features of Relief

14.4.1 Elevated Plateau and Marginal Escarpments

Surface morphology of the Jizerské hory is best described as a rolling upland, gently sloping to the south and truncated on the northern side by a high marginal escarpment. The plateau is not perfectly planar but characterized by alternating swells and wide troughs, with the amplitude of relief in the range of 100–200 m. However, slopes are gentle and rarely exceed 10° of inclination. Both the swells and the troughs predominantly follow WNW–ESE to NNW–SSE directions, sloping gently to the south-east. Thus, the northern part of the plateau is more elevated (Smrk 1124 m, Jizera 1122 m,

Černá hora 1085 m, Smědavská hora 1084 m, Holubník 1071 m), while in the south elevations within the range of 750–900 m a.s.l. are typical. In a few places the troughs widen and become flat-floored, giving rise to elongated basins (Fig. 14.2). The largest of them is Hala Izerska (its bigger part is on the Polish side), 7 km long and up to 1.5 km wide, followed by the Jizerka basin, which 4 × 1 km. In both extensive peat bogs occur, with the thickness of peat up to 3–4 m, whilst rivers change their pattern and develop alluvial channels of high sinuosity, with regular meanders and numerous paleochannels above the present-day floodplain.

The northern escarpment of the Jizerské hory above the town of Hejnice is the most impressive geomorphic feature of the area, visible from far in the north (Fig. 14.3). Over a distance of 1 km altitudes increase from 400 m to 900 m a.s.l., resulting in a slope of 30–35°, locally much more. In the upper section of the escarpment granite outcrops are numerous and bare rock slopes, interspersed with massive blocky talus, are hundreds of metres long (although mostly hidden by dense beech stands). The escarpment is considerably dissected and indented. The deepest valleys are oblique to the W–E trend of the scarp and follow SE–NW or SW–NE courses, as other valleys do in the southern part of the area. Their headwater areas located on the plateau and assume the form of wide and shallow troughs. Significantly, the presence of the escarpment is not related to lithological change and hence, possible variability of rock resistance. Granites occur both above, within, and below the scarp, which strongly suggests that the escarpment is fault-generated and hence, a primary tectonic landform.

The zig-zag course of the escarpment is notable. Its western part is elongated W–E, and then it sharply turns NW–SE at an acute angle, creating an embayment along the river Smědá, and after 7 km returns to the W–E course and continues eastward, beyond the Czech/Polish border. Here the escarpment is even higher, nearly 600 m at Mt. Smrk, but slope inclination is lower, possibly due to lithological change and replacement of granite by gneiss and schist of the metamorphic cover. Several Cenozoic volcanic outcrops in the foreland of the Jizerské hory are located in the westward prolongation of the latter part of the escarpment.

No analogous escarpment occurs to the south of the plateau, although a 3–4-km-wide belt of increasing incision and higher slope may be identified between the city of

Fig. 14.2 Jizerka. The plateau of the Jizerské hory consists of broad ridges and elongated basins, often with substantial peat accumulation (*Photo* J. Miklín)

Fig. 14.3 The steep northern escarpment of the Jizerské hory is of tectonic origin. The height of the scarp is about 600 m. Note numerous granite outcrops on a spur to the left of the deeply incised valley (*Photo* P. Migoń)

Liberec in the west and the towns of Tanvald and Desná in the east. The morphology of the western termination of the Jizerské hory plateau is complex. To the north of the city of Liberec an escarpment exists, although it is lower (up to 300 m high), wider and more dissected than the northern one. Further to the north-west an outlier of the plateau projects into the hilly foreland and takes the form of a narrow ridge with numerous rock slope sections, culminating in Špičák (724 m).

14.4.2 Valley Network and Its Controls

The main water divide in the Jizerské hory is shifted to the north, emphasizing the north–south asymmetry of the massif. The wide plateau reaches of north-flowing streams are very short, 1–2 km at most, and then continue as deeply incised valleys, with steep rocky sides, numerous waterfalls and boulder steps. The best examples are Velký Štolpich and Malý Štolpich just south of Hejnice, both with significant waterfalls (see Sect. 14.5.4). The valley network on the south-sloping side of the plateau is dominated by long straight reaches. The dominant directions are NNW–SSE (e.g. Černá Desná, Bílá Desná, Jedlová) and NW–SE (e.g. Jizerka, Kamenice, middle reach of Bílá Nisa). Less frequent are NNE–SSW flowing streams (e.g. upper reach of Černá Nisa). These directions are also followed by rivers flowing off the northern escarpment, both above and within it (e.g. Smĕda, Velký Štolpich). Interestingly, very few major rivers flow directly to the south, hence consistently with the general slope of the upland. These characteristics clearly indicate structural control on the drainage pattern. Dominant NNW–SSE to NW–SE fracture directions in the granite massif, as well as those perpendicular to them (Klominský 1969), must have influenced fluvial erosion since the inception of the drainage and have continued to exert a major influence despite up-faulting and apparent southerly tilting of the granite block.

Certain specific features of the drainage pattern in the Jizerské hory suggest that several captures may have taken place (Pilous 2009a). The drivers of capture were apparently streams capable of extending their drainage basins upstream from the escarpments through headward erosion. It is likely that the uppermost reach of Velký Štolpich has been captured by Černý potok, whilst at a much larger scale headward erosion of Jizera has likely reached the intramontane basin of Hala Izerska and diverted its drainage to the south. The original drainage would have been to the north-west, in the direction of the present-day Smĕda.

14.4.3 Rock-Controlled Ridges and Domes

Not everywhere have the Jizerské hory the appearance of a plateau. In the southern part of the area, a west–east trending granite ridge of Černostudniční hřbet indicates the boundary of the massif. It is associated with a specific variant of granite, known as the Tanvald Granite. The ridge is c. 10 km long and rises above the surrounding terrain by 250–300 m. Its summit, Mt. Černá Studnice, is 869 m a.s.l. To the north the ridge slopes down to a moderately incised hilly land underlain by the Liberec Granite, whereas to the south granites give way to schists and phyllites of the metamorphic country rock. Thus, the ridge clearly owes its existence to the increasing strength and resistance of the Tanvald Granite. Numerous tors and boulder fields add to the geodiversity of this part of the mountains.

The three highest elevations in the granite part of the Jizerské hory—Jizera, Černá hora, and Smĕdavská hora—are geomorphic features worth separate description. They are broad convex domes rising above the general plateau level, with a height of 100–150 m. Castellated tors are present on summits and particularly on slopes of these elevations, and being dominated by planar joints they emphasize the domical shapes of the hills. Analogous, bare granite domes have been often reported from arid and semi-arid lands (Twidale 1982). Those in the Jizerské hory are less impressive because of the presence of regolith and vegetation cover, but are structural features of the same kind.

14.5 Medium-Size and Minor Granite Landforms

14.5.1 Tors and Crags

Tors, understood as 'residual landforms rising from a regolith-veneered surface or rock platform, whether flat or sloping (…), which are composed of more than one individual compartment (boulder), but are too small to be distinguished as separate hills' (Migoń 2006: 87), are ubiquitous in the Jizerské hory and there are probably hundreds of them scattered across the plateau and the steep northern slopes (Pilous 2009b). They have different shapes, adjusted to locally dominant fracture pattern (Fig. 14.4). Most common are angular, castellated tors (castle koppies), developed in orthogonally fractured bedrock. Their height varies from as little as 2–3 m to more than 15 m, and they may be extensive laterally. In many locations castle koppies occur in clusters, close to one another and separated by

Fig. 14.4 Diversity of tor shapes. **a** rounded tor on the plateau in coarse-grained granite; **b** angular tor with talus around in the Tanvald Granite; **c** In some places tors are so close to one another that they combine to form more extensive 'rock cities'; **d** This subdued tor close to the town of Hejnice was probably remodelled by the Scandinavian ice sheet (*Photo* P. Migoń)

corridors (defiles), hence giving particular areas the appearance of 'rock cities' known mostly from sandstone areas. Polední kameny above the northern escarpment is a good example. Many tors show a clear presence of densely spaced horizontal joints (pseudobedding), although these tend to be lower than the castle koppies. Not all tors are necessarily angular and there are also examples of apparently chaotic piles of more or less rounded boulders, typically 2–5 m high. Solid rock outcrops are also common within the steep northern slope of the Jizerské hory, especially on valley sides. Here they emerge from the slopes as projecting rock spurs, only a few metres high on the upslope side, but up to 40–50 m on the downslope side. Nos in the valley of Černý potok and Krásná Maří above the valley of Velký Štolpich are good examples. In the metamorphic rock part of the mountains tors are scarce.

Tors occupy different geomorphic settings but are preferably associated with summits, crests and upper slopes rather than with lower slopes and valley floors. The most impressive examples can be found in the northern part of the massif, close to and within the northern escarpment, apparently in relation to more vigorous erosion and regolith removal in the area of higher relief. It is very likely that tors of the Jizerské hory have more than one origin. The two-stage model of tor origin through selective deep weathering and removal of loose weathering products to expose faintly weathered rock cores seems generally valid and occasional outcrops show granites differentiated into grus mantles with corestones. However, massive rock towers on the northern slope have gradually emerged from the slope, concurrently with its retreat and surface lowering and without a phase of antecedent deep weathering. On the other hand, many low angular tors and rock steps rising from the slope were probably exposed due to selective mechanical breakdown of jointed granite under periglacial conditions of the Pleistocene.

14.5.2 Weathering Features

Rock outcrops in the Jizerské hory are hosts to many curious minor features derived from selective rock breakdown. Preferential weathering along joints has led to the emergence of massive boulder-size compartments standing on narrow

Fig. 14.5 Two most famous weathering pits in the Jizerské hory. **a** Water-filled pit at Smržovka. **b** Large pit on a rotated boulder known as Kukaň (*Photo* P. Migoń)

stems ('rock mushrooms') and the origin of precariously perched, unstable boulders ('rocking stones'). Among surface microrelief weathering pits, i.e. circular or elongated flat-floored hollows on horizontal rock surfaces, are most abundant. They are typically a few tens of centimetres across and 20–30 cm deep, but the probably largest, and heart-shaped pit in the town of Smržovka is 1.5 m long and 50 cm deep (Fig. 14.5a). However, less regular basins may attain even greater dimensions, to 3 m long. Another natural curiosity is the large pit at Kukaň, once formed on a horizontal surface but due to subsequent boulder fall it is now incised into a side wall, resembling a tafone (Fig. 14.5b).

Other features of surface microrelief include runnels on gently sloping granite slabs, often draining and connecting weathering pits, series of flutes on vertical surfaces, and polygonal patterns of shallow cracks. Notable are examples of differential weathering of coarse granite versus fine-grained aplite veins. The latter typically project out of a rock face and are regularly fractured into decimetre-size cubes, contrasting with irregular, rough and pitted granite surfaces.

14.5.3 Caves

Differential, fracture-guided weathering of granite and gravitational processes have given rise to a number of roofed caverns within granite outcrops. Although rarely longer than 10 m and usually fully exposed to daylight, they are described as non-karstic caves. Two main genetic types may be distinguished. One is related to mass movements involved in degradation of granite tors. Caves of this type are mainly located along the northern escarpment, where high slope gradients and associated stresses facilitate processes such as joint opening, slab slides and boulder falls. Consequently, crevice-type caves, tunnels, and chambers within blocky talus are common. The other type is linked with removal of products of granite disintegration from between massive boulders. In this way, chambers within residual boulder tors originate. The origin of the largest underground system in the region, known as Valhala, is similar. This name is given to a system of interconnected passages, at least 150 m long, between boulders filling the valley floor of Černá Nisa. Here fluvial transport of fine material has been the main factor in opening the chambers.

14.5.4 Waterfalls

The region of Jizerské hory has probably more waterfalls than any other mountain massif in the Czech Republic (Pilous 2009a). Their abundance arises from the combination of two factors: the presence of steep terrain along the topographic margins of the plateau and the resistance of granite against fluvial erosion (Fig. 14.6). Consequently, waterfalls are preferentially located in two belts: within the northern escarpment and south of the main plateau, between the towns of Bedřichov and Desná. Velký vodopád in the valley of Velký Štolpich, which is 36.5 m high, is considered as the highest waterfall in the area, although in reality it consists of a series of steps and pools. The highest step is 9.8 m high. Dolní vodopád in the adjacent valley of Černý potok is 17 m high. In addition, long reaches of streams are bedrock channels, with frequent cascades, inclined slabs, pools, and boulder steps. Potholes are common, especially in the valley of Jizera.

Fig. 14.6 Waterfalls on the Černý potok which is incised into the northern escarpment of the massif (*Photo* P. Migoń and E. Pijet-Migoń)

14.6 Landscape Evolution

14.6.1 Deep Weathering and Planation

The contemporary physical landscape of the Jizerské hory is the result of protracted geomorphic evolution in changing tectonic and environmental conditions; hence, it is polygenetic and contains facets of various ages and lifetimes. It is assumed that uplift of the area to the present-day elevation has occurred relatively recently and both above and below the faulted escarpments remnants of paleolandscape can still be observed. The origin of these rolling to hilly pre-uplift surfaces in granites is usually linked with prolonged deep weathering that occurred in the Paleogene and most of the Neogene. Although in the Jizerské hory massif itself no remnants of thick saprolites have survived, they have been preserved in the northern foreland, and especially in the subsidence area around Zittau and beneath Oligocene–Miocene basalts (mainly in Lausitz, Germany). Kaolinitic saprolites, with clear evidence of differential weathering of granite, are locally up to 60 m thick (Franz 1969). Deep granular disintegration of granite, common within the plateau, could be root parts of these ancient weathering mantles but there is no solid evidence. Lithological and structural diversity of the granite pluton controlled the style and rates of weathering, and therefore the denudational surface was not a perfect planation surface. Instead, it was typified by alternating ridges and troughs following the main fracture systems, semi-closed basins, occasional domes on large and massive bedrock compartments, and probably a variety of tors and boulder fields, all interspersed with saprolite-covered surfaces. Subsequent removal of weathering products exposed the weathering front and has given rise to the regional etch surface (sensu Thomas 1989).

14.6.2 Uplift, Erosion and Regolith Removal

From the Oligocene onward the area has experienced increasing tectonic instability, along with the rest of the Bohemian Massif, and especially its northern part—the Eger Graben and its shoulders. The early stages of uplift were associated with volcanism and it is likely that the cone of Bukovec and volcanic bodies near Frýdlant date back to that period. The timing of major uplift of the Jizerské hory is not sufficiently constrained but the clarity of tectonics-related geomorphic features strongly suggests that it occurred in not too distant past, in Miocene–Pliocene times.

Late Cenozoic uplift of the massif took place along two main fault zones, both running broadly west–east. The northern one coincides with the imposing northern escarpment and is actually an eastern extension of a fault that limits the Zittau Graben from the south. The southern one is less evident as a topographic escarpment, but is emphasized by the line of waterfalls. Hence, large-scale tilting of the horst to the south may be inferred, consistent with the dominant drainage to the south. Several minor faults seem to offset different parts of the paleosurface, although their throw was

100–150 m at most. Their existence is hypothesized to the south-west from the intramontane basins of Hala Izerska and Jizerka, which would be half-graben rather than purely denudational features.

A side effect of uplift and enhanced erosion was widespread stripping of ancient weathering products and exposure of the weathering front. However, due to lack of data it is not possible to estimate the subsequent lowering of the plateau surface. Thus, it is safer to assume that the upland surface 'echoes' the early Cenozoic paleosurface rather than it represents its unchanged remnant, somehow 'frozen in time'.

14.6.3 The Pleistocene—Ice Sheets and Glaciers

During the Pleistocene the northern foothills of the Jizerské hory were reached by the Scandinavian ice sheets, most probably twice, during the Elsterian (Marine Oxygene Isotope Stage 12) and the Saalian (MIS 6) (Nývlt et al. 2011). The ice masses arrived from the north-west, across the adjacent Frýdlant Hilly Land, and were moving along the northern fault-generated escarpment of the massif and into the valleys dissecting the escarpment. Identification of the most probable vertical extent of the ice surface for 480–490 m a.s.l. (Fig. 14.1) is based on two lines of evidence. One includes the presence of outwash gravel to the south of the watershed Oldřichov Pass in the western part of the massif (Nývlt 2003), whereas in the other one emphasis is laid down on significant differences in the degree of weathering of granite surfaces below and above the altitude of 480–490 m a.s.l. which is hence interpreted as the Saalian trimline (Traczyk and Engel 2006; Černá 2011).

In front of the faulted margin of the Jizerské hory there occur numerous low granite elevations, from a few to 30–40 m high. The majority is elongated consistently with the former direction of ice sheet movement whilst discordantly to the predominant joint direction. Their surfaces are often smooth and subdued, without high upstanding compartments, typical for rock outcrops higher on the slope. Janásková (2009) considers these granite knobs as ice-remodelled hills—the roches moutonnées (Fig. 14.7). Ice sheet advance and decay in the front of the mountains has probably also caused drainage network changes and the origin of short epigenetic gorges linking wide, preglacial valleys (e.g. of Štolpich creek west of Hejnice).

The issue of possible mountain glaciation in the Jizerské hory is controversial. For long time they were considered to have been too far below the altitude of the Pleistocene snow line, reconstructed in the adjacent Krkonoše Mountains for c. 1200 m a.s.l. (Nývlt et al. 2011), for glaciers to occur. This view was reinforced by the absence of evident glacial geomorphological features such as cirques with rock walls, U-shaped troughs, and moraines. However, recent research in the enigmatic circular hollow of Rybí loučky, at a mere 850–950 m a.s.l., suggests that a small ice body capable of some erosional deepening of bedrock may have existed. Reasons for this abnormally low setting are sought in the peculiarities of local anemo-orographic systems and massive snow accumulation on the lee side of the ridge (Pilous 2006; Traczyk et al. 2008).

Fig. 14.7 Hilly granite landscape in front of the northern escarpment bears traces of glacial remodelling during an advance of the Scandinavian ice sheet (vicinity of the town of Hejnice) (*Photo* P. Migoń)

14.6.4 Present-Day Extreme Events

The protracted geomorphological history of the area and the considerable age of its main relief features do not mean that no landform change occurs nowadays. Although episodes of significant remodelling are rare, they nevertheless occur, in association with heavy rainfall that occasionally takes place, especially during summer. For example, on 29 July 1897 the highest ever daily precipitation in the entire Sudetes was recorded at Nová Louka station, on the upland surface—345.1 mm. Approximately, once per decade two-day precipitation events in excess of 200 mm are recorded, most recently on 6–7 August 2010 (Kulasová and Bubeníčková 2009; Pilous 2011). Rainfall of this magnitude results in widespread devastating floods and may also trigger mass movements.

The August 2010 event caused, besides floods and associated remodelling of river channels, two large and two minor debris flows on the very steep, NE-facing slopes of Mt. Smědávska hora—very rare phenomena on the forested slopes of the Jizerské hory (Pilous 2011). The longest debris flow track was 980 m long and 12–45 m wide, affecting approximately 60 % of the slope length (altitude range 915–580 m a.s.l.). Each flow was initiated by shallow debris slide and stripping of 1–2 m of regolith cover in rather indistinct slope hollows in the upper slopes. Then the sliding masses transformed into flows and further entrainment of loose regolith cover took place, so that granite bedrock has become exposed along most of the tracks (Fig. 14.8). Boulders as big as 2 m long travelled downslope as well as numerous tree logs, which then accumulated as massive log jams. Despite forced partial deposition on roads, a certain proportion of debris reached the channel of Směda river, indicating hillslope—valley floor coupling during extreme events.

14.6.5 Human Impact and Cultural Geomorphological Landscape

Direct anthropogenic influence on the geomorphological landscape of the Jizerské hory has been relatively minor. Granite quarrying used to be widespread due to high quality of the rock and its legacy includes many abandoned and a few working quarries, especially in the south-western part of the area, in the vicinity of Liberec—the biggest town. Some former quarries have been partially flooded and today they represent scenic components of the landscape. On the upland surface one can also find many former gravel pits, where products of deep-reaching granular disintegration of granite (grus) were excavated, mainly for road building purposes.

Granite has been widely used in construction and man-made structures built of granite typify the local cultural landscape. They range from small-scale elements such as monuments and stone crosses through dry stone walls on former arable land and imposing retaining walls on roads traversing steep slopes (mostly built in the late nineteenth century), to massive stone dams on local rivers. An increasing popularity of tourism since the second half of the nineteenth century prompted local tourist organizations to improve access to various places, especially to granite crags and waterfalls. Minor alterations of natural landforms (e.g. rock-hewn staircases) are therefore common.

Fig. 14.8 Exposed granite bedrock within the track of the 2010 debris flow below Mt. Smědavská hora (*Photo* P. Migoń)

14.7 Conclusions

The granite massif of Jizerské hory shows different controls on the geomorphic evolution, acting over a long time scale. They are related to both geology (influence of jointing, lithological differentiation of pluton, faulting) and climate. Within the Bohemian Massif it is probably the best example of this kind, of considerable geodiversity and with a particularly rich landform inventory. In the contemporary landscape one can find features of different origins and ages, spanning the period from the early Cenozoic. The most typical are broad convex elevations rising from the upland surface (domes), elongated basins with extensive peat accumulation, numerous tors and crags, boulder fields, bedrock channels, waterfalls, and weathering pits. Among the major relief features, the steep fault-generated northern escarpment stands out as a prominent geomorphic landmark. In the Pleistocene the Scandinavian ice sheet reached the footslopes of the mountains and remodelled bedrock elevations into roches moutonnées. On the plateau, very small glaciers may have existed on lee-slopes, fed by snow blown-in by westerly wind. Torrential floods and occasional debris slides and flows are main geomorphic processes acting nowadays.

All these geomorphic features can be easily seen by an interested visitor. The area is crossed by a dense network of marked trails for hikers, bikers, and cross-country skiers, whereas settlements offer ample accommodation opportunities. Numerous bare tor summits and observation towers elsewhere allow one to have an overview of larger areas. In several places educational trails have been set up. There remains little doubt that the area of Jizerské hory is one of the geomorphological highlights of the Czech Republic.

References

Balatka B, Pilous V (2009) Geomorfologické poměry Jizerských hor. In: Karpaš R et al, Jizerské hory. O mapách, kamení a vodě, Nakladatelství RK Liberec, pp 267–296

Černá B (2011) Reconstruction of the continental glaciation in the northern slope of the Jizera Mts. – Sbor. geol. Věd, Antropozoikum, 27, 23–38. Praha

Czudek T, Demek J, Marvan P, Panoš V, Raušer J (1964) Verwitterungs- und Abtragungsformen des Granits in der Böhmischen Masse. Peterm Geogr Mitt 108:182–192

Demek J (1964) Slope development in granite areas of the Bohemian Massif (Czechoslovakia). Z Geomorph Suppl 5:82–106

Demek J (1975) Planation surfaces and their significance for the morphostructural analysis of the Czech Socialist Republic. Studia Geogr 54:133–164

Franz H-J (1969) Die geomorphologische Bedeutung des Granitverwitterung in der Oberlausitz. Peterm Geogr Mitt 113:249–254

Janásková B (2009) The origin of rounded granite elevations in the northern foothills of the Jizera Mountains. Geomorph Slovaca Bohemica 9:7–16

Klominský J (1969) Krkonošsko-jizerský granitoidní masív. Sbor geol věd, Geol 15:7–133

Knotek Z (2009) Geologie Jizerských hor. In: Karpaš R et al, Jizerské hory. O mapách, kamení a vodě, Nakladatelství RK Liberec, pp 104–141

Kulasová A, Bubeníčková L (2009) Podnebí a počasí Jizerských hor. In: Karpaš R et al, Jizerské hory. O mapách, kamení a vodě, Nakladatelství RK Liberec, pp 342–383

Migoń P (2006) Granite Landscapes of the World. Oxford University Press, Oxford

Migoń P (2008) Main features of geomorphology of the Sudetes re-assessed in the light of digital elevation model. Geografie – Sborník ČGS 113: 400–416

Migoń P (2011) Geomorphic diversity of the Sudetes – Effects of global change and structure superimposed. Geogr Polon 84, SI Part 2:93–105

Nývlt D (2003) Geomorphological aspects of glaciation in the Oldřichov Highland, Nothern Bohemia, Czechia. Acta Univ Carol, Geogr 35:171–183

Nývlt D, Engel Z, Tyráček J (2011) Pleistocene Glaciations of Czechia. In: Ehlers J, Gibbard PL, Hughes PD (eds) Quaternary Glaciations – Extent and Chronology. A Closer Look. Developments in Quaternary Science 15, Elsevier, Amsterdam, pp 37–46

Pilous V (2006) Pleistocénní glacigenní a nivační modelace Jizerských hor. Opera Corc 43:21–44

Pilous V (2009a) Říční tvary a procesy w Jizerských horách. In: Karpaš R et al, Jizerské hory. O mapách, kamení a vodě, Nakladatelství RK Liberec, pp 418–446

Pilous V (2009b) Skalní útvary Jizerských hor. In: Karpaš R et al, Jizerské hory. O mapách, kamení a vodě, Nakladatelství RK Liberec, pp 297–341

Pilous V (2011) Povodňové mury v povodí horní Smědé v Jizerských horách. Sbor Severočesk Muz, Přír Vědy 29:3–40

Thomas MF (1989) The role of etch processes in landform development. Z Geomorph NF 33:129–142, 257–274

Tietz O, Büchner J, Suhr P, Goth K (2013) Field trip 3: Volcanology of the Lusatian Volcanic Field – New insights in old well-known. In: Büchner J, Rapprich V, Tietz O (eds) Basalt 2013. Cenozoic Magmatism in Central Europe. Abstracts and Excursion Guides, Czech Geological Survey Prague and Senckenberg Museum of Natural History Görlitz, pp 275–297

Traczyk A, Engel Z (2006) Maximální dosah kontinentálního zalednění na úpatí Ořešníku a Poledníku v severním svahu Jizerských hor. Geografie – Sborník ČGS 111: 141–151

Traczyk A, Engel Z, Janásková B, Kasprzak M (2008) Glacjalna morfologia wierzchowiny Gór Izerskich w świetle badań w rezerwacie "Rybí loučky" (Republika Czeska). Landform Anal 9:129–133

Twidale CR (1982) Granite Landforms. Elsevier, Amsterdam

Twidale CR, Vidal Romaní JR (2005) Landforms and Geology of Granite Terrains. Balkema, Amsterdam

Krkonoše Mountains: A Case Study of Polygenetic Relief

15

Vlastimil Pilous

Abstract

The Krkonoše Mts—the most elevated terrain in the Czech Republic—have been studied by German, Czech and Polish geomorphologists since the second half of the nineteenth century and in terms of geomorphology belong to the best investigated mountain ranges in Europe. The studies have focused on glacial and periglacial landforms and brought new findings, also in connection with the representation of landforms so far known only from subarctic locations. However, the main features of the range are genetically connected with protracted planation, neotectonic uplift and structural control exerted on the course of weathering, erosion and slope development. Therefore, the most characteristic landforms are extensive water divide planation surfaces, granite tors, deeply incised valleys, pointed ridges and peaks built of resistant hornfels of contact zone. The history of relief development goes back to at least the Late Cretaceous, with neotectonic uplift affecting the area in the Neogene and Quaternary. During the Pleistocene more than 10 glaciers existed in the summit parts, up to 5 km in length. Geomorphic legacy of glaciation includes cirques, U-shaped valleys, lateral and frontal moraines. Steep slopes of cirques and valleys are frequently modelled by debris flows, triggered by heavy rainfall. In the most recent times multiple human activities, mainly mining and agriculture, have modified geomorphology of the area.

Keywords

Krkonoše Mts • Planation surfaces • Neotectonics • Structural landforms • Glacial and periglacial landforms • Debris flows • Snow avalanches • Man-made landforms

15.1 Introduction

The Krkonoše Mountains are the highest mountain range (1603 m) of the Bohemian Massif that makes a part of the Paleozoic Variscan system. Although exceeding in height other Central European mountain ranges of this system (the Vosges and Schwarzwald in Western Europe, Šumava Mts, Králický Sněžník Mts, Hrubý Jeseník Mts in the Czech Republic) by mere 100–200 m, they are characterised by specific geomorphic features. These result from distinct geological structure made of combination of Paleozoic granite pluton, Upper Proterozoic crystalline strata, and resistant contact zone at the boundary of the two, neotectonic (Post-Oligocene) uplift (Chaloupský 1989) and subsequent erosion processes. Apart from neotectonic disturbances, erosion processes were also controlled by several important morphostructural, mainly passive structural factors. The development of glacial and periglacial landforms was significantly influenced by the presence of a continental Scandinavian ice sheet that approached the Krkonoše Mts most closely (7.5 km from the main ridge) of all high Variscan mountain ranges. The origin of relief of the Krkonoše Mts was also impacted by a perfectly developed, relief-conditioned

V. Pilous (✉)
Správa KRNAP, Dobrovského 3, 543 01 Vrchlabí, Czech Republic
e-mail: vlpilous@seznam.cz

T. Pánek and J. Hradecký (eds.), *Landscapes and Landforms of the Czech Republic*, World Geomorphological Landscapes, DOI 10.1007/978-3-319-27537-6_15

anemo-orographic system described for the first time from the Krkonoše Mts (Jeník 1961). Originally, this system was only considered to be an important ecological factor influencing living nature. However, recent research has shown that it considerably participated in the distribution of glacial, periglacial and Holocene landforms. An overview of geomorphological processes and landforms in the Krkonoše Mts has been recently presented by Pilous (2001) and Migoń and Pilous (2007).

15.2 Planation Surface

The processes of erosion and denudation gradually transformed the original Paleozoic Variscan mountain range into a planation surface (i.e. slightly protruding highlands with softly undulated landforms) that reflected structural conditions of the bedrock. The topography was characterised by wide, slightly inclined valleys and interjacent flat gentle-sloped elevations (Fig. 15.1). The origin of these landforms was fundamentally affected by deep chemical weathering in relatively warm and humid geological periods, particularly in the Paleogene, the result of which was deeply broken-up fine-grained weathering product. According to the latest research based on the thermochronological analysis of samples from the Krkonoše Mts (Danišík et al. 2010), between 100 and 75 Ma erosion removed rocks of the thickness of 3.6–6 km. Thus, a period of massive erosion and denudation (up to 300 m in a million of years) in the Late Cretaceous is inferred. On the other hand, erosion appears rather weak in the Late Neogene, despite strong uplift (see next section). Alternatively, the planation surface represents the vestige of a Permian peneplain that was covered by Mesozoic sediments and subsequently exhumed in the Late Cretaceous. This planation surface was broken and uplifted in blocks of various sizes as a result of Saxonian tectonic movements triggered by the Alpine orogeny. In this way the planation surface was exposed to intensive erosion and survived only at a few altitude levels in the summit parts of the Krkonoše Mts. It gives the whole mountain range a characteristic, yet relatively untypical appearance, with a prolonged main ridge of a relatively balanced altitude. This fact greatly affected the course of a wide range of subsequent geomorphic processes in the Pleistocene, as well as influenced the shape of resultant landforms. Given the important role of deep weathering in the development of the surface, it is appropriate to consider it as an etchplain (or etchsurface).

Fig. 15.1 Summit planation surface of the eastern Krkonoše Mts. In the Pleistocene it acted as deflation surface, feeding glaciers on the lee side. As a result of snow removal, most periglacial landforms originated on the summit surface. (View from the Sněžka Mt to the west) (*Photo* V. Pilous)

Etching and stripping of weathering mantle played a key part in the geomorphic evolution of medium-sized landforms. It is assumed that the first phase involved differential subsurface chemical alteration of local granite. Weathering exploited rock partings and caused bedrock differentiation into blocky compartments separated by weathered zones. More resistant rocks emerged at the surface in the second phase, in colder climatic periods, when surface erosion accelerated and massive rock compartments became exposed by water erosion and slope mass wasting processes. Rock formations of this type (e.g. castle coppies, tors) are more common on the Polish side of the range but some can be seen on the Czech side too, in the basin of the Mumlava River and in Sedmidolí.

15.3 Role of Neotectonic Processes in the Formation of the Mountain Range

Neotectonic activity in the Sudetes started as early as at the turn of the Neogene, whereas the activity peaked at the end of the Neogene and it continued relatively intensively in the Pleistocene as well (Migoń 1992). Research of neotectonic processes and resultant landforms has long been focused on the Polish side of the Krkonoše Mts (Migoń 1992). Its results proved older opinions (Ouvrier 1933) that the Krkonoše Mountains represent a typical faulted mountain range, or more precisely, a segmented horst. As for the Czech part of the Krkonoše Mts, tectonic processes have only been mentioned in relation to the landforms as a general background or in a very few case studies (Migoń and Pilous 2007). However, new research shows that their influence is more probably much greater than it has been supposed (Lysenko 2007).

Yet, the main factor in the development of landforms within the predominantly granite slopes on the Polish side remains to be tectonic activity (a distinct step-like topography with steep rectilinear slopes, parallel river systems, intramontane troughs parallel to the boundary faults, etc.) (Migoń 1992). On the other hand, passive structural factors prevail in the development of landforms in the Czech part of the Krkonoše Mts.

Altitudes of preserved levels of planation surfaces and benches/terraces in the deeply incised valley systems make it evident that tectonic uplift was vertically differentiated. Remnants of old planation surfaces can thus be found at several altitude levels. The horst of the Krkonoše Mts, and especially their ridge part, has a rectangular outline. The highest total uplift took place most likely in the northeastern part of the main massif (in the wider area of the Mt Sněžka), while the lowest one occurred on its diagonal margin (Rokytnice region in the basin of the Jizera River). The amount of uplift, however, cannot be unequivocally determined since tectonic factor is overprinted on the factor of unequal erosion and denudation, controlled by bedrock diversity. In this respect, the presence of a zone of hard orthogneiss in particular allows us to assume that the eastern half of the mountain range was less affected by erosion and denudation. In the Czech part of the mountain range faults are reflected in the landscape most clearly in the southeastern part of the mountain range (Fig. 15.2) and in the area of the Harrachov intramountain basin. The most important evidence of neotectonics are rectilinear, fault-controlled lower slope breaks and straight valley courses, as well as the occurrence of waterfalls (Mumlava, Jizera and Kamienna Rivers).

15.4 The Role of Passive Morphostructure

Several-hundred-metres-high neotectonic uplift revived erosion that led to the deepening of valley network and fast segmentation of the original planation surface. A vital role in controlling the course and rate of erosion was played by geological conditions, i.e. the factor of passive morphostructure. In this respect, it was the varied rock structure of the Czech part of the Krkonoše Mts that exerted an important role (Fig. 15.3), with (a) stretches of hard rocks of contact zone between granite and country rock passing through the whole western and central High Krkonoše Mts, (b) muscovite orthogneiss protruding close to the margins of the mountain range (in the whole central and eastern Krkonoše Mts), (c) short stretches of sericitic quartzites in the western Krkonoše Mts (in the basin of the Jizera River in the Rokytnice region) and (d) crystalline limestones (in the vicinity of Maršov municipality in the east). Similarly, a stretch of very hard, fine-grained biotite granite, coincidentally parallel to the contact zone, occurs within the granite pluton. This led to the origin of specific georelief in the Krkonoše Mts, namely two main summit ridges that are parallel to each other and approximately of the same height.

However, the most important factor in the development of erosional landforms was the contact zone, a 1-km-wide belt along the southern margin of the Krkonoše-Jizerské hory Mts granite pluton on whose hard rock the whole Český hřbet Ridge originated. Slowing down backward erosion from the south, this zone has left its mark in the appearance of the whole summit part of the mountain range. Owing to its presence, two summit plateaus, considered as remnants of an old planation surface, were preserved and played a decisive role in the geomorphic evolution of the massif in the Pleistocene. They acted as deflation zones supplying lee-side valley heads with voluminous snow masses, allowing glaciers to form, and provided a suitable setting for the formation of a wide range of periglacial

Fig. 15.2 Fault-generated southeast marginal slopes of the Krkonoše Mts between the towns of Vrchlabí and Svoboda n. Úpou (*Photo* V. Pilous)

landforms. Due to their height and continuous areal extent, in the Holocene the plateaux enabled the development of arctic and alpine tundra (Soukupová et al. 1995), with a range of subrecent to recent, cold climate landforms.

Other important landforms conditioned by the granite/country rock contact zone are subsequent valleys of the Mumlava River, the Labe/Elbe River, the Bílé Labe River and the Dlouhý potok Brook that segment the summit part of the mountain range. The valleys had a decisive influence on the development of landforms in glacial periods and the Holocene since their W–E direction corresponds with the prevailing wind flow, the so-called anemo-orographic systems.

Hard rocks of the contact zone also had an impact on the appearance of the very landforms they build. Their extremely steep slopes create a typical hogback shape in the area of Kozí hřbety ('goat back') Ridges and Čertův hřeben Ridge. As for the Kozí hřbety Ridges, these represent the longest (3 km) and the highest (up to 500 m) landform of this type in the Czech Republic (Fig. 15.4).

The second most important structural landform in the Krkonoše Mts is connected with the belt of hard and erosion-resistant muscovite orthogneiss. The gneiss belt is cut through by two rivers, the Labe River in Labská soutěska Gorge above the town of Vrchlabí and the Úpa River in the Horní Maršov-Temný Důl municipality. These sections are typified by significant narrowing of valleys, steepening of slopes, the occurrence of bedrock outcrops on the slopes, and in the case of the Labe River, the presence of rapids in the channel. Two landmarks of the southern margin of the Krkonoše Mts, the Žalý Mt (1019 m) and especially the joint massif of the Černá hora and Světlá hora (1299 m and 1244 m), markedly rising above their surroundings, owe their height to hard orthogneiss. A band of orthogneiss protruding between softer mica schists in the north and phyllites in the south played also a role in the development of the river system and probable old (Neogene or Early Pleistocene) river captures (Ouvrier 1933).

Distinct structural control can also be identified in sudden changes in the course of the Jizera River and its three incised meanders in the Rokytnice region in the western Krkonoše Mts. They are related to narrow stripes and insets of very hard sericite quartzite that cross the valley. Apart from that, quartzites create a group of individual rock outcrops exposed by selective erosion and weathering. Having been shaped by periglacial processes, they have a form of frost-riven cliffs. As a result of extraordinary hardness of quartzites, these rock formations are characterised by conspicuous sharp edges.

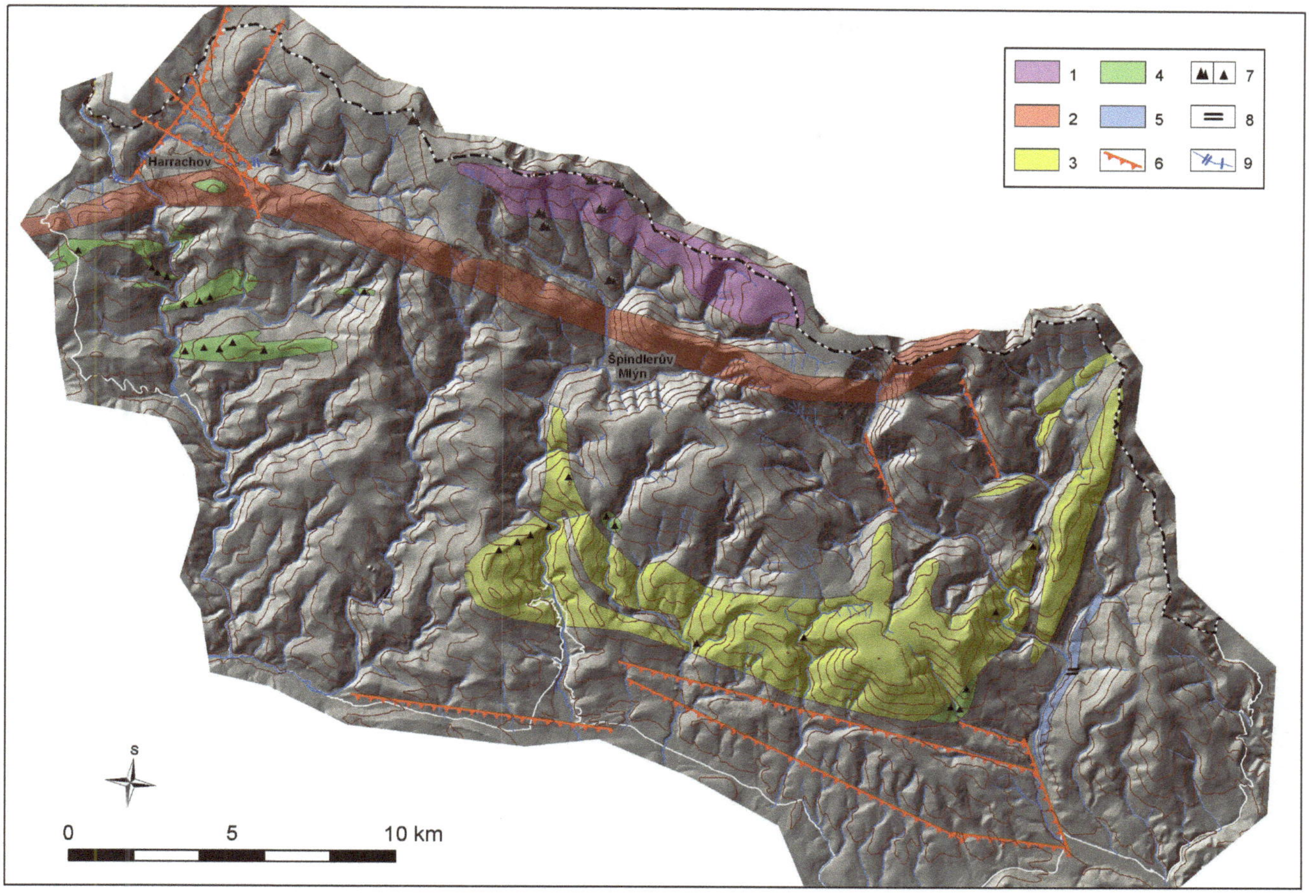

Fig. 15.3 Main structural and tectonic landforms in the Czech part of the Krkonoše Mts. *1* Hard, fine-grained granite, *2* hard rocks of contact zone (mainly quartzites), *3* muscovite orthogneiss, *4* sericitic quartzites, *5* crystalline limestones, *6* most conspicuous neotectonic faults, *7* granite tors and rock walls in crystalline rocks, *8* karst canyons, *9* tectonically and structurally conditioned waterfalls

The most peculiar rock formation here is a quartzite vein exposed by selective erosion and weathering, the so-called Končinská skalní zeď rock wall.

15.5 Glacial Landforms

Among all geomorphic features, glaciation of the Krkonoše Mts has been studied by geomorphologists most often. Partsch (1894), and later Migoń (1999) drew attention to an important role of preglacial landscape in the glaciation, particularly the extent and compactness of deflation zones on summit planation surfaces in the relatively low-altitude Sudetes mountain range. The latest and most complex findings on the extent of glaciation within the Krkonoše Mts have been presented by Engel (1997, 2003, 2007). However, other new findings show that some of the previous conclusions on the existence of a plateau ice field (Sekyra and Sekyra 2002) and the glaciation itself (Pilous 2012) need to be subject to new research, not only from the point of view of the role of anemo-orographic systems. Another issue that has to be reassessed is the number of glaciation periods. Whilst Riss (or even older) glaciation (>150 ka), in addition to the Last Glacial period, was proposed previously, recent evidence based on cosmogenic nuclide dating of moraine boulders suggests that all preserved accumulation landforms (moraines) come from different phases of the last glaciation period—Weichselian/Würm, mainly the most recent ones (25–10 ka). Contemporary research mainly involves cross-sectional profiling and longitudinal analysis, radiometric dating and macroscopic analysis of glacial sediments that, among others, are performed in order to contribute to the correlation of moraines on both sides of the mountain range (Engel et al. 2010, 2014).

Although the Krkonoše Mts are the highest mountain range of all central European Variscan mountain ranges, they were relatively little glaciated, especially if compared with the lower Vosges Mts and Schwarzwald. The latter can be characterised by a high degree of intervalley transfluence, whilst their highest parts were covered by an ice cap of an

Fig. 15.4 Kozí hřbety Ridges in the line of the contact zone represent the most conspicuous passive structural landform of the Krkonoše Mts (*Photo* V. Pilous)

incomparably larger extent. This was caused by more oceanic climate affecting these two mountain ranges and higher precipitation totals. Unresolved and most likely underestimated remains the role of the proximity of a Scandinavian ice sheet whose front almost reached the foot of the Krkonoše Mts during older glacial periods (Marine Oxygene Isotopic Stage 12 in particular)—a unique situation in the belt of high Variscan mountain ranges. Generally, it is supposed that the ice sheet cooled climate in the foreland area and thus in the Krkonoše Mts too. However, no studies have been made about the possible role of katabatic winds descending from the Scandinavian ice sheet. These winds are strong and dry, negatively affecting precipitation totals in the glacier foreland and at the same time they increase evaporation by adiabatic heating.

Opinions on the altitude of the snowline in the Krkonoše Mts are not uniform, with the variability of estimates being also related to different methods used by different authors. The most representative methods for three periods of the Würm glaciation are the Höfer and Penck–Brückner methods that specify the snowline altitude in the first period as of 1095 m (or 960 m using P–B method), 1055–1130 m in the second period (990–1075 m) and 1060–1170 m in the third period (1030–1100 m).

Given that very little of the Krkonoše Mts juts out over the snowline, a substantial effect on the distribution and size of glaciers was exerted by the location, extent and compactness of deflation zones in combination with the direction of dominant winds (i.e. anemo-orographic system). All larger glaciers originated in leeward, north- or east-oriented cirques with turbulent airflow, snow accumulation and formation of snow overhangs and largest avalanches in leeward positions. On the other hand, west-east trending valleys acted as funnels of accelerated airflow. Since the snow was constantly removed, the creation of glaciers in the west-facing valley heads was greatly limited.

Generally, glaciation of the Krkonoše Mts had a character of small solitary glacial units (Fig. 15.5). Despite this, joint erosion of the glaciers of Obří důl and Łomniczka valley on the Polish side led to the creation of the only horn summit of the Bohemian Massif (Mt Sněžka 1603 m). The Polish, shaded side was preferred for the formation of glaciers but

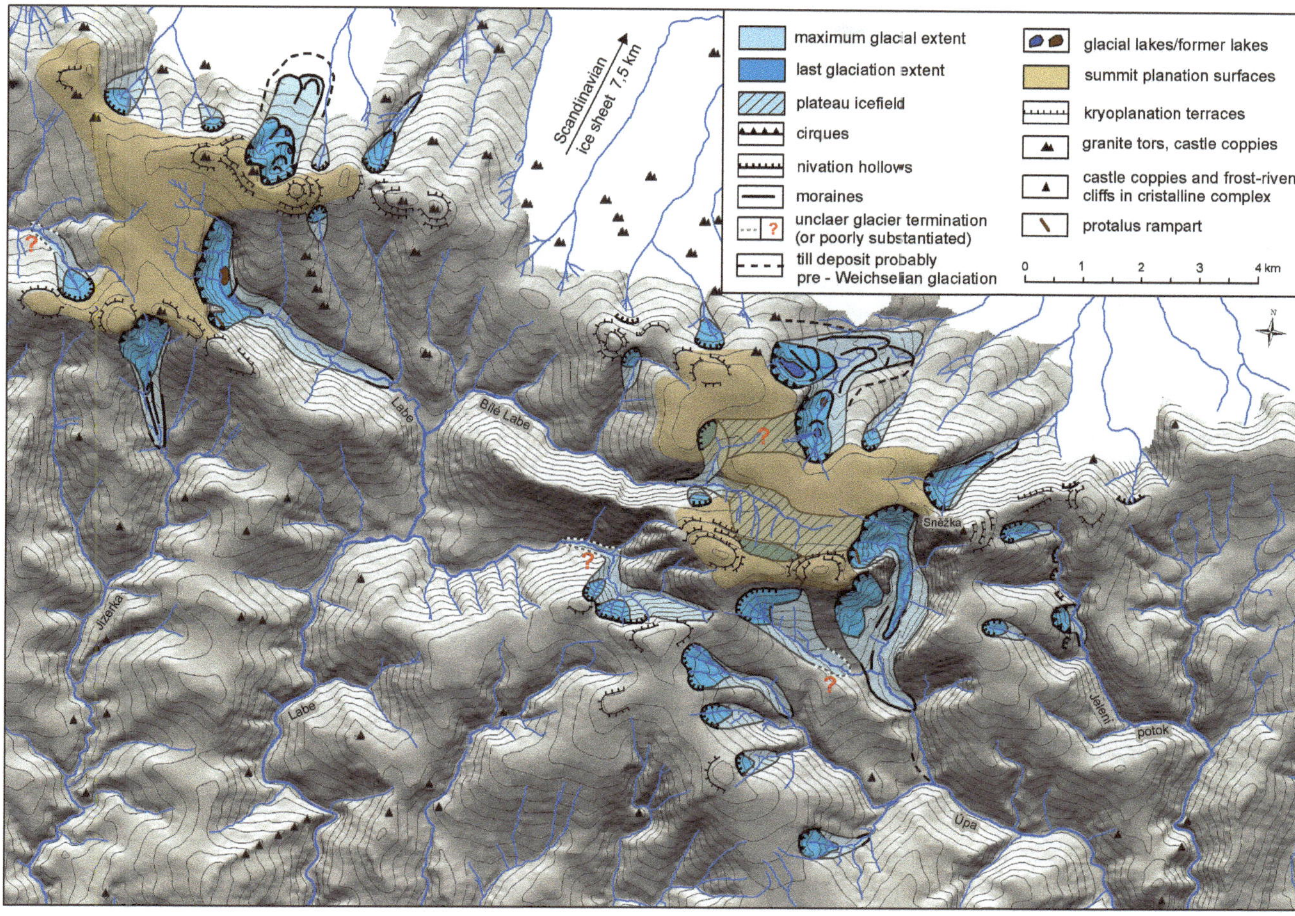

Fig. 15.5 Pleistocene glaciation in the Krkonoše Mts

its little segmented slopes were less favourable to the formation of longer valley glaciers. In the Krkonoše Mts there were three valley glaciers that created U-shaped valleys. Two of them, both about 5 km long and 100 m thick in the period of most extensive glaciation, are Labský důl and Obří důl troughs (Fig. 15.6), while a smaller one occurs on the Polish side in the Łomniczka valley. Even though other two glaciers created shorter ice tongues too (the glacier of Kotelní jámy and the glacier of the Wrzosówka valley in Poland), they reshaped the respective valleys below the cirques only marginally. Two largest glaciers on the Polish side fed by two cirques each (a 2-km-long glacier of Velká and Malá Sněžná jáma cirques and a 3.5-km-long glacier of the basins of Velký and Malý Staw lakes of the Łomnica stream) created wide ice tongues owing to the presence of preglacial landforms. The other glaciers of the Krkonoše Mts had a form of cirque glaciers and only a few largest of them extended as short (<1 km) ice tongues. Within the Czech mountains a specific feature of the Krkonoše Mts are remarkable hanging glacial cirques (Harrachova jáma cirque in the Labský důl trough and Velká a Malá Studniční jáma cirques in the Obří důl trough). Glacial erosion also had a great effect on the development of highest waterfalls in extra-alpine Central Europe. Two highest waterfalls are found on the edges of glacial cirques, in parts where tectonically uplifted planation surfaces meet glacially eroded landforms. These are the 148-m-high Pančavský vodopád Waterfall, and 129-m-high Horní Úpský vodopád Waterfall (Fig. 15.7). The most extensive moraines were left by two largest slope glaciers on the northern Polish side. As for the Czech side, the largest moraines can be found in the Obří důl trough, the only valley in which moraines have been preserved from all three glacial periods. Some moraines can also be identified in the Labský důl trough, below the Kotelní jámy cirques, below Vlčí jáma cirque and in the Jelení důl trough. The occurrence of moraines is related to glacial lakes whose number in the Krkonoše Mts was very small if compared with even less elevated Šumava Mts, Schwarzwald or the Vosges Mts. The only real cirque lakes are the Velký and Malý Staw lakes on the Polish side of the mountain range. The other lakes represent minor water bodies or even slowly disappearing bogs (7 of them within

Fig. 15.6 Glacial trough of the Obří důl Valley, with a U-profile (Sněžka Mt horn in the background) (*Photo* V. Pilous)

the moraine terrain below the Sněžné jámy cirques and one of them below the Kotelní jámy cirque, named Mechové jezírko Lake). The most important discovery of the latest years is the recognition of a former glacial lake in the headwater part of the Labský důl trough that existed from the period of the Late Glacial to Early Atlantic. This lake was later filled with sediments derived from an alluvial fan of the Labe River under the Labská rokle Ravine and consequently covered by humolites (Engel et al. 2010). On the other hand, landforms and gradient at some localities do not exclude a possibility that there may be more such vanished lakes in the Krkonoše Mts.

The latest findings of Engel et al. (2014) show that there were only small cirque glaciers during the last glaciation. The transition from glacial to periglacial conditions took place in the top parts of the Krkonoše Mts at the turn of the Early and Middle Holocene.

15.6 Periglacial Landforms

There is a wide range of periglacial landforms in the zone of relict planation surfaces (etchplains) which cover more than 10 km^2 in the highest part of the Krkonoše Mts. With regard to the type and extent, the diversity of these landforms greatly exceeds that recorded in other high Variscan ranges (both High Sudetes Mts—Králický Sněžník Mts and Hrubý Jeseník Mts, as well as Šumava Mts, Schwarzwald and the Vosges). Striking differences can be identified particularly when compared with the Vosges and Schwarzwald; the reason is related to growing continentalism eastwards and hence, the absence of glaciers in the top parts of the Krkonoše Mts. Other reasons include altitude, total extent and compactness of both etchplain remnants and perfectly developed anemo-orographic systems whose W–NW airflow removed the snow cover and thus enabled intensive frost action on regolith.

The occurrence but also the very identification of periglacial landforms is connected with the alpine timberline and dwarf mountain pine cover in locations above it. If landforms occurring below the timberline can nowadays be considered fossil, interpretation of landforms above the timberline is more complex. In the latter case, there are fossil landforms as well as those that have developed continuously, albeit very slowly, in the Holocene due to local tundra climate. In addition, the formation of others can be revived under certain circumstances, whilst some have developed only recently. In the latter case, their formation often takes

Fig. 15.7 Pančavský vodopád Waterfall in the Labský důl Trough is the highest in the whole extra-alpine central Europe (148 m) (*Photo* V. Pilous)

place due to combination of abiogenic and biogenic (vegetation) processes (Soukupová et al. 1995).

The first summarising overview of periglacial landforms in the Krkonoše Mts was presented by Králík and Sekyra (1969). Increased attention has been paid to these landforms and their genesis particularly in the last two decades (Sekyra and Sekyra 1995; Křížek 2007; Treml et al. 2010; Křížek and Uxa 2013). The most conspicuous and largest landforms are cryoplanation terraces occurring in various extent, length and perfection on almost all peaks of the Krkonoše Mts, although those in the altitude belt of 1300–1450 m a.s.l. are nowadays often hidden by vegetation (both forests and mountain pines). The landforms on several highest peaks (Vysoké Kolo, Luční hora, Studniční hora, Sněžka) jutting above the alpine timberline are perfectly developed and evident to the observer (Dvořák et al. 2004). They combine with stone polygons on the Luční hora Mt (Křížek and Uxa 2013). The summit part of the Luční hora Mt formed by cryoplanation terraces on all sides and in places arranged into a few levels separated by conspicuous frost-riven scarps (Fig. 15.8) is unique in the whole Central Europe.

Gentle slopes of planation surfaces in the top parts of the range are favourable for the development of periglacial landforms. On the other hand, steeper northern slopes of the Luční hora Mt show subsequent disturbance of periglacial landforms by gelifluction (solifluction) processes. Certain differences in the appearance of periglacial landforms are rock-controlled. Granites, with widely spaced block-like partings, disintegrate into coarse fragments (Vysoké Kolo Mt and Malý Šišák Mt), whilst schists break down into finer, platy fragments (Luční hora Mt, Studniční hora Mt and Sněžka Mt). Schists thus create more perfect forms of sorted patterned ground (stone polygons).

Frost sorted patterned ground in the form of polygons, stripes, networks and occasionally also circles are found almost exclusively above the alpine timberline, particularly on the Slezský hřbet Ridge and Český hřbet Ridge (Křížek and Uxa 2013). The most perfect forms can be observed on the Luční hora Mt (Fig. 15.9), Studniční hora Mt and Obří hřeben Ridge. On the other hand, less elevated parts of the mountain-top etchplain, covered with much finer grained weathering products and a large amount of humolites, contains earth and peat hummocks, peat and minerogenic pounus (Bílá louka and Pančavská louka Meadows); some solitary peat hummocks that resemble small palsas are characterised by an ice core that remains frozen up to several weeks longer than the surroundings.

Relatively more inclined slopes of top planation surfaces show cryogenic landforms affected by gravitation. Granites disintegrating into bigger fragments support the formation of block streams (Vysoké Kolo Mt and Malý Šišák Mt), while schists favour the origin of solifluction lobes and tongues (Luční hora Mt and Studniční hora Mt). Recent manifestations of cryogenic and solifluction processes include the so-called ploughing blocks that in the Krkonoše Mts are limited to granite bedrock since only granites with block-like partings create large boulders of sufficient weight. A wide range of small Holocene to recent microforms created by mutual participation of cryogenic processes and vegetation can also be observed on the surface of upland moorlands (Soukupová et al. 1995) or peat substratum. One of the last described forms is the so-called mountain beaded streams.

Products of frost weathering of bedrock in mountain-top locations are large boulder and block fields. Frost weathering

Fig. 15.8 Luční hora Mt, the second highest peak of the Krkonoše Mts and a part of the summit planation surface, is characterised by the highest concentration of periglacial landforms; remnant snow patches emphasise cryoplanation terraces and solifluction tongues (*Photo* V. Pilous)

processes on the valley slopes at lower locations in the Krkonoše Mts contributed to the origin of numerous frost-riven cliffs that are often structurally conditioned, i.e. connected with outcrops of more resistant rocks. Due to frost weathering, ridges are shaped into rock crags, sometimes accompanied by debris covers. However, cryoplanation terraces have not formed because of excessively steep slopes.

The summit parts of the Krkonoše Mts also contain small Holocene, subrecent and recent nivation forms, particularly in places of extraordinary accumulation of snow by wind. As a result of nivation processes acting on bedrock by compacting and pushing firn masses, minor nivation hollows have formed. These, however, need to be distinguished from the fossil and much larger Pleistocene nivation hollows in valley heads. Small nivation forms can be found e.g. on Vysoké Kolo Mt, Kotel Mt, Luční hora Mt and in the incised reach of the Bílé Labe River below the Luční bouda challet. Yet, the most conspicuos form is on the southern slopes of the Studniční hora Mt. Nivation forms also include protalus ramparts that can be found on steeply inclined valley bottoms or on cirque steps of some of the cirques (Harrachova jáma cirque, Krakonošova zahrádka area in Obří důl trough).

15.7 Relief Development in the Holocene

A dominant geomorphic process during the Holocene is water erosion whose effects can be observed after each flood. In the context of Czech mountains it is the local evidence of river braiding that needs to be pointed out and that can particularly be studied in the cirques of the Labský důl trough and Obří důl trough (Figs. 15.10 and 15.11). Except for a single case, these areas have nowadays become stabilised and overgrown. The exceptional case is represented by joint erosion of the Rudý potok Brook on its outwash plain connected with the bottom of the U-shaped valley of the Obří důl trough downstream from the Dolní Úpský vodopád Waterfall. At this location, signs of older but temporarily stabilised (up to 1970s) river braiding had been observed before but during a flood in 1974 river braiding was revived. As a result, the braiding section zone is

Fig. 15.9 Periglacial stone stripes in the eastern part of the summit plateau of the Luční hora Mt. The view towards the east reveals relief diversity of the eastern summit planation surface; Modré sedlo Saddle and Bílá louka Meadow in the centre as a part of the planation surface; the Studniční hora Mt with cryoplanation terraces and a three-sided horn of the Sněžka Mt in the background, the only peak of this type in the Bohemian Massif (*Photo* V. Pilous)

extended with every single flood as a boulder-rich alluvial deposits within a forest, of an extent of several hectares.

There are two types of Holocene (up to recent) processes that are important in the Krkonoše Mts and in terms of intensity cannot be compared with any other mountain range in the Bohemian Massif. The first type, debris flow, is fast slope movement (Fig. 15.10) whose resulting forms were first described and genetically differentiated by Pilous (1973, 1975, 1977). Even though since that time debris flows have been recognised or newly observed in other five mountain ranges of the Czech Republic, their number and role in the development of landforms is undoubtedly most important in the Krkonoše Mts. It is reasonable to assume that debris flows have been occurring throughout the whole Holocene. However, considering how fast debris flow landforms can vanish, only those younger than 150 years can be identified. Most of them and at the same time the largest of them originated during catastrophic floods in the year 1882 and especially in 1897, when they destroyed two houses in the Obří důl trough and killed 7 people. Nowadays, there are over 250 debris flow tracks recognised, among them about 180 in the Czech part of the mountain range and 70 in the Polish part of the range. The largest debris flow tracks on the slopes of the Sněžka Mt are 900 m long and cover c. 5 ha (Fig. 15.12). The total area affected by debris flows identified so far can be estimated for c. 100 ha and the volume of material they moved for c. 300 000 m^3. Debris flows often occur in clusters, which is conditioned by strictly local concentration of torrential rainfall. The most important localities are Obří and Dlouhý důl troughs in the Czech

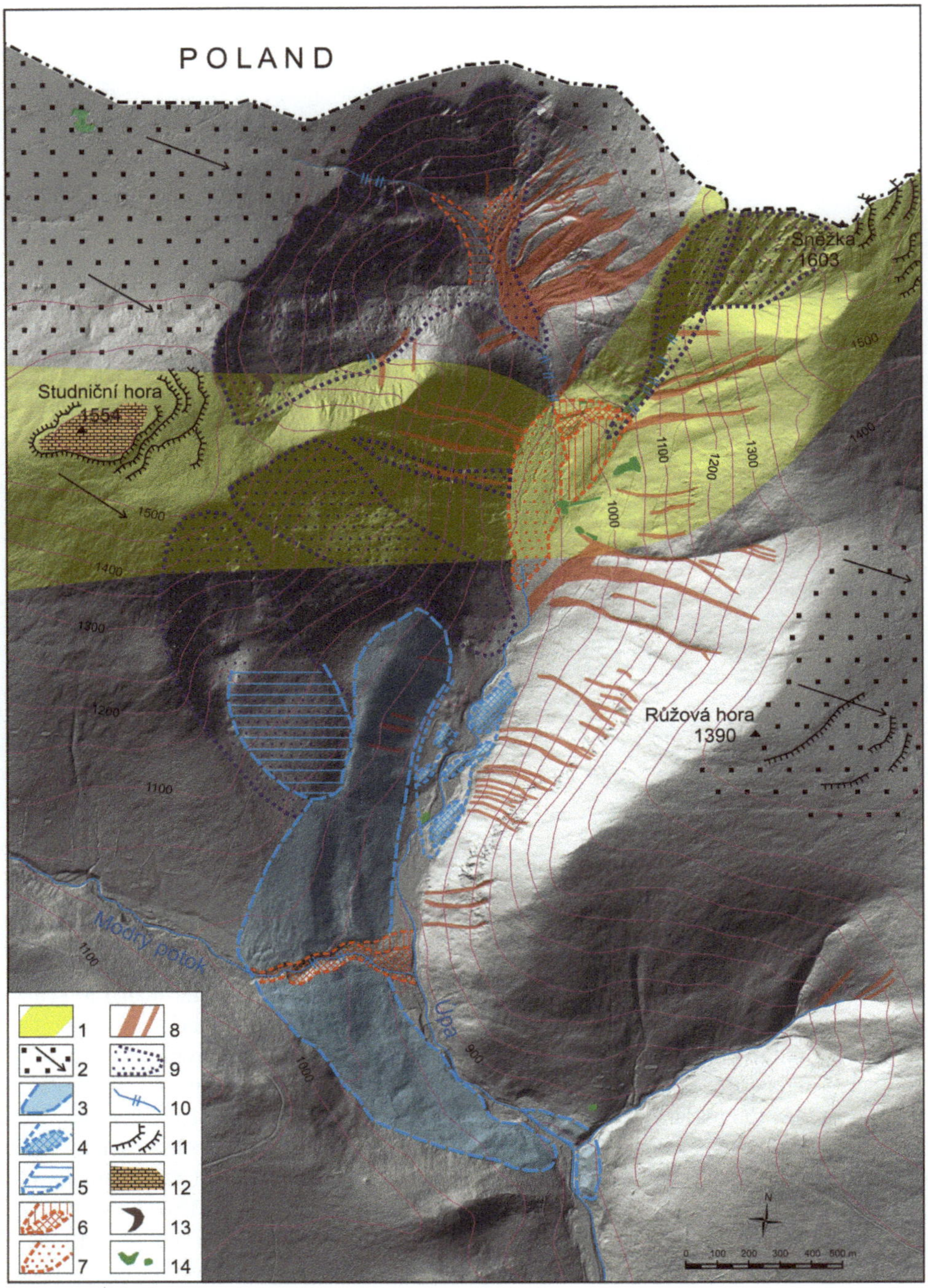

Fig. 15.10 The Obří důl Trough, the largest glacial trough of the Krkonoše Mts is a perfect example of polygenetic Pleistocene and Holocene development of the mountain range including intensive recent geomorphological processes. *1* Structural forms built by hard rocks of the contact zone, *2* deflation areas on planation surfaces, with the direction of prevailing winds, *3* moraines of maximal glaciation, *4* moraines of the last glaciation, *5* unsorted glacier accumulation (rock glacier?), *6* fossil and active alluvial fans, *7* braided river, *8* debris flow tracks, *9* avalanche paths, *10* waterfalls, *11* cryoplanation terraces, *12* patterned ground, *13* protalus rampart, *14* mining landforms

Fig. 15.11 Recently braiding channel of the Úpa River in the Obří důl Trough (*Photo* V. Pilous)

Krkonoše Mts and the Łomniczka Valley in the Polish part of the Krkonoše Mts.

Considering their height and shape, the Krkonoše Mts are known for a large number of avalanche paths (51 in the Czech part and 50 in the Polish part). On average, 50 avalanches slide down the avalanche paths every season. Monitoring and research in the last few tens of years (Kociánová et al. 2004) showed that some snow avalanches, particularly ground avalanches, are a significant geomorphic agent as they entrain and transport weathered material. Research has brought new findings such as the occurrence of "jumping" rocks hurled as far as tens of metres, or the occurrence of avalanche tables—a small analogy of glacier tables—that originate during thawing of an avalanche mass.

15.8 Man-Made Landforms

Man started to penetrate the interior of the Krkonoše Mts in the Fourteenth century, mainly in connection with the extraction of ores. Although the occurrence of ores and raw materials (particularly limestones) was extensive, only a few deposits were of greater significance (Svatý Petr Valley, Černý Důl and Obří důl trough). Minor deposits were exhausted relatively quickly and mines ceased to exist. Yet, a large number of mining landforms, mainly waste dumps and quarries, can still be found all over the mountain range. The mountain range as a whole was only locally affected. The only exception was gold extraction that culminated in the sixteenth century. Extraction was mainly carried out at

Fig. 15.12 Slopes of the Sněžka Mt above the cirque of the Úpská jáma cirque are the place of highest concentration of debris flow activity, with the longest and widest debris flow tracks in the Krkonoše Mts (*Photo* V. Pilous)

the surface by washing gold-bearing weathered material. This left numerous noticeable landforms in the form of hundreds-of-metres long and several metres deep mine trenches (Fig. 15.13). Apart from that, gold mining took place in shallow pits along water streams by means of panning and in a few cases in the form of deep mining. The area affected by surface extraction exceeds 50 ha and was concentrated at the foot of the eastern Krkonoše Mts, in a strip extending from the Černý důl trough, via Rudník, Javorník and Svoboda n. Úpou to Sklenařovice at the foot of the Rýchory Massif (Pilous 1986; Tásler et al 2003). In some localities water ditches were built in order to bring water necessary for placer mining from mountain streams situated at higher altitudes.

A very specific phenomenon of the Krkonoše Mts is represented by agrarian landforms. The original forests were almost cut down in the Sixteenth century to satisfy the needs of silver mines in Kutná Hora (central Bohemia) to which timber was floated on the Labe and Úpa rivers. After the forests were cut down, local people started agricultural activities by establishing numerous meadow enclaves. In order to enhance fertility they collected rocky fragments and stones from the fields and accumulated them at selected places to create stone piles and embankments along the enclaves. Larger meadows were frequently divided with inner stone piles and solitary simple and compound embankments. Due to their enormous extent, these agrarian landforms have become an important relief component and along with scattered settlements, scrub and tree vegetation that covers them, they are one of major landscape attributes of the Krkonoše Mts.

15.9 The Krkonoše Mts in Tourism, Sport and Art

The Krkonoše Mts as the highest mountain range of continental Europe to the north of the Alps (except Scandinavia) became a popular destination at the very beginning of tourism. Original agricultural and pastoral settlements scattered all over the range up to ridge locations (called "boudy"—chalets) provided tourists with board and lodging; first, as a secondary activity but later the settlements changed directly into accommodation units. Since the end of the nineteenth century the Krkonoše Mts have been the most visited mountain region in the Czech lands, with the Sněžka Mt, the Labe River spring and the Mumlavský vodopád Waterfall being the most favoured locations among visitors. In the last few tens of years tourists have preferred to visit the range for winter sports, mainly downhill skiing. In this respect, the Krkonoše Mts played a vital role in the beginning of skiing in the whole Central Europe: the first group of people to use skis was forest personnel of Earl Harrach in 1892. Tourism and winter sports gave rise to fast growth of infrastructure (construction of new roads, ski slopes, ski lifts and cableways), by which they get into conflict with landscape protection organisations, mainly because the Krkonoše Mts as being a relatively small mountain range are at the same time the first and most important national park in the Czech Republic.

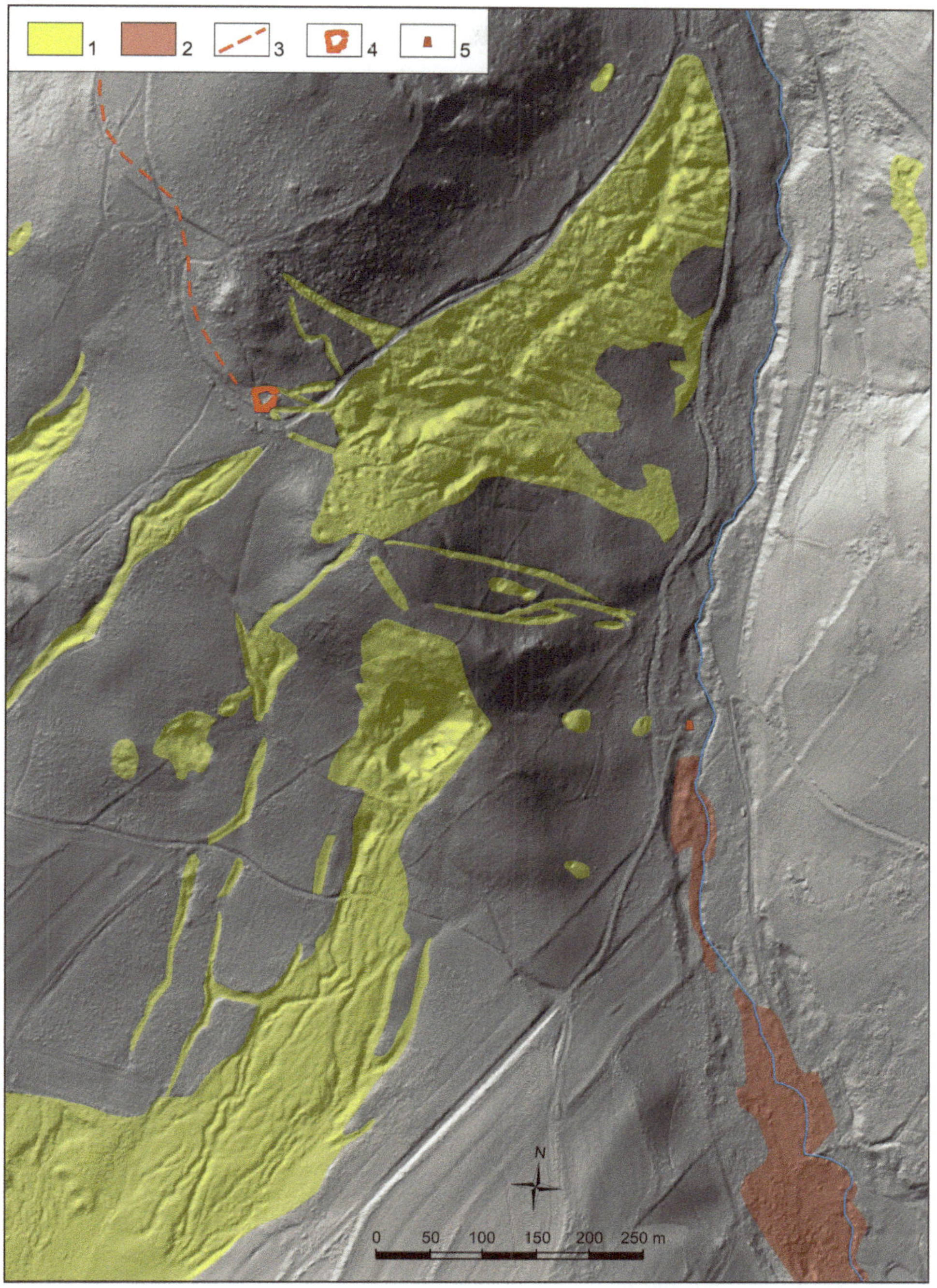

Fig. 15.13 Bartův les Forest and Sklenařice Municipality at the foot of the Rýchory Ridge. An example of relief strongly remodelled by surface ore mining. *1* surface mining areas, *2* placer fields, *3* water ditch, *4* water reservoir, *5* mouth of hereditary shallow pit

The attractiveness of nature in the Krkonoše Mts also attracted artists, particularly painters. Romantic or classicist depictions of local nature belong to the oldest ones in the Czech territory. Painting concentrated on landscape scenes and attractive geomorphological forms such as cirques, waterfalls and "idylic" life of local highlanders. The most

Fig. 15.14 View of the deepest cirque of Úpská jáma, with the Horní Úpský vodopád Waterfall on the cirque wall (in the background) and Dolní Úpský vodopád Waterfall across a rock step (in the foreground). This coloured etching by Anton Balzer from the year 1794 shows local morphology with astounding detail (*Photo* from the collection of the Krkonoše Museum of the Krkonoše NP Administration Office in Vrchlabí)

famous collection of coloured etchings is that of Anton Balzer from the year 1794 (Fig. 15.14); other renowned painters are František Kaván and Fritz Hartmann. Another important role was played by glass-making industry represented especially by the glass factory in Harrachov. Apart from that, the Krkonoše Mts became one of literary topics. In this respect, a specific position is that of fairy tales and legends about Krakonoš, a good and just spirit of the Krkonoše Mts.

15.10 Conclusion

The Krkonoše Mts are the highest mountain range of the Bohemian Massif. Beside the altitude itself, it was their specific geological and tectonic structure that played an essential role in the development of landforms that are different from those of neighbouring high mountain ranges. As a result, we can observe processes and landforms which other Variscanmountain ranges either miss or include at a much smaller scale or in less typical representation. Therefore, the Krkonoše Mts comprise polygenetic relief containing a mosaic of different landforms within a relatively small area. They include Tertiary planation surfaces—etchplains, tectonic and erosion slopes, structural landforms, as well as glacial, periglacial and karst landforms. Geomorphic diversity is enhanced by contemporary erosion and avalanche processes, as well as combination with extremely strong action of orographically conditioned winds in the form of the so-called anemo-orographic system. All this enriches the mountain range with individual meso-, micro- and nanoforms, many of which also substantially influenced the variety of living nature (e.g. cirques, summit etchplains). This was also the reason why the Krkonoše Mts were declared as the first and most precious national park of the Czech Republic.

References

Chaloupský J (ed) (1989) Geologie Krkonoš a Jizerských hor. ÚÚG, Praha 288 p

Danišík M, Migoń P, Kuhlemann J, Evans NJ, Dunkl I, Frisch W (2010) Thermochronological constraints on the long-term erosional history of the Karkonosze Mts., Central Europe. Geomorphology 117:78–89

Dvořák IJ, Kociánová M, Pírková L (2004) Příklad využití technologií GPS a GIS při studiu kryoplanačních teras na Luční a Studniční hoře. Opera Corcontica 41:18–24

Engel Z (1997) Pleistocénní zalednění české části Krkonoš. Przyroda Sudetów Zachodnich 6:223–234

Engel Z (2003) Současný stav poznatků o pleistocénním zalednění české části Krkonoš. Geografie – Sborník ČGS 102:288–302

Engel Z (2007) Late Pleistocene glaciations in the Krkonoše Mts. In: Goudie AS, Kalvoda J (eds) Geomorphological Variations. P3 K, Praha, pp. 269–285

Engel Z, Nývlt D, Křížek M, Treml V, Jankovska V, Lisá L (2010) Sedimentary evidence of landscape and climate history since the end of MIS 3 in the Krkonoše Mountains, Czech Republic. Quat Sci Rev 29:913–927

Engel Z, Braucher R, Traczyk A, Laetitia L, Team Aster (2014) ^{10}Be exposure age chronology of the last glaciation in the Krkonoše Mountains, Central Europe. Geomorphology 206:107–121

Jeník J (1961) Alpínská vegetace Krkonoš. Kralického Sněžniku a Hrubého Jeseníku. Teorie anemo-orografického systemu, Academia, Praha 460 p

Kociánová M, Špatenková I, Tondrová A, Dvořák IJ, Pilous V (2004) Základové a smíšené laviny ve vztahu k přemísťování svahovin a dynamice vegetace. Opera Corcontica 41:86–99

Králík F, Sekyra J (1969) Geomorfologický přehled Krkonoš. In: Fanta J (ed) Příroda Krkonošského národního parku. SZN, Praha, pp 59–87

Křížek M (2007) Periglacial landforms above the alpine timberline in the High Sudetes. In: Goudie AS, Kalvoda J (eds) Geomorphological Variations. P3 K, Praha, pp 313–337

Křížek M, Uxa T (2013) Morphology, sorting and microclimates of relict sorted polygons, Krkonoše Mountains, Czech Republic. Permafrost Perigl Proc 24:313–321

Lysenko V (2007) Morfotektonická analýza Krkonoš pomocí metod dálkového průzkumu Země. Opera Corcontica 44(1):37–42

Migoń P (1992) Tektoniczne formy rzeźby na północnym stoku Karkonoszy. Opera Corcontica 29:5–24

Migoń P (1999) The role of preglacial relief in the development of mountain glaciation in the Sudetes, with the special reference to the Karkonosze Mountains. Z Geomorph NF Suppl 113:33–44

Migoń P, Pilous V (2007) Geomorfologie. In: Flousek J et al (eds) Krkonoše – příroda, historie, život. Baset, Praha, pp 103–124

Ouvrier H (1933) Beiträge zur Morphologie des Hohen Riesengebirges. Veröff Schles Ges Erdkunde, Breslau 88p

Partsch J (1894) Die Vergletscherung des Riesengebirges zur Eiszeit. Forschungen zur deutschen Landes- und Volkskunde 8(2):103–194

Pilous V (1973, 1975, 1977) Strukturní mury v Krkonoších – I. I. III. část. Opera Corcontica 10:15–69, 12:7–50, 14:7–94

Pilous V (1986) Antropogenní montánní tvary reliéfu v Krkonošském národním parku – III. část. Opera Corcontica 23:5–52

Pilous V (2001) Giant Mountains – the story of rock and ice. Vrchlabí, 31 pp

Pilous V (2012) Zalednění Jeleního dolu ve východních Krkonoších ve vztahu k anemo-orografickým systémům. Opera Corcontica 49:101–120

Sekyra J, Sekyra Z (1995) Recent cryogenic processes. In: Soukupová L et al., Arctic-alpine tundra in the Krkonoše, The Sudetes. Opera Corcontica 32:5–88

Sekyra J, Sekyra Z (2002) Former existence of a plateau icefield in Bílá Louka meadow, Eastern Giant Mountains: hypothesis and evidence. Opera Corcontica 39:35–43

Soukupová L, Kociánová M, Jeník J, Sekyra J. (eds.) (1995) Arctic-alpine tundra in the Krkonoše, The Sudetes. Opera Corcontica 32:5–88

Tásler R et al (2003) Těžba zlata v okolí Svobody nad Úpou. Svoboda n. Ú., 50 pp

Treml V, Křížek M, Engel M (2010) Classification of patterned ground based on morphometry and site characteristics: A case study from the High Sudetes, Central Europe. Permafrost Perigl Proc 21:67–77

Bohemian Paradise: Sandstone Landscape in the Foreland of a Major Fault

16

Jan Mertlík and Jiří Adamovič

Abstract

The region of Bohemian Paradise in NE Bohemia is the most varied sandstone landscape in the Bohemian Cretaceous Basin. This is caused by the complex stratigraphy of Cretaceous (Upper Turonian to Coniacian) sediments, abrupt lateral changes in the development of sandstone bodies, intrusions of volcanic rocks and, most of all, by the different response to stresses mediated by the Lusatian Fault in the NE. The effects of thrusting along the Lusatian Fault include rotation of blocks in its foreland, faulting and brecciation of Cretaceous sandstones, grain cataclasis and silica cementation, and deformation banding. Farther from the Lusatian Fault, stress resulted in the formation of an orthogonal system of vertical joints—a necessary prerequisite for the development of ruiniform relief within the so-called *sandstone rock cities*. A dozen of sandstone districts can be distinguished, representing patches of rugged wilderness in an agriculturally utilized land. They became attractive for vacationers and tourists as early as in the mid nineteenth century, and for climbers from the early twentieth century. The Bohemian Paradise was proclaimed a Protected Landscape Area in 1955 and entered the network of European Geoparks of UNESCO in 2005.

Keywords

Bohemian Paradise • Cretaceous sandstones • Lusatian Fault • Landform evolution • Solution caves • Rock climbing

16.1 Introduction

The region known as the Bohemian Paradise (*Český ráj* in Czech) is limited by the ridge of Kozákov (parallel to the Lusatian Fault) in the north–east, by the Jizera River in the north–west and by the Klenice stream in the south. Its southern part is dominated by low-positioned plateaus with relicts of fluvial terraces. Although no precise geographical limits of the Bohemian Paradise can be drawn, it is generally understood as a region between Turnov and Jičín, dominated by complex relief developed on Upper Cretaceous quartzose sandstones. In the regional geomorphological subdivision, it constitutes the districts of Vyskeřská vrchovina (=Hill country) and Turnovská stupňovina (=Stepped area) (Balatka and Kalvoda 2006).

The aesthetic values of the region motivated the establishment of the Bohemian Paradise Protected Landscape Area in 1955. It became the first protected landscape area in the former Czechoslovakia, even preceding the establishment of national parks. In 2005, the wider region of the Bohemian Paradise was included into the network of European Geoparks of UNESCO as the first such area from the territory of the Czech Republic. In 2015, it entered the UNESCO Global Geoparks network. A dozen sandstone

J. Mertlík (✉)
Administration of the Bohemian Paradise PLA, Turnov, Czech Republic
e-mail: mertlik@cesky-raj.info

J. Adamovič
Institute of Geology CAS, v.v.i., Prague, Czech Republic
e-mail: adamovic@gli.cas.cz

T. Pánek and J. Hradecký (eds.), *Landscapes and Landforms of the Czech Republic*, World Geomorphological Landscapes, DOI 10.1007/978-3-319-27537-6_16

districts can be identified within the region, each showing specific geological conditions and specific forms of relief and microrelief (Adamovič et al. 2006).

Landscape of the Bohemian Paradise constitutes a very fine mosaic of habitats differing in their temperature, humidity and nutritional value of soil. This implies a high variation in the vegetation cover. About half of the area is covered with forests, mostly with altered tree species representation. Plateaus with loess loams are used as agricultural land. Valleys with streams of sufficient discharge host ponds, which served for fish production or for water storage at grain mills and saw mills. The largest towns are Jičín (population of 16,300), Turnov (14,300) and Mnichovo Hradiště (8,200). The rugged relief in the central part of the region rather promoted the existence of small communities and farms on the individual plateaus. Many of the rural wooden houses date to the eighteenth–nineteenth centuries and are protected as cultural monuments (e.g. Vesec village near Sobotka). Natural and cultural values of the country, together with the lack of employment opportunities for local people, have resulted in the present-day use of most dwellings for recreational purposes only.

16.2 Emergence of the Name of Bohemian Paradise

The early nineteenth century was characterized by marked changes in the Czech society. Under growing industry, Napoleonic Wars provided a possibility of earnings on war supplies. Traditional noble families started to sell their land and real estates. An increasing interest in the recognition of Bohemian landscape spread among town people. In 1805, Zacharias Römisch purchased the Malá Skála dominion and started the construction of the Pantheon—a memorial to outstanding personalities—around the ruins of the Vranov Castle. In 1821, the Hrubá Skála dominion was purchased by Jan Antonín Lexa of Aehrenthal, and water-curing spa was founded by Antonín Vincenc Šlechta at nearby Sedmihorky in 1840 under the support of the Aehrental family. The new resort, together with the Hrubá Skála Chateau, soon became frequented by Czech artists.

After 1848, interests of minor nations in the Austro-Hungarian Monarchy became suppressed, which led to the aggravation of relations between the Czech and German ethnics. Germans living in the borderzone of Bohemia established their *Paradies von Böhmen* (Paradise of Bohemia) in the area between Litoměřice and Teplice in the České středohoří Mts. This fertile land was close to the meaning of the word 'paradise', but was inhabited almost solely by German-speaking population. The introduction of a parallel 'paradise' was initiated in 1886: Václav Durych started to publish tourist guidebooks and was the first to refer to the region between Jičín, Železný Brod, Mnichovo Hradiště and Libáň as the Bohemian Paradise. Both 'paradises' then persisted side by side until the end of World War II. After the expulsion of the German population the term *Paradies von Böhmen* became forgotten, being replaced by the name *Zahrada Čech* (Garden of Bohemia).

In this respect, the currently used term Bohemian Paradise was coined as a brand to be used for the promotion of tourism. Compared to the 10,000 years lasting inhabitation of this land, the history of this name is very short.

16.3 Geology

Bohemian Paradise is located in the Bohemian Massif, one of the exposures of the exhumed Variscan mountain belts in Europe. In this area, Variscan-consolidated crystalline rocks are overlain by a max. 1800 m thick fill of the Late Paleozoic Mnichovo Hradiště Basin and a maximum of 400 m thick sedimentary fill of the Bohemian Cretaceous Basin. In the NE, the extent of basinal sediments and volcanics is limited by the Lusatian Fault (Fig. 16.1).

The Lusatian Fault is one of major structures of the Elbe Fault Zone (Scheck et al. 2002). During the collision of the African plate and Iberian microplate with Europe in the latest Cretaceous and Paleocene, major thrusting of the NE block over the SW block occurred along the fault, with the total displacement of >2.5 km (Coubal 1990; Danišík et al. 2010; Coubal et al. 2014). The steep fault plane was orientated almost perpendicular to the acting stress and functioned as a bulldozer blade, transmitting most energy to its foreland.

As a result of this process, basinal blocks in the foreland of the Lusatian Fault show moderate to steep dips towards the SW, and a succession of progressively younger rocks is exposed at the present erosional surface in this direction. Cretaceous sandstones in a belt ca. 1 km broad along the fault show many features of brittle deformation, ranging from deformation bands with only a mm-sized displacement to large-scale striated fault planes (Coubal et al. 2014). Sandstones adjacent to the fault locally contain secondary silica cement precipitated due to tectonic and hydrothermal processes, and show therefore higher resistance to weathering.

Crystalline rocks to the NE of the Lusatian Fault comprise the Lower Paleozoic sedimentary successions intercalated with basic volcanics and their tuffs, subjected to epizonal metamorphism in the Carboniferous. This process gave rise to chlorite-sericite phyllites, metadiabases and basic metatuffites. Upper Carboniferous and Permian rocks

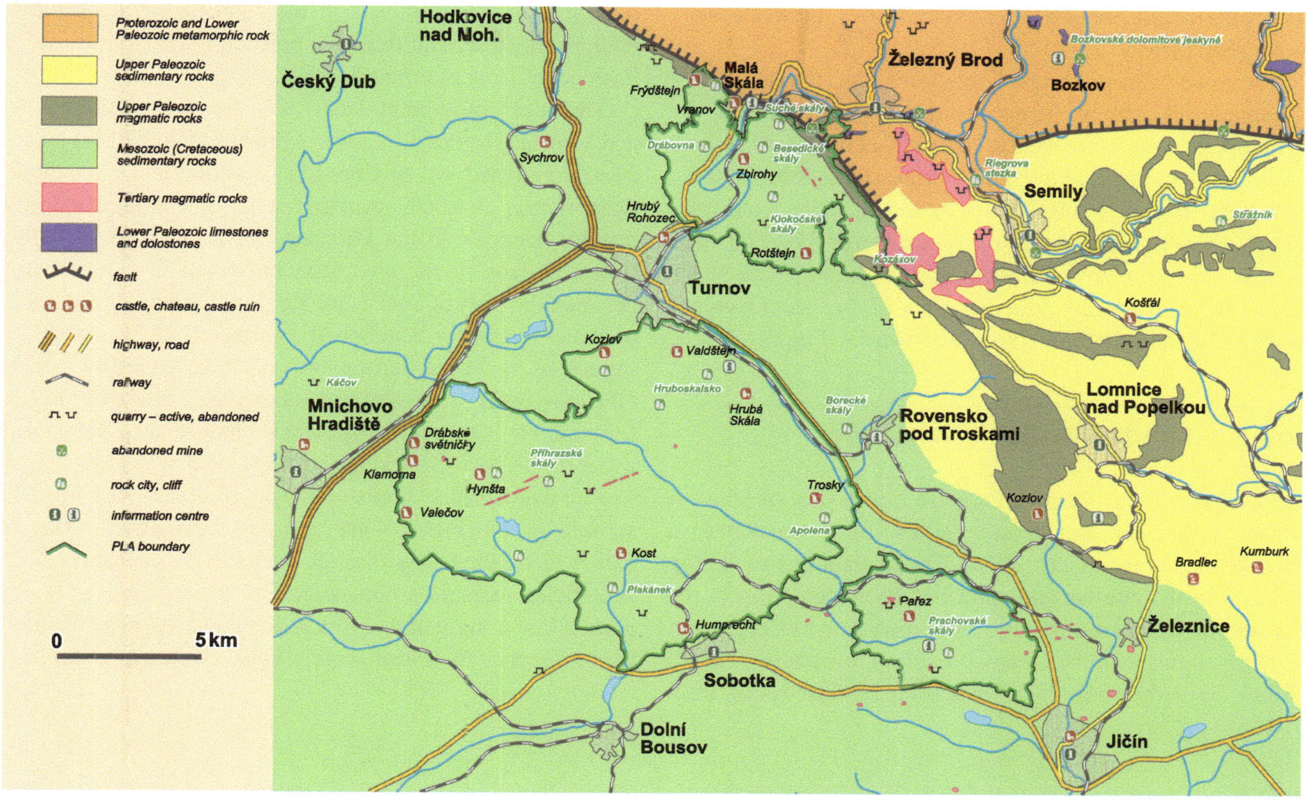

Fig. 16.1 A simplified geological map of the Bohemian Paradise. The distribution of Permo-Carboniferous rocks and Cretaceous sediments is limited by the Lusatian Fault in the north

SW of the Lusatian Fault include mudstones, sandstones and conglomerates deposited in fluvial and lacustrine environments. They are intercalated with effusions of acidic (rhyolites) and intermediate volcanics (andesites). Andesitic lava flows are best exposed in the quarries at Bezděčín, NW of Malá Skála, and at Kozákov ridge, showing a variety of void-filling minerals (chalcedony, opal, amethyst, calcite, zeolites) which laid the foundations of the local jewellery industry.

Shallow-marine sediments of the Bohemian Cretaceous Basin cover vast majority of the Bohemian Paradise region. They span the interval of 95–87 Ma (Mid-Cenomanian to Late Coniacian) and belong to five formations defined by Čech et al. (1980): the Peruc-Korycany, Bílá hora, Jizera, Teplice and Březno formations. The most important of these is the Teplice Formation (Upper Turonian to Coniacian) which includes an areally extensive body of quartzose sandstone 80–140 m thick, generally referred to as the *Hrubá Skála Quader* (Krejčí 1869). It was deposited in a subaquatic delta which prograded from the seashore (now beyond the N basin margin) towards the SE and progressively filled the whole depth of the water column (Uličný 2001; Uličný et al. 2003). Sandstones of the *Quader* contain frequent cross-bedded sets max. 8 m thick but mostly tens of centimetres thick, erosional surfaces, and most notably gently dipping surfaces interpreted as delta foresets (Uličný 2001). The upper part of the *Quader* shows a succession of upward-coarsening cycles topped with conglomerate beds. The Teplice Formation sandstones in some districts (Klokočské skály, Borecké skály; *skály* in Czech = cliffs, prominent rock outcrops) previously attributed to the *Hrubá Skála Quader* have been recently proved to represent a separate body underlying the *Quader* and prograding to the W (Čech 2013).

The Teplice Formation is separated from the underlying Jizera Formation (Middle to Upper Turonian) by an interval of calcareous claystone 10–15 m thick. The Jizera Formation is arranged into 3–4 upward-coarsening cycles. Lower and middle portions of the cycles are formed by silty calcareous fine-grained sandstones to siltstones, and their tops are formed by fine-grained, locally coarse-grained quartzose sandstones. The best exposures of the Jizera Formation are accessible in the incised Jizera river valley between Malá Skála and Turnov.

The lowermost preserved Cretaceous unit, i.e. the Peruc–Korycany Formation (Cenomanian) is represented by cross-bedded quartzose sandstones with a conglomerate bed, maximum 2 m thick, at the base. Rocks of this formation are preserved in a belt sub-parallel to the Lusatian Fault and usually show prominent tectonic dips towards SW. The areas of best exposure in the Bohemian Paradise include 'Pantheon' and Suché skály on either bank of the Jizera River at Malá Skála, and the SW slopes of Kozákov ridge.

16.4 Landscape Evolution

Landscape of the Bohemian Paradise is a mosaic of relics of structural plateaus and cuestas, subjected to erosion to a variable degree (e.g. Čech 2013). The inclinations of the top surfaces of the cuestas are controlled by tectonic dip in the Cretaceous sediments, ranging from moderate to steep (even vertical) angles near the Lusatian Fault to low angles (ca. 4°) farther from the fault (Fig. 16.2). These dips result from rotation of tectonic blocks in the foreland of the Lusatian Fault during the main thrusting episode (latest Cretaceous to Paleocene; Coubal et al. 2014).

The most prominent landmarks are exhumed conduits and erosional relics of Lower Miocene (20–16 Ma) volcanoes composed of olivine-rich basanite and nephelinite (Rapprich et al. 2007). Another stage of basanitic volcanic activity occurred in the Pliocene (5 Ma; Cajz et al. 2009). Relics of lava flows have been preserved in a broad strip between the hills of Sokol and Kozákov, the basanitic stocks of which probably functioned as magma ascent paths. As suggested by brittle deformation of these lava flows, these effusions were followed by later-stage thrusting at the Lusatian Fault and tilting of blocks which shaped the present relief of the NW–SE-elongated ridge of Kozákov.

The two basanite spires of Trosky hill (elev. 488 m)—the symbol of the region—are volcanic conduits 17 Ma old which made their way through a cinder cone. Relics of the superficial volcanic products are still preserved, which suggests a low to moderate uplift and denudation rates from the Mid-Miocene to Pleistocene. This contrasts with high

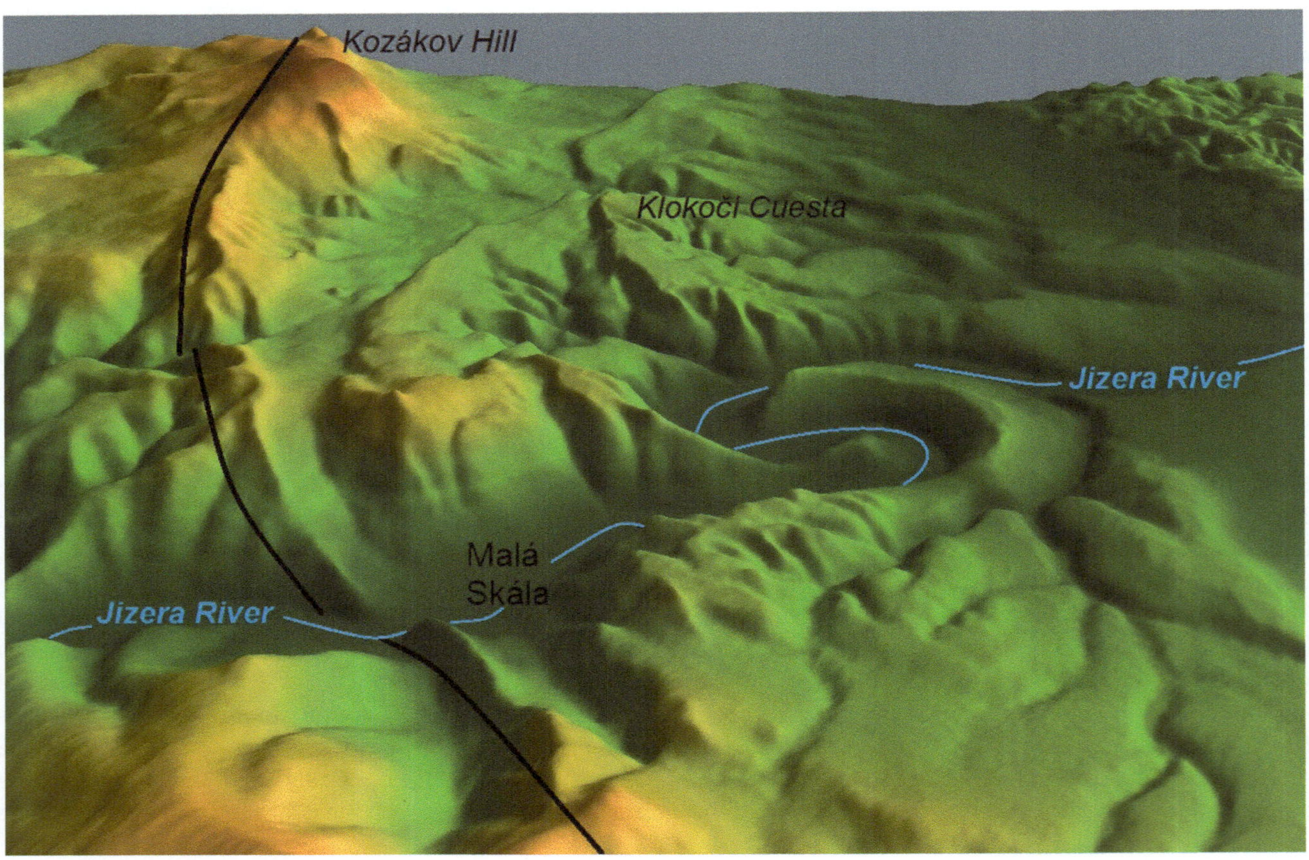

Fig. 16.2 Digital elevation model of the Malá Skála area showing the course of the Lusatian Fault (*black line*) and relief controlled by southwesterly tectonic dips of strata in the Bohemian Cretaceous Basin. A view towards the southeast

Pliocene uplift rates inferred by Rapprich et al. (2007) for the Kozákov ridge area near the Lusatian Fault.

The Pliocene paleo-Jizera river (or possibly several rivers fed by alluvial fans along the Lusatian Fault scarp) was flowing from the Turnov area towards the SE, crossing the sandstone plateau between Hrubá Skála and Vyskeř hill and depositing gravels now found in relics at topographic highs around Sobotka at 380–390 m a.s.l. In the Early Pleistocene, the Jizera channel was deflected towards the south–west and eroded a prominent valley. With the ongoing westerly shift of the Jizera river course, its left-sided tributaries of Libuňka, Žehrovka and Klenice streams were becoming more important in shaping the relief of the Bohemian Paradise.

In the Late Pleistocene, the area was covered by aeolian sediments: loess and loess loams, locally exceeding 10 m in thickness. Aggradation of fluvial sediments at the Pleistocene/Holocene transition resulted in the formation of shallow lakes and lowland moors. The thicknesses of humolites reach 8–9 m in some valleys. Streams sourced in areas of calcareous sandstones and marlstones deposited calcareous tufa mounds in the Holocene.

16.5 Suché Skály: Tectonically Dragged Sandstone Strata

The Suché skály are formed by a belt of quartzose sandstones of the Peruc–Korycany Formation (Cenomanian) whose strata were tilted and locally even overturned during the main thrusting at the Lusatian Fault. The sandstones are coarse-grained, passing to conglomerates. Sandwiched between Permian andesites and Lower Turonian marlstones, Cenomanian sandstones were an ideal channelway for hydrothermal fluids released during the thrusting and during later stages of volcanic activity. Their pore spaces were therefore partly filled with secondary silica, mainly having the character of quartz overgrowths on detrital grains. Besides hydrothermal silicification, sets of deformation bands with cataclasis and silica precipitation are very common. The presence of paired systems of deformation bands halved by bedding planes suggests that this semi-brittle deformation preceded the rotation of strata, hence also the main thrusting. Three types of fault planes are observed: 1. faults striking NW–SE, both reverse and normal, with slickensides, 2. faults and fracture zones striking NNE–SSW, with dilated fractures, only rarely showing strike-slip movement and 3. faults striking E–W, activated as strike-slip faults.

The rotated block of resistant sandstone at Malá Skála gave rise to a continuous rocky crest of the Suché skály (Fig. 16.3), which continues to the right bank of the Jizera River towards Frýdštejn (the crest of "Pantheon"). The crest is about 80 m broad at its base, ca. 1000 m long, with SSW cliff faces 40–80 m in height in its central part. It is segmented into four main "spires" by saddles at points of its intersection with transverse faults. The prominent crest of the Suché skály is a landmark of the northern limit of Bohemian Paradise and a frequent motive for paintings and photographs.

Fig. 16.3 A view of the Suché skály with houses of the Malá Skála community in the front (*Photo* J. Adamovič)

16.6 Calcareous Sandstones in the Jizera River Valley

Between Malá Skála and Turnov, the Jizera river is incised into calcareous sandstones of the Jizera Formation (Middle to Upper Turonian). The type section of the Jizera Formation of basin-wide validity lies in a road cut at Dolánky, on the right bank of the river (Čech et al. 1980). Unlike quartzose sandstones of the Teplice Formation, calcareous sandstones to limestones of the Jizera Formation are rich in fossils including oyster beds and bivalves *Pinna* in life positions. They disintegrate to centimetre- to metre-sized fragments which accumulate in talus cones at cliff bases.

Calcareous sandstones of the Jizera Formation provide substrate favourable for the origin of karstic caves, in addition to true pseudokarst caves. The most important karst system with ponors, blind and semi-blind valleys lies on the right bank of the Jizera river to the north of Turnov—the Ondříkovice System is the only documented karst system with an active subterranean stream in Bohemia. The resurgence is the cave of Bartošova pec with its explored length of 225 m, accessible with diving equipment only (Balatka and Sládek 1975; Bruthans et al. 2006).

Small valleys on the left bank of the Jizera river north of Turnov host streams which start on quartzose sandstone but sink underground as soon as they reach calcareous sandstone. Resurgence waters at cliff bases are used to supply the town of Turnov. Several crevasse caves were formed in the Jizera Formation calcareous sandstone on the W slope of Sokol Hill within an extensive landslide along NNE–SSW-striking fractures. The crevasses are often buried beneath blocks of quartzose sandstone from the Teplice Formation. Walls of the Semikraska Cave are covered with a dehydrated moonmilk coating.

16.7 Klokočské skály: A Textbook of Sandstone Geomorphology

The cliffs are developed in quartzose sandstone of the Teplice Formation (Upper Turonian) on a crustal block with a tectonic dip of ca. 5° towards the SW. This block forms a notable cuesta (Klokočí Cuesta) whose head has the character of an escarpment with sandstone pillars up to 30 m high. The cuesta is limited by the deeply incised valley of the Jizera River in the NW and by the valley of the Stebenka stream in the SE. Drainage of the cuesta is towards the SW, through canyon-like valleys, mostly with no watercourses.

The tectonic dip relates to the position of the site ca. 3 km from the Lusatian Fault. Fault proximity also increases the variety of meso- to micro-scale weathering forms. Tectonic phenomena on the cuesta include minor faults on its periphery, fracture zones, and deformation bands. The latter are very frequent but unevenly distributed along the whole cuesta. They are often associated with grain cataclasis, their bundles are typically 5–30 mm thick and several metres long, of two prevailing strikes: NNW–SSE and E–W (Fig. 16.4). Their morphological expression depends on the slope of the cliff face. A mere contrast in the shade can be observed on horizontal surfaces while grooves as much as 20 cm deep are developed on vertical faces, and low ribs are developed on overhanging faces. Riedel shears accompanying the deformation bands on horizontal surfaces locally host deep pit karren with rhombic cross sections.

Fig. 16.4 Intersecting deformation bands of two principal orientations beneath the Pětichlapka Cliff, Klokočské skály (*Photo* J. Adamovič)

Spherical concretions 0.2–2 m in diameter concentrate to a level 5–15 m below the top of the sandstone body (Mertlík and Adamovič 2005). Some of them form positive relief; others are filled with loose sand which becomes easily evacuated. The cementing agent has been mostly dissolved; poor silica cement has been found in their interiors (Adamovič et al. 2013) but a former presence of carbonate cement cannot be excluded. Surfaces of some of these concretions show shear planes (Fig. 16.5), which probably originated in a narrow zone between the hard, undeformed concretion and the ambient sandstone mostly deformed via grain re-orientation. Some of the shear planes are several square metres in area and have the character of 'tectonic mirrors'.

A special type of solution caves has developed at sites with large concentrations of cavities left after destruction of spherical concretions (Vítek 1987; Mertlík and Adamovič 2005). The septa among the cavities may get destructed due to cavity enlargement or tectonic fracturing, producing caves maximum 30 by 15 m in horizontal dimension (Postojna Cave, Fig. 16.6). Some of the caves in the Klokočské skály were formed along sub-horizontal fracture zones. Talus caves are mostly of limited length, and typical fissure caves are rare.

The rich variety of microforms of sandstone relief includes ubiquitous honeycomb pits and frequent wandkarren: series of vertical grooves separated by sharp ribs. Although quartzose sandstones with kaolinite admixture do not provide ideal conditions for the preservation of body fossils, casts of inoceramid bivalves and ferruginized wood fragments are occasionally present. Trace fossils are, however, abundant: the most conspicuous is the boring trace of ichnogenus *Planolites* 3 m long and over 10 cm in diameter.

The Klokočí Cuesta is transected by the Zelený důl (=-valley), which is over 60 m deep and reaches to the cuesta head. Inverse character of microclimate in this valley is favourable for the fern *Blechnum spicant*, a species typical of montane fir-beech forests. The lower reach of this valley hosts numerous springs. Dry top of the cuesta was originally

Fig. 16.5 A cavity after an ellipsoidal concretion with shear planes along its circumference. Klokočské skály near Rozumov (*Photo* J. Adamovič)

Fig. 16.6 The Postojna Cave is the largest cave in the Bohemian Paradise, 30 m in length (*Photo* J. Mertlík)

covered by acidophilous oak forests and relict pine forests; these were later partly replaced by artificially introduced tree species.

16.8 Hrubá Skála Area: Castles and Chateaus on Sandstone

The structural plateau lying at ca. 400 m a.s.l. south of Turnov is covered by relics of Pliocene fluvial terraces and topped by a Lower Miocene volcano of Vyskeř (466 m). Between Hrubá Skála and Turnov, this plateau is dissected by valleys drained by the Libuňka stream in the NE and the Kacanovský potok in the SW. This area of complex relief, ca. 2 by 3.5 km in size, is referred to as the Hrubá Skála rock city. The rock city is developed in the Hrubá Skála Quader—the upper one of the two immediately superimposed quartzose sandstone bodies attributed to the Teplice Formation (Upper Turonian—Coniacian). The sandstone body is cut by an orthogonal system of joints (NW–SE and NE–SW), probably originated in response to thrusting at the Lusatian Fault. The rock city is a perfect example of a fully mature sandstone landscape. It comprises hundreds of solitary pillars up to 60 m in height; its central part is referred to as *Skalák* by rock climbers (Fig. 16.7). Its SW part is called *Údolíčka* (Valleys) with pillars and cliff faces seldom exceeding 20 m in height.

The Hrubá Skála area has a long cultural history. The highest points of sandstone relief served as sites of fortified settlements and, later, of medieval castles. The chateau of Hrubá Skála (Fig. 16.8) is a prominent landmark: it lies at the site of a former castle. A curative water spa of Sedmihorky has been founded under the chateau in mid nineteenth century, making use of cold water issued at the base of the sandstone body above an impermeable claystone layer.

Fig. 16.7 The central part of the Hrubá Skála rock city known as *Kapela* (The Band) (*Photo* J. Adamovič)

The group of rock pillars called *Čertova ruka* (Devil's Hand) was the site of a Celtic fort and a timber-made medieval castle. The castle of Kavčiny was also made of wood and occupied a few rooms carved in sandstone. It was founded in the fourteenth century, conquered in 1440, and destructed shortly thereafter. It functioned as an advanced fortification to the larger Valdštejn Castle. The latter was situated on several pillars interconnected with bridges. Its ruins were adapted to meet the demands of growing tourist activities in the nineteenth century. The remains of dwellings at the rocky ridge of Chlum (Kozlov) farther to the west can be also considered a former castle or a refuge for local people during wartime.

16.9 Příhrazy Area: Tensional Joints with Ferruginous Lining

The top surface of the structural plateau around Příhrazy lies at 370 m a.s.l. It is dominated by the centrally positioned volcanic conduit of olivine nephelinite at Mužský hill (464 m), dated to 19 Ma (Rapprich et al. 2007). The exhumed conduit is surrounded by a relic of calcareous claystones overlying the Teplice Formation, baked to form porcellanites at contact with the magma. The western and northern rims of the plateau are prominent landmarks, rising some 150 m above the lowland underlain by the Jizera Formation calcareous sandstone. The cliffs are formed by

Fig. 16.8 An aerial view of the Hrubá Skála Chateau and the group of pillars known as *Dračí skály* (Dragon's Cliffs) (*Photo* J. Adamovič)

quartzose sandstones of the Hrubá Skála Quader, showing numerous sets of giant-scale cross-bedding.

Joint system striking NW–SE, parallel to the Lusatian Fault, is the most common but joints ENE–WSW are also present. The joints bear clear signs of tension: plume structures, twist hackles and twist hackle fringes (Adamovič et al. 2006). In the small rock city of *Železné věže* (Iron pillars) 1 km SW of Příhrazy, corresponding morphologies are visible on opposite faces of crevasses, indicating dilation by 1 m or more. Joint faces of all orientations in this rock city are covered by hydrothermal ferruginous linings (Mertlík et al. 2002), see Fig. 16.9. This suggests an updoming of the area around Mužský Hill during magma emplacement and hydrothermal fluid circulation.

Tensional character of joints along the plateau rim is mostly due to gravitational sliding of blocks. Mass movements are enhanced by the presence of a layer of claystone beneath the 100 m thick sandstone body, i.e. at the base of the Teplice Formation. The best described event was the landslide near Dneboh in 1926 which destroyed several houses of the Kavčina community. Slow gravitational creep is being constantly monitored by extensometers in outcrops and in boreholes.

A pseudokarst abyss 21.5 m deep lies at *Krásná vyhlídka* near the plateau rim. It evolved from a sinkhole located on a series of joints in sandstone. These joints have not been dilated by gravity as suggested by their strike perpendicular to plateau rim and by their tight character at depth. Instead, they have been widened by erosion by rainwater swallowed by the sinkhole on a plateau otherwise covered by impermeable loess. Many other sinkholes and semi-blind valleys can be found in the vicinity (Balatka 1980).

16.10 Plakánek Valley: Tufa Cascades

The Plakánek Valley lies on the southern margin of the Bohemian Paradise, NW of Sobotka. It contains the generally south-flowing Klenice stream, incised in the Hrubá Skála Quader. The body of quartzose sandstone is 55 m thick, featuring SSW-dipping clinoform surfaces interpreted as foresets of a prograding subaquatic delta body (Uličný 2001), see Fig. 16.10. The transverse profile of the valley with steep cliff faces and a flat bottom reflects the stages of erosion and subsequent aggradation: sediments up to 9 m thick have been documented from other valleys in this area since the end of the Last Glacial. The valley also hosts several springs, where water issues into pools with sandy bottoms. Ellipsoidal cavities tens of centimetres in diameter are a notable element of the sandstone microrelief; they are visible on vertical faces and often concentrate to a specific bedding plane. Their origin by dissolution of pre-existing carbonate or silica cement confined to concretions is probable (Adamovič et al. 2013).

Fig. 16.9 Ferruginous linings on tensional joints in the *Železné věže* (Iron pillars) rock city near Příhrazy. The joint in the front shows high-relief twist hackles with fringes, a clear sign of extension (*Photo* J. Adamovič)

The Kost Castle lies in the upper reach of the Plakánek Valley. It was built on a sandstone butte in the 13th century from sandstone ashlars extracted from local quarries. A reclaimed sandpit lies ca. 1 km WNW of the Kost Castle. Accumulations of Lower Pleistocene gravel and sand lie ca. 40 m above the bottom of the Plakánek Valley and reach 12 m in thickness (Valečka 2013). They were found to fill cavities in sandstone cliffs lining the valley, thereby preserving old microrelief (Balatka 1987).

The Vesecký Plakánek is a left-sided tributary valley of the Plakánek Valley. A water mill, now not in operation, was partly built in the rock massif. In its proximity, several mounds of calcareous tufa, both modern and fossil, are present. One of the fossil ones lies 200 m upstream of the mill, in the stream bed. Tufa accumulations were extracted in the past for the production of lime to be used in agriculture. It can be speculated, however, that burnt lime from this source was also used in the construction of the Kost Castle. One of the exploited tufa mounds lies in a slope above the stream west of the Semtiny Farm.

16.11 Prachovské skály: A Century of Czech Rock Climbing

The Prachovské skály evolved from a plateau of the Hrubá Skála Quader, slightly inclined to the south. The plateau is now dissected to a fully mature rock city 2.5 by 1.3 km in size, with hundreds of pillars. These are confined by an orthogonal set of vertical joints striking WNW–ESE (spacings in metres) and perpendicular joints NNE–SSW (spacings in tens of metres). The highest point of Svinčice (451 m) is an exhumed basaltic stock.

The aesthetic value of the Prachovské skály is suggested by the visit of the Austrian Emperor Franz II of the rock city in 1813, during his stay in Jičín within the military campaign against Napoleon. The first tourist paths were signposted in the 1870s with the help of Vojta Náprstek, Czech ethnographer and founder of the Czech Tourist Club. The first climbing club *Prachov* was founded in Jičín in 1907, and the area can be considered the birthplace of sandstone climbing in Bohemia. In 1933, the rock city was proclaimed a nature reserve.

Passages between the pillars are typically narrow and some of the pillars are leaning or have fallen, giving rise to false arches and talus caves in block accumulations (Vítek

Fig. 16.10 Southerly dipping bedding planes (clinoforms) in the Plakánek Valley are interpreted as a head of a submarine sand delta body prograding into the basin. The clinoforms are truncated by a subhorizontal flooding surface at their top (*Photo* J. Adamovič)

1980). The longest cave is the *Kladivo* (Hammer) Cave with a total length of over 40 m. Caves in the Prachovské skály host the tallest root stalagmites in the Bohemian Paradise.

16.12 Caves and Cavities in Quartzose Sandstone

Bohemian Paradise boasts the highest number of pseudokarst caves of all sandstone areas in the Czech Republic. Most of them are located on the Klokočí Cuesta. Although the Czech legislation provides no definition of a cave, the Slovak Act No. 543/2002 Coll. defines the cave as a natural space accessible to man, whose length or depth exceeds 2 m and is larger than the size of its entrance. About 500 caves have been registered within a speleological survey in the Bohemian Paradise, even though spaces less than 2.5 m in length were attributed rather to cavities. This figure represents about one-half of all pseudokarst caves in the Czech Republic.

The character of the caves depends on the degree of tectonic deformation, the number and size of spherical concretions, and the vulnerability to gravitational sliding of blocks. Generally, all types of caves defined by Vítek (1981) are present in the Bohemian Paradise, but most of them do not exceed 10 m in length. The longest caves are of fissure or crevasse type. Many of the alleged bedding caves were in fact formed at subhorizontal fracture zones, with locally developed slickensides and 'tectonic mirrors'. It is the combination of cement dissolution and tectonic deformation which is the most effective in speleogenesis, producing caves of hundred square metres in areal extent, such as the Postojna Cave (Fig. 16.6) in the Klokočské skály. Its name was adopted from that of the famous Slovenian karstic cave and reflects the spirit of Slavic integrity within the former Austro-Hungarian Empire.

Evacuation of sand from the caves was not governed by gravity as most of the caves are horizontally elongated or inclined away from the cave entrance. Flowing water could have contributed to sand transport where fractures are present (Fig. 16.11), but sand digging by animals definitely

Fig. 16.11 The Průtočná Cave in the Klokočské skály has a tunnel-like shape. During torrential rains, water flows through the cave (*Photo* J. Mertlík)

played an equally important role. The caves and rock-shelters in the Bohemian Paradise have been utilized by man since the Mesolithic, as evidenced by numerous archaeological finds (e.g. Šída and Prostředník 2007).

16.13 Conclusion

The variety of landforms of all scales preserved in the Bohemian Paradise due to the favourable geological conditions makes it a perfect natural laboratory for the study of recent geomorphic processes. The crest of Suché skály with vertical strata of silicified sandstone is an excellent example of multiple sediment deformation near a major fault. Cuestas and strongly dissected plateaus contain rock cities with complex relief of hundreds of rock pillars, including many examples of arches and mushroom rocks. Various types of pseudokarst caves and cavities occur within hundreds of metres from true karstic caves. Microrelief on vertical as well as horizontal cliff faces is controlled largely by salt weathering and frost weathering but the resulting forms are affected by structural inhomogeneities of many kinds: sedimentary structures including trace fossils, concretions, silicified and ferruginized fractures, tensional fractures, and also by deformation bands which are specific for this sandstone region.

Acknowledgements This contribution was prepared with the support of Institutional Research Plan RVO 67985831 of the Institute of Geology CAS, v.v.i.

References

Adamovič J, Mikuláš M, Cílek V (2006) Sandstone districts of the Bohemian Paradise: emergence of a romantic landscape. Geolines 21:1–100. http://geolines.gli.cas.cz/index.php?id=volume21

Adamovič J, Mikuláš R, Mertlík J (2013) Origin of regular cavities in European sandstones: field evidence for dissolution of carbonate and silica cement. In: Migoń P, Kasprzak M (eds) Sandstone landscapes. Diversity, ecology and conservation. Proceedings of the 3rd international conference on sandstone landscapes, Kudowa Zdrój (Poland), 25–28 April 2012, Wrocław, pp 13–18

Balatka B (1980) Landforms of the Příhrazy Plateau in the Bohemian Paradise PLA. Pam Přír 5(1980):554–559 (in Czech)

Balatka B (1987) Fossil forms of sandstone weathering and erosion in the Kostecká pahorkatina. Sbor Čes Geogr Spol 92:60–63 (in Czech)

Balatka B, Kalvoda J (2006) Geomorphological regionalization of the relief of Bohemia. Kartografie Praha, Praha, 80 p

Balatka B, Sládek J (1975) Pseudokarst phenomena in the eastern part of the Českodubská pahorkatina. Ochr Přír 30:211–212 (in Czech)

Bruthans J, Zeman O, Vysoká H (2006) Geology and hydrology of the Bartošova pec Cave and vicinity. In: Jenč P, Šoltysová L (eds) The sandstone phenomenon of the Bohemian Paradise, ZO ČSOP Křižánky, Turnov, pp 79–91 (in Czech, English abstract)

Cajz V, Rapprich V, Schnabl P Pécskay Z (2009) A proposal on lithostratigraphy of Cenozoic volcanic rocks in Eastern Bohemia. Zpr geol Výzk v Roce 2008:9–14 (in Czech, English abstract)

Čech S (ed) (2013) Explanatory text to geological map CR 1:25,000, 03-342 Rovensko pod Troskami. Czech Geological Survey, Praha (in Czech)

Čech S, Klein V, Kříž J, Valečka J (1980) Revision of the Upper Cretaceous stratigraphy of the Bohemian Cretaceous Basin. Věst Ústř úst geol 55(5):277–296

Coubal M (1990) Compression along faults: example from the Bohemian Cretaceous Basin. Miner Slov 22:139–144

Coubal M, Adamovič J, Málek J, Prouza V (2014) Architecture of thrust faults with alongstrike variations in fault-plane dip: anatomy of the Lusatian Fault, Bohemian Massif. J Geosci 59:183–208

Danišík M, Migoń P, Kuhlemann J, Evans NJ, Dunkl I, Frisch W (2010) Thermochronological constraints on the long-term erosional history of the Karkonosze Mts, Central Europe. Geomorphology 117:78–89

Krejčí J (1869) Studien im Gebiete der böhmischen Kreide-Formation. I. Allgemeine und orographische Verhältnisse, sowie Gliederung der böhmischen Kreide-Formation. Archiv naturwiss Landesdurchf Böhmen, Bd I, Sect II. Fr. Řivnáč, Praha, pp 39–179

Mertlík J, Adamovič J (2005) Some significant geomorphic features of the Klokočí Cuesta, Czech Republic. Ferrantia 44:171–175

Mertlík J, Adamovič J, Nešporová M (2002) Český ráj. In: Adamovič J, Cílek V (eds) Ironstones of the Bohemian Cretaceous Basin. Knih Čes speleol spol 38, Praha, pp 105–127 (in Czech). http://adamovic.euweb.cz/katalog.pdf

Rapprich V, Cajz V, Košťák M, Pécskay Z, Řídkošil T, Raška P, Radoň M (2007) Reconstruction of eroded monogenic Strombolian cones of Miocene age: a case study on character of volcanic activity of the Jičín Volcanic Field (NE Bohemia) and subsequent erosional rates estimation. J Geosci 52:169–180

Scheck M, Bayer U, Otto V, Lamarche J, Banka D, Pharaoh T (2002) The Elbe Fault System in North Central Europe—a basement controlled zone of crustal weakness. Tectonophysics 360:281–299

Šída P, Prostředník J (2007) The Late Palaeolithic and Mesolithic in the Bohemian Paradise: perspectives for a study of the region. Archeol rozhledy 59:443–460 (in Czech, English abstract)

Uličný D (2001) Depositional systems and sequence stratigraphy of coarse-grained deltas in a shallow-marine, strike-slip setting: the Bohemian Cretaceous Basin, Czech Republic. Sedimentology 48:599–628

Uličný D, Čech S, Grygar R (2003) Tectonics and depositional systems of a shallow-marine, intra-continental strike-slip basin: Exposures of the Český ráj region, Bohemian Cretaceous Basin. Geolines 16:133–148. http://geolines.gli.cas.cz/index.php?id=volume16

Valečka J (ed) (2013) Explanatory text to geological map CR 1:25,000, 03-341 Kněžmost. Czech Geological Survey, Praha (in Czech)

Vítek J (1980) Pseudokarst forms in Prachovské skály Cliffs. Českosl Kras 31(1979):45–56 (in Czech)

Vítek J (1981) Morphogenetic typification of pseudokarst in Czechoslovakia. Sbor Českosl geograf spol 86:153–165 (in Czech, English abstract)

Vítek J (1987) Pseudokarst forms in sandstones of Klokočské skály Cliffs. Českosl Kras 38:71–85 (in Czech)

Adršpach-Teplice Rocks and Broumov Cliffs—Large Sandstone Rock Cities in the Central Europe

17

Jan Vítek

Abstract

Structural plateaus and cuestas underlain by sedimentary formations of Late Cretaceous age are a part of Intra-Sudetic Basin in the northeastern Bohemia. Processes of erosion, weathering, and gravity-driven slope movements created important rock cities which originated in differently resistant and tectonically jointed sandstones. Rock cities are systems of rock towers, pillars, ridges, and cliffs separated by deep canyons, narrow gorges, and crevices. Some of these rock formations are as much as 100 m high. Mesoforms and microforms of weathering and denudation, especially rock perforations (rock arches, windows), various types of pseudokarst caves, tors, mushroom rocks, rock hollows, rock basins, and karren are common. The largest and the most visited rock cities are the Adršpach-Teplice Rocks (Adršpašsko-teplické skály) and the Broumov Cliffs (Broumovské stěny) which belong to Broumovsko Protected Landscape Area.

Keywords

Adršpach-Teplice Rocks • Broumovsko • Sandstones • Late Cretaceous • Intra-Sudetic Basin • Rock cities

17.1 Introduction

Rock cities and other rock formations built of Late Cretaceous sandstones are a part of the submontane landscape which occurs east of the highest mountain range of the Czech Republic—Krkonoše Mts. The term "rock city" depicts a complex ("labyrinth") of solitary rock towers and cliffs which are separated by narrow gorges, cracks, or crevasses one from another. The rock cities in northeastern Bohemia are part of *Broumovsko Protected Landscape Area*, named after the nearby town of Broumov. The most important rock cities are located within the Adršpach-Teplice Rocks (Czech *Adršpašsko-Teplické skály*) and the Broumov Cliffs (Czech *Broumovské stěny*). They are characterized by a huge variety of landforms and are easily accessible for tourists, belonging to the most visited places of the Czech nature. These sandstone formations continue in Poland in the Góry Stołowe National Park, where similar sandstone rock formations can be found.

J. Vítek (✉)
Department of Biology Faculty of Science, University of Hradec Králové, Rokitanského 62, 500 03 Hradec Králové, Czech Republic
e-mail: jan.vitek@uhk.cz

17.2 Geological and Geomorphologic Setting

Broumovsko Protected Landscape Area with its rock cities is a part of the geomorphologic unit known as Broumovská vrchovina (=Highland) (Demek et al. 2006). In terms of regional geological division of the Bohemian Massif, this area is part of the Intra-Sudetic Basin which is a long active sedimentary basin between older (Precambrian and Lower Paleozoic) crystalline complexes of Krkonoše Mts. and Orlické hory Mts. (Chlupáč et al. 2002). The Intra-Sudetic Basin contains Carboniferous and Permian sediments, to a lesser also paleovolcanites of from these periods, followed then by and sediments of Mesozoic age (Triassic and especially Late

T. Pánek and J. Hradecký (eds.), *Landscapes and Landforms of the Czech Republic*, World Geomorphological Landscapes, DOI 10.1007/978-3-319-27537-6_17

Cretaceous period) (Tásler et al. 1979). Carboniferous sediments (conglomerates and arkoses with coal seams) create the southwest part of the Intra-Sudetic Basin, with an asymmetric mountain ridge of Jestřebí hory (Žaltman, 739 m). It is delimited by tectonic fault (Hronov-Poříčí Fault) against the Krkonoše-piedmont Basin. Permian sediments (mostly "red sandstones") fill the northeastern part of the Intra-Sudetic Basin, along the shallow valley of Stěnava River. Permian volcanic rocks include rhyolites, andesite, and basalts. They create rugged highlands along the borders with Poland, with the highest peaks approaching 900 m a.s.l. (Královecký Špičák, 880 m, in Vraní hory, and Ruprechtický Špičák, 880 m, in Javoří hory). In the overburden of Permian sediments sandstones of Triassic age appear in some places which is the only occurrence of Triassic rocks in Bohemian Massif.

The central part of the Intra-Sudetic Basin is created by the Police Basin (named after the town Police nad Metují), where lithologically diverse sediments of Late Cretaceous age occur (stratigraphic unit Cenomanian, Turonian, and Coniacian). The thickness of Cretaceous sediments reaches up to 500 m (Dvořák 1968). Two main types of sediments may be distinguished. The fine-grained sediments (marls, claystones, calcareous siltstones etc.) create the major part of the area. These sediments can be found, e.g., in rock outcrops on slopes of the river Metuje valley and its tributaries. The second main type is quartzose and arkosic sandstones. Important landform assemblages, including 'rock cities', came into existence primarily in the so-called block sandstones whose breakdown typically proceeds along two sets of vertical joint, perpendicular to each other, and horizontal partings, which may follow bedding surfaces. For sandstones of the Police Basin, distinctive diagonal bedding is typical (Adamovič et al. 2010).

The Police Basin has a brachysyncline structure, with an axis elongated in NW–SE "Sudetic direction" (Fig. 17.1). Both sides of the basin are typified by stepped morphology and the occurrence of homoclinal (cuesta) ridges, whose gentle slopes are inclined toward the middle of the basin according to the dip of the strata, whereas rocky escarpments occur on the opposite steep escarpments (Fig. 17.2). The evolution of Police Basin and its landforms was controlled by the presence of tectonic faults established during the Paleozoic. In the Cenozoic faults were revived under the influence of Saxonian tectonics and thereby some parts of the area moved about tens or even hundreds of meters in

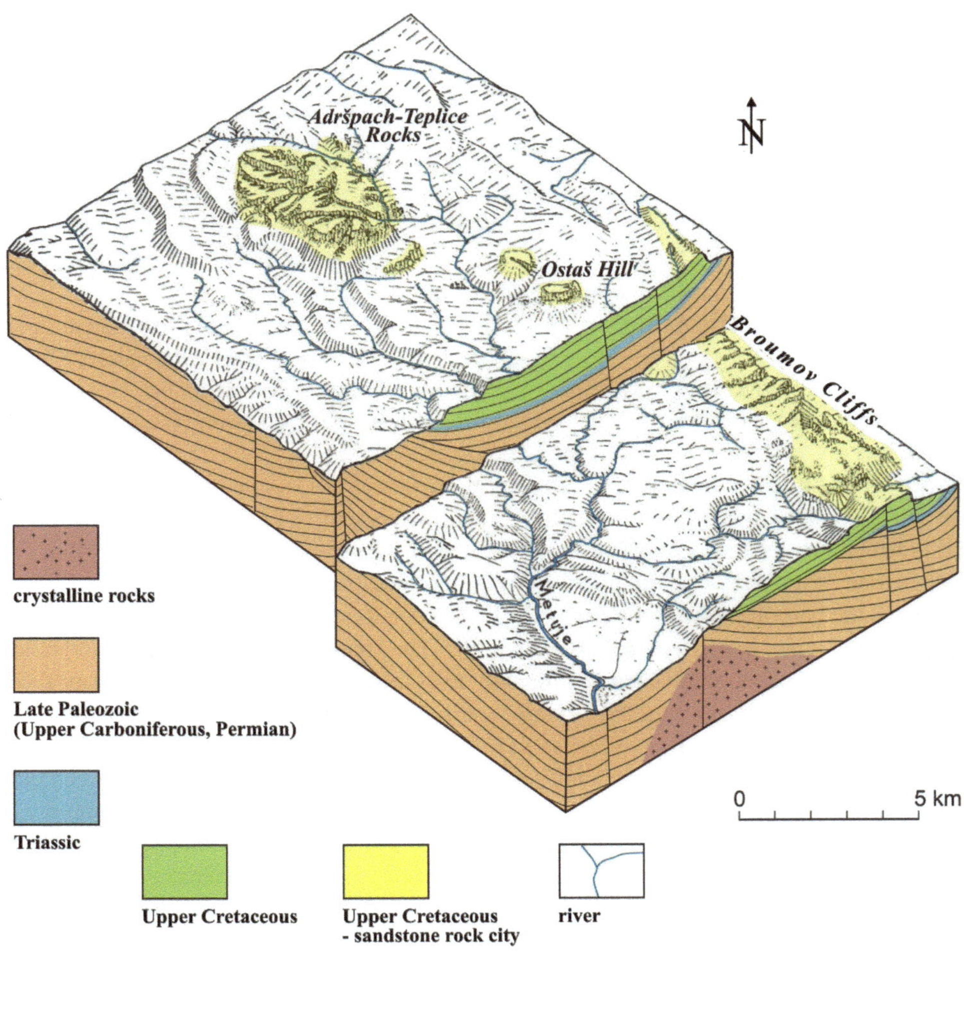

Fig. 17.1 Position of the rocky areas of Adršpach-Teplice Rocks and Broumov Cliffs in Police Basin (Modified after Bína and Demek 2012)

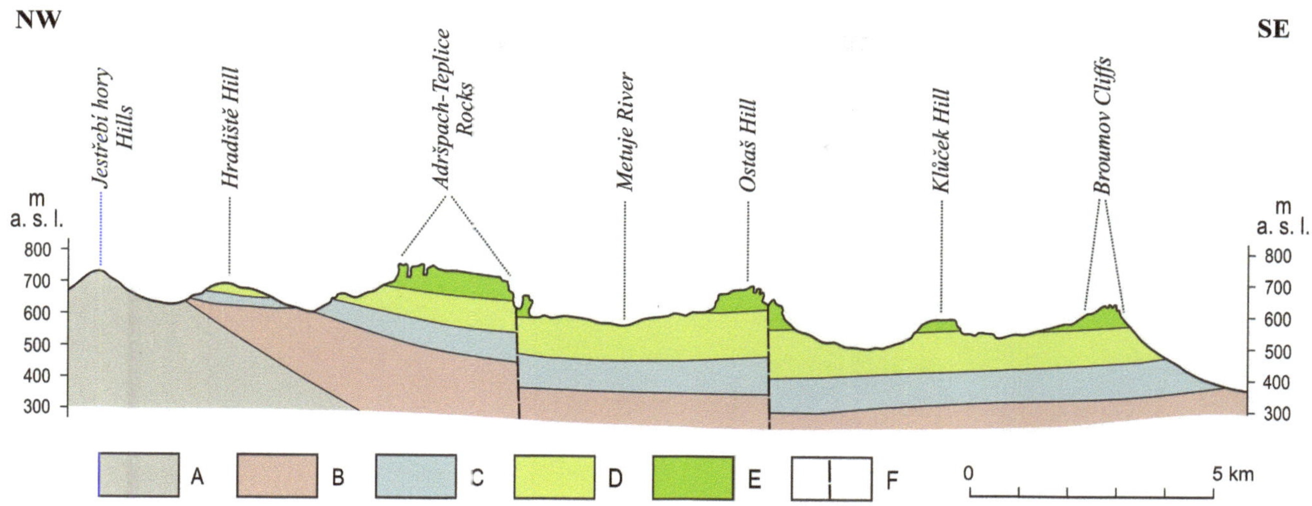

Fig. 17.2 Profile of the central part of the Intra-Sudetic Basin in the northeastern Bohemia. *A* Upper Carboniferous, *B* Permian, *C* Triassic, *D* Late Cretaceous, *E* Late Cretaceous—block sandstones with rock cities, *F* Signifiant faults (Modified after Tásler et al. 1979)

horizontal and also in vertical direction (Tásler et al. 1979; Vejlupek 1986). The most distinctive fault is the Police Fault (or Police Fault zone) running in NW–SE direction and the most distinctive cross fault is the Skály fault in perpendicular SW–NE direction. Joint systems in sandstones and other compact rocks have practically identical directions. The brachysyncline of the Police Basin is an important reservoir of drinking water which is accumulated in the Late Cretaceous sediments.

17.3 Plateaus and Cuestas

Rock cities in the thick-bedded sandstones of the Broumov Highland have evolved within tableland plateaus or on smaller mesas, or they are parts of asymmetric ridges—cuestas (Fig. 17.3). Most of these elevations create significant dominants of the landscape. From the morphogenetic point of view tablelands, mesas, and cuestas are denudational landforms whose origin and evolution was largely controlled by structural (lithological and tectonic) conditions. They were formed by activity of stream erosion, weathering processes, and gravitation slope movements (Balatka and Sládek 1984; Demek and Kopecký 1994; Härtel et al. 2007). Sandstone tablelands and mesas, which appear in the central part of Police Basin, are relicts of presence of the youngest layers of Late Cretaceous sediments (stratigraphic unit Turonian—Coniacian) in the Intra-Sudetic Basin (Dvořák 1968; Vejlupek 1986). Quartzose sandstones—with grains of diameter of 0.1–0.5 mm big—with clayey and also ferruginous cement are predominant. Ferruginization of sandstones happens particularly near by faults (Adamovič et al. 2002). Inclination of sandstone layers is low in flat, central part of the Police Basin (on average 5° to the axis of the basin), and it increases toward its margins. Proper conditions for the formation of rock cities occurred especially along the jointed edges of table plateaus and mesas (Fig. 17.4).

The most extensive tableland (almost 20 km^2) is the plateau of *Adršpach-Teplice Rocks* situated in the northern part of the Police Basin. From the northern up to the eastern edge of this tableland, the Metuje River incised into the plateau and separated smaller sandstone rocky hills (mesas): Křížový vrch Hill (668 m), Lada Hill (623 m), Lysý vrch Hill (611 m) etc. The narrow sandstone ridge named Skalský hřbet (698 m) was dislocated by tectonic fault (Skály fault) from the southern slope of tableland of Adršpach-Teplice Rocks.

The mesa of Ostaš (700 m), exceeding its surrounding by about 200 m, forms a distinct elevation in the middle part of the Police Basin. The structural plateau on the top is 500 × 400 m in dimensions and inclined in SW-SSW direction, following the dip of sandstone layers. This plateau is bordered by rocky walls on all sides. These rocky walls can be up to 40–60 m high. The NE slope is of tectonic origin and the block of Kočičí skály (rock city Dolní bludiště—Lower Labyrinth) was downthrown along the Police Fault due to faulting in the Neogene. In the apical part of the structural platform of Ostaš there are frequent forms of weathering and denudation of siliceous sandstones (rock city Horní bludiště—Upper Labyrinth, with tors, mushroom rocks, rock windows, rock hollows, and small caves). The western edge of the plateau and the adjoining part of the slope are affected by rock creep processes, with creation of deep crevices (including crevice-type caves), a system of deflected and collapsed rock blocks (Vítek 1979; Demek and Kopecký 1994). These processes, conditioned by inclination

Fig. 17.3 Southeastern part of Broumov Cliffs with its asymmetric table hills of Božanovský Špičák (on the *left*) and Koruna (on the *right*) (*Photo* J. Vítek)

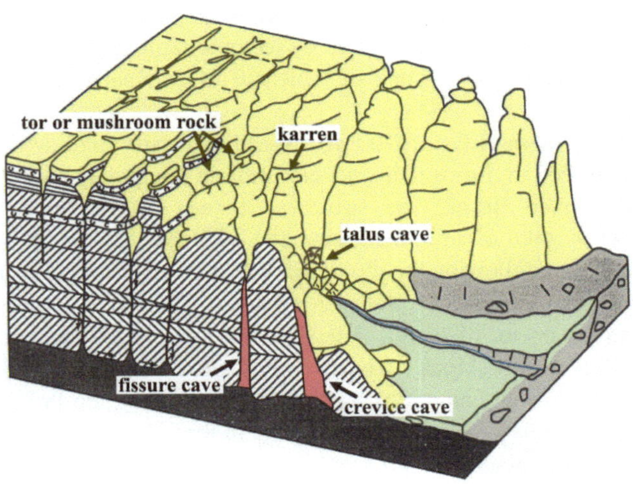

Fig. 17.4 Scheme of development of a sandstone rock city and pseudokarst forms (Modified after Čech and Gawlikowska 1999)

of layers of block sandstones, are recently active (Stemberk et al. 1994).

The marginal parts of the Police Basin are composed of discontinuous cuesta ridges in the NE, NW, and SW part of the basin. Inward backslopes of cuestas, broadly concordant with the inclination of layers, are low angle surfaces, while front escarpments are steep, with rocky rims in some places. Slopes of cuestas are dissected by canyon-like valleys which are filled with fallen boulders. Distinct rock cities of block sandstones (stratigraphic unit Middle Turonian, Dvořák 1968) were created particularly in the Broumov Cliffs (Broumovské stěny). In the SW part of this area, largely on the territory of Poland, they are followed by the Góry Stołowe mountains, protected in the Góry Stołowe National Park (Witkowski et al. 2008). The top of the sandstone table mountain of Szczeliniec Wielki (919 m) is the highest peak in the whole area of the Bohemian Cretaceous Basin.

17.4 Adršpach–Teplice Rocks

The best known and also the largest rock city occupies the table plateau of Adršpach-Teplice Rocks (Adršpašsko-teplické skály). It covers an area of 20 km^2 and is the most extensive, continuous complex of sandstone rocks in the Czech Republic (Mikuláš et al. 2007). The present-day tableland is a relict of an originally more extensive area made up from quartzose block sandstones in the northern part of the brachysynclinal Police Basin. The axis of the basin slopes slightly from the northeast to the southwest, whereas sandstone layers are inclined toward the axis line of the basin, to SW and NE on respective side of the syncline. The highest hill—Čáp (786 m)—is located in southwestern part of the plateau. The thickness of the sandstone layer (stratigraphic unit Turonian–Coniacian) reaches up to the 120 m (Dvořák 1968; Čech and Gawlikowska 1999). Distinct sandstone units with diagonal bedding may be seen in the stratigraphic column exposed in the rock cities (Adamovič et al. 2010). The plateau is disrupted by tectonic faults and fissures which have had huge importance for the development of landforms, including rock cities. The area is split into two parts. The smaller part in the north is called the Adršpach Rocks (*Adršpašské skály*), named after the nearby village of Adršpach. The Teplice Rocks (Teplické skály), named after the town of Teplice nad Metují, are located in the south. A canyon-like valley named Vlčí rokle crosses the table plateau in WNW–ESE direction and defines a 6-km-long border between these two rock cities. The whole area of Adršpach-Teplice Rocks is largely covered with spruce forests, also birch trees and European mountain ashes are frequent. The remains of the native pine vegetation are preserved on the exposed rock outcrops.

Adršpach-Teplice Rocks National Nature Reserve—Národní přírodní rezervace Adršpašsko-teplické skály (with an area of 18.03 km^2) was established in 1933.

17.4.1 Adršpach Rocks—Breathtaking Rock City

A typical rock city is represented particularly by the Adršpach Rocks, located in the north of the table plateau (Fig. 17.5). The sandstone plateau is dissected by deep canyons and gorges; its development was mainly controlled by the presence of subvertical fissures or joints systems. The narrow parts of gorges are locally called "streets" and the wider parts are called "squares". River erosion has influenced the development of gorges, whilst mechanical weathering (mainly gelivation) of sandstones in the zones of increased frequency of subvertical joints significantly contributed to the evolution of gorge walls (Vítek 1979) (Fig. 17.6). The upper reach of Metuje River is located in the

Fig. 17.5 Northeastern edge of the plateau of the Adršpach-Teplice Rocks (rock group "Království"—Kingdom) (*Photo* J. Vítek)

Fig. 17.6 Narrow canyons and fissure caves are formed by weathering and erosion of densely fissured sandstones ("Sloní náměstí"—Elephant Square in Adršpach Rocks) (*Photo* J. Vítek)

main valley of the Adršpach Rocks. In the central part of the rock city, the river makes two waterfalls, with the Upper Waterfall—Horní vodopád, open to public, being 16-m-high and hidden in a rocky shaft. The gorge above the waterfall has been modified to a man-made small lake where tourists can enjoy boat rides.

As a result of erosion and weathering processes as well as gravity-driven slope movements, the northern edges of the sandstone plateau and narrow rock ridges are divided into solitary towers and groups of pillars which create a typical rock city (Fig. 17.7). The most massive towers are 70–100 m high. Most of them are close to tourist trails. The towers have various names, e.g. Milenci (Lovers) (Fig. 17.8), Starosta a Paní starostová (Mayor and Mrs. Mayor) (Fig. 17.7), Král (King) (Fig. 17.5), Homole cukru (Sugar Loaf), etc. The rock cities are favorite destination of rock climbers.

Many minor landforms were formed by processes of differential (mechanical and chemical) weathering and denudation which exploited unequal resistance of different units within the sandstones. These processes formed various rock perforations (Fig. 17.8)—rock arches and windows (e.g. Čertův most—Devils` Bridge, Džbán—Pitcher, etc.). The surface of rock walls has a rich mosaic of microforms—rock ledges on harder bedding planes (some of them highlight the progress of diagonal bedding), small rock hollows, rock basins (also weathering pits), and karren. Unlike other areas of sandstone in the Bohemian Cretaceous Basin microforms genetically related to salt weathering, such as honeycombs, are rare (Mikuláš et al. 2007).

17.4.2 Mysterious Teplice Rocks

The Teplice Rocks occupy a bigger, southern part of the plateau of the Adršpach-Teplice Rocks. A distinct ferruginization of sandstone cement occurred close to the tectonic faults, particularly secondary cementation by iron oxyhydroxides (Adamovič et al. 2002). Massive rock walls are typical for the area of Teplice Rocks. These walls are

Fig. 17.7 Adršpach rock city with its rock towers "Starosta a Paní starostová" (Mayor and Mrs. Mayor) (*Photo* J. Vítek)

hundreds of meters long and up to the 60 m high (e.g. Chrámová stěna and Martinská stěna in the southernwest part of the area). More expensive rock cities are rare and can be found, e.g., around the perimeter of Skalní ostrov, in Anenské údolí, Supí skály, Skalský hřbet, and in other places.

The sandstone plateau is dissected by a system of canyons and gorges. Their orientation follows the orientation of fissures and joint systems. Major valleys are Vlčí rokle and the canyon of Skalní potok. A number of side gorges runs to these valleys. Some of these valleys are connected with each other at almost right-angles; they are very narrow and tens of meters deep, e.g., gorges named Siberia (Sibiř) and Underworld (Podsvětí), Large and Small Temple Square (Velké a Malé Chrámové náměstí), etc. Many of these gorges are inaccessible because their bottom is filled with collapsed blocks and boulders. In some gorges, talus caves exist in gaps between collapsed boulders. These caves are usually hardly accessible or require crawling. For example, the of cave Teplická jeskyně (Teplice Cave), situated in boulder chaos at the bottom of one canyon, is 1065 m long and it is the longest pseudokarst cave in the Czech Republic (Cílek and Kopecký 1998). Weathering of sandstones creates narrow and deep fissures along the joint systems, while gravitation movements of rock blocks transform them into open crevices and crevice-type caves (Vítek 1979); the example of such a cave is Skalní chrám—Rock Dome (45 m long, 50 m high and 1–2 m wide) which is close to the marked hiking trail. The total length of "pseudokarst system" in the central part of the Teplice Rocks is estimated to be 20 km (Mlejnek et al. 2009), but the research is not complete so far. In one smaller pseudokarst cave of the Teplice Rocks, the presence of the so-called root stalagmites (Fig. 17.9) was recorded for the first time in the Czech Republic (Vítek 1980).

Mesoforms and microforms of selective weathering and denudation of sandstones are abundant in the Teplice Rocks. Rock formations (tors) stand out from the surface of the sandstone plateau in some places. They are characterized by

Fig. 17.8 Rock formation "Milenci" (The Lovers), situated in the middle of the Adršpach Rocks, belongs to the highest rock towers (over 100 m) in the sandstone rock cities of the Czech Republic. Diagonal bedding is noticeable (*Photo* J. Vítek)

Fig. 17.9 Root stalagmite at the bottom of a talus cave in the Teplice Rocks (*Photo* J. Vítek)

diverse shapes. Following variable resistance of adjacent layers, mushroom rocks are created. The "cap" of a mushroom rock is formed from more resistant layers and the "stem" is built by less-resistant layers, usually more porous, and/or thinly bedded. Other surface microforms are rock hollows, karren (e.g. on the top of the formation called Rock Crown–Skalní koruna, Fig. 17.10), rock basins (weathering pits), etc. Contoured shapes of ferruginous incrustations close to the Police Fault in the southern part of the Teplice Rocks are remarkable (Adamovič et al. 2002, 2010).

17.5 Broumov Cliffs

A distinctive sandstone area in the northwest Bohemia is the Broumov Clifs (Broumovské stěny) which are made of quartzose sandstones and arkosic sandstones of Middle Turonian age (Dvořák 1968; Čech and Gawlikowska 1999). The asymmetric ridge (cuesta) of Broumov Cliffs takes an area of 10 km^2 and divides the Police Basin in the south-west from the Intra-Sudetic Basin in the northeast (Fig. 17.2). The ridge is 12 km long in NW–SE direction and gets higher in the same direction (from Honský Špičák, 662 m, at the NW edge to Božanovský Špičák, 773 m, at the SE extremity). The sandstone area continues from the pass of Machovské sedlo (620 m) in the southeasterly direction and makes Góry Stołowe Mts. in Poland (Migoń 2008). The southwestern slope of the cuesta follows the dip of sandstone layers pointing into the middle of the Police Basin. The opposite northeast-facing front slope creates the escarpment of Broumov basin up to 250 m high (Fig. 17.3). Somewhat different situation is in the southeast, highest part of this area (Adamovič et al. 2010). Slightly inclined structural plateaus (hills Velká kupa. 708 m, Signál, 708 m, Koruna, 769 m, Božanovský Špičák, 773 m) are set aside according to the tectonic faults. The total height of the escarpment exceeds 300 m. The whole area is covered by forest, with predominance of spruce, birch, and pine.

Both the front and the backslope of the cuesta of Broumov Cliffs are not continuous but dissected by a series of rocky gorges. Consequent valleys, which incise into the

Fig. 17.10 Teplice Rocks—“Skalní koruna” (Rock Crown) rock tower, with sandstone karren at the top (*Photo* J. Vítek)

gently inclined southwestern slope, are longer and occupied by left-side tributaries of river Metuje (itself a tributary of the Labe (Elbe) river). The most attractive example is a 1.5 km long and up to 70-m-deep canyon named Kovářova rokle (Blacksmith’s Gorge) (Fig. 17.11) between the tourist cottage Hvězda and the village of Hlavňov. Obsequent rocky gullies, which intersect the steep northeastern slope, are shorter and contribute runoff to the river Stěnava (Odra drainage basin). For example, Zaječí rokle, Třešňová rokle and others are highly rugged and up to 70 m deep. The bottom of some gorges is filled with collapsed sandstone blocks and boulders. Hardly accessible talus caves were created in gaps between boulders (e.g., 400-m-long cave Jeskyně pod Luciferem). Gravitational deflection of rock blocks in the edge area of the cuesta creates narrow crevices and pseudokarst crevice caves. An impressive rock formation is 5-m high rock gate Kamenná brána (Rock Gate) (Fig. 17.12) on the top of Velká kupa hill.

Typical rock cities, with their characteristic assemblages of high rock towers and pillars, are not plentiful in the Broumov Cliffs. There are only small clusters, e.g., in the upper part of the canyons of Kovářova rokle (Skalní divadlo —Rocky Theater) and Zaječí rokle, in the segmented edge of the cuesta (Strážná hora, 689 m, Modrý kámen, 686 m, Signál, 708 m). Weathering and denudation of sandstone layers dissimilar resistance underpins various rock shapes, including rugged tors in the upper parts of plateaus and ridges. The so-called mushroom rocks, present among others above the villages of Slavný and Bělý, on the plateaus of

Fig. 17.11 Consequent valley in Broumov Cliffs—sandstone canyon “Kovářova rokle” (Blacksmith’s Gorge) (*Photo* P. Migoń)

Fig. 17.12 Rock formation "Kamenná brána" (Rock Gate) has come into existence by disintegration of the rocky edge of the cuesta of Broumov Cliffs (*Photo* J. Vítek)

Fig. 17.13 Sandstone tor (mushroom rock) on the plateau of Broumov Cliffs near the village of Slavný (*Photo* J. Vítek)

Božanovský Špičák, Signál, and in many other places, are the most characteristic examples (Fig. 17.13). The top surface of the Božanovský Špičák hills is particularly rich in emerging rock formations of diverse shapes, often resembling animals. Hence, the whole area is informally named 'Skalní zvěřinec" (Rock Menagerie), with individual formations such as Velbloud—Camel, Želva—Turtle, Varan—Monitor, Kočička—Cate, Kačenka—Duck.

Rock basins (weathering pits) and karren are the most common examples of rock microforms (Vítek 1979). For example, in a part of Strážná hora hill (688 m) at the northwestern edge of the Broumov Cliffs rock basins up to 2 m wide and 0.5 m deep are cut into the rock surface. Some of them are filled with soil and overgrown. In the same locality and in other places (e.g., in Božanovský Špičák), one can also find 10–40 cm deep rillen karren.

In the central part of the Broumov Cliffs, in a place called Hvězda (Star), a baroque chapel from the year 1733 and stylish inn are located and contribute to the cultural heritage of the area (Fig. 17.14).

Fig. 17.14 Baroque chapel "Hvězda" (*Star*) in the central part of Broumov cliffs (*Photo* J. Vítek)

17.6 Rock Cities and People

Rock cities in the Broumovsko Protected Landscape Area—especially the Adršpach-Teplice Rocks and the Broumov Cliffs—belong to the most attractive and the most popular natural areas in the Czech Republic. Humans were interested in these rocks already from the Middle Ages (Cílek et al. 1998; Härtel et al. 2007; Adamovič et al. 2010; Migoń and Latocha 2013). Three castles were built on the marginal rocky cliffs: Adršpach was built on Starozámecký vrch next to the village of Dolní Adršpach, Střmen was built in the entry part to the Teplice Rocks and Skály (Katzenstein) was built on Skalský hřbet in the vicinity of Teplice nad Metují. The purpose of these castles was to guard the borders of the land. Nowadays, we can see only small rests of these castles. Historical data tells us that people began to permeate the hardly accessible rocky terrain relatively late. First access roads were made for the purpose of tree felling and timber transport. From the first half of the eighteenth century, these paths were rearranged to provide easier access for visitors. Already in the year 1790, the German poet and geologist J. W. Goethe visited the Adršpach Rocks and the Teplice Rocks. Today, the most interesting places are accessible via marked hiking trails and educational trails. About 250,000 tourists visit the Adršpach Rocks and about 50,000 tourists visit the Teplice Rocks every year (entry into the both rock cities is chargeable).

Sandstone rock formations in the Broumovsko Protected Landscape Area belong to the most valuable climbing terrains in the Czech Republic, offering also the most demanding climbing. Over 1800 rock towers were climbed in the Adršpach-Teplice Rocks (Adamovič et al. 2010). Rock climbing is also possible in selected terrains of the Broumov Cliffs, on the mesa of Ostaš, in the Kočičí skály Rocks and on Křížový vrch hill.

17.7 Conclusion

The Adršpach-Teplice Rocks and the Broumov Cliffs belong to the most precious and most visited geomorphological sites in the Czech Republic. They are strictly protected and are listed as National Nature Reserves. Both localities are representative examples of sandstone relief which has developed on the sharply demarcated plateaus and asymmetric ridges (cuestas). The course and pace of geomorphological processes, especially erosion and gravitational slope movements, were significantly controlled by lithological, structural, and tectonic conditions. Some sandstone landforms in these rocky areas have record dimensions within the sandstone areas of Central Europe. For example, the Adršpach Rocks are the most extensive continuous rock cities and some rock towers are over 100 m high. Another example is the longest pseudokarst talus cave (Teplická Cave). It is situated at the bottom of the craggy canyon of Teplice Rocks and is 1065 m long. Also some mesoforms and microforms of sandstone relief, e.g., rock perforations, mushroom rocks, rock basins (weathering pits), various types of karren, which attract the attention of the visitors. It is unquestionable that many extraordinary geological formations are still waiting for discovery in the rugged and hardly accessible rocky terrain.

References

Adamovič J, Cílek V et al (2002) Ironstones of the Bohemian Cretaceous Basin. Knihovna České speleologické společnosti. Praha, 38, 170 pp

Adamovič J, Mikuláš R, Cílek V (2010) Atlas pískovcových skalních měst České a Slovenské republiky. Academia, Praha, 460 pp

Balatka B, Sládek J (1984) Typizace reliéfu kvádrových pískovců české křídové pánve. Rozpravy Československé akademie věd, Řada matem. přír. Věd. Academia, Praha, 94, 6, 80 pp

Bína J, Demek J (2012) Z nížin do hor. Geomorfologické jednotky České republiky. Academia, Praha, 344 pp

Cílek V, Kopecký J et al (1998) Pískovcový fenomén: klima, život a reliéf. Knihovna České speleologické společnosti, 32. ČGS, Praha, 174 pp

Čech S, Gawlikowska E (1999) Góry Stołowe, mapa geologiczno-turystyczna. Adršpašsko-teplické skály, geologická mapa pro turisty. Państwowy Instytut Geologiczny, Český geologický ústav. Warszawa, Praha

Demek J, Kopecký J (1994) Geomorphological processes and landforms in the southern part of the Polická vrchovina Highland (Czech Republic). GeoJournal 32(3):231–240

Demek J, Mackovčin P et al (2006) Zeměpisný lexikon ČR. Hory a nížiny. Agentura ochrany přírody a krajiny ČR, Praha, Brno, 582 pp

Dvořák J (1968) Stratigrafie, litologie a podloží svrchní křídy ve vnitrosudetské pánvi. Věstník Ústředního ústavu geologického 43:423–430

Härtel H, Cílek V, Herben T, Jackson A, Williams R (eds) (2007) Sandstone Landscapes. Academia, Praha, 496 pp

Chlupáč I et al (2002) Geologická minulost České republiky. Academia, Praha, 436 pp

Migoń P (2008) Rzeźba i rozwój geomorfologiczny Gór Stołowych. Przyroda Parku Narodowego Gór Stołowych. In: Witkowski A, Pokryszko B, Ciężkowski W (eds) Przyroda Parku Narodowego Gór Stolowych. PNGS, Kudowa Zdrój, pp 49–69

Migoń P, Latocha A (2013) Human interactions with the sandstone landscape of Central Sudetes. Appl Geogr 42:206–216

Mikuláš R, Adamovič J, Hájek A, Spíšek J (2007) Adršpašsko-teplické skály Cliffs and Ostaš Hill (Czech Republic). In: Härtel H et al (eds) Sandstone Landscapes. Academia, Praha, pp 332–335

Mlejnek R, Ouhrabka V, Růžička V (2009) Poseidon – a complex system of underground spaces in sandstone in the Czech Republic. NSS News 67(8):4–7

Stemberk J, Košťák B, Kopecký J (1994) Deformations in sandstones due to table hill desintegration. Zeszyty Naukowe Akademii Rolniczej we Wrocławiu 255(7):187–193

Tásler R et al (1979) Geologie české části vnitrosudetské pánve. Ústřední ústav geologický, Academia, Praha, 296 pp

Vejlupek M (1986) Strukturní stavba polické a svatoňovicko-hronovské pánve. Věstník Ústředního ústavu geologického 61(3): 139–148

Vítek J (1979) Pseudokrasové tvary v kvádrových pískovcích severovýchodních Čech. Rozpravy Československé akademie věd, Řada matem. přír. Věd. Academia, Praha, 89, 4, 58 pp

Vítek J (1980) Die Wurzelstalagmiten auch in Sandsteinhöhlen Böhmens. Der Höhlernforscher, 12, 1, pp 1, 12

Witkowski A, Pokryszko B, Ciężkowski W (2008) Przyroda Parku Narodowego Gór Stolowych. PNGS, Kudowa Zdrój, 404 pp

Žďárské Vrchy Highland—Geomorphological Landscape in the Top Part of the Bohemian-Moravian Highland with the Unique Crystalline Rocks Forms

18

Karel Kirchner

Abstract

Žďárské vrchy Highland is built by metamorphic rocks and is situated in the upper part of Bohemian-Moravian Highland. A typical feature of Žďárské vrchy is the occurrence of isolated groups of rocks or rock formations (walls, towers, small rock cities), which dominate on flat upland ridges. These ridges are separated by open valleys with flat headwaters. Rock formations are limited by vertical walls and often significantly protrude above the surrounding landscape, occasionally reaching heights well above 30 m, although 15 m is the typical height value. The Žďárské vrchy Highland is a part of the Žďárské vrchy Protected Landscape Area and many rock formations were declared as natural monuments. Landforms of the Žďárské vrchy Highland have originated through polygenetic development since the Cretaceous and are in many ways unique within the highlands in the Variscan Europe.

Keywords

Žďárské vrchy highland • Rock forms • Metamorphic rocks • Rock towers and walls • Weathering pits

18.1 Introduction

The Žďárské vrchy (=Highland) represents a compact geomorphological sub-unit, which forms the second highest part of the Bohemian-Moravian Highland, situated in the central part of the Czech Republic. The geomorphological evolution of this area has been very long and traces of the pre-Cretaceous planation, remnants of Cretaceous sediments and Tertiary planation surfaces are preserved there. In the Pleistocene, the area was significantly modelled by cryogenic processes, whilst the Middle Ages colonization caused an anthropogenic transformation of natural landscape and individual landforms. The main morphostructural features of relief were conditioned by tectonic movements during the Alpine orogeny. A unique feature of the central part of the Žďárské vrchy Highland is the occurrence of isolated groups of rocks or rock formations, which often have the character of towers that dominate the flat ridges and complement the diversity of the landscape. The region is an important water divide and headwater area. The principal European Elbe–Danube Water Divide runs along the watershed ridges. Some significant Czech and Moravian rivers—Svratka, Sázava and Chrudimka spring here (Fig. 18.1). Landforms and other natural elements, combined with long-term cultivation of land have given rise to the varied mosaic of forests, meadows, rocky pastures, ponds and small settlements that form a harmonious cultural landscape (Fig. 18.2). This picturesque landscape with rock formations has often become the subject for significant Czech painters (Fig. 18.3). Thanks to those qualities, the Žďárské vrchy Highland has become the core of the Protected Landscape Area (PLA) Žďárské vrchy, established in 1970. The PLA's mission is to preserve the harmoniously balanced cultural landscape with significant natural phenomena. These include many geological and geomorphological sites, especially rock outcrops which affect and underpin valuable ecosystems. Therefore, a

K. Kirchner (✉)
Institute of Geonics, Czech Academy of Sciences, v.v.i. Branch Brno Drobného 28, 602 00 Brno, Czech Republic
e-mail: kirchner@geonika.cz

T. Pánek and J. Hradecký (eds.), *Landscapes and Landforms of the Czech Republic*,
World Geomorphological Landscapes, DOI 10.1007/978-3-319-27537-6_18

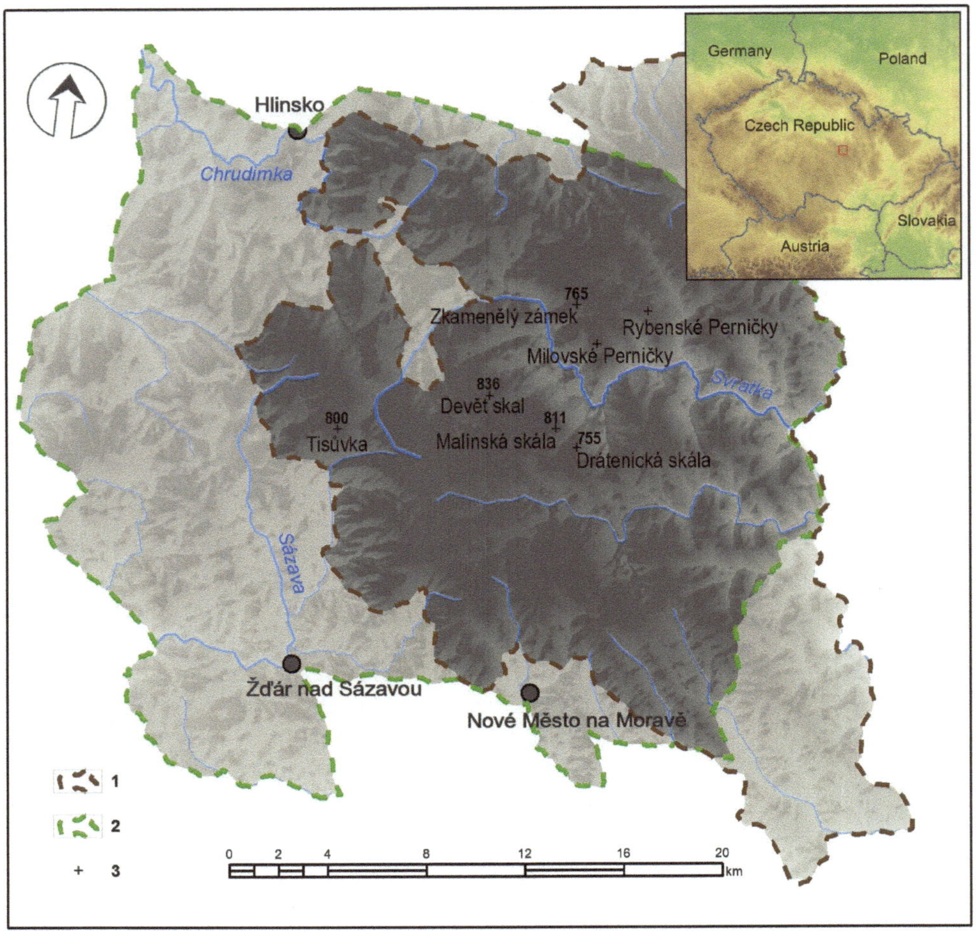

Fig. 18.1 Schematic map of the study area. Explanation: *1* border of the geomorphological sub-unit Žďárské vrchy Highland, *2* border of the Protected Landscape Area Žďárské vrchy, *3* geomorphological sites

significant number of these rock formations were declared as natural monuments, e.g., Natural Monument Devět skal (=Nine Rocks), Natural Monument Milovské Perničky etc. The sites that are described in the following text are also declared as natural monuments or natural reserves and they characterize the diversity of landforms (especially rock forms) in the area of the interest.

18.2 Localization

The Žďárské vrchy region is situated in the northeastern part of the Bohemian-Moravian Highland. The top parts of Žďárské vrchy reach the heights of 700–800 m a.s.l.—Mt. Devět skal (836 m a.s.l.) is the highest peak in the area and the second highest point of the Bohemian-Moravian Highland. Other important peaks are Křovina (830 m), Kopeček (822 m), Malinská skála Rock (811 m), and Žákova hora (810 m). The basement consists mainly of Variscan crystalline rocks of the Bohemian Massif, which spreads in the area of Bohemia and the western part of Moravia. The area of interest lies in the central part of the Bohemian Massif and extends into several geological units. The majority of the basement is built by the metamorphic rocks of Svratka Crystalline and Polička Crystalline. Only the southern part is formed by the metamorphic rocks of the Strážek Moldanubicum (Hanžl et al. 2011).

18.3 Geology and Morphostructures

The Svratka Crystalline is located in the central and northeastern part of the area. The two-mica orthogneiss and migmatites prevail, alternating with mica schist and mica schist gneisses. In these rocks, there are interbeds of amphibolites and skarns, in the past widely exploited as deposits of iron ore. The majority of rock formations in the highest parts of the Žďárské vrchy are built of rocks of the Svratka Crystalline. The northeastern part of Žďárské vrchy is built by the Polička Crystalline which consists primarily of fine-grained biotite gneiss with amphibolite interbeds, crystalline limestones and quartzites. The southern part of the area is built by the Strážek Moldanubicum. This unit is represented mainly by sillimanite—biotite paragneisses,

Fig. 18.2 Harmonic cultural landscape in the top part of Žďárské vrchy (*Photo* K. Kirchner)

which are also very migmatized or granitized. The paragneisses often contain interbeds of exotic rocks mainly amphibolite, orthogneiss, but also crystalline limestone, sporadically serpentinites, quartzites and skarns (Melichar et al. 2004).

Relics of Cretaceous sediments (sandstones with brown-grey clays) represent an evidence of marine transgression in the Early and the Late Cretaceous that probably covered most of the area. These sediments can be found in a depression between the municipalities Herálec and Svratka. During the tectonic updoming of the Žďárské vrchy between the Early and Late Tertiary, the basin was formed and the Cretaceous sediments were locked in, as well as in troughs on the south-western edge of the area, e.g., Dlouhá mez Cretaceous in the Dářská brázda (=Graben) (Demek 2002).

The Žďárské vrchy area has a character of a flat highland, which is the result of close relations between landforms, geologic structure and tectonics. From the morphostructural point of view, the Žďárské vrchy are a post-Cretaceous asymmetric horst limited by the parallel faults of NW–SE extension. Orientation of the main wide flat ridges is influenced and conditioned by the fold tectonics and the occurrence of geomorphologically more resistant metamorphic rocks—migmatites and orthogneiss. These ridges are separated by valleys with flat shallow valley ends (called Žďár type of relief, Demek et al. 1965). Due to the dome-shaped character of relief, watercourses go radially, and the radial drainage pattern has probably evolved in response to the young tectonic domal uplift. Due to the existence of transverse tectonic faults, the river network also acquired rectangular character. The tectonic conditions and the effects of geomorphologic resistance of metamorphic rocks are reflected in the character of the Svratka valley and its tributary Fryšávka. The narrow water gaps alternate with riverine basins where alluvial plains with the naturally meandering channel occur, e.g. the floodplain of the Svratka River in Milovská kotlina (=Basin) or the floodplain of the Fryšávka River near the village of Kuklík.

Fig. 18.3 Painting "To spring near the Krátká village" by Josef Jambor (1947)—Typical landscape scenery in the central part of the highland (published with permission of the Municipal Authority Tišnov)

18.4 Main Relief Features and Geomorphic Development

A typical feature of the Žďárské vrchy landscape is the presence of groups of rock formations, which have a character of both isolated rock outcrops (walls, towers) as well as small rock cities. These rock formations dominate the flat top ridges and adjacent flat slopes and are conditioned by the geomorphologically resistant metamorphic rocks—orthogneiss and migmatites. They are limited by vertical walls and often significantly protrude above the surrounding landscape; therefore, they are locally called as pulpits. Traditionally, these rock formations are connected with various legends, which have been registered in the folk literature. The highest among them are Drátenická skála (=Rock) (35 m), Čtyřpaličatá skála (= Rock) (30–31 m) and Malínská skála (= Rock) (20 m). Usually, the height of the rock formations oscillates around 15 m.

In 1930, after a large-scale wind calamity and deforestation, these rock formations were well visible from afar and dominated the surrounding landscape (Fig. 18.4). However, they have gradually disappeared in the re-established forest since then. At present, the PLA Administration considers the possibility of uncovering selected rock formations again to let them return to be a significant feature of the landscape. The PLA Administration will propose transformation of spruce monocultures to open canopy forests with diverse species composition. In this activity, the visual axis should be released to let the rock formations dominate.

The geomorphological evolution of rock formations is complicated and related to the overall development of the wider area, which has gone through several phases. The periods of tectonic activity and phases of planation alternated. The basic features of relief were developed through the post-Cretaceous period in consequence of tectonic movements related to the folding and thrusting of the Carpathians. Uplifts and subsidence are linked to the transgression of the Cretaceous sea, which probably reached the majority of the area. The Paleogene planation surface in the upper parts of the area began to form and later, due to

Fig. 18.4 Deforested ridge of Žďárské vrchy south-west of the Křižánky village with well-visible rock forms of Bílá skála (=White Rock, 743 m a.s.l.) at the turn of the twentieth century. The site looked in a similar way after wind calamity in 1930 (source—Administration of Protected Landscape Area Žďárské vrchy)

favourable climatic conditions it assumed the character of a peneplain, sometimes with a considerable cover of weathered material.

During the subsequent tectonic dome uplift of a horst at the turn of the Late Tertiary, gradual erosion of weathered mantle and denudation of the Cretaceous sediments occurred. When the basal surface of weathering was revealed, a planation surface of etchplain type originated. On the exposed basal surface of weathering some resistant rocks stood out as isolated rock formations and groups of tors. There is also a possibility that high isolated rock formations may have originally developed in the context of deep weathering processes such as inselbergs (Ivan and Kirchner 1999). In the cold periods of the Quaternary, rock outcrops were transformed by frost weathering. Rock walls of frost-riven cliffs and cryoplanation terraces with extensive debris accumulations, rocky and boulder fields and streams were formed. The surfaces of the walls of rock formations are covered by small weathering forms (honeycombs, weathering pits and rock niches, pseudolapies). Caves and tunnels were created along fissures in the rock masses by gravitational and mechanical weathering processes. Within the thick debris covers, some talus caves are situated.

Since the Middle Ages, the landscape and landforms in the area were influenced by economic activity, resulting in various forms of anthropogenic origins. The agrarian heaps and terraces are the typical shapes associated with colonization. They represent important landscape elements and serve as a proof of the historical development of the territory. These features act harmoniously and complete the cultural landscape, e.g. near small settlements of Samotín, Krátká and Blatiny (Bajer et al. 2014).

18.5 Selected Unique Geomorphological Sites

18.5.1 Devět Skal

The largest rock formation, not only in the Žďárské vrchy Highland, but in the whole Bohemian-Moravian Highland, is located at the altitude between 780–836 m about 3 km southwest of the Křižánky village. The site consists of many rock formations and together they form a small and unique rock city. Rock formations are situated both in the top part of the ridge of the Devět skal (=Nine Rocks) and in the steep upper part of the eastern slope. They are built of the two-mica orthogneiss and migmatites of the Svratka Crystalline. The extensive destruction landforms are characteristic for this site, especially rugged ridges and isolated rocks (tors). There are also accumulation landforms, the most important one being the block stream heading to the south. The upper part of the site consists of three separate massive rock ridges which locally have character of a rock wall (Fryšavský hřbet, Hlavní hřbet (=Central Ridge), and Křižánecký hřbet) and which are divided into nine large and three smaller blocks. The site is dominated by the Hlavní hřbet (Fig. 18.5), 90 m long, with the height of vertical walls oscillating between 5 and 19 m. The Hlavní hřbet is divided into the three sections by two passes. These particular sections are characterized by the occurrence of rock towers as Žďárská věž (=Tower) and Větrná věž (=Windy Tower), with a beautiful view of the surrounding rock city. Southwest of the Hlavní hřbet, the Fryšavský hřbet is situated (length 70 m, height 6–8 m). This one has an irregular shape and consists of five smaller towers (e.g. Strmá věž (=Steep Tower), Malá věž (=Small Tower)), separated by a few distinctive passes. East of the Hlavní hřbet, the Křižánecký hřbet is located (about 60 m long, the walls reach up to 16 m). Close to the ridges, there are some isolated rock formations (e.g. Trůn Throne Rock with a height up to 16 m). In the lower part, the rocky wall of the Trůn Throne Rock passes into a semi-circular depression, probably a nivation hollow, partially filled with a block heap (or block accumulation) which continues further down as a block stream. Based on the measurements of joints and fissures, the relationship between morphology and structural elements of migmatites was confirmed (Fig. 18.6). The outline of the nivation hollow, as elsewhere in regions of the Czech Republic, built in crystalline rocks (e.g. Votýpka 1974), follows the geometry of joints and foliation planes. Morphological characteristics of rock formations (elongation, fragmentation into sub-blocks, occurrence of the passes) depend on the direction of joint systems in the underlying rocks and the position of rock formations in respect to the slope or plain. The joint system is dominated by two main directions: the primary system is NE-SW, followed by NNW-SSE and the third one is N-S. These fissures divide

Fig. 18.5 North-eastern, 19-m-high rock wall at Devět skal rock group (*Photo* K. Kirchner)

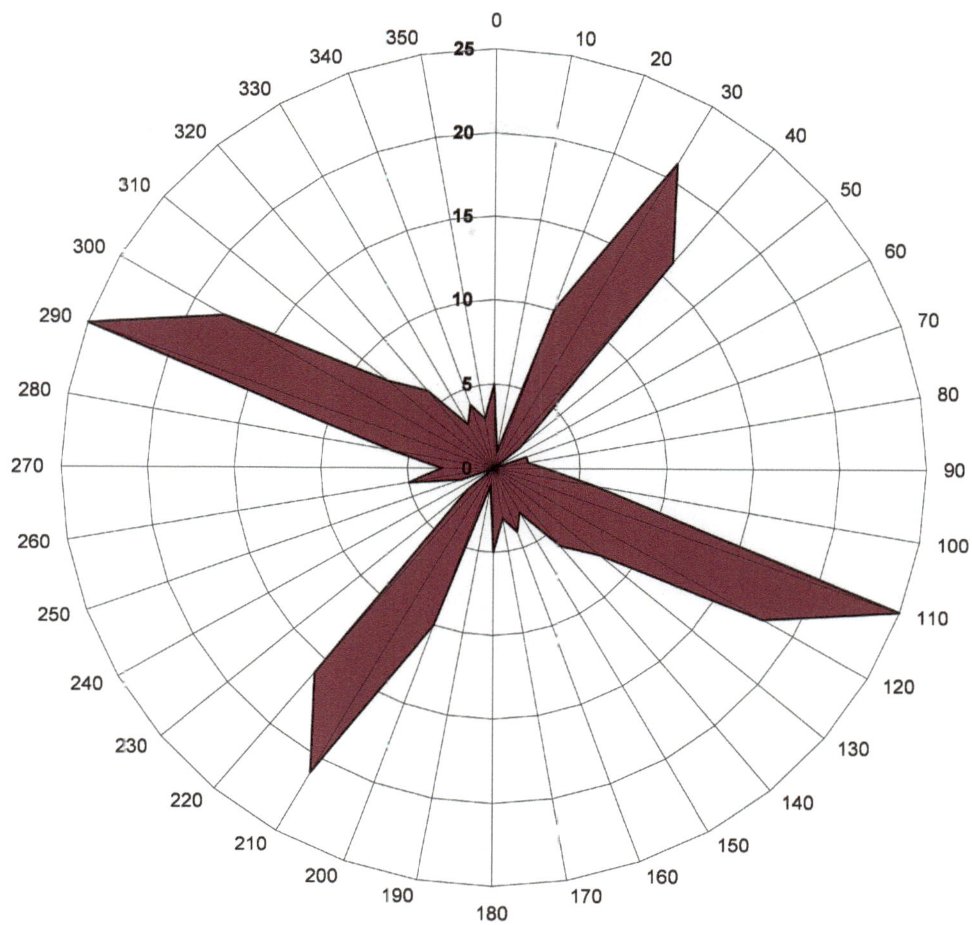

Fig. 18.6 Joint diagram for the Trůn Throne Rock, at Devět skal locality (axis Y—number of measurements)

the rock masses into segments of variable height. The WNW-ESE and NW-SE directions form a secondary joint system (Romportl 2003). The diversity of landforms at this site is complemented by the extensive accumulations of debris and block streams up to 100 m long, which are linked to the rock formations. The surfaces of rock formations (e.g. at Větrná věž) are covered with weathering microforms—rock niches and honeycombs, which are linked to joint directions and foliation surfaces; typically, there are rock windows in the upper parts of the rock wall of the Hlavní hřbet.

18.5.2 The Tisůvka Rock

The top rock wall of Tisůvka is located about 1.5 km northwest of the Cikháj village and reaches a maximum altitude of 800 m. It is built of the migmatized gneiss of the Svratka Crystalline. The massive rock wall rises up from the top part of a flat ridge (Fig. 18.7). This significant and mysterious rock formation has always attracted people's attention and is popularly called the Devil's stone. The 60-m-long walls extend in the NW-SE direction and their maximum height is 15 m. The wall is divided into two parts by a transversal fissure. Elongation of the rock wall is determined NNW- SSE joints. The ENE-WSW and N-S joint systems and surfaces of foliation then influence the division into sub-segments and underping the occurrence of weathering microforms through selective weathering effects. At the base of the rock walls, large niches of abri type originated. The vertical walls of rock formations were modelled by cryogenic processes during the cold periods of the Pleistocene and have the character of structurally conditioned frost-riven cliffs. Below the rock walls, massive debris piles were formed (up to 3 m high) and these are followed by a large cryoplanation terrace, covered with weathered blocks. The morphological diversity of the site is complemented by rock ledges, rock windows, and the unique occurrence of honeycombs 3–16 cm across and up to 5 cm deep.

18.5.3 Drátenická Skála

The site is located about 200 m northwest of the Blatiny village, at the altitude of 715–775 m. On the top part of the ridge, there is a massive 200-m-long rock wall of N–S extension and a few further isolated rock formations.

Fig. 18.7 Rock wall of Tisůvka emerges from the top part of flat ridge (*Photo* K. Kirchner)

Fig. 18.8 Orlí věž in the southern part of rock ridge at Drátenická skála is 28-m-high (*Photo* K. Kirchner)

Drátenická skála belongs to the highest peaks in the area. It is built by two-mica orthogneiss and migmatites of the Svratka Crystalline. Following the course of tectonic fissures (NE–SW, NW–SE), the rock wall is divided into several blocks separated by cols. These sub-blocks are dominated by significant rock towers. The most remarkable ones and visible from afar have been called "pulpits" by local residents. Among them, the 28-m-high Orlí věž (=Eagle Tower) (Fig. 18.8) and particularly the dominant Sokolí věž (=Falcon Tower), 35 m high, are most impressive (Doležal et al. 2006). In the talus below small debris caves occur, including a tunnel (width 1.5 m, height 1.5 m, length about 5–7 m) in the middle of the wall. It was formed by physical weathering in a 1-m-thick tabular layer of gneiss, distinctively divided into sub-layers. At the base of the rock wall, debris/block piles are situated which continue downslope as block streams which then merge with the block field on the slopes.

18.5.4 Malínská Skála

Malínská skála outcrop is situated about 2 km northwest of the Blatiny village at altitudes of 720–811 m. The site which occupies a flat ridge, is characterized by the occurrence of many rock towers and walls that form an arc more than 200 m long, open to the southwest. Rock formations are built of two-mica orthogneiss and migmatites of the Svratka Crystalline. The top rock formation, called Zubří skála (=Bison Rock), has the shape of a rock ridge and reaches the height of 20 m (Fig. 18.9). In the topmost part, a mushroom rock has formed along the joints. At the foot of the rock towers, rock niches (abri) have formed. From the steep southern and south-eastern slopes of Malínská skála, several distinctive rocky ridges rise. There is also a large depression with amphitheatrical shape where a large block stream begins. This depression is considered to be a partly destroyed nivation depression (cirque). In a nearby massive rock formation called Výspa (height 21 m), a typical nivation depression occurs, whose shape is adjusted to the joint system (Kirchner and Roštínský 2007). Maximum height of the rock wall in the middle of the depression is 12 m, while along the fissures a small cave has developed (depth 2–3 m). The site is significant especially because of the occurrence of block and debris accumulations. On the southeastern and eastern slope there are five block streams, the longest one being up to 80 m long, which merge with the block field in the lower part of the slope.

Fig. 18.9 Rock towers at Malínská skála locality (*Photo* K. Kirchner)

Fig. 18.10 Zkamenělý zámek with its highest rock tower, 22 m high. A fissure cave is located at the base of the rock formation—entrance is seen in the right part of the photograph (*Photo* K. Kirchner)

18.5.5 Zkamenělý Zámek

The site is located 1 km east of the Svratka village, on a flat ridge and adjacent slopes, at 720–765 m a.s.l. Bedrock is built by two-mica orthogneiss and migmatites of the Svratka Crystalline. The site is composed of six distinct rock formations arranged in a horseshoe arch, open to the northwest. The main formations are located on the western edge of the top part of the site. The inner space of the horseshoe is c. 40 m × 16 m. Rock formations are separated by transversal joints trending NE–SW. The highest and dominant rock formation of the site is called the Zámecká tower (22 m high, Fig. 18.10). Exfoliation phenomena along surfaces of foliation may be observed in the south-eastern and eastern wall of the tower, as well as a small fissure cave, 3 m long and 0.65 m wide. From the south-western foot of the rock formation, a huge debris/block pile extends which then transforms into a block stream. Other rock formations nearby are up to 10 m high, with the exception of one called Pivovar ("Brewery"), situated about 30 m west of the main group, which reaches a height of 15 m. Rockfall along the joint surfaces has created a huge overhang 5.5 m deep. Vertical walls of rock formations are rich in small weathering forms, honeycombs and abri, whereas at the top of the Pivovar occur two destroyed weathering pits measuring 43 and 35 cm long, respectively.

Unique landforms, which show the first traces of human activity in the top part of the Žďárské vrchy, are the remains of a Slavic settlement from the sixth century to the seventh century, situated in the eastern and southern part of the site. Here one can find remains of ramparts and deep trenches, up to 1.5 m deep and in places up to 3 m width. Natural rock formations were also included in the fortification. The settlement area was about 85-m-long and 55 m wide, surrounded by double ramparts. Later in the Middle Ages it could have served as a watch fort.

18.6 Typical Small Weathering Rock Forms

Vertical and horizontal surfaces of rock formations in the Žďárské vrchy abound in weathering forms that increase the diversity of the sites. Besides the already discussed small weathering forms, honeycombs, rock niches and ledges, there are also some rare features in metamorphic

Fig. 18.11 Unique weathering pits in metamorphic rocks at Milovské Perničky (*Photo* K. Kirchner)

rocks—weathering pits. They are found in greater concentrations (tens of pits) at the sites of Rybenské Perničky (1.5 km west of the Pustá Rybná village) and Milovské Perničky (2 km northeast of the Křižánky village). The biggest weathering pit is located at the former locality, measuring 50–60 cm across and 20 cm deep (Vítek 1975). On the other site, eight well-developed weathering pits have been documented. There is also a considerable number of initial forms that often intersect one another (Fig. 18.11). In the past, weathering pits were popularly known as 'gingerbread' forms, because their shape resembled gingerbread mould. These minor weathering forms have given the name to their host rock forms ("Perničky" = ginger biscuits). But there are other interpretations of the name and origin. Folk legends often attributed the divine origin and purpose to the weathering pits, or it was believed that they were created for cult purposes or that they were used for ignition of guard fires, to grind grain, etc.

18.7 Problems of Nature Protection and Human Impact

The area of Žďárské vrchy is located on the border of Bohemia and Moravia. During the early Middle Ages the area was covered with large forests. The initial settlement of the wider area started in the thirteenth century from the lowlands, during the period of colonization under the rule of Přemyslids dynasty. The development of settlement network was associated with burning the forests to gain land for agriculture. The monasteries, especially Cistercian monastery in Žďár, founded in 1252, and aristocratic estates became the centres of this first wave of colonization. Slavic settlement on Zkamenělý zámek is a local phenomenon and so far an exception. Substantial economic development of the area occurred during the second wave of colonization at the turn of the sixteenth century, when aristocracy founded ironworks and glassworks, ponds and big farming estates. The last wave of the so-called pasture colonization, which reached the most elevated areas, started in the eighteenth century. The most picturesque villages and settlements such as Krátká, Samotín, Blatiny, located very close to the rock formations presented earlier, originated in this period. At the beginning of the nineteenth century, the area was a major producer of iron in the Czech lands and products of glassworks of Herálec and Milovy were known throughout Europe. As a result of exhaustion of wood resources in the local forests production gradually declined. The mentioned period of colonization is related to growing influence of the economic activities. In the meantime, agricultural activities created characteristic landforms of agricultural terraces and heaps of stones collected from fields that complement the predominantly forest landscape mosaic. The unique landscape of the Žďárské vrchy began to change rapidly with the socialist agricultural production in the 1950 and 1960s. Consolidation of the pieces of land, elimination of the indigenous ways of farming and traditional field boundaries, implementation of surface drainage schemes and regulation of small water flows quickly changed the original appearance of the landscape. Therefore, founding of the PLA Žďárské vrchy in 1970 also meant preservation of the residues of the harmonious cultural landscape and the area of the interest has become the core of the protected area, with its rock formations now protected as nature monuments. Currently, the PLA management activities are towards restoration of long-distance views of the rock formations in the top parts of the area because most of them are now surrounded by tall spruce forests, which obscure visibility.

18.8 Conclusions

The Žďárské vrchy Highland is a part of the Bohemian Massif (Variscan Europe) and its landforms have developed over a very long period. Basic landform features were formed after Cretaceuos marine regression. A new phase of planation began, culminating in the development of the Paleogene planation surface, on which a weathered mantle, mostly kaolinitic and often than 100 m thick, was created (Czudek and Demek 1970). Owing to neotectonic movements, the territory was uplifted and weathering mantles were removed, the basal surface was exposed and a younger planation surface of etchplain type originated. The geomorphic evolution of area were accomplished in two stages (Czudek 2005). On the exposed basal surface of weathering some resistant rocks stood out as isolated rock formations and tor groups. These tors were then gradually modelled by cryogenic processes in the Pleistocene, consistent with the two-phase development model for tors (Demek 1964).

Landform development of the Žďárské vrchy Highland has been very complicated and its relief is polygenetic. Within the Variscan Europe we can look for similar landforms, e.g., in the German highlands (Rheinisches Schiefergefirge, Schwarzwald, Odenwald) or in France in the Massif Central (Král 1999). Unlike the aforementioned examples, the Žďárské vrchy Highland (especially its central part) is a unique area, because many tall rock formations are located on ridge tops in the gently undulating Variscan crystalline highland. These rock formations contribute to the peculiar character of the area and are not present in other Hercynian Mountains of the Europe (Embleton 1984).

Acknowledgments Geomorphological research in the Žďárské vrchy Highland was supported by a long-term conceptual development support of research organisations (Institute of Geonics, Czech Academy of Sciences, v.v.i.) RVO: 68145535.

References

Bajer A, Hlaváč V, Kirchner K, Kubalíková L (2014) Za skalními útvary CHKO Žďárské vrchy. Mendelova univerzita v Brně, Brno 88 p

Czudek T (2005) Vývoj reliéfu krajiny České republiky v kvartéru. Moravské zemské muzeum, Brno 238 p

Czudek T, Demek J (1970) Některé problémy interpretace povrchových tvarů České vysočiny. Zprávy Geografického ústavu ČSAV: 1970 7 (1):9–28, Brno

Demek J (1964) Formy zvětrávání a odnosu granodioritu v Novohradských horách. Zprávy Geografického ústavu ČSAV: 1964 (9): 6-15, Opava

Demek J (2002) Vývoj georeliéfu. In: Čech L, Šumpich J, Zabloudil V a kol. Jihlavsko. Chráněná území ČR VII, pp 36-37, AOPAK ČR a EkoCentrum Brno, Praha

Demek J Balatka B, Czudek T, Láznička Z, Linhart J, Loučková J, Panoš V, Raušer J, Seichterová H, Sládek J, Stehlík O, Štelcl O, Vlček V (1965) Geomorfologie Českých zemí. Academia Praha, 335 p

Doležal F, Bořil P, Zavadil A, Ouda J, Bořecká K, Glajc P, Hartmann P, Trefulka F, Vogt J, Juda J, Trefulková M, Stejskal D (2006) Žďárské vrchy. TJ Vysočina, Žďár n.S, Průvodce po horolezeckých terénech Vysočiny, 172 p

Embleton C (ed) (1984) Geomorphology of Europe. Macmillan Reference Books, London 465 p

Hanžl P, Melichar R, Buriánek D, Krejčí Z, Kociánová L (2011) Geology of the Žďárské vrchy area: a review. Travaux Géophysiques 40: 24, Institute of Geophysics of ASCR, Praha

Ivan A, Kirchner K (1999) Morfostrukturní charakteristika Žďárských vrchů. Geologické výzkumy na Moravě a ve Slezsku v roce 1998, VI:17–18, Brno

Kirchner K, Roštínský P (2007) Geomorfologická inventarizace vybraných skalních útvarů v centrální části CHKO Žďárské vrchy. Sborník Prací Přírodovědecké fakulty Ostravské univerzity, Geografie, Geologie 237(10):48–64

Král V (1999) Fyzická geografie Evropy. Academia, Praha 348 p

Melichar R, Břízová E, Buriánek D, Buriánková K, Čurda J, Fürych V, Hanžl P, Kirchner K, Lysenko V, Mrnková J, Roštínský P, Rýda K, Skácelová Z, Vít J (2004) Vysvětlivky k základní geologické mapě České republiky 1:25000, list 24-111 Sněžné. Česká geologická služba, 58 p

Romportl D (2003) Geomorfologické poměry centrální části CHKO Žďárské vrchy. Univerzita Karlova, Praha, Magisterská práce 95 p

Vítek J (1975) Mikroformy na skalách Perničkách. Ochrana přírody 7:209–210

Votýpka J (1974) Vznik a vývoj mezoreliéfu a mikroreliéfu Sedmihoří. Acta Universitatis Carolinae 1974, Geographica 2:17–34

The Dyje Canyon-like Valley: Geomorphological Landscape of Deep Valley at the Eastern Part of the Marginal Slope of Bohemian Massif

19

Karel Kirchner

Abstract

Southeastern margin of the Bohemian-Moravian Highland is cut by deep river valleys with incised meanders. The Canyon-like Valley of the Dyje River belongs to the most important and attractive of them. The valley is embedded into the Paleogene regional planation surface (etchplain) within the hilly land relief, creating sharp morphological contrasts, unique in the Bohemian Highland and in the adjacent part of Austria. The deep valley of Dyje River has become a strategic river since the Middle Ages, because the valley traditionally formed the border between Austrian countries and the Kingdom of Bohemia. In the period of the Cold War, the area was closed to the public for more than 40 years as it was a part of the "iron curtain". After political changes in the Czech Republic Europe in 1989, the border area was open to the public. Due to rare landscape phenomenon (including landforms), the Dyje valley with its surroundings was declared a national park in 1991 (Podyjí National Park). On the Austrian side of the Dyje River valley, the national park Thayatal was declared in 2000. Three basic relief types have been defined in the study area: (i) Polygenetic regional planation surface, (ii) Fluvial canyon-like valley of the Dyje River and its tributaries and (iii) Polygenetic marginal slope of Bohemian Highland. Within these relief types, a few unique sites are located, related to fluvial erosion, periglacial remodelling of rock slopes, selective weathering and denudation. Special attention is devoted to site with Ice caves within the largest rock slide in the area, which conditioned the origin of the longest pseudokarst cave system in crystalline rocks in the Czech Republic.

Keywords

Podyjí National Park • Canyon-like valley • Meanders • Rock formations • Crevice-type caves

19.1 Introduction

The southeastern edge of the Variscan geomorphological subsystem of the Bohemian-Moravian Highland consists of two geomorphological units: the southern highlands of Jevišovická pahorkatina Hillyland followed by Bobravská vrchovina Highland to the northeast. Towards the eastern borders, these geomorphological units neighbour with a young relief of the Fore-Carpathian depressions, which belongs to the Outer Western Carpathians. Southeastern margins of Jevišovická pahorkatina Hillyland and Bobravská vrchovina Highland are cut by deep river valleys (rivers Svratka, Jihlava, Jevišovka and Dyje/Thaya), typified by incised meanders. The canyon-like valley of Dyje, with many incised meanders (situated in the Jevišovická pahorkatina Hillyland), cut into the Paleogene planation surface, is one of the unique geomorphological phenomena (Fig. 19.1). The valley of the Dyje River forms the border between Austria and the Czech Republic and the area was closed to the public more than 40 years as it was a part of the border

K. Kirchner (✉)
Institute of Geonics, Czech Academy of Sciences, v.v.i. Branch Brno Drobného 28, 602 00 Brno, Czech Republic
e-mail: kirchner@geonika.cz

T. Pánek and J. Hradecký (eds.), *Landscapes and Landforms of the Czech Republic*, World Geomorphological Landscapes, DOI 10.1007/978-3-319-27537-6_19

Fig. 19.1 Western part of the Dyje River valley in the Podyjí National Park. Bedrock meanders of the Dyje River are incised into the regional planation surface (*Photo* P. Lazárek)

zone. Due to political changes in Europe in 1989, the Dyje valley with its surroundings was declared a national park in 1991 (Podyjí National Park, Podyjí NP), where valuable components of the natural environment and landscapes only slightly disturbed by human activities are present. On the Austrian side of the Dyje/Thaya River, the Nationalpark Thayatal was declared in 2000. Both National Parks create a bilateral cross-border protected area. In this chapter, attention will be paid to landforms of the Dyje valley and the surrounding area as well as to the unique geomorphological sites in the Podyjí NP, which include incised meanders, large rock outcrops and the site with the longest pseudokarst caves in crystalline rocks in the Bohemian Highland.

19.2 Geological and Geomorphological Setting

The geological basement of the canyon-like valley of Dyje and the surrounding area is built of crystalline rocks, both igneous and metamorphic; these rock complexes distinctly dominate in the deeply eroded SE part of the Bohemian Massif. The territory belongs to the classic area of Variscan nappe tectonics. Three high-order geological units participate in the geological structure of the Podyjí NP and adjacent terrain. These are, from the east to the west, the Brunovistulicum, the Moravicum (or newly Moravosilesicum) and the Moldanubicum. The Brunovistulicum is composed of intrusive rocks of the Dyje Massif (Cadomian) (Scharbert and Batík 1980), believed to be, together with the Moravicum, the complicated structure of the Dyje Dome. Petrographically, the Dyje Massif is rather homogenous in the interested area. The prevailing biotite granite, also named the Dyje granite, occurs close to contacts with the Moravicum which is schistosed and mylonitized (Fig.19.2).

The Moravicum is extraordinarily complex by its structure, composed mainly of metasediments and gneiss. It consists of three units. In the west, there is the eastern marginal part of the Vranov unit built of paragneiss with intercalations of amphibolites and crystalline limestone. Under these rocks, the Bíteš unit is composed of two mica orthogneiss. The lowest, Lukov unit consists of phyllites, schists with crystalline limestones (marbles), erlans and quartzites (Batík 1993).

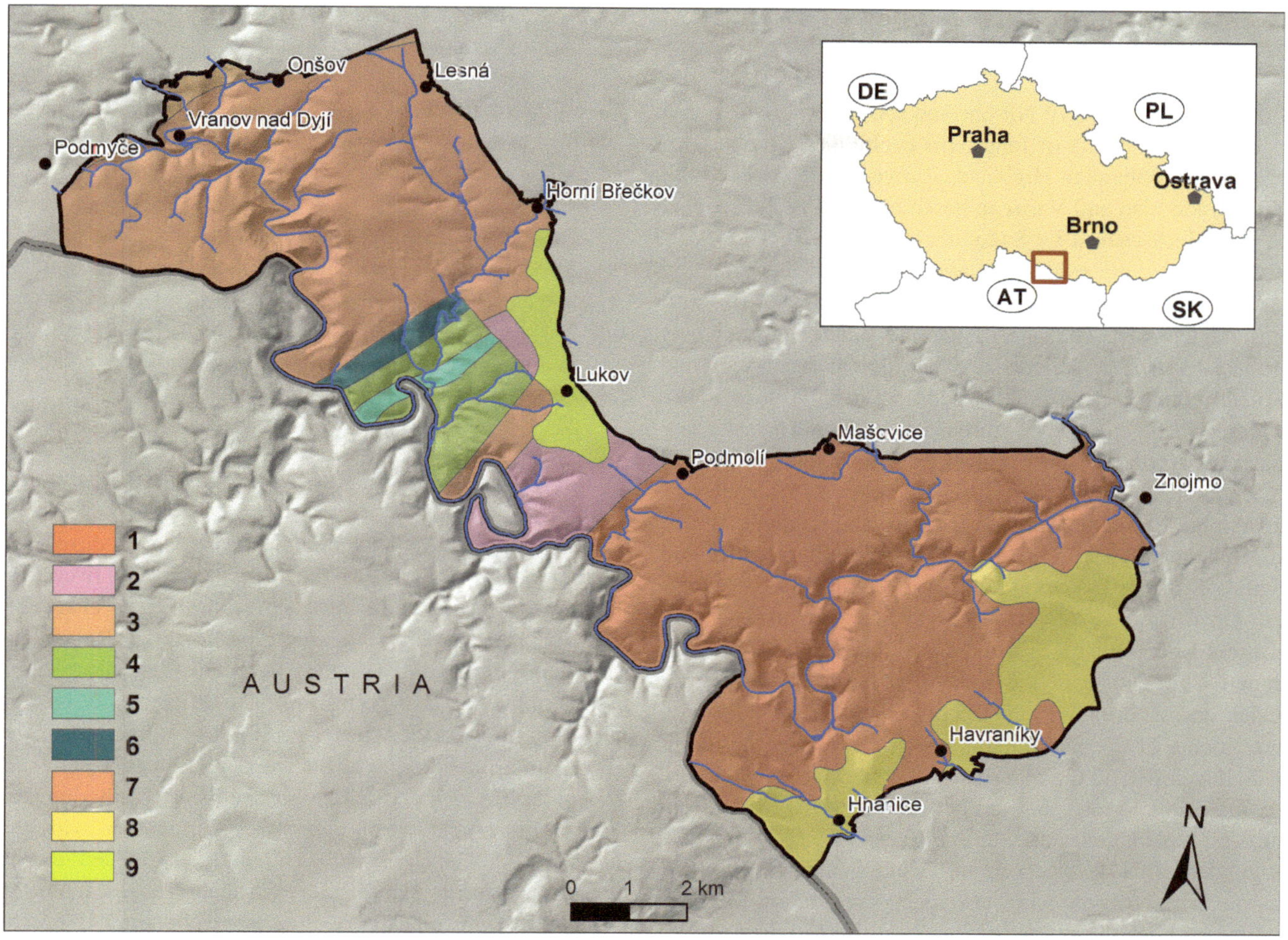

Fig. 19.2 Schematic geological map of the Podyjí National Park (arranged after—Cháb et al. 2007): Geologická mapa ČR, Czech Geological Survey. http://mapy.geology.cz/geovedni_mapy500/?center=-655117,-1189131&scale=100000 Explanation: Dyje Massif *1* biotite and two-mica granites and granodiorites, locally deformed and metamorphosed, Moravicum *2* gneisses and migmatites, *3* mica schists and gneisses, *4* phyllites, *5* marbles, *6* calc-silicate rocks (erlans and quartzites), *7* two-mica orthogneisses, Miocene sediments *8* clays, marls, minor sands, gravel, (marine facies), *9* sands, gravel, clays

To the west and north of the study area, different types of paragneiss and ortogneiss are the prevailing Moldanubian rocks. On the east, it borders with the Jevišovická pahorkatina Hillyland and the Carpathian Foredeep, filled with the Egerian, Eggenburgian, Ottnangian, Karpatian and Badenian sediments, predominantly marine.

The remnants of Miocene sediments are located also in the eastern part of the area. The polymictic gravels and sands are supposed by Batík (1993) to be of Ottnangian age. On plains and flat slopes loess and loess loam are widespread (Šušolová 2005), in the lower parts of the valley slopes there are some remnants of the Dyje river terraces (Roetzel et al. 2005; Roetzel 2010).

The Dyje valley forms the core zone of Podyjí NP and is located on the southeastern edge of the Jevišovická pahorkatina Hillyland geomorphological unit and the altitudes oscillate in the interval 301–450 m. The lowest point—the bottom of the Dyje valley on the eastern edge Podyjí NP is 215 m a.s.l., maximum altitude is reached in the western part of the area—Býčí hora, 536 m above sea level. Almost 20 % of the area is covered by slopes with an inclination higher than 15°. The meandering flow of the Dyje River has a length of 41.6 km within Podyjí NP, but the straight distance is only 17 km (coefficient of the flow development is quite high—2.4). In the landscape of NP Podyjí there are considerable contrasts. The cross section shows the sharp contrast between the deep valley of the Dyje and the flat planation surface. The longitudinal profile shows crucial contrast between the western margin of the NP Podyjí, where the Dyje flows through the depression filled by the Vranov dam, then it creates a deep canyon valley in Podyjí NP and on the eastern margin, the Dyje flows into wide shallow depression of the Dyjsko-svratecký úval Graben.

19.3 Geomorphic Development of Study Area

The relief of the Podyjí NP and the surrounding wider area represents the effects of long-term evolution of the eastern edge of the Bohemian Highland. It reflects deep denudation of the Cadomian and Variscan basements and its subsequent modifications caused by Saxonian tectonics. After the Variscan orogeny, the relief was developed towards a planation surface that was formed under the conditions of a warm seasonal climate. The age of the old planation surface is probably Mesozoic–Paleogene. The tropical weathering had probably the effect on the development of landforms in the Middle Paleogene, when the deep kaolinitic weathering materials around Únanov and Mašovice villages originated (Roetzel et al. 2005). In relation to tectonic movements of the Alpine-Carpathian orogeny, transport of the weathered materials and exposure of the basal weathering surface and evolution of the planation surface of etchplain type occurred. On weathered rock outcrop and their eluvia, formation of the silicites occurred in the Late Miocene, e.g. findings from Čížov village in the western part of the Podyjí NP (Cílek and Andrejkovič 1999). In connection with flexural movements of the Bohemian Massif, the escarpment etchplain developed on the marginal slope, during denudation of the planation surface. In parallel the river network was cut into this planation surface, hence an incised etchplain originated. This development was complicated by Miocene marine transgression and subsequent erosion of Miocene sediments after the retreat of the sea. Characteristic development and long sequence of the incised meanders of the Dyje River probably testifies to the epigenetic origin of the valley.

Neogene sediments have not been found in the Dyje valley and therefore the valley development took place in the Quaternary (Brzák 1998). Based on the analysis of other valleys in eastern margin of the Bohemian Massif, some authors (e.g. Batík and Šebesta 1996; Demek 2007) presented the opinion that the Dyje valley is pre-Miocene in age. However, the relics of gravel at different relative heights above the contemporary Dyje valley (e.g. polymict sandy gravels—considered as Late Pliocene—at the site of Nový Hrádek, on the contact with the Dyje valley) indicate gradual incision during the Pleistocene (Roetzel et al. 2005). The tectonic development of the eastern margin of the Bohemian-Moravian Highland is probably related to a significant meandering of the Dyje River associated with the origin of a number of incised meanders, but the problem has not been fully resolved.

19.4 Relief Types of the Podyjí National Park

The relief of the Podyjí NP is dominated by three basic landform assemblages, which differ in geological basement, morphology and genesis:

(i) Polygenetic regional planation surface,
(ii) Fluvial canyon-like valley of the Dyje River and its tributaries
(iii) Polygenetic marginal slope of the Bohemian Highland. Within these main types, 9 subtypes have been defined (Fig. 19.3).

19.4.1 Polygenetic Regional Planation Surface

The flat or slightly undulated platforms which are developed on the rocks belonging to the Moravicum and Brunovistulicum represent an erosional surface, which is characteristic for the whole southeastern part of the Bohemian-Moravian Highland. The oldest sediments that allow direct dating of the surface come from the Miocene. It is very likely that the surface was formed already in the Mesozoic. The kaolinite weathering crusts are also of great importance. The processes of deep kaolinitic chemical weathering and denudation of the weathering material together with the uncovering the basal weathering surface has led to the development of an etchplain-type planation surface. Within the concept of planation surfaces (etchplains) (Thomas 1974), the regional planation surface of the wider surroundings of Podyjí NP belongs to the category of incised etchplains that are characterized by the sporadic occurrences of the old weathered material with a deeply incised valley network. Migoń (2004) ranks this type of surface as one of the options of the complex etchplain. Thomas (1994), but also Migoń (2004) adopt a broader concept of an etchplain in which etching processes and etchplanation include also the sandy types of weathering in the areas of middle and even higher latitudes. Typical surface landforms in the area of interest are ruwares and tors.

19.4.2 Fluvial Canyon-like Valley of the Dyje River and Its Tributaries

The deeply incised, canyon-like valley of the Dyje River valley is the main and unique landscape feature of the Podyjí NP. In the section of the Dyje River valley on the territory of

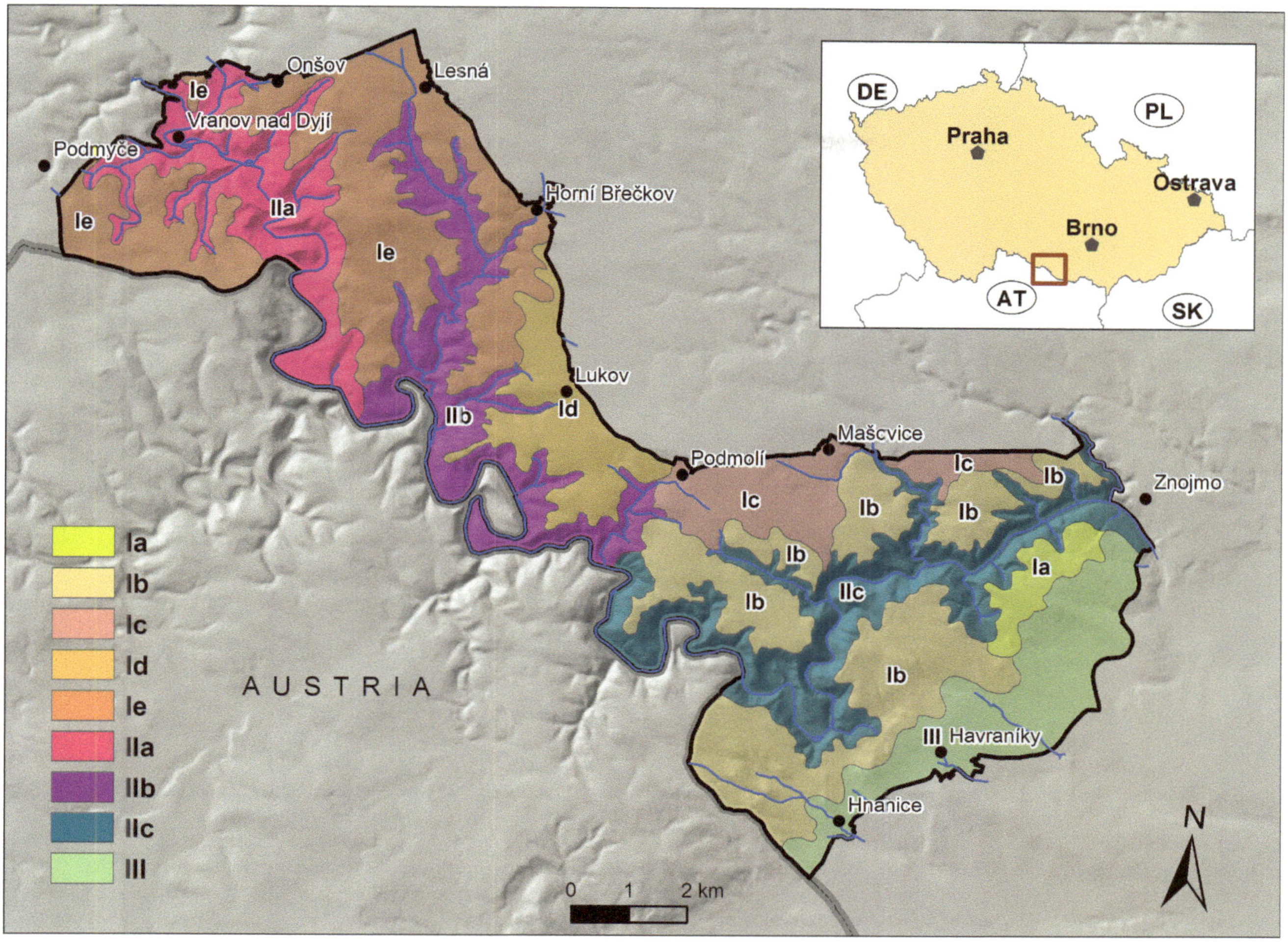

Fig. 19.3 Geomorphic types of the Podyjí NP landscape Explanation *I*. Polygenetic regional planation surface *Ia* on granitoid rocks of the Dyje Massif with ruwares and cover of quartz gravel (Ottnangian) *Ib* on granitoid rocks of the Dyje Massif with ruwares *Ic* on granitoid rocks of the Dyje Massif with cover of loess and loess loam, in some places with remnants of kaolinic weathering crust *Id* on metamorphic rocks of Moravicum with prevailing mica schists and crystalline limestones, in some places with cover of loess and loess loam as well as quartz gravel (Ottnangian) *Ie* on metamorphic rocks of the Moravicum with prevailing two-mica orthogneiss (Bíteš orthogneiss), sporadic occurrence of ruwares II. Fluvial canyon-like valley of the Dyje River and its tributaries *IIa* deeply incised Dyje valley (depth–235 m) and its tributaries into the Bíteš orthogneiss, steep rocky slopes with abundant gravitational deformations, rectangular incised meanders, symmetrical cross valley profile, the Dyje River is anti-dip stream *IIb* deeply incised Dyje valley (depth 120–150 m) and its tributaries into mica schists with crystalline limestones, smooth meanders, including completely cut off, with asymmetrical cross profile, undercut meander slopes with rock formations, flood plain is developed *IIc* deeply incised Dyje valley (depth from 120–160 m) and its tributaries into granitoid rock of the Dyje Massif, symmetrical and asymmetrical meanders, steep valley slope with rock formations, block fields and streams, manifestation of gravitational processes and exfoliation *III*. Polygenetic marginal slope of Bohemian Highland—tectonically conditioned—on granitoid rocks of the Dyje Massif, in some places with cover of Miocene sediments as well as loess and loess loam, occurrence of small weathering forms on granite

Podyjí NP three subtypes are distinguished: IIa, IIb, IIc. (Fig. 19.3).

The western section of the Dyje River between Vranov nad Dyjí and Hardegg towns, classified as subtype IIa, is developed in the Bíteš orthogneiss (Moravicum). As a result of high resistance of orthogneiss, the incised meanders are rather rectangular than regularly arched. The cross-section of the canyon in the meander bends is relatively symmetric (entrenched type). In relation to the position and inclination of foliation planes in gneiss (mostly 20–40° to the NW), the Dyje River is an up-dip stream. The valley is the deepest in this section (over 235 m). The valley slopes are dotted, especially in the upper parts, with many rock formations (rock towers, crags, ridges). The nature of this area is expressed in a detailed geomorphological map (Demek and Kopecký 1996) (Fig. 19.4). Gravitational slope processes and cryogenic weathering processes are the most important agents in this area.

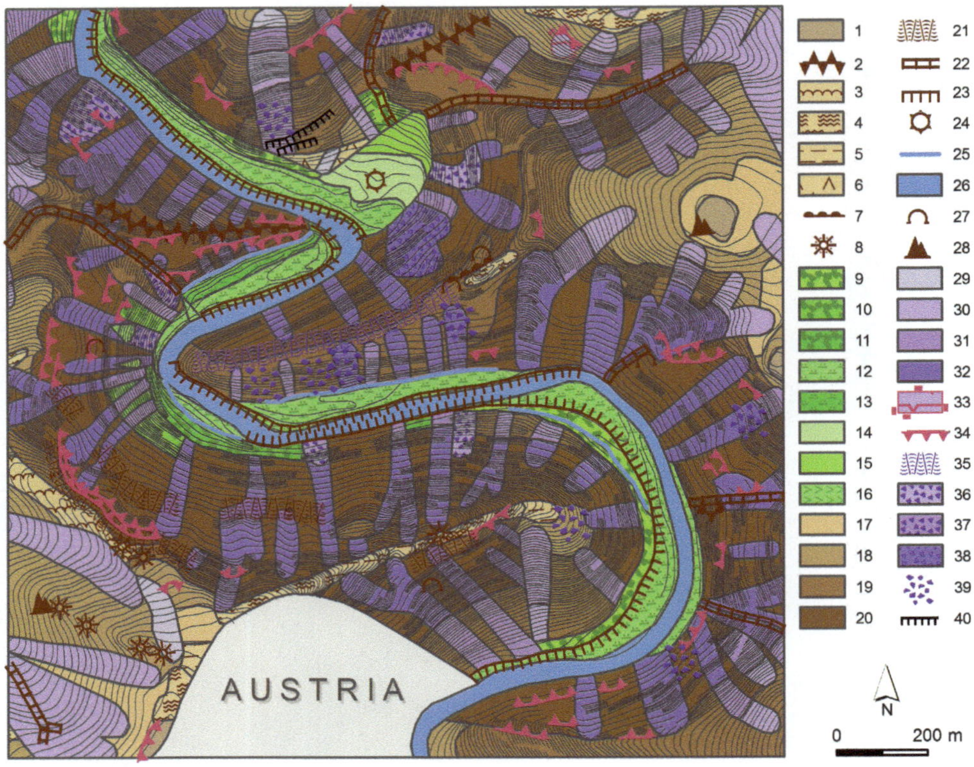

Fig. 19.4 Geomorphological map of western part of the Dyje River valley in the Podyjí NP southward of the Vranov nad Dyjí town with the incised meander of Dyje River. The site Ice Caves is situated in meander spur (arranged after Demek et al. 2009). Legend to the geomorphologic map. Explanations: *1* remnant of the planation surface (etchplain), *2* structural range, *3* narrow and rounded ridge developed by the intersection of slopes, *4* broad and rounded ridge developed by the intersection of slopes, *5* spur, *6* structural ridge, *7* scarp–landslide scar, *8* monadnock, *9* talus pile inclined 5–15°, *10* talus pile inclined 15–25°, *11* talus pile inclined 25–35°, *12* low floodplain (Holocene), *13* high floodplain (Quaternary), *14* surface of alluvial cone with inclination 2–5°, *15* surface of alluvial cone with inclination 5–15°, *16* debris cone, *17* erosional–denudational slope inclined 2–5°, *18* erosional–denudational slope inclined 5–15°, *19* erosional–denudational slope inclined 15–25°, *20* erosional–denudational slope inclined 25–35°, *21* erosional-denudational slope inclined more than 35°, *22* gully, *23* scarp due to lateral erosion of watercourse, *24* meander core, *25* abandoned river bed, *26* river bed, *27* pseudokarst cave entrance, *28* tor, *29* dell inclined 2–5°, *30* dell inclined 5–15°, *31* dell inclined 15–25°, *32* dell inclined 25–35°, *33* cryoplanation terrace, *34* frost-riven cliff, *35* rock wall modelled by cryogenic processes, *36* block stream (angular block inclined 5–15°), *37* block stream (angular block inclined 15–25°), *38* block field with inclination 25–35°, *39* angular block, *40* agricultural baulk

The central section of the Dyje valley between Hardegg town and the estuary of the Žlebský potok creek (subtype IIb) is composed of less resistant metamorphic rocks of the Lukov unit (Moravicum). The meanders have characteristic asymmetric transverse profile (type ingrown). The depth of the valley oscillates between 120–150 m. In the concave banks bedrock outcrops are abundant (Ivan and Kirchner 1994).

The eastern section of the Dyje, belonging to subtype IIc, is built by granites of the Dyje Massif (Brunovistulicum). It reaches the depth of 120 m, sporadically up to 160 m. The valley sides are steep, with rock formations, extensive boulder fields and block streams, with manifestations of gravitational processes and exfoliation. The incised meanders are both symmetric (entrenched) and asymmetric (ingrown). On the lower parts of the slopes and on the valley bottom there occur remains of fluvial gravel accumulation. To the west of the town of Znojmo, in the area of King's Throne, the rest of the gravel river terrace about 6 m thick was discovered (basis in the 30 m above the today Dyje valley bottom). Based on the comparison with gravel-sandy accumulations in Dyjsko-svratecký úval Graben, correlation with the younger gravel-sandy cover is possible, in the city of Brno area then with Tuřany terrace and dating to the lower Pleistocene (Kirchner et al. 1996). These findings have relevance for the understanding of denudation chronology of the Dyje River (see Sect. 19.3).

19.4.3 Polygenetic Marginal Slope of Bohemian Highland

The genesis of the eastern marginal slope of the West European platform in SW–NE direction, which limits the Carpathian Foredeep, is related to young tectonic movements (Fig. 19.5). These tectonic movements date to the turn of the Neogene with later renewals during Lower Badenian. The Lower Miocene Sea probably transgressed already on the pre-existing slope and covered the slope with marine sediments. The marginal slope included a number of inselbergs which are also typical for the foreland (Ivan and Kirchner 1998). It is probable that the slope and inselbergs within it were covered by Miocene sediments and after marine regression these deposits were removed and the pre-Miocene denudational surfaces cut in bedrock were exhumed (Roštínský and Roetzel 2005). On the inselbergs themselves, affected by exfoliation, there occur rounded granite boulders; occasionally microforms (weathering pits, pseudolapiés) can be found (Fig. 19.6). According to classification of Thomas (1994), the area can be identified as an escarpment etchplain.

19.5 Selected Unique Geomorphological Sites

In Podyjí NP, there are a number of unique geomorphological sites which are linked to the steep slopes of the Dyje valley and the eastern marginal slope. Valuable ecosystems are linked to the abiotic features. In terms of nature protec-

Fig. 19.5 Eastern marginal slope of the Bohemian-Moravian Highland between Hnanice and Havraníky villages. Rock outcrops built of granite and core stones with small weathering forms are characteristic for this area (*Photo* K. Kirchner)

Fig. 19.6 Weathering pits in the granite outcrop near the Havraníky village (eastern part of the Podyjí NP) (*Photo* K. Kirchner)

tion, the Podyjí NP is divided into three zones, designated regarding natural values. In each zone, the methods and ways of protection are graded. The strictest protection regime is established for the first zone where the most valuable geomorphological sites are located. For the purposes of nature conservation, Podyjí NP was divided into 19 landscape-ecological segments. Among them, twelve are linked to the Dyje valley phenomenon, whilst most of the rest is located on the eastern edge of the park (marginal slope of the highlands), and one landscape-ecological segment includes a waterlogged depression with a valuable habitat on the planation surface (more in Škorpík et al. 2007).

Spacious sites with rock landforms and large scree and boulder accumulations are located in the eastern part of the Dyje Valley (subtype IIc—basement built of granites of the Dyje Massif) and in the western sector of the Dyje Valley (subtype IIa—basement built of Bíteš orthogneiss) (Kubalíková 2009). The following text briefly describes selected geomorphological sites that are characteristic for the area of interest, especially for the relief subtypes mentioned above.

19.5.1 The Kamenná Moře—Block Fields

The locality is situated on the right steep slope of the Dyje River valley north of Paper Mill, about 6 km southwest of the town of Znojmo. Bedrock is formed by granites of the Dyje Massif. Due to unloading of granite and subsequent frost weathering in the Pleistocene ice-ages, very manifold rocky landforms are created there. Rocky rugged ridges, pillars and tors belong to the most significant and largest destructive forms (Kirchner and Demek 2009). In the upper part of rock ridges, crevice-type caves originated due to loosening along joint surfaces. The largest cave is located on the rocky ridge Výří skála (Owl Rock)–with extent: height up to 10 m, length 3–4 m and width up to 1.5 m.

Block fields and block streams (which reach the length up to 120 m) are located in the middle and lower part of locality (Fig. 19.7). In the lower part of block fields small debris caves occur. Pseudokarst lapiés and weathering pits developed on granite corestones and bedrock outcrops.

19.5.2 The Šobes Meander

The locality lies on a slip-off spur of an impressive entrenched meander and neighbouring abandoned meander Lipina with a rocky outlier of the Dyje River, approximately 7 km southeast of the town of Znojmo. Bedrock is formed of granite of the Dyje Massif. Man settled this area already 30,000 years B.P. and continuous settlement was established around 4300 years B.C. There are many forms of weathering of granite (large corestones, block fields, block streams). Original forest steppe and xerothermic vegetation were mostly replaced by vineyards on agricultural terraces, already in the Roman times–First Century AD. In this part of the valley, nine mills have been erected situated since 1497, but floods destroyed 3 mills already during the sixteenth century (Fig. 19.8).

19.5.3 The Liščí Skála—Fox Rock

The site is located on a left steep rocky slope of the Dyje valley, exposed to the south, about 4 km southwest of Mašovice village. The Fox Rock is a large rock formation

Fig. 19.7 Site Kamenná moře—Block fields in the eastern part of the Podyjí NP. Gravitational processes and unloading of granite caused rugged ridges with rock pillars and crevice-type caves (*Photo* K. Kirchner)

Fig. 19.8 Šobes meander lies on a slip-off spur of an impressive entrenched meander of the Dyje River. Original xerothermic vegetation was mostly replaced by vineyards on agricultural terraces (already since Roman times–First Century A.D.) (*Photo* P. Lazárek)

consisting of several stepped ridges which probably originated due to a large rock slide. The bedrock consists of biotite, schistosed granite of the Dyje Massif. The site is characterized by high diversity of small rocky landforms (ribs, pillars, rock mushrooms). Manifestations of exfoliation (exfoliation sheets 10–15 cm thick) are also notable as well as other forms of weathering and erosion of granite (tafoni, rock windows, mushroom rocks, boulder streams, etc.). Due to southern aspect which results in intense insolation of the steep slopes, temperatures of the rock surface during summer reach the high values. Thermic weathering enhances disintegration of the granite and the development of tafone-type cavities.

19.5.4 The Kozí Stezky—Goat Paths

The site is located on the left steep outer bank of the incised meander of the Dyje River, about 2 km south from Čížov village; the basement is built mainly of mica schist of the Lukov unit of Moravicum. The steep rocky slopes of the Dyje valley abound in cryogenic rock forms with interesting shapes such as spherical niches. In the surroundings of the site, there are remains of medieval mining depressions and shafts. Northeast from a rock formation at the mouth of the Klaperův potok creek a soil complex in loess was found, dated probably to the last interglacial (Cílek et al. 1996). Bedrock in this part is composed of crystalline limestone, there is a small karst system with a karst spring Loucký and karst caves.

19.5.5 The Vraní Skála—Crow Rock

The site is dominated by a rugged cliff 80 m high on the left slope of the Dyje valley, located about 1.5 km southwest of Lukov village. The amphitheatrically curved rock wall was formed in the outer bank of the incised meander of the Dyje

Fig. 19.9 Rock wall (up to 80 m high) of the Vraní skála–Crow Rock creates dominant form in the undercut slope of an incised meander of the Dyje River (*Photo* P. Lazárek)

River and it is characteristically dissected by mainly vertical cracks into sub-blocks due to the loosening of the rock cliffs. The latter are built mainly by mica schist, with the subordinate role of stripes of orthogneiss of Lukov unit (Fig. 19.9). At the foot of the hill, boulder scree slopes are situated. The steep slopes are also cut by deep gullies (in the upper part of the slope up to 7 m deep), which pass into significant ravines in the lower parts of the slopes.

19.5.6 The Ledové Sluje—Ice Caves

The site is one of the most important and the best-known sites in Podyjí NP. It is situated on a spur of the incised meander of the Dyje River, with the rest of the core in Bíteš orthogneiss, about 2.5 km southeast of Vranov nad Dyjí town. An extensive set of surface rock forms, pseudo-sinkholes, crevice-type caves and large block accumulations occur at this site. They were formed by gravitational slope processes which were conditioned by tectonic discontinuities and weathering. Fissures and underground spaces are linked to the main scars of the rockslides and landslides (Fig. 19.10). Due to specific microclimatic conditions, the underground spaces have ice decoration, which in some years persists through the summer period. The pseudokarst caves have been known since the nineteenth century due to occurrence of ice in caves (Roth 1863; Jarz 1882). Detailed speleological mapping was carried out in the 1990s. As many as 17 caves been documented; with the Brněnská jeskyně cave being the longest, about 400 m (Demek and Kopecky 1996). So far, the mapped underground spaces reach a length of about 2,000 m (Jiří Kopecký, oral communication). These underground spaces form the longest pseudokarst system in the crystalline rocks in the Czech Republic. However, triggering mechanisms which led to a large rock movement remain unknown.

Fig. 19.10 Crevice-type cave at the 'Ledové sluje' site. Caves' origin is due to gravitational processes, which occur along longitudinal fissures (*Photo* K. Kirchner)

Lateral erosion of the Dyje River and undercutting of the slopes are usually discussed. Influence of tectonic activity (earthquakes) in combination with lateral erosion is also mentioned.

19.6 Human Activities and Landforms

Conspicuous Dyje River valley formed the historic border between Austria and the Czech lands. In the eastern part of the Podyjí NP, which is in contact with fertile Dyjsko-svratecký úval Graben, the terrain has long been used in residential and agricultural activities. The first evidence of human settlement in the area come from the older Paleolithic (Svoboda et al. 1994; Neruda 2007). The impact of humans on the landforms gradually increased, but a deep wooded valley of the Dyje constituted a barrier to colonization process towards the west. At the time of the Great Moravian Empire, one of the most important centres of the area existed in Hradište near the town of Znojmo. After connecting Moravia to Premyslide State (first half of the eleventh century), the town around the Znojmo Castle began to develop (the castle was founded in 1017–1037). The town of Znojmo and its surroundings were transformed into a centre of economic, social and cultural life. After stabilizing the southern border of Moravia in the eleventh century, Dyje became a strategic river. Along the Dyje River valley, several castles were founded (e.g. Bítov, Vranov, Znojmo). Important castles were linked with towns (e.g. Znojmo). The origin of anthropogenic landforms was related to communication, military, defensive and residential activity. Water mills were other anthropic features that influenced the morphology of the Dyje valley and the surrounding hillsides in historical time. Around them, the whole range of the man-made forms was created, e.g. mill races, weirs, agricultural terraces, small quarries and communication lines. In the thirteenth century, colonization (partly German) increased in Bohemian countries, accompanied by substantial expansion in agricultural production, trade, mining, crafts and transport. These changes were a part of an extensive economic remodelling of the European continent (Žemlička 1998; Klápště 2005). From this period onward, more significant effects of economic activity began to appear, which manifested themselves in the emergence of fragmented cultural landscape. Many anthropogenic forms (residential, mining, agricultural such as terraces for growing grape wine and water-management forms) thus have historical foundation. In the western part of Podyjí NP during the nineteenth century, the noble families of Mniszek and Stadnický, owners of the Vranov estate contributed to landscape and relief changes. In the vicinity of the Vranov chateau (Fig. 19.11), a large natural landscape park was established and consisted of the series of targeted modifications of the relief.

The location of the Podyjí NP on the state border influenced the relief by military activities and construction of technical buildings. Remnants of defensive military construction (concrete bunkers) from the period before the World War II (1937–1938) can be found both in the Dyje valley and on the surrounding plains. The considerable disturbance of the landscape and relief was brought by the construction of the border zone with engineer-technical barriers during the Cold War (1951) (Fig. 19.12). The abolition of the border zone occurred in 1990 and today, its morphological distinctiveness is negligible. It can be said that historic anthropogenic landforms (in both extent and

Fig. 19.11 Western part of the Dyje River valley in the surrounding of the Vranov nad Dyjí town, with the dominant castle of Vranovský zámek (*Photo* L. Reiterová)

character) do not represent the disturbance of the landscape of the Podyjí NP, but have become an organic part of the relief as well as they document the development of the area in the past.

19.7 Conservation

After the Velvet Revolution, the unique quality of the canyon-like valley of the Dyje River led to the establishment of Podyjí NP in 1991 and subsequently, on the Austrian side, the Nationalpark Thayatal was declared in 2000. A unique bilateral cross-border protected area was formed. Human impact on the landforms in the past did not affect the character of the relief and vice versa, historic landforms have become an organic part of the relief of the Podyjí NP and they also constitute the legacy of the past development of the area. The unique natural environment of the Podyjí NP and the high quality of nature protection was reflected already in 1993 when the National Park was classified in the category II of the worldwide list of the protected areas under the International Union for Conservation of Nature. The Podyjí NP is a member of the Europark Federation (the pan-European non-governmental organization which associates large protected areas). In 2000, the Podyjí NP received the European Diploma of the Council of Europe as it met the strict criteria of quality and management of the protected area; this diploma has been held by the NP Administration until now even after the repeated evaluations.

Fig. 19.12 Engineering-technical barriers constructed in the period of Cold War in the border zone in the valley bottom of Dyje River near Šobes meander (*Photo* J. Vlasák)

Acknowledgments Geomorphological research in the Podyjí NP was supported by long-term conceptual development support of research organization (Institute of Geonics, Czech Academy of Sciences, v.v.i.) RVO: 68145535

References

Batík P (1993) Geologická mapa Národního parku Podyjí 1 : 25.000. Český geologický ústav, Praha, Geodézie, Brno, Geodézie, České Budějovice

Batík P, Šebesta J (1996) Vývoj toku řeky Dyje mezi Vranovem nad Dyjí a Znojmem a jeho vliv na vznik "Ledových slují". Věstník Českého geologického ústavu 71(3):297–299

Brzák M (1998) Příspěvek k vývoji údolí Dyje mezi Vranovem a Znojmem na základě morfografické analýzy a výzkumu fluviálních sedimentů. Geografie – Sborník České geografické společnosti 1998, 103, 1:31–45

Cháb J, Stráník Z, Eliáš M (2007) Geologická mapa ČR. Czech Geological Survey. http://mapy.geology.cz/geovedni_mapy500/?center=-655117,-1189131&scale=100000

Cílek V, Andrejkovič T (1999) Silicity z Čížova: příspěvek ke geomorfologii miocenní krajiny Podyjí. Thayensia (Znojmo) 2:3–13

Cílek V, Hradilová J, Ložek V (1996) Sprašová sedimentace v západní části NP Podyjí. Příroda - sborník prací z oboru ochrany přírody Sv. 3, Výzkum lokality Ledové sluje u Vranova nad Dyjí (NP Podyjí), pp 73–81, Praha

Demek J (2007) Geomorfologické unikáty Národního parku Podyjí. Thayensia (Znojmo) 2007(7):37–48

Demek J, Kopecký J (1996) Slope failures in metamorphic basement rocks of the Dyje river valley, Podyjí National Park Czech Republic. Moravian Geographical Reports 4(2):2–11

Demek J, Havlíček, M, Kopecký, J, Roetzel R (2009) Reliéf na krystaliniku – údolí řeky Dyje. Geomorfologická mapa - měřítko 1:15 000. In Hrnčiarová T, Mackovčin P, Zvara I. et al.: Atlas krajiny České republiky. Oddíl 4. Přírodní krajina, Pododdíl 4.4. Reliéf. Mapa č 65, pp 121. Ministerstvo životního prostředí ČR, Výzkumný ústav Silva Taroucy pro krajinu a okrasné zahradnictví, v.v.i, Praha

Ivan A, Kirchner K (1994) Geomorphology of the Podyjí National Park in the southeastern part of the Bohemian Massif. Moravian Geog Rep 2(1):2–25

Ivan A, Kirchner K (1998) Granite Landforms in South Moravia (Czech Republic). Geografia Fisica e Dinamica Quaternaria 21 (1):23–26

Jarz K (1882) Die Eishohlen bei Frain in Mahren. Petermanns Geographische Mitteilungen, 28:170–176, Gotha

Kirchner K, Ivan A, Brzák M (1996) K rozšíření kvartérních fluviálních sedimentů v údolí Dyje v NP Podyjí. Geologické výzkumy na Moravě a ve Slezsku v roce 1995: 21–23, Brno

Kirchner K, Demek J (2009) Relief of the Podyjí National Park and Geomorphologic Aspects of its Protection (Czech Republic). Memorie Descrittive Della Carta Gelogica D'Italia. Geomorphology and Cultural Heritage Vol. LXXXVII, 2009: 91–98. Istituto Superiore per la Protezione e la Ricerca Ambientale, Servizio Geologico d' Italia, Roma

Klápště J (2005) Proměna Českých zemí ve středověku. Nakladatelství Lidové noviny, Praha 624 pp
Kubalíková L (2009) Block Accumulations in the Western Park of the Podyjí National Park (Czech Republic): Preliminary Analysis of their Distribution. Moravian Geographical Reports 17, 2009, 1:49–55
Migon P (2004) Etching, etchplain and etchplanation. In: Goudie, A. S. ed. Encyclopedia of Geomorphology. Routledge, pp 345–347
Neruda P (2007) Starší doba kamenná v Podyjí – současný stav a perspektivy. Thayensia (Znojmo) 2007(7):291–303
Roetzel R, Fuchs G, Havlíček P, Übl Ch, Wrbka T (2005) Erläuterungen zur geologischen Karte der Nationalparks Thayatal und Podyjí. Geologie im fluss, Geologische Bundesantstalt, Wien 92 p
Roetzel R (2010): Geologie und Geomorphologie im Nationalpark Thayatal-Podyjí. In: Wiss. Mitt. Niederösterr. Landesmuseum - 21, 2010: 35–66, St. Polten
Roštínský P, Roetzel R (2005) Exhumed Cenonoic landforms on the SE flank of the Bohemian Massif in the Czech Republic and Austria. Zeitschrift für Geomorphologie 49(1):23–45
Roth A (1863) Die Eishohlen bei Frain in Mahren. Programm des k.k. Gymnasiums in Znaim am Schlusse des Schuljahres, Znojmo, pp 3–17
Scharbert S, and Batík P (1980) The age of the Thaya (Dyje) pluton. Verh. Geol. Bundesanst. 3:325–331, Wien
Svoboda J, Czudek T, Havlíček P, Ložek V, Macoun J, Přichystal A, Svobodová H, Vlček E (1994) Paleolit Moravy a Slezska. Dolnověstonické studie, sv. 1, Archeologický ústav AV ČR, Brno, 209 pp
Škorpík M, Demek J, Grulich V, Kirchner K, Kříž H, Plánka L, Petruš J (2007) Chráněná území NP Podyjí. In Mackovčin P, Jatiová M, Demek J: Chráněná území ČR – Brněnsko, svazek IX. Agentura ochrany přírody a krajiny ČR a EkoCentrum Brno, Praha, 88 pp
Šušolová J (2005) Spraše na území Národního parku Podyjí. Acta Musei Moraviae, Scientiae Geologicae 90(2005):155–170
Thomas MF (1974) Tropical geomorphology. MacMillan, London 332 pp
Thomas MF (1994) Geomorphology in the tropics. Wiley, Chichester 460 pp
Žemlička J (1998) Století posledních Přemyslovců. Melantrich, Praha, 412 p

The Moravian Karst: An Interconnection Between Surface and Subsurface Natural Sceneries

20

Jaroslav Kadlec and Petr Neruda

Abstract

The Moravian Karst belongs to the famous and frequently visited area with characteristic karst morphology and the largest cave systems in the Czech Republic. The karst evolution history covers a time span between Early Cretaceous and present. In particular, the Cenozoic processes, driven by climatic oscillations, and tectonic activity at the close contact between the Bohemian Massif and Carpathian orogenic range, have controlled the geomorphological processes and the present morphology of the area. Shallow valleys, and the first horizontal cave passages were formed by streams following a N-S drainage pattern during the latest Oligocene. The Carpathian nappe thrusting over the Bohemian Massif eastern margin occurred during the Early Miocene increased the stream gradients and accelerated fluvial erosion in the whole territory. The Moravian Karst limestone belt was dissected by deep canyon-like valley which was subsequently filled with marine clayey deposits during the Mid Miocene transgression. The present karst valleys and canyons were cut after the sea retreat. The large cave systems, formed by subsurface streams, were developed in relation to the valley and canyon incision.

Keywords

Moravian Karst • Morphology • Tectonics • Climate • Macocha Abyss • Caves • Prehistoric settlement

20.1 Introduction

The Moravian Karst belongs to the Drahanská Highland, extending at the Bohemian Massif eastern periphery. The limestone body forms a 25 km long and 2–5 km wide N–S trending belt covering an area of ca. 78 km^2. The karst surface is dissected by valleys locally changing to narrow glen-like canyons. The valleys of Lažánecké and Křtinské údolí running across the limestone belt divide the karst area into northern, central and southern segments (Fig. 20.1).

J. Kadlec (✉)
Department of Geomagnetism, Institute of Geophysics CAS, v.v. i., Prague, Czech Republic
e-mail: kadlec@ig.cas.cz

P. Neruda
Anthropos Institute, Moravian Museum, Brno, Czech Republic
e-mail: pneruda@mzm.cz

The Moravian Karst represents an area with the spectacularly developed karst morphology including half-blind and blind valleys, karst canyons, large sinkholes or deep abysses. The subsurface rivers formed the largest cave systems in the Czech Republic. Five show-caves yearly attract a large number of visitors. The romantic karst relief was a possible reason why this area was called the Moravian Switzerland in nineteenth century. The modern name was proposed by Procházka (1899) in his papers describing the character and origin of the Moravian Kart features. Both surface and subsurface morphological evolution of the karst area was controlled by limestone structure, regional tectonic activity and climatic conditions.

T. Pánek and J. Hradecký (eds.), *Landscapes and Landforms of the Czech Republic*, World Geomorphological Landscapes, DOI 10.1007/978-3-319-27537-6_20

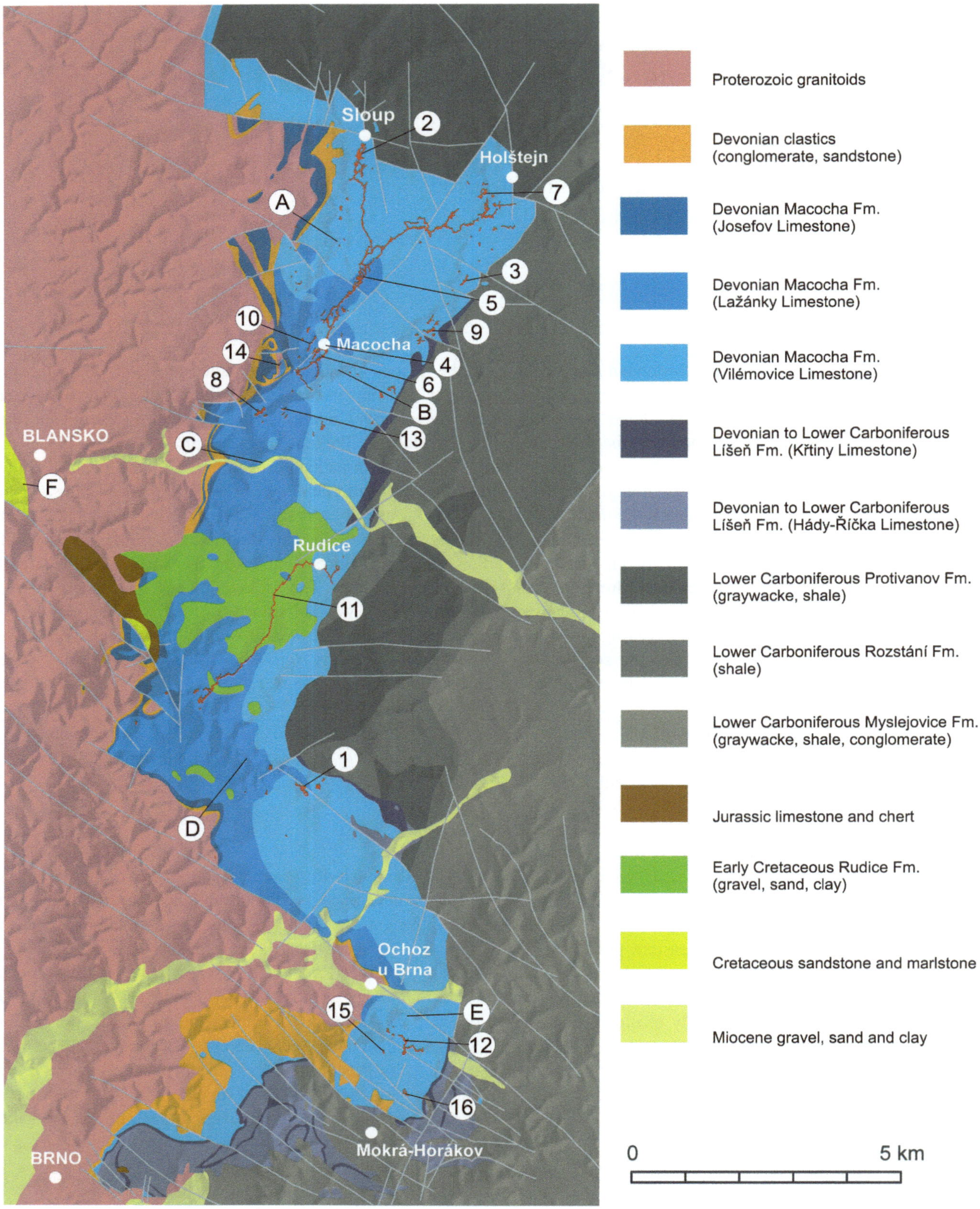

Fig. 20.1 Geological map of the Moravian Karst and its adjacent areas, main cave systems in red (constructed by I. Balák). Main valleys: *A* Pustý Canyon, *B* Suchý Canyon, *C* Lažánecký Canyon, *D* Křtinské Valley, *E* Říčka Valley, *F* Blanenský Graben; caves mentioned in the text: *1* Výpustek Cave, *2* Sloupsko-šošůvské Caves (including the Kůlna Cave), *3* Císařská Cave, *4* Macocha Abyss, *5* Amatérská Cave, *6* Punkevní Caves, *7* Holštejnská Cave, *8* Kateřinská Cave, *9* Balcarka Cave, *10* Zazděná Cave, *11* Rudické propadání–Býčí skála Cave, *12* Ochozská Cave, *13* Rytířská Cave, *14* Pod hradem Cave, *15* Pekárna Cave, *16* Mokerská Cave

20.2 Concise History of the Moravian Karst Exploration

Long tradition of scientific exploration of surface and subsurface natural phenomena of the Moravian Karst began in the early seventeenth century. The first reports notifying of the Pleistocene fauna bones found in the Výpustek Cave were written by O. Crolius in 1608, and one year later by Flemish physician and mineralogist B. de Bodt (Zajíček and Hromas 2013). The most remarkable report of the Sloupské, Kůlna and Císařská Caves, including the Macocha Abyss description, was prepared by J. A. Nagel, an Austro-Hungarian court mathematician and physicist in 1748. The manuscript is completed by remarkable drawings by K. Beduzzi (Fig. 20.2). The next outstanding person who conducted large archaeological and paleontological projects in the Moravian Karst underground was local physician H. Wankel. He summarized his results in a number of papers and in the science popularization book (Wankel 1882). Scientific activity accelerated at the turn of the twentieth century. Several specialists, namely V. J. Procházka, M. Kříž and K. Absolon, left profound traces in the history of Moravian Karst research. K. Absolon was Wankel's grandson and is the most famous Moravian Karst explorer and researcher of the first half of the last century. He described his remarkable projects, mentioned in Sect. 20.4.1, in his two-volume monograph (Absolon 1970).

The recent research history is connected with geomorphological investigations conducted in the 1960s, mainly by members of the Brno geographic school (e.g. Panoš 1964; Štelcl 1964). However, the first dating of fluvial and speleothem sequences dominating cave passages in the Moravian Karst, and which were necessary for reliable reconstruction of karst processes, were available as late as in the mid 1990s (Šroubek and Diehl 1995; Hercman et al. 1997). Subsequent research activities focused mostly on fluviokarst cave system history have revealed a number of exciting relationships between surface and subsurface evolution of karst morphology (e.g. Kadlec et al. 2001). However, some questions are still waiting for a definitive answer.

Fig. 20.2 Nagel's exploration of the Císařská Cave as recorded by K. Beduzzi in 1748. Adopted from Absolon (1905–1911)

20.3 Geological and Geomorphological Evolution

Limestone of the Moravian Karst was deposited from the Middle Devonian to Early Carboniferous. Early Devonian basal clastic member is overlain by reef carbonate complex several hundred metres thick, called the Macocha Formation (Middle Devonian to Frasnian age, Bábek et al. 2007) characterized by limestone alternating with stromatoporoid-coral reef banks and bioherms in several megacycles (Hladil 1983). Limestones intercalated with calcareous shales overlying the Macocha Formation belong to the Líšeň Formation (Famennian to Lower Visean age) and are several tens of metres thick. The overlying non-karstic marine deposits, called the Drahany Culm facies (Tournaisian to Upper Visean), were deposited by turbiditic currents and are composed from up to several-kilometre-thick sequence of shales and greywackes, and gravity flow conglomerates (Chlupáč et al. 2002). The Paleozoic limestone and non-karstic complex of strata were folded and thrust over the Proterozoic granitoid Brno Massif during the Variscan orogeny (Hladil et al. 1999). The karst area is tectonically bordered against the non-karstic rocks (Fig. 20.1).

The platform sedimentary record is represented by not very thick relics of Late Jurassic (Callovian to Oxfordian) marine calcareous sandstone and cherty limestone (Kettner 1960). Subsequently, the Moravian Karst area was likely completely covered by quartzose sandstones and calcareous mudstones during the Late Cretaceous (Cenomanian to Turonian) marine transgression. The last marine transgression affected the Moravian Karst in the Middle Miocene (Badenian). The sea expanded from the Carpathian Foredeep basin across the eastern margin of the Bohemian Massif, including the Moravian Karst area (Brzobohatý and Cícha 1993). With the exception of the above-mentioned marine invasions, the limestone surface has long been exposed to subaerial weathering processes of different intensity and duration. In particular, processes active in the Cenozoic, driven by climatic oscillations, and tectonic movements at the close contact between the Bohemian Massif and the Carpathian orogenic range, have controlled geomorphological evolution and the present morphology of the area.

The Cenozoic stage was preceded by the above-mentioned Cretaceous sea transgression/regression cycle and subsequent erosion of marine sediment during the latest Cretaceous and Paleogene (e.g. Dvořák et al. 1993). The contemporary shallow depression of the Moravian Karst in respect to the surrounding non-karstic landscape probably originated during the Late Paleogene as a result of fluvial and lacustrine processes (Kettner 1960). Shallow valleys were formed by streams following the N–S drainage pattern during the latest Oligocene (Panoš 1964). The first horizontal cave passages related to these shallow valleys were developed as indicated by the cave, called the Mokerská Cave, exposed during limestone mining in the Mokrá Quarry, close to the Moravian Karst southern termination. The cave passage is completely filled with fluvial and flood cave deposits with a supposed age of Oligocene/Miocene transition. The above-situated vertical cracks, genetically younger than the horizontal cave passage, were filled mostly with lacustrine clays containing Early Miocene (Ottnangian) vertebrate fauna remains of small mammals, turtles, frogs and reptiles (Ivanov et al. 2006).

Thrusting of the Carpathian nappes over the eastern margin of the Bohemian Massif occurred during the Early Miocene (Karpatian), resulting in an increase of the stream gradients and accelerated fluvial erosion in the whole territory (Brzobohatý and Cícha 1993). The Moravian Karst area was dissected by deep valley, trending generally towards the adjacent Carpathian Foredeep (Kettner 1960). As evidenced by recent research, the glen-like Lažánecký Canyon was incised by a stream running from the Blanenský Graben in the west, across the Moravian Karst limestone belt to the Carpathian Foredeep in the east (Černý et al. 2013). We can assume that local cave systems that developed north of the Lažánecký Canyon (e.g. caverns along the Macocha Abyss) were drained to the canyon bottom during the Early/Middle Miocene transition (Panoš 1964; Kadlec et al. 2001). Badenian marine clays and marls subsequently filled the karst canyons and valleys and covered the surface of the entire karst area. This prominent transgression event, lasting ca. 1 million years, interrupted all previous karst processes.

The Svitava River, running west of the Moravian Karst, became a dominant drainage base for the karst streams after the Badenian sea regression. This river incised a new valley in granitoids of the Brno Massif, flowing from the Blanenský Graben to the south, into the Carpathian Foredeep basin. The backward stream erosion formed the modern valleys cutting the limestone body (Kettner 1960). Changes in local hydrography caused abandonment of the Lažánecký Canyon. This pre-Badenian drainage route has remained filled with marine clay up to 160 m thick (Schütznerová-Havelková 1958; Kadlec et al. 2001), blocking forever the supposed karst springs at its bottom. The Badenian marine deposits are also preserved in several smaller valleys and karst depressions in the southern segment of the Moravian Karst. The marine clays, usually reworked, were found very rarely inside the caves, indicating that modern large fluvial cave systems were developed for most part of the Moravian Karst after regression of the Badenian sea.

20.4 Geomorphological Characteristics

20.4.1 Northern Segment

The Pustý and Suchý canyons dominate in morphology of the northern segment. The surface of karst plateau located between the canyons is scarred by sinkhole depressions, usually following tectonic zones, connected by vertical shafts with horizontal cave passages. The largest depression—the Macocha Abyss—was opened to the plateau surface (see Sect. 20.4.2).

The main cave system draining this segment of the Moravian Karst is called the Amatérská Cave, after amateur cavers who penetrated inside in 1969. The cave is connected by sumps with additional cave systems both upstream and downstream. The total length of all passages of these caves is more than 35 km, representing the largest cave system in the Czech Republic.

The cave system is run through by the Punkva River arising from the Sloupský and Bílá voda streams confluence in the Amatérská Cave. The streams are sinking underground in the half-blind ponor valleys at Sloup and Holštejn, respectively. The Punkva River running 3 km through the cave passages resurges in the Punkevní Caves close the Macocha Abyss (Fig. 20.1). Other minor streams join the cave system from the east.

The half-blind Sloupské Valleyhalf-blind Sloupské Valley, where the Sloupský Stream enters the Sloupsko-šošůvské Caves, was incised and filled with fluvial deposits in relation to the underground drainage evolution. At present, the valley is filled with Pleistocene gravel and sand up to 60 m thick (Kadlec et al. 2001). The Sloupsko-šošůvské Caves, almost 5 km long, developed mostly along tectonic dislocations in close contact between Devonian limestone and Lower Carboniferous greywacke, consists of two cave levels connected with chasms up to 70 m deep. Although the Sloupské Caves were accessible for ages, most passages have been officially opened to visitors since 1881. The Šošůvská Cave was discovered at the beginning of the twentieth century and included into the show-cave system (Zajíček and Hromas 2013). The cave passages are partly filled with cave deposits recording a number of aggradation/degradation cycles controlled by subsurface fluvial dynamics related to climatic and paleoenvironmental changes. A remarkable section, containing a large quantity of Pleistocene fauna bones with dominating cave bear, was exposed by Wankel's team in the Kostní (Bone) Passage during paleontological excavations in 1850–1852. The position of bones in the mixture of clastic deposit indicates that the animals died during catastrophic floods (e.g. Musil 2014), with estimated timing of these events as the Late Pleistocene (e.g. Kadlec et al. 2007). Paleomagnetic dating of fine flood deposits exposed in the Sloupsko-šošůvské Caves, completed with U-series dating of speleothems, allows us to reconstruct the Pleistocene geomorphological history of the half-blind ponor valley evolution (Kadlec et al. 2001, 2002).

The 'sister-like' half-blind ponor valley is developed 5 km east from the Sloupské Valley, at the Holštejn village, where the Bílá voda Stream enters the cave system. An analogous two-level cave drained the valley during the Cenozoic. The upper cave level, called the Holštejnská Cave, was almost completely filled with fluvial deposits during the Pleistocene and Holocene, as evidenced from paleomagnetic, U-series and radiocarbon datings (Kadlec et al. 2001).

The Amatérská Cave itself is developed in two downstream divergent levels (Hromas et al. 2009). The lower level is mostly formed by phreatic cave tubes, whereas the upper flood level has large passages with course fluvial sandy gravel covering the passage floor (Fig. 20.3). Flowstone decorations are not very abundant and usually limited to places located higher than the flood water can reach.

The underground Punkva River is resurging from the Punkevní Caves opened to the Pustý Canyon. The caves were discovered by the legendary karst explorer K. Absolon and his team in 1909. The fossil cave passages, decorated with beautiful speleothem forms, were modified into a show-cave in 1910. The bottom of the exciting 138 m deep Macocha Abyss was made accessible through the Punkevní Caves in 1914. Absolon's outstanding projects continued via mining

Fig. 20.3 Bílá voda Stream in the Amatérská Cave (*Photo* I. Audy)

Fig. 20.4 The Punkva River during an extreme flood in 1997, running across the visitor path in the Macocha Abyss (*Photo* P. Zajíček, adopted from Kadlec (2010))

modification of the active water cave passage allowing the visitors to pass through on boats since 1933 (Absolon 1970), except for the Punkva River flood events (Fig. 20.4).

The next show-cave located at the junction of the Suchý and Pustý canyons is called the Kateřinská Cave. A large chamber close to the cave entrance was known for several centuries (Fig. 20.5). Distal passages decorated with thin stalagmites were discovered in 1909 and open to visitors the following year. The cave is separated from the Punkva River drainage system. The last show-cave opened in the northern segment of the Moravian Karst is the Balcarka Cave, located close to the eastern margin of the karst area (Fig. 20.1). The cave was discovered between 1923 and 1948 (Zajíček and Hromas 2013). Due to the fact that the passages were not eroded by running water, the cave walls and ceilings are decorated with magnificent speleothem sceneries.

20.4.2 Origin and Geomorphological Evolution of the Macocha Abyss

The 138 m deep Macocha Abyss has been known for many centuries (Fig. 20.6). The abyss name is derived from the old legend describing a scary story about a bad step-mother ('macocha' in a local dialect) who threw her scorned step-son into the depth of the abyss. However, the boy luckily caught the tree branches and the brave wood-cutters saved him. After the step-mother heard it, she ran to the abyss and jumped down.

For long time the Macocha Abyss was the only place where the Punkva River, running through unknown cave systems, could have been observed. The explanation of the origin of this vertical cavity is a key point in the reconstruction of the cave system evolution formed at the Punkva River resurgence. The first documented descent to the abyss bottom was conducted in 1723 (Zajíček and Hromas 2013). Key scientific knowledge about the interior of this phenomenal cavity brought expeditions to the abyss organized by H. Wankel and J. Kříž in the late nineteenth century, and in particular by K. Absolon at the beginning of the twentieth century. Most Moravian Karst researchers have explained the origin of the abyss via the collapse of a limestone roof above the large cavity developed below a sinkhole (e.g. Absolon 1970). Sediments forming a cone steeply descending to the abyss bottom were supposed to be a relic of the cavity roof (Fig. 20.7).

Fig. 20.5 Hlavní Chamber in the Kateřinská Cave (*Photo* P. Zajíček, adopted from Zajíček and Hromas et al. (2013))

Fig. 20.6 Macocha Abyss by F. Kaliwoda in 1857 (Archive of the Cave Administration of the Czech Republic)

Fig. 20.7 Lake on the Macocha Abyss bottom. The sedimentary cone covered by vegetation is visible in a background. (*Photo* P. Zajíček, adopted from Zajíček and Hromas et al. (2013))

The architecture and morphology of the sedimentary cone were elucidated using two geophysical techniques: (1) shallow refraction seismic and (2) vertical electrical sounding (Kadlec et al. 2001). Figure 20.8 shows the results of the geophysical survey. Most of the cone volume is formed by sediments with specific resistivity on the order of hundreds of Ωm corresponding to sand and sandy gravel. Sand and sandy gravel are coated by a layer 1–2 m thick, with resistivity values on the order of thousands of Ωm. This upper layer corresponds to coarse limestone scree deposited on the surface of the cone. The limestone bedrock steeply inclines towards the bottom of the Macocha Abyss.

We assume the beginning of the Macocha Abyss evolution in the Early Cretaceous period, when the cockpit-like karst depressions were formed (see Sect. 20.4.3). The abyss could have had the form of a sinkhole drained by vertical karst routes (Panoš 1964). The next stage of cavity evolution is traceable from the latest Paleogene period, when the shallow karst valleys were formed in the Moravian Karst. Morphology of the Macocha depression changed to a small blind valley termination where the stream, coming from the shallow karst valley, was sinking underground (see Fig. 20.9, Kadlec et al. 2001). In the Early Miocene the cave passages draining the Macocha Abyss area to the Lažánecký Canyon were probably formed. These passages remained permanently flooded by groundwater after the Badenian marine transgression. The next phase of the Macocha Abyss evolution is connected with the incision of the deep canyon-like valleys in the post-Badenian periods. The Punkevní Caves draining the abyss were developed in three levels, following the incision of the nearby Pustý Canyon.

A crucial event affecting the present morphology of the Macocha Abyss was the collapse of the limestone roof above the underground chamber developed in the abyss. The limestone scree choked up the drainage paths at the abyss bottom and induced the underground stream level rise up to 50 m, as documented by fluvial deposits filling the fossil Písečná Cave passage (Fig. 20.9). The same fluvial sediments were deposited in the Zazděná Cave located 200 m from the Macocha Abyss (Fig. 20.1). The Zazděná Cave sandy gravel shows an architecture documenting an anomalous vertical transport of sediments from underground stream passage located higher up. The end of fluvial deposition in the Zazděná Cave passage is marked by a flowstone bed deposited on the fluvial sediment surface at the beginning of the last glacial stage, 114–100 ka ago (Kadlec et al. 1996). We therefore assume that the collapse of the Macocha Abyss roof happened during the last interglacial, before 114 ka, causing the underground stream flow in the higher passage via the Písečná Cave to the Macocha Abyss. The supposed original cylindrical shape of the abyss was elongated in northwesterly direction by the effect of headward stream erosion in response to the steep gradient formed by 50 m difference between the Písečná Cave and the abyss bottom. A steeply inclined erosional slope was developed in the limestone bedrock from the Písečná Cave to the abyss bottom (Fig. 20.8). The slope was covered with fluvial sandy gravel and sand transported by the stream resurging from the Písečná Cave. The surface of fluvial sediments was later overlain with a bed of coarse limestone scree fallen from the adjacent vertical rocky walls of the abyss. The running water erosion subsequently unchoked the sedimentary plug at the

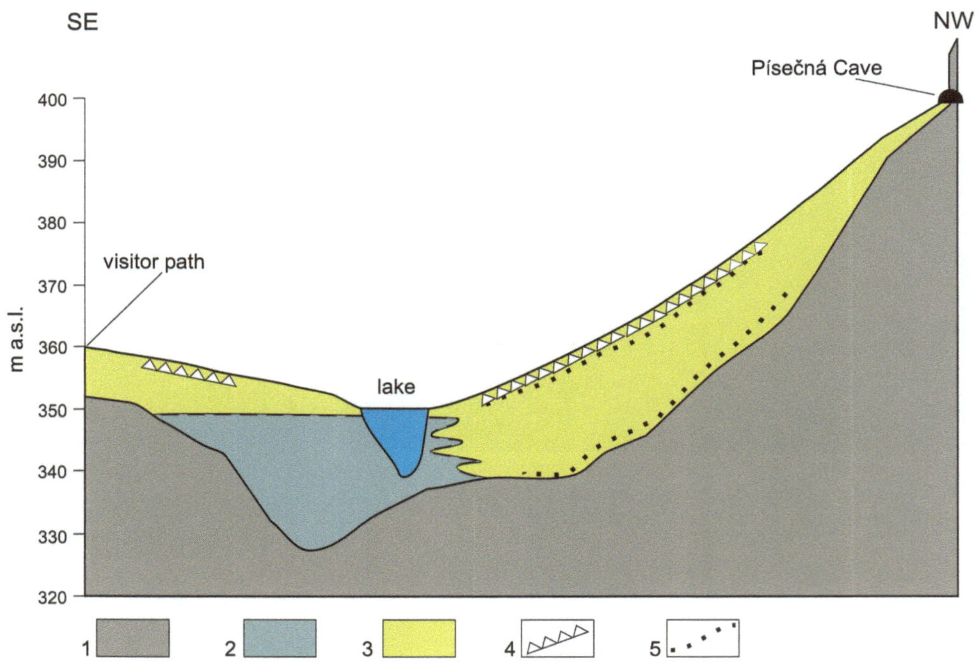

Fig. 20.8 Vertical section through the Macocha Abyss bottom, after the geophysical survey (modified after Kadlec et al. 2001). *1* limestone, *2* water saturated limestone scree, *3* sand and sandy gravel, *4* blocky limestone scree, *5* lithological boundary detected by shallow seismic survey

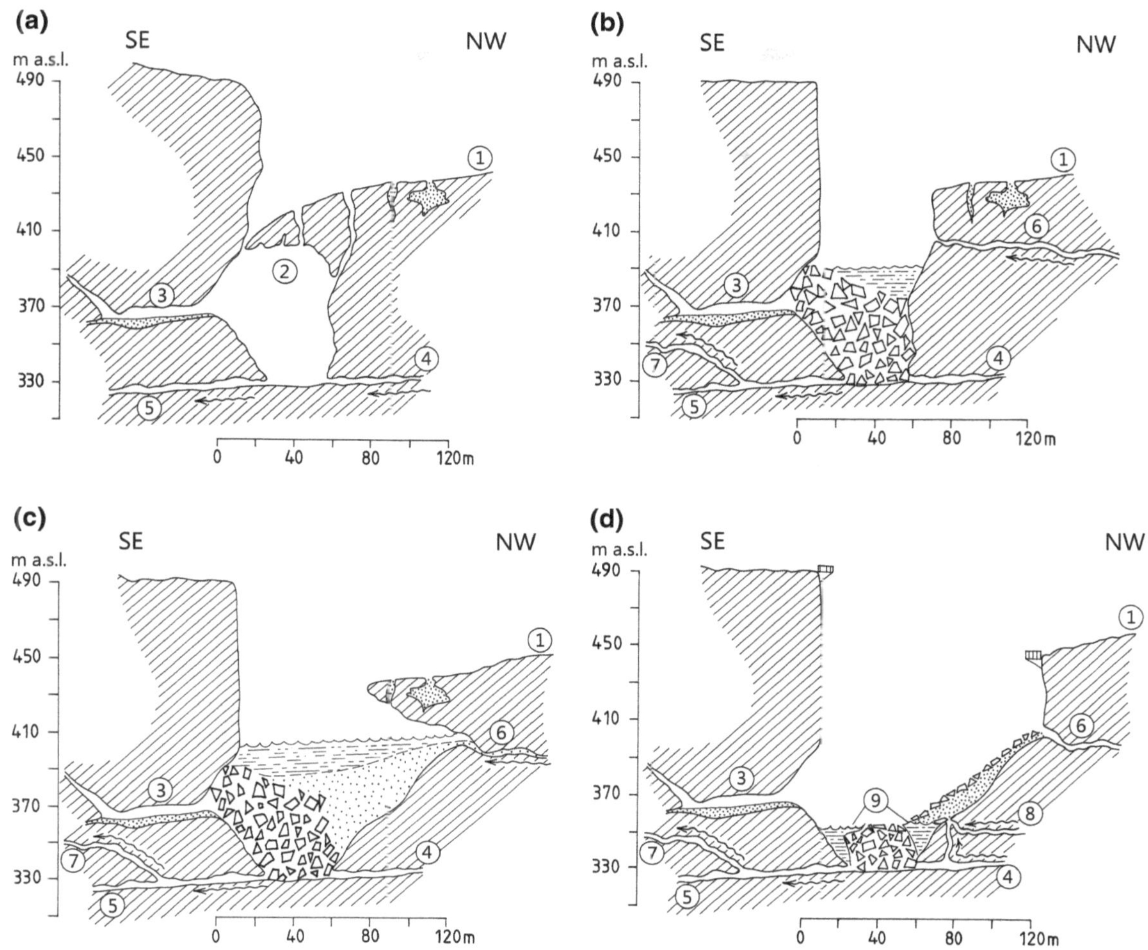

Fig. 20.9 Vertical sections showing main stages of the Macocha Abyss evolution (modified after Kadlec and Beneš (1996)). **a** Late Paleogene and Neogene forming of the vertical cavity and drainage cave passages (expressed by arrows), **b** last interglacial collapse of the cavity roof, **c** elongation of the cavity due to headward stream erosion and subsequent infilling, **d** present state after erosion of the sedimentary fill and a rest of the cavity roof. *1* blind ponor valley, *2* underground cavity, *3* Erich's Cave, *4* and *5* Early Miocene drainage routes, *6* last interglacial drainage passage coming to the Písečná Cave, *7* and *8* Pleistocene drainage routes, *9* lakes at the abyss bottom

abyss bottom formed after the roof collapse. The stream left the Písečná Cave and returned to the Macocha Abyss bottom where we see it today.

20.4.3 Central Segment

The area underwent intense tropical karstification during the Jurassic/Cretaceous transition and in the Early Cretaceous period. Cockpit-like depressions originated at that time were syngenetically filled with sandy gravel and sand described as the Rudice Formation (e.g. Bosák et al. 1979). This paleokarst phenomenon stayed preserved in the central segment due to subsidence along faults crossing the limestone belt in the southerly continuation of the Blanenský Graben (Kettner 1960). The iron oxide and hydro(oxide) ores were precipitated from the sediment pore fluids near contacts of the clastic fills and the limestone walls of the depression. The iron ore has been mined here occasionally since early Iron Age (Wankel 1882). Intense progress in iron mining and smelting began in ninth century during the first Slavonic state period, when the local iron supplied the whole Moravian population (Souchopová et al. 2002). The morphology of the filled karst depressions is known from mining maps and sections constructed in the eighteenth and nineteenth centuries when iron was exploited in a total of 270 mines (Fig. 20.10, Balák et al. 1997).

The Křtinské Valley, representing the southern limit of the central segment, was incised in the same periods as the Pustý and Suchý canyons in the northern segment, being controlled by incision of the Svitava River. The valley is crossing the limestone belt in SE-NW direction. The central segment is drained by a 13 km long cave system, called the Rudické propadání–Býčí skála Cave. The cave is generally

Fig. 20.10 Morphology of the tropical karst depressions located below the Rudice Village. Slightly modified after Balák et al. (1997)

formed as a single stream passage created by the Jedovnický Stream. The initial segment of the cave, located close to the eastern border of the karst area, is descending from a ponor blind valley by vertical passages 90 m deep into a lake chamber and then follows a horizontal underground canyon. Although several karst shafts wriggle upward from the horizontal passage, only one was connected by local cavers with the karst surface. It is called the Rudická Chasm, and with depth of 153 m it represents the deepest chasm above groundwater level in the Czech Republic. The Rudické propadání Cave is separated by a central sump from the downstream cave passage called the Býčí skála Cave. The underground stream, running through the cave system, resurges at the Křtinské Valley bottom close to the western karst limit.

20.4.4 Southern Segment

The geomorphological evolution of this karst segment is identical with the processes recorded in the northern and central segments of the Moravian Karst. The southern

Fig. 20.11 View to the half-blind Sloupské Valley with the Kůlna Cave entrance (*Photo* P. Neruda)

Fig. 20.12 Archaeological excavations in the Pekárna Cave entrance conducted by K. Absolon in 1920s (the Anthropos Institute archive)

segment is the smallest in size and karst features are condensed here. The area is intersected by the Říčka Valley, trending generally from NE to SW (Fig. 20.1). Most of the local caves, concentrated along the valley, were developed in relation to the valley incision. The main cave system, called the Ochozská Cave, is 1750 m long and was formed by a subsurface stream sinking in the half-blind Hostěnické Valley located at the eastern border of the karst area (e.g. Himmel and Himmel 2012). The cave consists of active, partly undiscovered conduits and a large flood passage filled with thick sequences of fluvial and flood sediments, locally covered by flowstone crusts and stalagmites deposited during the Last Glacial stage and in the Holocene (Doláková 2000; Kadlec et al. 2000). The stream resurgence is again situated at the western periphery of the limestone area close to the border with the Devonian clastics.

20.5 The Moravian Karst as a Human History Archive

The caves of the Moravian Karst were occupied by people many times. A unique, Late Pleistocene natural archive has been exposed in the Kůlna Cave, which is a part of the

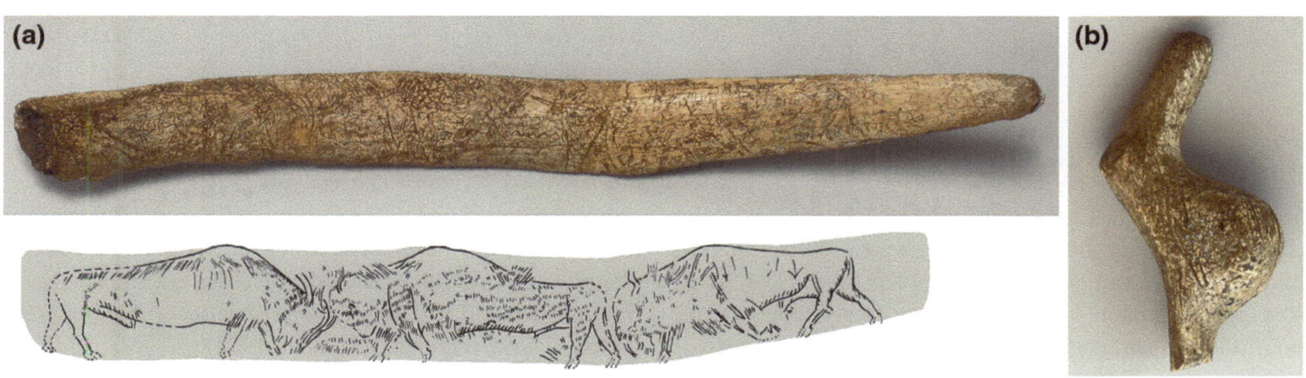

Fig. 20.13 Magdalenian bison engraving into a horse rib (**a**) and a female sculpture carved into ivory (**b**) found in the Pekárna Cave Late Glacial deposits (*Photo* K. Jursa)

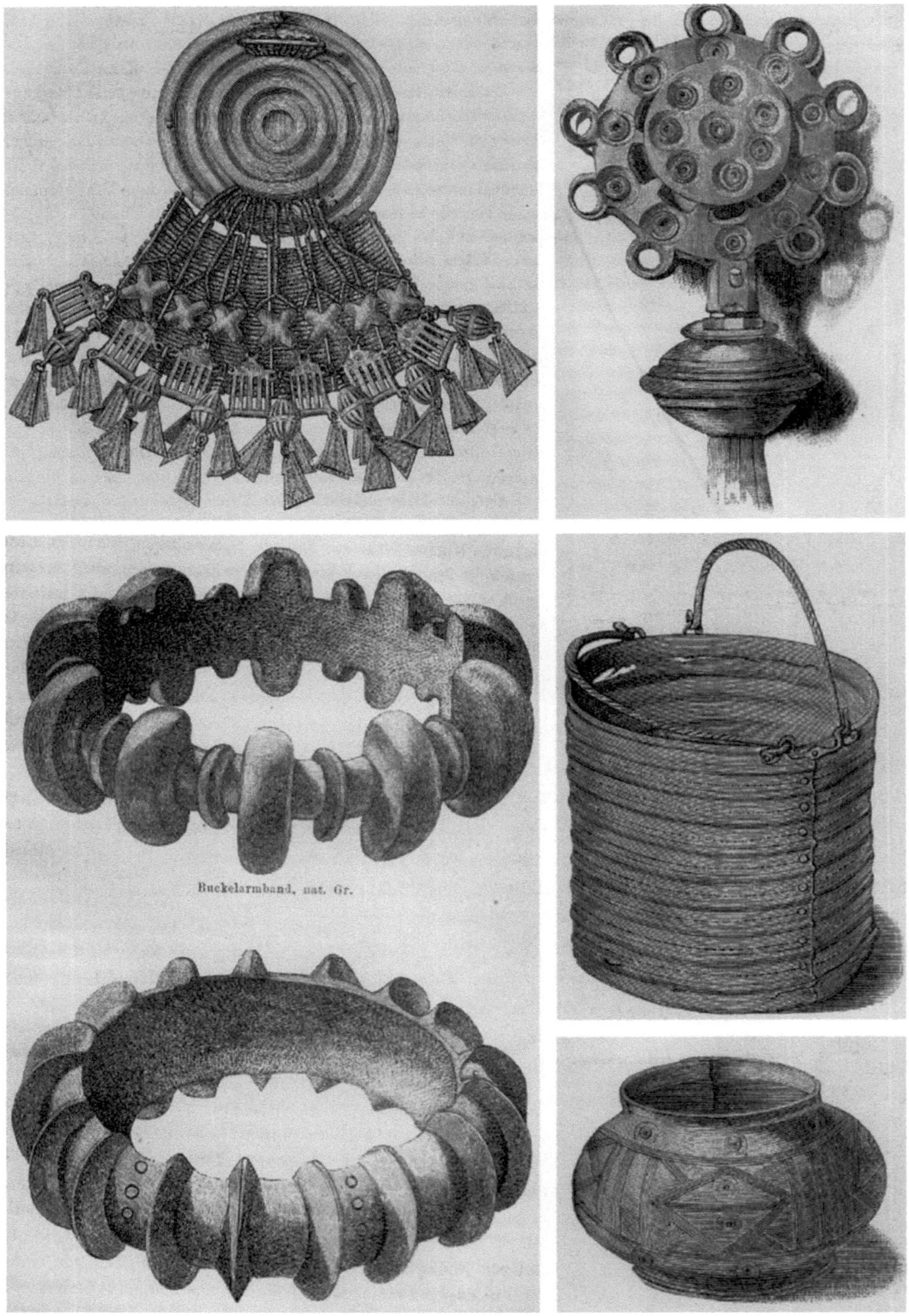

Fig. 20.14 Selected Iron Age finds from the Býčí skála Cave. Adopted from Wankel (1882)

Sloupsko-Šošůvské Caves (Fig. 20.11). Up to 14 horizons have been distinguished and studied by K. Valoch and his team in the cave sedimentary fill (Valoch 1988). The lowermost layer containing Levallois artefacts likely belongs to the end of the Saalian glacial stage. Remarkable are the findings attributed to the Taubachian period from the last interglacial. Uppermost Middle Paleolithic layers yielded Micoquian industry with bifacial tools made by the Neanderthal dwellers that lived in the cave between 90 and 50 ka (Neruda and Nerudová 2014). The Neanderthal maxilla, occipital bone and milk teeth found in this sediments document this important period of cave occupation (Jelínek 1988).

During the early Upper Paleolithic humans began to sparsely visit the karst area, as evidenced by leaf points of the Szeletian period found in the Rytířská Cave in the Suchý Canyon (Valoch 1965), or in the Pod hradem Cave located in the Pustý Canyon, where also Aurignacian artefacts have been identified (e.g. Wright et al. 2014). Rare presence of the Gravettian hunters, who expanded in Moravia along the mammoth migration paths ca. 27 ka ago, are only documented from the Kůlna Cave deposits (Nerudová and Neruda 2014).

A breakthrough in cave settlement strategies was driven by the Late Glacial climatic oscillations, when the Magdalenian reindeer hunters often lived in the caves of the Moravian Karst (e.g. Svoboda 2000; Valoch 2001). In addition to the Kůlna Cave, a large human herd settled the Pekárna Cave opened to the Říčky Valley in the southern segment (Fig. 20.12). This tunnel-shaped cave has yielded a unique collection of Magdalenian mobile art, including a woman statuette carved into ivory, or a horse rib decorated with engraved bisons (Figs. 20.13a, b). Apart from this site, several smaller reindeer hunting stations were identified, e.g. in the Býčí skála and Balcarka caves. In addition, sparse artefact collections from some other sites indicate that people often used the caves as short-term refuges. However, hunters sometimes preferred the karst plateau surface located at the Moravian Karst southern termination, overlooking the nearby Carpathian Foredeep lowland. The Magdalenian hunters left the area concurrently with the retreat of the Late Glacial fauna, following climate amelioration.

Our knowledge regarding the post-Paleolithic settlement of the Moravian Karst is fragmentary. For instance, a burial in the Výpustek Cave, comprising of vessels of the Linear Pottery Culture (Wankel 1871), suggests that the Neolithic people also visited the Moravian Karst for non-utilitarian reasons. Early Bronze Age people carried out a wide spectrum of activities in the cave environment. Both settlement sites and evidence of burial activities were found. In particular, the Hallstatt period became famous after a discovery in the Předsíň (Vestibule) Chamber in the Býčí skála Cave, where in 1882 H. Wankel and his team uncovered human remains showing evidence of violent death together with an enormous quantity of weapons, parts of 3–4 wheeled chariots, decorative items, parts of horse tackles and other utility objects (e.g. Nekvasil 1993, Fig. 20.14). Interpretation of this mysterious finding remains enigmatic. Scarce finds from the Kůlna Cave deposits indicate the local Late Iron Age and the Roman Era human activities (Podborský 2011).

The diverse morphology of the Moravian Karst, with deeply incised valleys and canyons hosting many cave entrances, has excited the fantasies of visitors since medieval times, especially during the Romantic Period at the turn of the nineteenth century.

The Výpustek Cave, opened to the in the central segment, shows an extraordinary human usage of the cave environment. The two floor cave system, with the total length of passages reaching almost 2 km, represents a fossil ponor cave. Until the beginning of the twentieth century, the large upper passage was filled with cave deposits up to several metres thick. The phosphate cave sediments, abundant in Pleistocene fauna bones, were mined in the 1920s and used as agriculture fertilizer. Many bones of cave bears, lions, hyenas and other Pleistocene animals were found during mining operations (Musil 2010). The completely mined-out cave passage was later modified and used for military purposes, and later still as an underground military factory during World War II. The last stage of this cheerless story occurred in 1960s, when a top secret anti-atomic bunker was constructed for the Czechoslovak army high leadership inside the cave. In the 1990s, the army left the cave and the complete facilities have served as a special show-cave since 2006.

20.6 Conclusions

Geomorphological history of the Moravian Karst reveals a succession of processes driven by climatic and tectonic forcing since the Early Cretaceous epoch. These processes formed a spectacular landscape with characteristics karst canyons, half-blind or blind valleys and sinkholes. The geomorphological evolution of the surface karst features was closely related to the underground drainage pattern and large fluviokarst cave system formation. The cave deposits record the changes of paleoenvironmental conditions including the oldest traces of human settlement in our territory. However, others natural karst archives, e.g. stalagmites recording climatic oscillations, are still waiting for future appreciation.

Acknowledgments We are grateful to the photographers Peter Zajíček, working for the Cave Administration of the Czech Republic, and Igor Audy for the cave picture provision, Ivan Balák for construction of the Moravian Karst geological map and Vendulka Kadlecová for editing of the cartoon figures. We highly appreciate discussions with Jiří Bruthans, a hydrogeologist from Charles University in Prague, contributing to fine-tuning ideas concerning the Moravian Karst Cenozoic evolution

References

Absolon K (1905–1911) Kras Moravský a jeho podzemní svět. Wiesner, Praha, 218 pp

Absolon K (1970) Moravský kras, vol. 1 and 2. Academia, Praha, 761 pp

Bábek O, Přikryl T, Hladil J (2007) Progressive drowning of carbonate platform in the Moravo-Silesian Basin (Czech Republic) before the Frasnian/Famennian event: facies, compositional variations and gamma-ray spectrometry. Facies 53(2):293–316

Balák I, Baldík V, Klejzarová A, Kovařík M, Kožoušková H, Leitgeb I, Ocetková L, Podborský V, Pokladník J, Souchopová V, Stloukal P, Štefka L, Varner D, Vybíhal K, Zouharová K (1997) Rudická plošina v Moravském krasu. Městská knihovna Blansko, 94 pp

Bosák P, Glazek J, Gradziński R, Wójcik Z (1979) Genesis and age of sediments of the Rudice type in fossil-karst depressions. Čas Mineral Geol 24(2):147–154
Brzobohatý R, Cícha I (1993) Karpatská předhlubeň. In: Přichystal A, Obstová V, Suk M (eds) Geologie Moravy a Slezska. Sborník příspěvků k 90. výročí narození Prof. dr. Karla Zapletala. Morav. Zem. Muzeum a Sekce přír. věd PřF MU, Brno, pp 123–128
Chlupáč I, Brzobohatý R, Kovanda J, Stráník Z (2002) Geologická minulost České republiky. Academia, Praha 436 p
Černý J, Otava J, Melichar R (2013) Tektonická mapa, 24-411 Jedovnice. Archív ČGS, Praha
Doláková N (2000) Palynological studies from the Ochozská Cave and from the Šošůvka part of the Sloupsko-Šošůvská Cave (Moravian Karst). Geolines 11(1):172–174
Dvořák J, Štelcl O, Demek J, Musil R (1993) Geologie a geomorfologie Moravského krasu. In: Musil R (ed) Moravský kras. Labyrinty poznání, GEO program, Adamov, pp 32–76
Hercman H, Laurizen SE, Glazek J, Vít J (1997) Uranium-series dating of speleothems from Amaterska and Holstejnska Caves, Moravian Karst, Czech Republic. In: Proceedings of 12th Int Cong Speleol, Basel, pp 45–47
Himmel J, Himmel P (2012) Jeskyně v povodí říčky. ČSS ZO 1-05 Královopolská, Brno, 58 pp
Hladil J (1983) The biofacies sedimentation of Devonian Limestones in the central part of the Moravian Karst. Sbor geol věd, Ř G 38:71–94
Hladil J, Melichar R, Otava J, Galle A, Krs M, Man O, Pruner O, Čejchan P, Orel P (1999) The Devonian in the Easternmost Variscides, Moravia: A holistic analysis directed towards comprehension of the original context. Abh Geol B-A 54:27–47
Hromas J a kol (2009) Jeskyně. In: Mackovčin P, Sedláček M (eds) Chráněná území ČR, sv. XIV. Agentura ochrany přírody a krajiny ČR a EkoCentrum Brno, Brno, 608 pp
Ivanov M, Musil R, Brzobohatý R (2006) Terrestrial and marine faunas from the Miocene deposits of the Mokrá Plateau (Drahany Upland, Czech Republic) - Impact on palaeogeography. Beiträge zur Paläontologie 30:223–239
Jelínek J (1988) Anthropologische Funde aus der Kůlna-Höhle. In: Valoch K (ed) Die Erforschung der Kůlna-Höhle 1961–1976. Anthropos, 24, Brno, pp 261–283
Kadlec J (2010) Macocha - propast s pohnutou minulostí. Od druhohor po čtvrtohory. Vesmír 89(5):300–303
Kadlec J, Beneš V (1996) Jak vznikla Macocha? Speleo 23:5–17
Kadlec J, Hercman H, Beneš V, Šroubek P, Diehl JF, Granger D (2001) Cenozoic history of the Moravian Karst (northern segment): Cave sediments and karst morphology. Acta Mus Morav Sci Geol 86:111–160
Kadlec J, Hladíková J, Žák K (1996) Isotopic study of cave carbonates from Moravian Karst. In: Lauritzen SE (ed) Conf Climate Change: The Karst Record. Ext Abst, Bergen, pp 67–71
Kadlec J, Pruner P, Venhodová D, Hercman H, Nowicki T (2000) Stáří a geneze sedimentů v Ochozské jeskyni. Geol výzk na Moravě a ve Slezsku 7:19–24
Kadlec J, Pruner P, Venhodová D, Hercman H, Nowicki T (2002) Stáří a geneze sedimentů v Šošůvské jeskyni (Moravský kras, Česká republika). Acta Mus Mor Sci Geol 87:229–243
Kadlec J, Šroubek P, Diehl JF, Hercman H, Nowicki T, Pruner P, Venhodová D (2007) How old are cave deposits abundant in Pleistocene fauna preserved in the Bone Passage in the Sloupsko-šošůvská Cave (Moravian Karst)? Scrip Fac Sci Natur Univ Masaryk Brun 35:37–41
Kettner R (1960) Morfologický vývoj Moravského krasu a jeho okolí. Čs kras 12:47–84
Musil R (2010) Výpustek, bájná jeskyně u Křtin. Acta Speleol, 1/2010, Průhonice, 115 pp
Musil R (2014) Morava v době ledové. Prostředí posledního glaciálu a metody jeho poznávání. muniPRESS, Brno 228 pp
Nekvasil J (1993) Chronologie moravského halštatu. In: Podborský V et al (eds) Pravěké dějiny Moravy. Vlastivěda moravská - země a lid. Muzejní a vlastivědná společnost v Brně, Brno
Neruda P, Nerudová Z (2014) New radiocarbon data from Micoquian layers of the Kůlna Cave (Czech Republic). Quat Inter 326–327:157–167
Nerudová Z, Neruda P (2014) Chronology of the Upper Palaeolithic sequence in the Kůlna Cave (okr. Blansko/CZ). Arch Korresp 44 (3):307 324
Panoš V (1964) Der Urkarst in Ostflügel der Bohmischen Masse. Z Geomorphol N F 8(2):105–162
Podborský V (2011) Pravěké a středověké nálezy z jeskyně Kůlny a okolí. In: Valoch K a kol (eds) Kůlna Historie a význam jeskyně. Acta speleologica, Správa jeskyní České republiky, Průhonice, pp 139–148
Procházka VJ (1899) Moravský kras. Sbor Čes spol zeměvěd, Praha 51 p
Schütznerová-Havelková E (1958) Mocnost tortonských sedimentů v Lažáneckém údolí v Moravském krasu. Čs kras 11:180–182
Souchopová V, Merta J, Truhlář J, Balák I, Štefka L (2002) Cesta železa Moravským Krasem. Technické muzeum v Brně, Brno 123 pp
Svoboda J (2000) The Eastern Magdalenian: Hunters, Landscape and Caves. In: Price HA, Peterkin GL (eds) Regional approaches to adaptation in Late Pleistocene, Western Europe. BAR, Oxford, pp 179–189
Šroubek P, Diehl JF (1995) Paleomagnetické/environmentálně magnetické studium jeskynních sedimentů Moravského krasu. Knih Čes spel spol 25:29–30
Štelcl O (1964) Geomorfologické poměry jihozápadní části Drahanské vrchoviny. Sbor Čs spol zem 69:21–45
Valoch K (1965) Paleolitické nálezy z Rytířské jeskyně v Moravském krasu. Anthropozoikum, řada A 3:141–155
Valoch K (1988) Die Erforschung der Kůlna Höhle 1961-1976. Anthropos, 24, Brno, 204 pp
Valoch K (2001) Das Magdalénien in Mähren: 130 Jahre Forschung. Jahrbuch des Romisch-Germanischen Zentralmuseums Mainz 48:103–159
Wankel H (1871) Prähistorische Alterthümer in den mährischen Höhlen. Selbstverlag der Verfassers, Wien 36 pp
Wankel H (1882) Bilder aus der Mährischen Schweiz und ihrer Vergangenheit. Wien, 422 pp
Wright D, Nejman L, d'Errico F et al (2014) An Early Upper Palaeolithic decorated bone tubular rod from Pod Hradem Cave. Czech Republic. Antiquity 88(339):30–46
Zajíček P, Hromas J et al (2013) Show caves of the Czech Republic. Cave Administration of the Czech Republic, Průhonice 208 pp

Region of the Rychlebské Hory Mountains—Tectonically Controlled Landforms and Unique Landscape of Granite Inselbergs (Sudetic Mountains)

21

Petra Štěpančíková and Jakub Stemberk

Abstract

The Rychlebské hory Mountains are situated in the north-eastern part of the Bohemian Massif. They are part of Sudetic Mountains, which are divided from Sudetic Foreland by the Sudetic Marginal Fault. Controlling the mountain front for a length of 130 km, the fault is one of the morphologically most striking features in the Bohemian Massif and has been studied by numerous geologists, geomorphologists, and geophysicists for several last decades. Its Pleistocene seismicity was proved by paleoseismological survey and its recent potential seismic threat is shown by minor historical earthquakes. In contrast to the mountainous relief, the adjacent Žulovská pahorkatina (Hilly Land) forms a unique granite landscape of gently undulated basal weathering surface of etchplain with numerous low exfoliation domes, isolated inselbergs and rock landforms. Middle Pleistocene continental ice-sheet, which reached the area twice—in Elsterian 1 and Elsterian 2, influenced the development of rock forms and caves, the latter ones being shaped by its meltwater into characteristic "heart-like" profile Post-glacial (post-Saalian 1) uplift of the area, which was most probably related to glacioisostatic rebound, resulted in removal of glacial deposits and valley deepening. Striking geomorpho-diversity and its scientific value of the entire area remain attractive for nature-lovers regardless their profession.

Keywords

Rychlebské hory • Žulovská pahorkatina • Tectonic landforms • Fault scarp • Inselberg • Granite landforms

21.1 Introduction

The region of the Rychlebské hory (hory = mountains) is a part of Sudetes Mountains situated in the north-eastern part of the Bohemian Massif and with the adjacent Žulovská pahorkatina (=Hilly Land) forms a unique landscape with strikingly rich geomorpho-diversity and mineral resources related to different lithologies and tectonic evolution (Figs. 21.1 and 21.2). The northern limit of Sudetes Mountains is controlled by one of the morphologically most pronounced structure in the Bohemian Massif—Sudetic Marginal Fault, which is traceable in morphology at a length of 130 km. Due to subsidence within the Sudetic Foreland, the region is also one of a few areas with preserved Tertiary deposits and buried deep saprolite. Moreover, complex geological history can be read from Pleistocene glacial sediments deposited by repeatedly overriding continental ice-sheets, which stopped at the mountain front, as well as from alluvial fan deposits in front of the Rychlebské hory. The latter ones were truncated by the fault. Such diversified geology and morphology have brought also economic benefits due to mineral extraction since medieval times.

P. Štěpančíková (✉) · J. Stemberk
Department of Neotectonics and Thermochronology, Institute of Rock Structure and Mechanics, Czech Academy of Sciences, Prague, Czech Republic
e-mail: stepancikova@irsm.cas.cz

J. Stemberk
e-mail: kuba.stemberk@gmail.com

J. Stemberk
Department of Physical Geography and Geoecology, Faculty of Science, Charles University, Prague, Czech Republic

T. Pánek and J. Hradecký (eds.), *Landscapes and Landforms of the Czech Republic*, World Geomorphological Landscapes, DOI 10.1007/978-3-319-27537-6_21

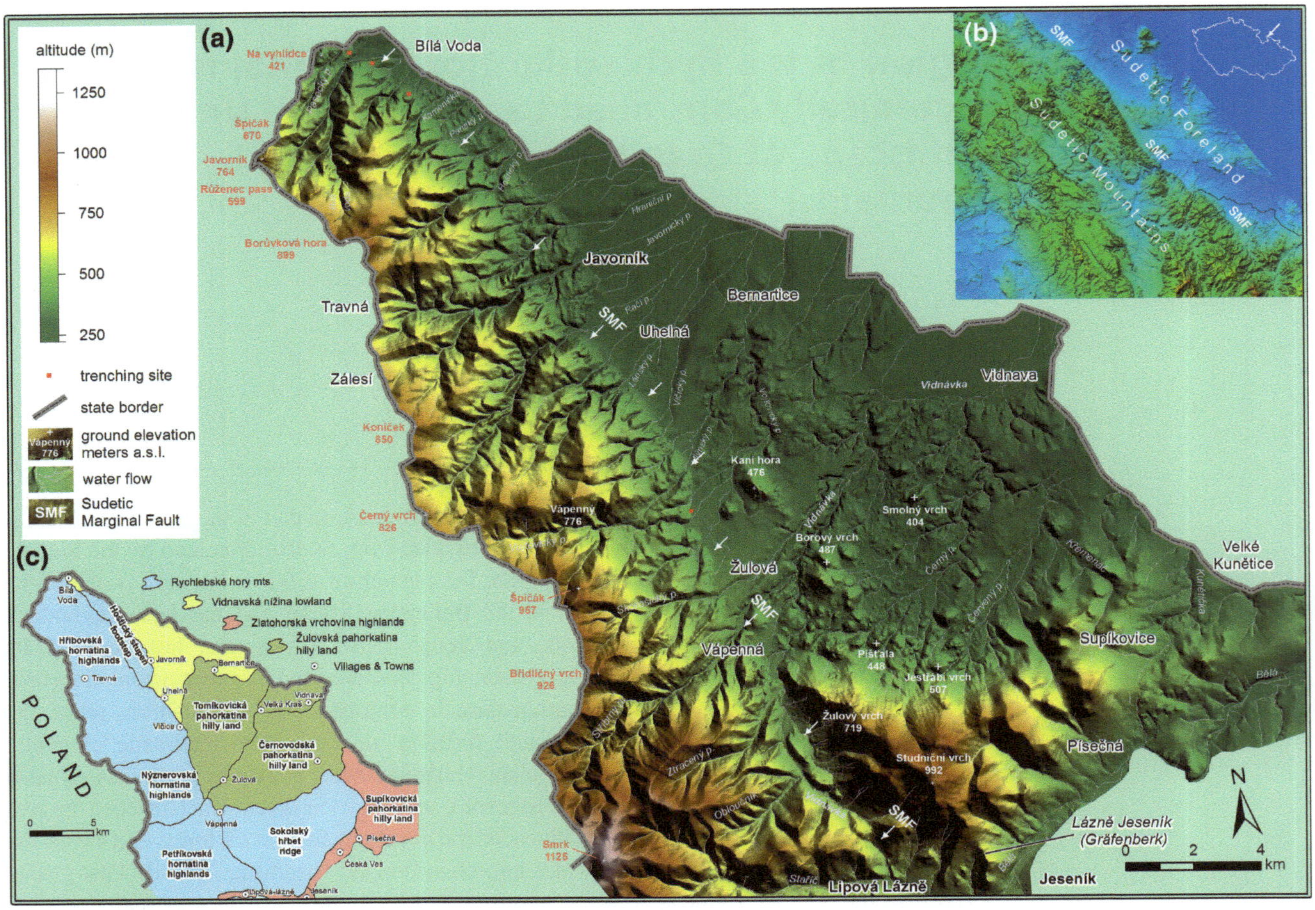

Fig. 21.1 **a** The Rychlebské hory Mts and topographic situation on the digital elevation model based on 2 m contour lines, **b** digital elevation model of the Sudetic Mountains bordered by the Sudetic Marginal Fault towards the Sudetic Foreland, **c** geomorphological units

21.2 Different Geology: The Key to Sudetic Versus Fore-Sudetic Landscape Diversity

Our area of interest is situated in the north-eastern part of the Bohemian Massif, whose crystalline basement is dissected by faults into a mosaic of blocks uplifted to different elevations. The major structure running through the area is NW-SE trending Sudetic Marginal Fault (SMF), steeply dipping to the NE. The fault divides two distinct geological units with different morphologies: an elevated Sudetic block with Lugicum Unit represented by mountain ranges with broad ridges and deeply dissected uplands (average altitudes 400–800 m near the fault) and relatively subsided Fore-Sudetic block with Silesicum unit, which is typified by gently undulated relief with scattered groups of hills or slightly dissected uplands (Fig. 21.1). During geological history vertical movements along the SMF alternated and the Fore-Sudetic block is considered to be over about 5 km more eroded than the Sudetic block although it occupies lower altitude today (Cwojdziński and Żelaźniewicz 1995). This is also the reason why such different lithologies occur next to each other and control the present-day morphology. Geomorphological units which occupy the differentially uplifted blocks are as follows: the Rychlebské hory, their north-eastern part Sokolský hřbet (=Ridge), the Žulovská pahorkatina and the Vidnavská nížina (=Lowland) (Fig. 21.1). The Rychlebské hory within the Sudetic block, as well as its NE part Sokolský hřbet, are horst-like ridges asymmetrically uplifted over up to 600 m in respect to the Žulovská pahorkatina block. Their Proterozoic to Paleozoic metamorphic sedimentary shales and igneous rocks comprise gneisses, mica shists, amphibolites, marbles, quartzites, graphitic shales, metamorphic volcanites (Fig. 21.2). The morphology of the mountains is controlled by neotectonic uplift, subsequent enhanced erosion and valley incision. In contrast, the adjacent Žulovská pahorkatina in the Fore-Sudetic block represents the basal weathering surface of an etchplain (Demek et al. 1964). It comprises granitoids of the Žulová granite pluton of Variscan age, which

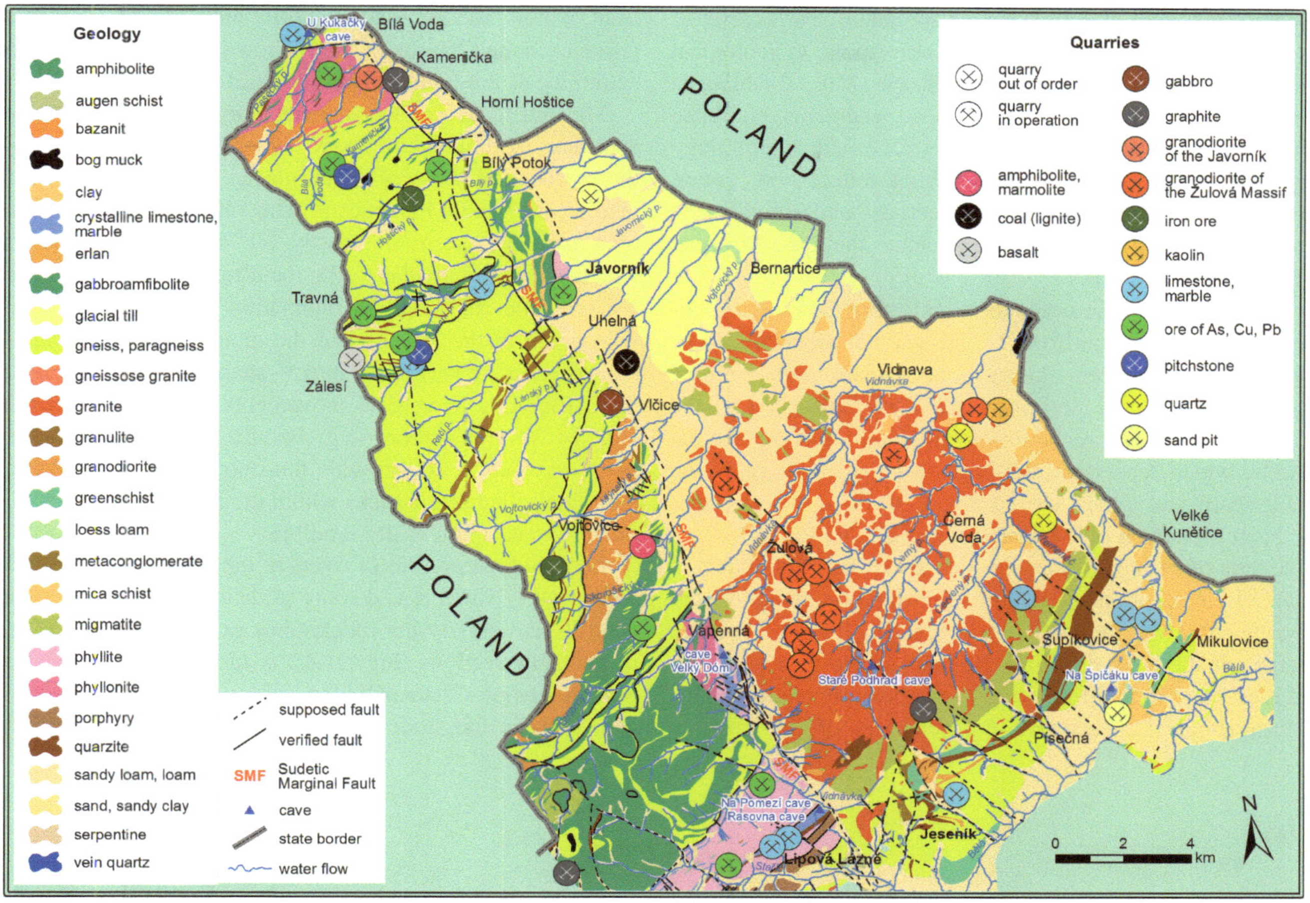

Fig. 21.2 Geological map with localisation of quarries (modified after Paleček et al. 2013)

represents an apical part of a vast granitic body. The unique granite landscape of this hilly land is mostly characterized by gently undulated surface hosting numerous low exfoliation domes and isolated inselbergs, exposed from Tertiary kaolinite-rich saprolite. During the Middle Pleistocene, the surface of the Žulovská pahorkatina has been directly covered by continental ice-sheet, which reached the mountain front of the Rychlebské hory twice—in Elsterian 1 (620,000–635,000 years) and Elsterian 2 (400,000–460,000 years) (Žáček et al. 2004). In the north, the down-thrown block of the Vidnavská nížina has a flat surface with slightly entrenched valleys. It is a part of Neogene Paczków graben filled with fluvial to lacustrine deposits up to 400 m thick. These deposits cover a granitic basement kaolinized to a depth of about 50 m (Badura et al. 2004).

The differential relief development in the adjacent morphostructural units of the Rychlebské hory (Sudetic block) and Žulovská pahorkatina (Fore-Sudetic block) has resulted not only in contrasting landscapes but in a systemic disparity between rock landforms in both units, despite their common structural and lithological control. The Sokolský hřbet (NE part of Rychlebské hory) is characterized by numerous frost-riven cliffs and cryoplanation terraces, extensive blockfields, and limited depth of granitic grus weathering. The extensive granitoid block accumulations developed on the marginal slopes are a result of exposure of intensively disintegrated rocks due to uplift. In contrast, rock landforms observed in the Žulovská pahorkatina are characterized by numerous rock steps and tors exposed at the basal weathering surface, deeper grus weathering profiles on residual hills, the limited extent of blockfields and the absence of frost-riven cliffs (Štěpančíková and Rowberry 2008).

21.3 Tectonic Landforms of the Rychlebské Hory Mountains

The SMF controls the pronounced mountain front of Sudetes, which is higher in the south-eastern part (120–300 m) and less elevated in the north-western part (50–180 m), however with large differences among individual segments. In the Rychlebské hory it reaches 250–300 m. To some extent, varying height contrasts between the Sudetes and the

Sudetic Foreland along the whole mountain scarp correlate with the lithology; the highest parts correspond to geological units built mainly of gneisses (Sowie Mts, Rychlebské hory/Złote Mts), whereas the lowest segments comprise poorly consolidated Permian clastics. The fault morphology and the height of the mountain front reflect both young tectonic activity and lithological contrasts and are assumed to represent the cumulative uplift since the Late Miocene (Badura et al. 2007).

Based on variable orientation and morphotectonic properties of the mountain front, 15 segments of the SMF can be distinguished between Złotoryja and Jeseník town, eight of them in the Czech portion. Particularly important is the presence of clearly recognizable triangular and trapezoidal facets, which represent inter-basin areas at the base of the fault scarp and which were formed due to uplift and subsequent dissection of the scarp by gullies. A flight of two to five tiers of triangular facets, which show their differentiated state of preservation and degree of erosional remodelling are recognizable on the fault segments (Fig. 21.3). Average height of these facets in the Rychlebské hory are: 28, 60, 111, 173 and 275 m for successively older generations. This tiering suggests at least five uplift episodes of the mountain front, which probably started shortly after 31 Ma, since they postdate faulted basalts of the Sichów Hills area, at the NW extremity of the fault (Badura et al. 2007).

The uplift of the Rychlebské hory has controlled the evolution of drainage network, which was also affected by the presence of the continental glacier in the Sudetic Foreland in the Middle Pleistocene. The valley network consists of a parallel system of mostly NE–SW-oriented valleys, which are perpendicular to the main ridge and with general NE direction controlled by inclination of the area since the Paleogene. In the NW part of the mountains small streams are forced to actively erode into V–shaped valleys within the SMF zone due to vertical movements documented on this fault by several authors (e.g. Badura et al. 2007). Bigger rivers have up to 200 m wide flat-floored valleys, probably fault-controlled as the rocks within the fault zone are heavily fractured and more easily erodible.

Longitudinal profiles of rivers of the Rychlebské hory are considerably variable and some of them reflect also tectonic control, particularly the knickpoints occurrence near places of fault intersections (Fig. 21.4). A morphometric indicator Stream Length-Gradient index (SL) was applied to reveal anomalies in longitudinal profiles due to its sensitivity to

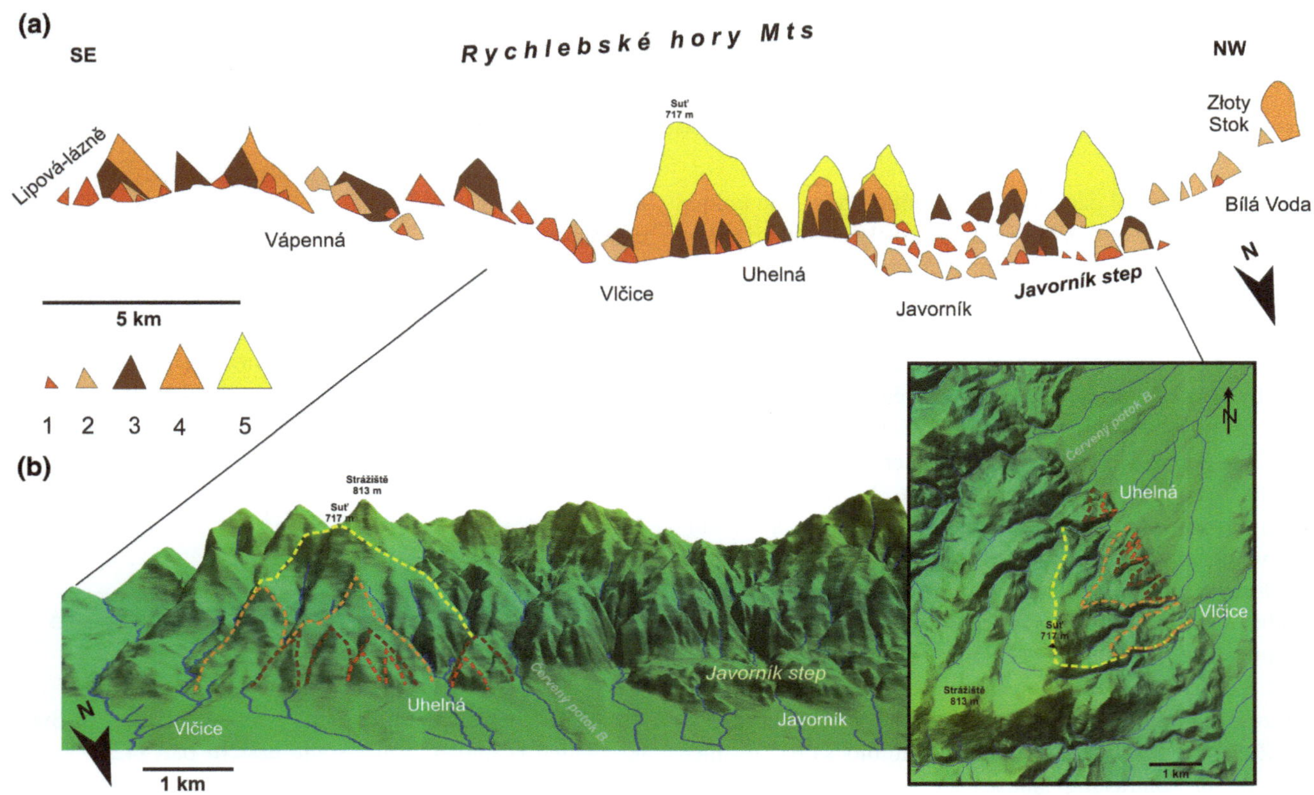

Fig. 21.3 a Perspective drawing of staircase-like arranged faceted spurs along the Czech portion of the Sudetic Marginal Fault between Bílá Voda and Lipová–lázně, *1–5* sequence of triangular facets from youngest to oldest (adapted from Badura et al. 2007), **b** 3D view of digital elevation model with facets around Uhelná village, 5× exaggerated

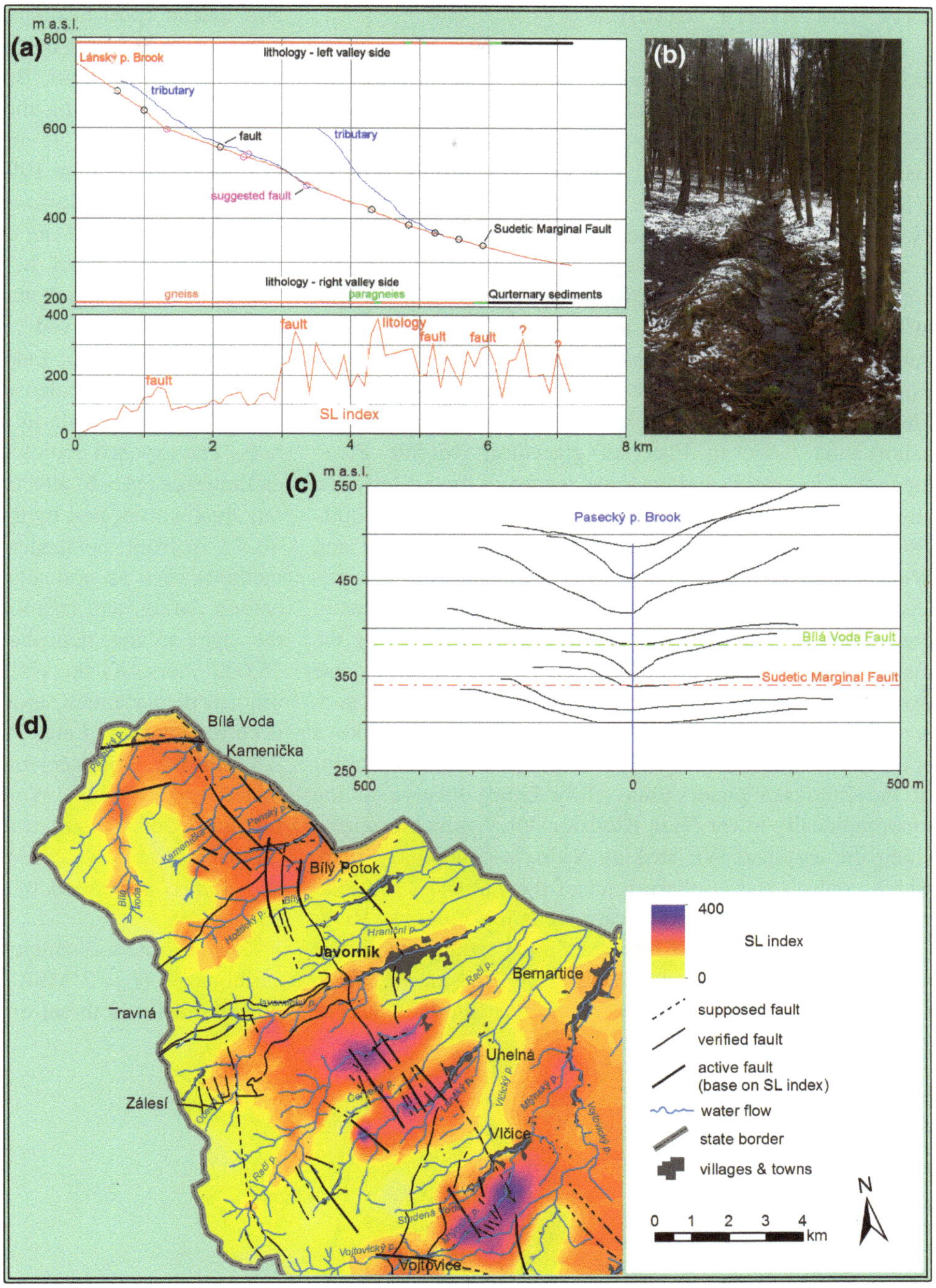

Fig. 21.4 **a** Longitudinal profile of the Lánský potok with position of crossing faults and simplified lithology, and SL index, **b** flat valley of Pasecký potok, **c** cross-section of the Pasecký potok valley with position of major faults, **d** SL index map with faults expressed by high SL indices

changes in slope in the valley bottom. After comparison of the local maxima of SL index with geological structures and after excluding lithological transitions, sites with rock outcrops occurrence, and places of streams confluence followed by increased erosive power, the tectonic control could be considered. Based on the presence of local maxima of SL index four groups with the occurrence of active faults based on their azimuths were delineated: the area near Bílá Voda village with Bílá Voda fault and parallel fault with azimuth 80°; the area between Kamenička brook and Panský potok (=Brook), and between Račí potok and Studená voda, where a great number of faults parallel with the SMF in direction 140°–150° occur; the faults with azimuth 90° near bifurcation of the Vojtovický potok and Mlýnský potok; the area near Horní Hoštice village with curved faults in direction 0° with increased erosion.

21.4 Quaternary Tectonics

Evidence for Quaternary faulting is found in morphology along the SMF. It includes relatively linear mountain fronts, deflected drainages, hanging wine-glass valleys, and deep dissection of the footwall of the Sudetes Mountains (Fig. 21.1). Middle and Late Pleistocene activity is proved by river terraces that are truncated by the fault and that show 5–20 m high scarp in their longitudinal profiles in the Polish portion of the fault (Krzyszkowski and Pijet 1993). Incision of the Žulovská pahorkatina and glaciofluvial accumulation flats in front of the Rychlebské hory and removal of debris from mountain valleys also point to uplift in the Late Pleistocene. The continental ice-sheet reached the Žulovská pahorkatina twice in Elsterian glaciation (Middle Pleistocene). After deglaciation, three levels of fluvial terraces/alluvial fans were deposited, during Saalian 1 (240,000–280,000 years), Saalian 2 (130,000–180,000 years) and Weichselian (10,000–80,000 years), respectively. Their relative heights above the channel of Vidnávka River attain greater values than terrace levels of the same age along the main Nysa Kłodzka River. The differences between these two rivers attain 20 m at the highest level 1, at least 8 m at level 2, and up to 2–3 m at level 3 (Fig. 21.5) (Štěpančíková et al. 2008). These discrepancies imply post-Saalian 1 uplift of the Žulovská pahorkatina (Hilly Land) relative to the topographically lower Nysa Kłodzka valley, which occurred along the assumed WE striking Vidnava-Głuchołazy fault, which controls the northern limit of the Žulovská pahorkatina by a 30–40 m high fault scarp along the Vidnávka River (Fig. 21.1). Based on the affected fluvial terraces the uplift rate that diminished towards the Late Pleistocene, from approximately 20 to 3 m, could be inferred. Similar values of post-glacial (post-Saalian) uplift of 20–35 m, which also decreases to 2–5 m in the Late Pleistocene, are reported from the Sudetes Mountains (Badura et al. 2004). The total uplift along the SMF during the Middle and Late Pleistocene is then estimated to be 20–30 m up to 60–80 m (Krzyszkowski and Pijet 1993; Migoń 1993). It is supposed to be a result of combination of glacioisostatic rebound and tectonics during which the mountain front of Sudetes and many blocks within the Fore-Sudetic block have been uplifted. As a result, fluvial erosion increased and a large amount of Quaternary sediments was removed.

Direct evidence of Late Pleistocene activity of the SMF was documented in paleoseismological trenches excavated at the locality Bílá Voda in the north-west of the Rychlebské hory. The exposed geological data revealed at least 4–5 large earthquakes (M > 6) affecting deposits of an alluvial fan. The fan apex is truncated by the SMF and left-laterally offset by 30–45 m from the feeder channel (Fig. 21.6). The dating methods such as optically stimulated luminescence, radiocarbon dating, and cosmogenic nuclide dating (^{10}Be) yield the age of the deposits close to the fan apex about 25,000 years. As the Holocene deposits do not show significant displacement, most of the recorded slip must have occurred during Late Pleistocene, with slip rate 1.8–2.8 mm/year. The acceleration of slip rate was probably due to ice-loading of the Weichselian ice-sheet, which had its margin about 150 km from the locality at ~20,000 years (Štěpančíková et al. 2013). The Late Pleistocene left-lateral slip is also confirmed by horizontal offset of valley sides (Fig. 21.7).

Sparse microseismicity and rare historical moderate earthquakes (I = 7 MSK) suggest that the SMF is still a potential seismic threat to the region. The largest historical

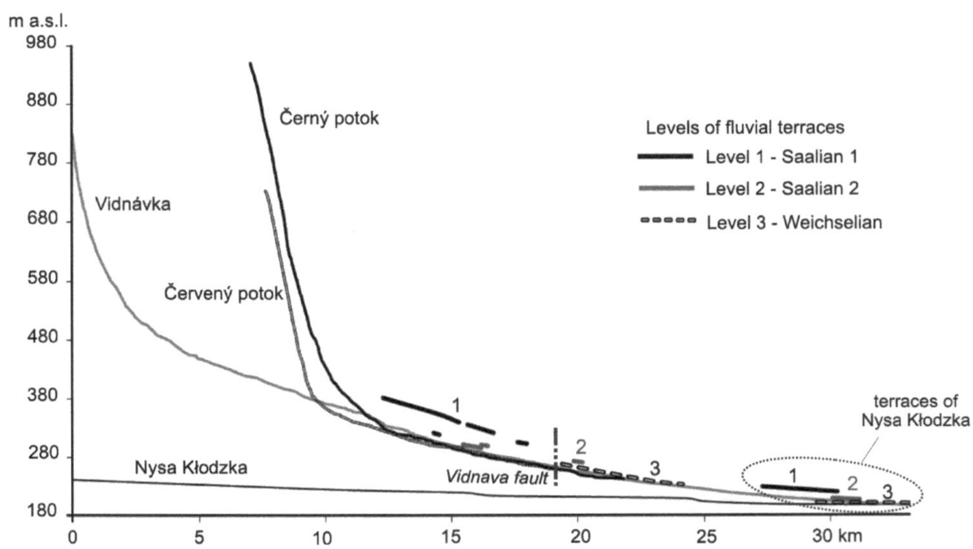

Fig. 21.5 Longitudinal profiles with fluvial terraces levels of the Vidnávka River, the Černý and Červený potok and the Nysa Kłodzka River at the confluence area. Terraces/alluvial fans are of the following ages: level *1* Saalian 1 (240,000–280,000 years), *2* Saalian 2 (130,000–180,000 years), *3* Weichselian (10,000–80,000 years) (adapted from Štěpančíková et al. 2008)

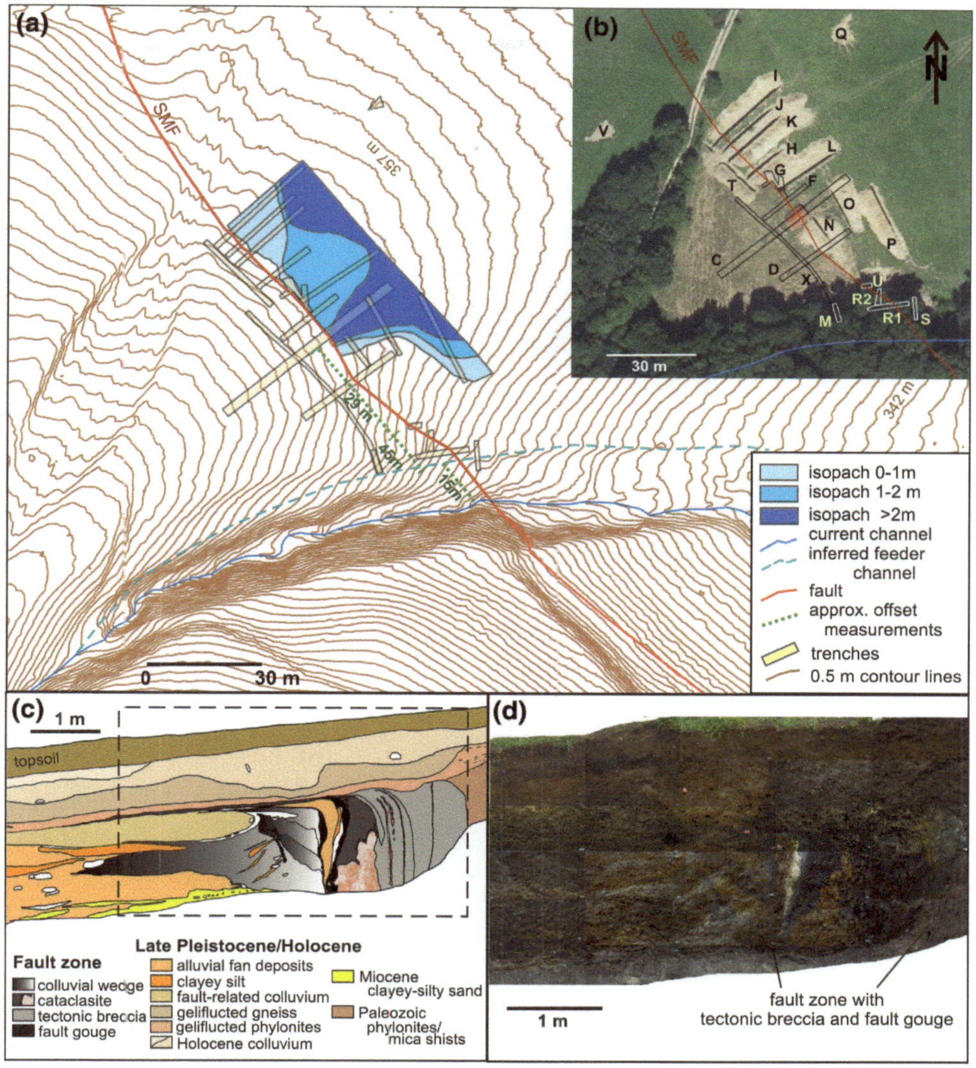

Fig. 21.6 a Isopach map of the truncated alluvial fan deposits inferred from the trenches at the locality Bílá Voda and estimated offsets of the alluvial fan apex, b position of the trenches on the aerial photograph (for situation see Fig. 21.1); Sudetic Marginal Fault (SMF) is interpolated based on the data from the trenches and geophysical survey, c geological profile of trench I; dashed rectangle shows the extent of photomosaic, d photomosaic of the zone of SMF in trench I (*Photo* P. Štěpančíková)

earthquake in Lower Silesia was also felt in Bílá Voda on 11th June 1895, where it reached I = 4.5 MSK (Sponheuer 1952).

Ongoing uplift of the Rychlebské hory is confirmed by the spatial pattern of enhanced erosion as well as by monitoring of displacements on tectonic structures (Štěpančíková et al. 2008). Deformeters TM71, which record microdisplacements on faults three-dimensionally, are installed within the SMF zone in caves Na Pomezí, Na Špičáku, Rasovna, and in a trench in Bílá Voda. The results show aseismic microdisplacements within the SMF zone with a rate of hundredths to tenths of a millimetre per year since 2001. The movements imply the NNW-SSE stress field, which is in accordance with the present-day stress field in the Bohemian Massif (Peška 1992).

21.5 Inselbergs of the Žulovská Pahorkatina Hilly Land

The term "Žulovská" pahorkatina (Hilly Land) comes from the Czech word 'žulová', which means 'granitic'. Various landforms typical for granite weathering make its unique landscape. These landforms have been formed since the Paleogene, after exposure of the Žulová granite pluton from a depth of several kilometres. Relief formation was controlled by deep selective chemical weathering with its characteristic product of kaoline, which was favoured along numerous joint and fault zones and facilitated by humid or semi-humid tropical or subtropical climate (Fig. 21.8). The age of chemical weathering products is supposed to be Paleogene to Middle Miocene (Milický et al. 1985). As the

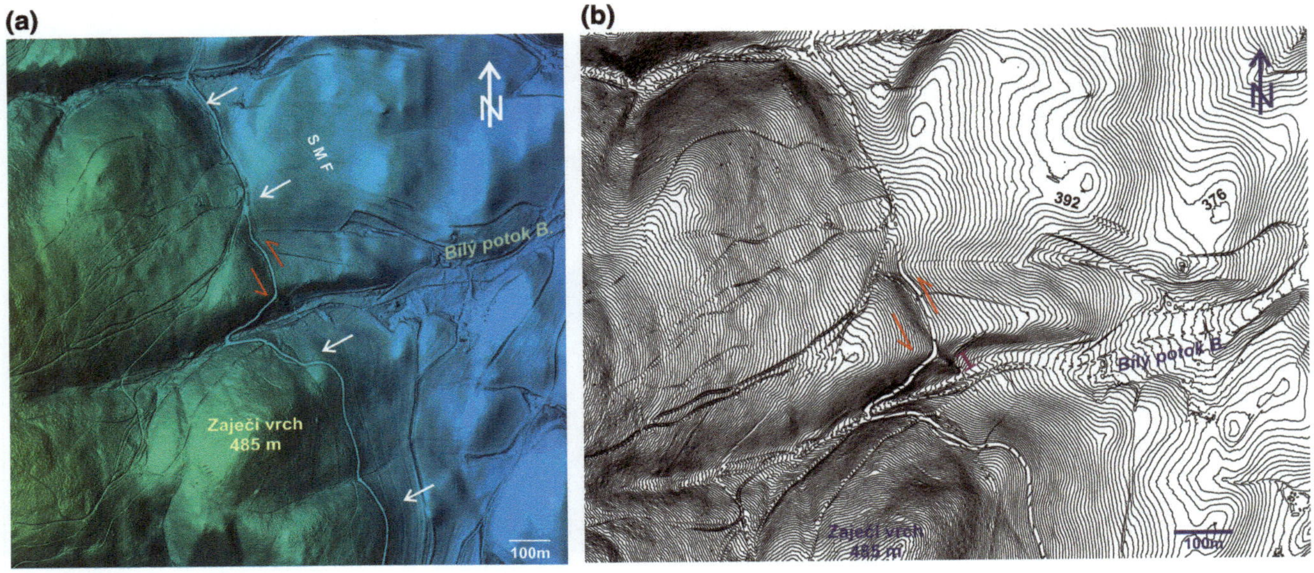

Fig. 21.7 **a** Left-laterally offset valley side of the Bílý potok Brook along the Sudetic Marginal Fault on digital elevation model based on LiDAR data (data produced by GEODIS Brno, s.r.o), **b** offset valley side on one meter contour lines of the area; *red arrows* show the kinematics, the *purple line* shows about 30 m horizontal offset

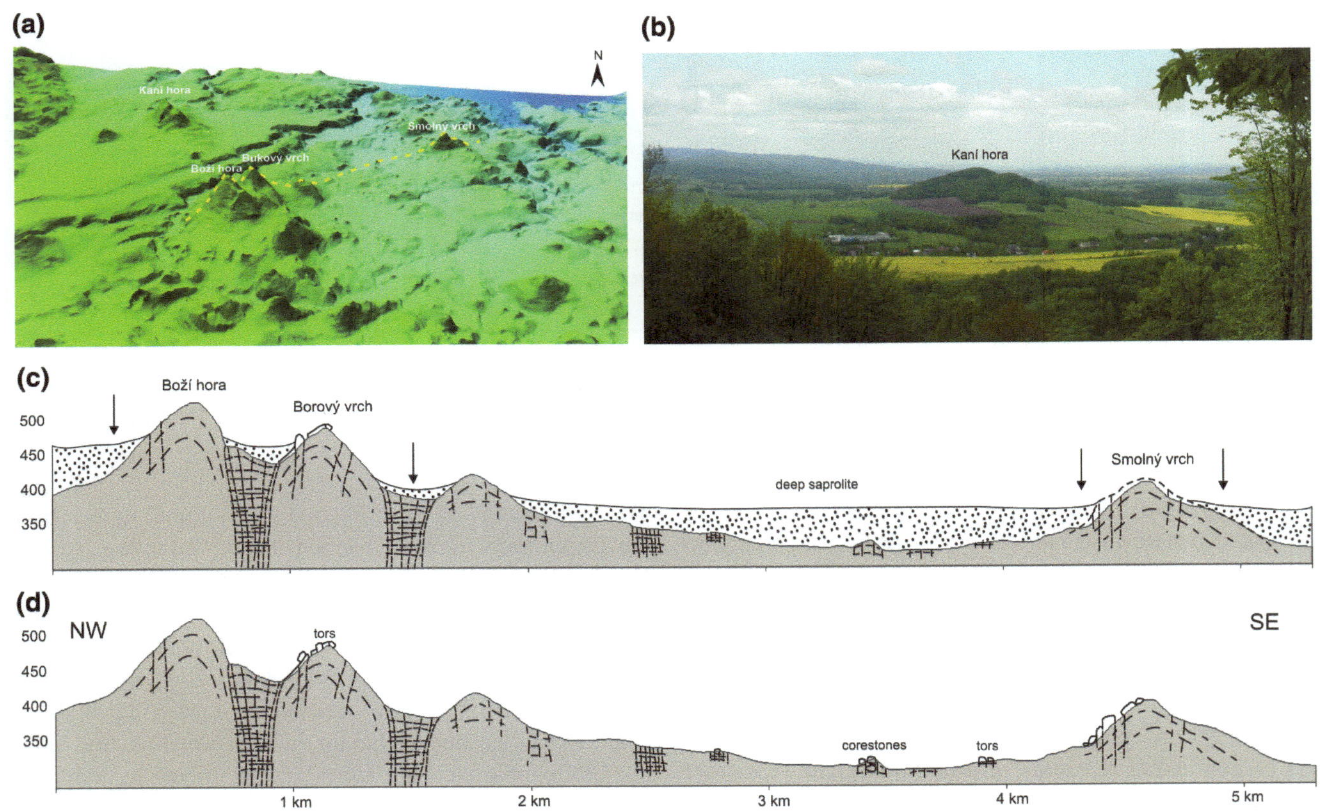

Fig. 21.8 **a** 3D relief of the Žulovská pahorkatina (Hilly Land) with granite inselbergs and low exfoliation domes, **b** inselberg of Kaní hora Mt, **c** development scheme of the inselbergs, *upper* figure—deep weathering conditioned by different jointing, periodic stripping (*arrows*) and exposure of hills, **d** subaerial degradation of inselbergs and their remodelling with formation of residual rock forms (for the cross-section position see the *yellow dashed line* in (**a**)) (*Photo* P. Štěpančíková)

Fig. 21.9 The etchplain of the Žulovská pahorkatina with small residual hills as exposed irregularities of the stripped basal weathering surface 1 km east of Smolný vrch; the Sokolský hřbet in the background (*Photo* P. Štěpančíková)

kaolinite-rich saprolite was mostly stripped off, basal remnants were recorded only in boreholes in depressions between residual hills (e.g. Demek et al. 1964). Residual kaoline up to 90 m thick is exposed in a pit close to Vidnava town and preserved due to its position in a tectonically subsided block (Milický et al. 1985). The surface of the hilly land was then remodelled to a gently undulated relief with protruding isolated hills due to etchplanation and landsurface lowering.

Numerous low exfoliation domes (ruwares) with gentle convex slopes have relative relief 20–30 m, while isolated residual hills, so called inselbergs, represent mostly high exfoliation domes with relative relief of 100–150 m. There are 34 inselbergs in the Žulovská pahorkatina, which conspicuously rise out of the surrounding sub-horizontal surface of low relief. These hills are structurally and lithologically controlled, and are often elongated in the NW-SE direction such as for example Mt Kaní hora (476 m a.s.l.), which is formed within the oldest intrusion of the pluton—quartz monzonite and bounded by a NW-trending fault (Figs. 21.1 and 21.2). Some of the inselbergs show dome-like pattern of sheet fractures with castellated rocks on the top such as for example Borový vrch (487 m a.s.l.) and Smolný vrch (404 m a.s.l.). These hills are composed of biotite granodiorite with up to 1 m thick slabs, which are destructed into isolated sheets and residual rocks on the convex slopes (Fig. 21.8). As the relative heights of the inselbergs exceed the estimated thickness of kaolinite-rich saprolite (up to 50 m), numerous phases of inselberg exposure in the broader region of the Sudetic Foreland were inferred. Besides structural and lithological control, and related selective weathering, long-term development with alternation of tectonic and climatic stability and short instability episodes have resulted in the present-day shape and height of the inselbergs (Migoń 1997).

Low exfoliation domes and small residual hills south of Smolný vrch had long been considered as roche moutonnées sculpted by glacial erosion (Demek 1976) (Fig. 21.9). However, recent detailed morphometrical analysis rather suggests that these granitoid elevations are built of exposed irregularities of the stripped basal weathering surface. Elongation of the elevations shows strong structural control but does not show any micro or macro-morphological characteristics that would be typical for roche moutonnées such as an asymmetric shape consistent with the direction of glacier advance or glacial striations. Moreover, the topographical position of the area suggests their younger age than the presence of the last ice-sheet (Ivan 1983; Víděňský et al. 2007).

21.6 Touristic Geomorphological Sites

21.6.1 Granite Rock Forms

The unique granite landscape of the Žulovská pahorkatina also hosts attractive residual rock forms, which have been exposed at the basal weathering surface. They usually form the tops of castellated residual hills that protrude above the surrounding denudation surface, except for tors, which could be located in a variety of positions within the slope profile. The rock forms were formed by complex processes of weathering and denudation, controlled not only by mechanical and chemical, but also biochemical processes. Due to their scientific and aesthetic values these rock forms are protected at various levels.

Residual rock forms occur at the top of a residual hill (398 m a.s.l.) at the SW limit of the Žulovská pahorkatina. The castellated rocks, rock walls, and tors in medium-grained biotite granodiorite with microforms on their surfaces are rather isolated and well-rounded (Fig. 21.10). The absence of debris or extensive blockfields suggests their etched origin according to a two-phase evolution model, which consists of a period of differential weathering followed by regolith stripping, rather than origin due to frost-shattering in periglacial environment. Thus, here the tors represent corestones exposed from saprolite. Instructive example of corestones can be also found on the NW slope of the Sokolský hřbet (Fig. 21.11).

Fig. 21.10 Residual rock landforms composed of granodiorite in the Žulovská pahorkatina **a** Tors and rock walls with weathering microforms to the N of Jestřábí vrch, **b** weathering pits and tafoni located on the top tor on inselberg Píšťala, **c** isolated precariously balanced rocks on the slope of inselberg Borový vrch, **d** tors on the slope of Borový vrch, **e** rock slabs on the SW slope of Borový vrch, **f** rockwall with pseudokarren and weathering pit on the SW slope of inselberg Smolný vrch, **g** weathering pit on Smolný vrch (*Photo* P. Štěpančíková)

Fig. 21.11 **a** Stages of boulder development by differential joint-controlled subsurface weathering and spheroidal disintegration of granites followed by removal of debris (modified after http://www.yosemite.ca.us/library/geologic_story_of_yosemite/images/43.jpg), **b** initial phase of corestone exposure; rounded shapes contrast sharply with the paragneiss overlaying the granite (top right hand-side), **c** the corestone with onion-skin weathering. The quarry is located near Mt Zelená hora in the Sokolský hřbet, see Fig. 21.1 (*Photo* P. Štěpančíková)

Inselbergs Píšťala (447 m a.s.l.), Borový vrch and Smolný vrch host various rock forms: tors with abundant microforms such as weathering pits, often connected by draining channels, tafoni, pseudokarren, honeycombs, rock cavities, etc. Moreover, Borový vrch is famous for tors on the slopes and isolated rock slabs, and Smolný vrch for its weathering pits, which were considered as artefacts in sites of pagan sacrificial rituals and called "Venus bowls" (Fig. 21.10). A group of tors occurs also to the SE of Kaní hora Mt in fine-grained biotite granodiorite, where open joints control their shape. Rock fragments with differing lithologies such as metagabbro, quartz, and quartz diorite were found deep inside these joints. This suggests that these fragments have been transported by the continental glacier, thus the tors are of pre-glacial origin (Štěpančíková and Rowberry 2008).

21.6.2 Karst Caves

Karst in the Rychlebské hory is formed in Devonian crystalline limestone. During the Middle Pleistocene the continental glacier and climatic cooling affected the area and significantly influenced cave development in the area.

One of the most attractive caves in the area is Na Špičáku cave (Fig. 21.2). The cave is formed under the Velký Špičák hill (482 m a.s.l.), with marble castellated rocks controlled by tectonics. The cave was modelled by meltwater from the Middle Pleistocene (Elsterian 2?) continental ice-sheet that shaped the corridors into distinctive heart-shaped cross-section, which are typical for this cave. This origin and cave modelling occur quite rarely in central Europe. Beside Na Špičáku cave we can find it also in the Staré Podhradí cave, which is formed within the contact of marble and granite (Fig. 21.12) (Král 1958). Moreover, Na Špičáku cave is the oldest mentioned cave in central Europe, noticed by Anthonius Wale in 1430. The cave provided a hiding place from the Middle Ages, which is evidenced by four thousands of epigraphic inscriptions on the walls, the oldest one being from 1519. Not far from the cave, near Supíkovice town, relicts of typical tropical tower karst "mogotes" were excavated from fluvioglacial deposits. Depressions among karst towers are filled by reddish kaoline-rich clay and sandy deposits. The karst towers can be compared to granite inselbergs in the Žulovská pahorkatina, which were formed during the same Paleogene periods typified by tropical climate (Czudek and Demek 1960).

The touristic Na Pomezí caves are the longest and most extensive karst system formed in crystalline limestones in the Czech Republic. The high chemical purity of the limestone, favourable bedding dip with several variably oriented systems of fissures and faults were the primary controlling factors of cave development. The Quaternary cave evolution was also influenced by cooling and the presence of the

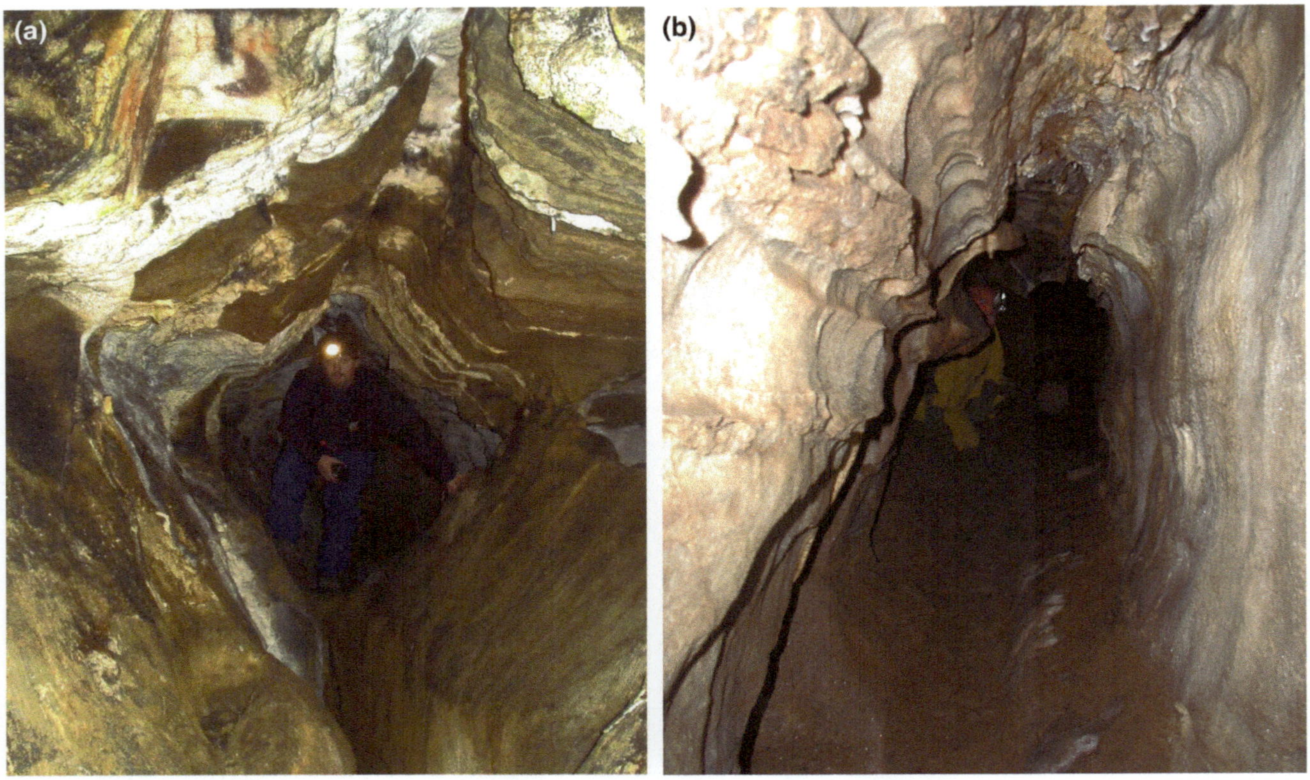

Fig. 21.12 Caves modelled by glacier meltwater, which resulted in sub-horizontal corridors: **a** heart-shaped corridor in the Na Špičáku cave with epigraphic inscriptions on the wall (top left corner), **b** Staré Podhradí cave formed within the contact of marble and granite. For location see Fig. 21.2 (*Photo* P. Štěpančíková)

ice-sheet in the immediate proximity, when the limestone massif was modelled by periglacial processes. It was torn by ice in cracks up to a depth of 20 m. Seeping water widened the fissures by its corrosive and erosive effects resulting in a system of joined corridors (Král 1958). The cave system has been partly known from as early as 1936, but it was only in 1949 when the caves were fully explored. Typical for the Na Pomezí caves are cascades and large stalactites, one of them being shaped and known as "a heart".

21.7 Social and Economic Value of Unique Geological Variability

21.7.1 Times of Prosperous Mining

The area of the Rychlebské hory, Žulovská pahorkatina and the surrounding area have abundant mineral resources (Fig. 21.2), which have brought economic benefit to the region since medieval times. A number of deposits of metal ore could be found here such as gold, silver, copper, lead, zinc, iron ore, arsenic ore, formations of bismuth-cobalt-nickel-uranium ore, molybdenum and wolfram. Mines and quarries for graphite, kaoline, clays, quartz, peat and especially for crystalline limestone and granite were opened. Deposits of sand and gravel were also quarried. Mining of iron ore and non-ferrous metals was carried out with slight fluctuations since the beginning of the settlement, mainly from the sixteenth to nineteenth century. During the nineteenth century and the first half of the twentieth century there was stagnation of ore mining, which resulted in its gradual disappearance. The only exception was mining of uranium and associated ores in the second half of the twentieth century near Zálesí village (Skácel 2006). During the nineteenth century high-quality marble was discovered near Supíkovice and Velké Kunětice villages. It was quarried for lime burning, but mainly as a decorative stone for statues, decorative elements of buildings, monuments, etc. Beautiful marble products that come from this region have become a part of many important European buildings (e.g. Pergamon Museum in Berlin). That is the reason why Supíkovice village is also called "Silesian Cararra".

Names of several villages, places or peaks that evidence mineral resources, remain a curiosity of the region. We can find the village of Vápenná ("Limestone village"), Žulová ("Granitic town"), Písečná ("Sand village"), Uhelná ("Coal village"), Zlaté hory ("Golden mountains town"), peaks of

Stříbrník ("Silver hill"), Vápenný vrch ("Lime Hill"), Žulový vrch ("Granitic Hill"), Čedičový vrch ("Basalt hill"), etc.

21.7.2 Priessnitz Spa: Birthplace of Famous Hydrotherapy

The Rychlebské hory were also a home of a peasant, Vincent Priessnitz (1799–1851), who was born in the village of Gräfenberg (recently Lázně Jeseník Spa, Fig. 21.1), in former Austrian Silesia. Despite not being a doctor, he became wold famous for pioneering modern hydrotherapy. When he was 16-years old he sustained injury during an accident when a horse carriage run over his chest and Priessnitz received numerous bruises and rib fractures. Local healer gave him no hope of recovery, but Priessnitz tried cold water as the same remedy he used in the treatment of the domestic animals for which he cared. He was completely cured in a short time. As the news on this remarkable cure without using any drugs or lotions spread, many wounded and chronically ill persons came to him for relief. Priessnitz was soon fully occupied with their treatment. He was not a quack; he worked honestly and openly, and was earnestly and sagaciously developing the great principles which he recognized. His cure based on perspiration followed by cold applications, using cold mountain spring water, outdoor exercise and manual labour, had great success. Although his methods seemed to be crude he helped to recover a great number of chronic invalids whose affections were practically incurable by the medical methods common those days. No wonder that he drew general attention to the previously little appreciated importance of applying water as a remedial agent used in various simple methods. Despite original scepticism of scientific physicians the Austrian emperor granted him a diploma and gold medal after an official investigation of his work and methods. The French government sent the head of the army medical department to study his methods, so the hydrotherapy was introduced into the French military service, later also to other countries and promptly disseminated (Kellogg 1910). In 1822 Vincenz Priessnitz founded the first hydrotherapy institute in the world. He built also a house for accommodation of patients in Gräfenberk, which is now the Museum of Vincenz Priessnitz, attracting a number of tourists. Progressive expansion of the spa influenced the development of the poor mountainous village and its surroundings, so it became a sought-after balneal place also for rich and noble clientele such as archduke Franz Karl, writer Nikolai Gogol, sculptor Hermann Bissen, physicians Elisabeth Blackwell, Sir Charles Scudamore, etc. During history the spa resort experienced different transformations reflecting also the world events such as wars. At the end of the nineteenth century hydrotherapy became a great fashion and the reputation of Gräfenberk spa continuously rose. In the present-day the Priessnitz Medical Spa plc serves as a prominent climatic spa with tens of cold water springs, which yearly attracts more than 10,000 patients by its health benefits as well as by its emplacement in the middle of gorgeous mountains. In 1999, the 200th anniversary of Vincenz Priessnitz´s birth was included in the UNESCO list of world´s cultural anniversaries (Kočka and Kubík 2006).

21.8 Conclusion

The area of the Rychlebské hory (Mountains) and Žulovská pahorkatina (Hilly Land) provides striking landscape diversity, which is a result of a long polygenetic history within the area. Due to different tectonic evolution of both morphostructural units related to faulting along the Sudetic Marginal Fault, which is one of the morphologically most pronounced features in the Bohemian Massif, diverse relief has been formed. The Rychlebské hory have been uplifted mainly during late Cenozoicand their relief rejuvenated by increased river incision. They sharply contrast with the Žulovská pahorkatina composed of granitoids, which represents one of the most remarkable parts of the Bohemian Massif with its gently undulated surface with numerous inselbergs, exposed irregularities of the stripped basal weathering surface, and granite rock forms, which remind a subtropical or tropical relief. Post-glacial (post-Saalian 1–240,000 years) uplift of the region related most probably to glacioisostatic rebound resulted in removal of most of remaining weathering products and Quaternary deposits, and in river valley deepening.

Due to diversified lithology, numerous quarries and sand pits are spread over the area, some of the abandoned ones used for water recreation. Except for quarrying the mostly forested area has never been affected by any industry and with its unique landscape, clean and untouched nature, and remote position, it is a sought-after place by those who prefer calm ambient to relax or do sports or even to heal in the local climatic Priessnitz Spa with long and unique history.

Acknowledgments The long-term research in the area was supported by the Czech Science Foundation projects No. 205/08/P521 and No. P210/12/0573, Czech Ministry of Education, Youth and Sports project No. LH12078, and grant foundation of Charles University GAUK 862213. This work was carried out also thanks to the support of the long-term conceptual development research organisation RVO: 67985891.

References

Badura J, Przybylski B, Zuchiewicz W (2004) Cainozoic evolution of Lower Silesia, SW Poland: a new interpretation in the light of sub-Cainozoic and sub-Quaternary topography. Acta Geodyn Geomater 1 3(135):7–29

Badura J, Zuchiewicz W, Štěpančíková P, Przybylski B, Kontny B, Cacoń S (2007) The Sudetic Marginal Fault: a young morphotectonic feature at the NE margin of the Bohemian Massif, Central Europe. Acta Geodyn Geomater 4 4(148):1–23

Cwojdziński S, Żelaźniewicz A (1995) Podłoże krystaliczne bloku przedsudec-kiego. Geologia i ochrona środowiska bloku przedsudeckiego. Przew. 66 Zjazdu PTG, 11–28

Czudek T, Demek J (1960) Formy fosilního krasovění v podloží glaciálních usazenin u Supíkovic ve Slezsku. Přírodovědný časopis slezský 21:588–591

Demek J (1976) Pleistocene continental glaciation and its effects on the relief of the northeastern part of the Bohemian Highlands. Studia Societatis Scientiarium Torunensis 8. Sectio C (Geographica et Geologia) 4–6:63–74

Demek J, Marvan P, Panoš V, Raušer J (1964) Formy zvětrávání a odnosu žuly a jejich závislost na podnebí. Rozpravy ČSAV, řada mat.-přír. 74 9:1–59

Ivan A (1983) Geomorfologické poměry Žulovské pahorkatiny. Zprávy GgÚ ČSAV 20(4):49–69

Kellogg JH (1910) Rational hydrotherapy: a manual of the physiological and therapeutic effects of hydriatic procedures, and the technique of their application in the treatment of disease. 4th rev. ed., F. A. Davis company, Philadelphia, 21–1247

Kočka M, Kubík A (2006) Vincenz Priessnitz: Světový přírodní léčitel. 1. vyd. Pavel Ševčík – Veduta, Štíty, 183 pp

Král V (1958) Kras a jeskyně Východních Sudet. Acta Universitatis Carolinae. Geologica 2:159 pp

Krzyszkowski D, Pijet E (1993) Morphological effects of Pleistocene fault activity in the Sowie Mts., southwestern Poland. Zeitschr. Geomorph. N. F. Suppl.-Bd. 94:243–259

Migoń P (1993) Geomorphological characteristics of mature fault-generated range fronts, Sudetes Mts., Southwestern Poland. Zeitschr. Geomorph. N. F. Suppl.-Bd. 94:223–241

Migoń P (1997) Crystalline Rock Inselbergs in Southwestern Poland, Origin and Paleoenvironmental Significance. Studia Geograficzne LXVI, Wyd, Uniw. Wrocl 102 p

Milický V, Kabát F, Křelina B (1985) Kaolín z Vidnavy a jeho tradiční využití. Sbor. geol. věd. Technologie, geochemie 20:203–232

Paleček M, Pospíšil V, Hanžl P, Krejčí Z (2013) Mapová aplikace geologická mapa 1 : 50 000. Praha. http://mapy.geology.cz/geocr_50/

Peška P (1992) Stress indications in the Bohemian Massif: reinterpretation of the borehole televiewer data. Stud Geophys Geod 4:307–324

Skácel J (2006) Historický pohled na poznávání a těžbu zdrojů nerostných surovin na Jesenicku. VI. Svatováclavské česko-polsko-německé setkání v Jeseníku, sborník referátů, Vlastivědné muzeum Jeseník, pp 9–30

Sponheuer W (1952) Erdbebenkatalog Deutschlands und der angrenzenden Gebiete für die Jahre 1800 bis 1899. Akademie-Verlag GmbH, Berlin 195 p

Štěpančíková P, Rockwell T, Hartvich F, Tábořík P, Stemberk J, Ortuño M, Wechsler N (2013) Late Quaternary Activity of the Sudetic Marginal Fault in the Czech Republic: A signal of Ice Loading? In: C. Grutzner, A. Rudersdorf, R. Peréz-Lopéz, K. Reicherter, eds.: Seismic Hazard, Critical Facilities and Slow Active Faults. 4th International INQUA Meeting on Paleoseismology, Active tectonics and Archeoseismology, Aachen, Germany, pp 259–262

Štěpančíková P, Rowberry M (2008) Rock landforms that reflect differential relief development in the north-eastern sector of the Rychlebské hory and the adjacent area of Žulovská pahorkatina (SE Sudeten Mts, Czech Republic). Acta Geodyn Geomater 5 3 (151):297–321

Štěpančíková P, Stemberk J, Vilímek V, Košťák B (2008) Neotectonic development of drainage networks in the East Sudeten Mountains and monitoring of recent fault displacements (Czech Republic). Geomorphology 102(1):68–80

Vídeňský A, Nývlt D, Štěpančíková P (2007) Příspěvek k otázce vzniku granitoidních elevací v západní části Černovodské pahorkatiny, žulovský batolit. Geol Výzk Mor Slez v r 2006:35–39

Žáček V, Čurda J, Kočandrle J, Nekovařík Č, Nývlt D, Pecina V, Skácelová D, Skácelová Z, Večeřa J (2004) Základní geologická mapa České republiky 1:25 000 s Vysvětlivkami, 14-222 Vidnava. Česká geologická služba, Praha 46 p

Periglacial Landforms of the Hrubý Jeseník Mountains

22

Marek Křížek

Abstract

This chapter deals with the occurrence, morphology and activity of periglacial landforms in the Hrubý Jeseník Mts. Redistribution of snow during the Last Glacial period, cold climate and the presence of extensive planation surfaces at high elevations have created favourable conditions for the formation and evolution of periglacial landforms, some of which are preserved to this day. Most of these landforms are relict (tors, frost-riven cliffs, cryoplanation terraces, blockfields, sorted polygons and nets, and large solifluction steps), and only a small part of climatically less demanding periglacial landforms are active (ploughing blocks, earth hummocks, small sorted circles, nivation hollows and small solifluction lobes). Special attention is paid to patterned ground, which provides information about current and past freeze-thaw effectiveness. Earth hummocks, found at wind-swept sites, on frost-susceptible, fine-grained regoliths, are the most interesting type of patterned ground. Evidence of present-day activity of earth hummocks are distorted soil horizons, vertical and horizontal displacement of clasts, cryoexpulsion features and cracks on crests of earth hummocks, frequent freeze-thaw cycles and long-term freezing. The origin of earth hummocks has been identified as being at the break of the Subboreal/Subatlantic. The occurrence and activity of earth hummocks, sorted circles and ploughing blocks at several sites above the alpine timberline of the Hrubý Jeseník Mts. allows us to regard these areas as parts of the mountain periglacial zone.

Keywords

The Hrubý Jeseník Mountains • The High Sudetes • Planation surfaces • Cirque • Periglacial landforms • Patterned ground • Nivation hollow

22.1 Introduction

22.1.1 Geographical Setting

The Hrubý Jeseník Mts. in the north-eastern part of the Bohemian Massif (Fig. 22.1) constitute the second highest mountain range in the Czech Republic after the Krkonoše Mts. (see Chap. 15). Mt. Praděd (1,492 m a.s.l., 50°4′59″N; 17°13′51″E) is the highest peak in the mountain range. The total area of this range is 520 km^2. This Variscan fault-block mountain range is composed of metamorphic rocks (gneisses, phyllites, mica schists and quartzites).

The mountains are situated in a cold climate region (Quitt 1971), which is characterized by short, cool and wet summers, cold springs and long cold winters with very long duration of snow cover. The mean annual air temperature on Mt. Praděd is +1.7 °C (1960–1990; Coufal et al. 1992). The mean air temperature of the warmest month (July) is 9.7 °C, whilst the lowest mean air temperature goes down to −7.5 °C in January. Mean annual precipitation is 1,231 mm (1947–

M. Křížek (✉)
Department of Physical Geography and Geoecology Faculty of Science, Charles University in Prague, Albertov 6, 128 43 Praha 2, Czech Republic
e-mail: marek.krizek@natur.cuni.cz

T. Pánek and J. Hradecký (eds.), *Landscapes and Landforms of the Czech Republic*, World Geomorphological Landscapes, DOI 10.1007/978-3-319-27537-6_22

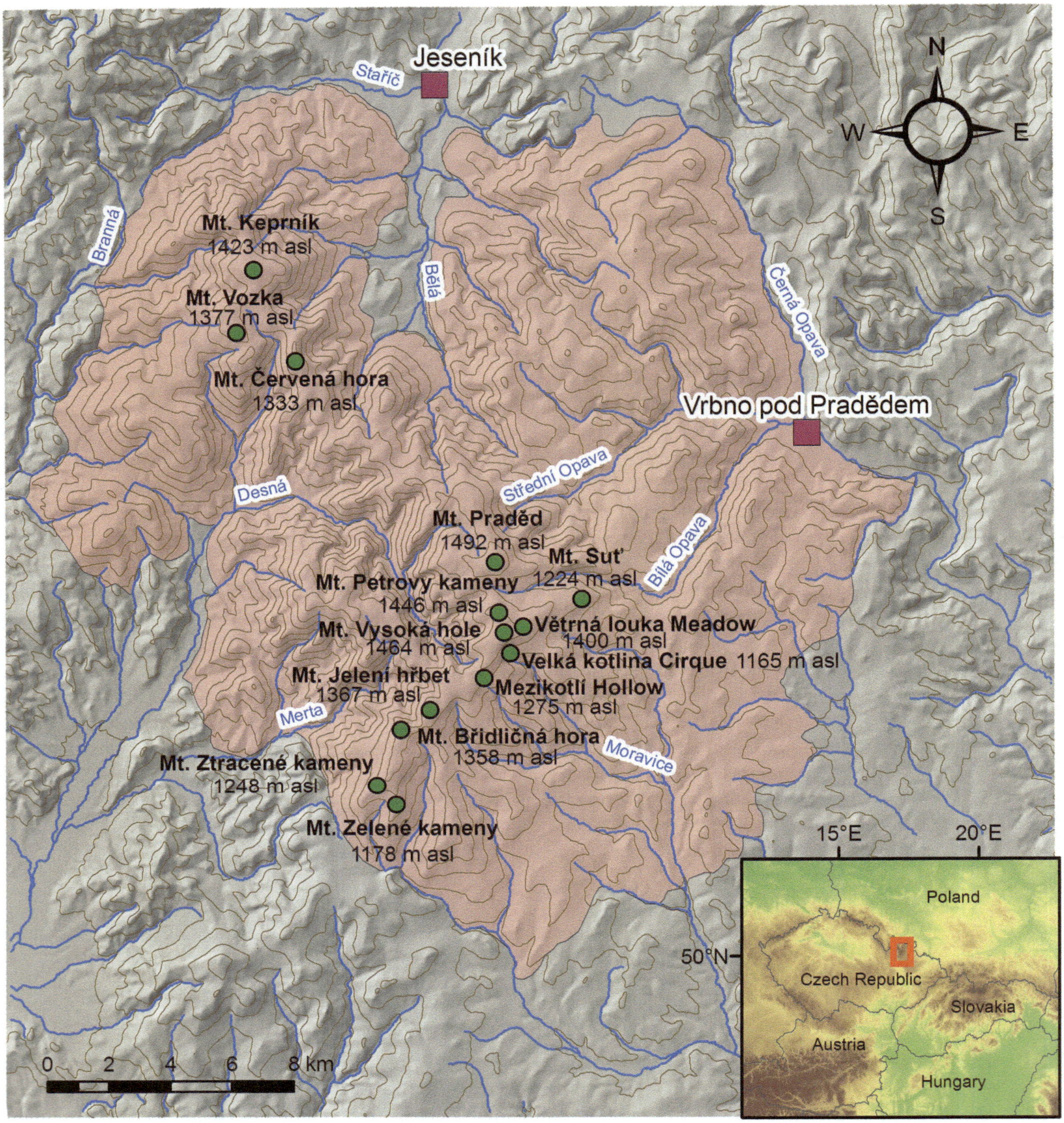

Fig. 22.1 Geographic position of the Hrubý Jeseník Mts. and location of described periglacial and glacial phenomena

1985; data of the CHKO Jeseníky). July and August are months with the highest precipitation, while the lowest precipitation is in February. Although snowfall may occur in the highest parts of the mountain range at any time during the year, the mean number of days with snow cover is 171 per year (data of the CHKO Jeseníky). The mean annual maximum snow depth is 195 cm on Mt. Praděd and the greatest snow depth occurs in early March. Podzol and cryptopodzol soils are dominant (Tomášek 2003), and cambisols are developed in the lower parts of the valley slopes.

The Hrubý Jeseník Mts. includes montane, subalpine and alpine vegetation areas (Neuhaüselová et al. 1998). The flora is predominantly montane, and thermophilic species are very rare. Most of the range (approx. 80 %) is covered by secondary forests. Spruce trees predominate from the lowest areas to the alpine timberline, which occurs at an average altitude of 1,310 m a.s.l. (Treml and Banaš 2000). The alpine belt of the Hrubý Jeseník Mts. is of natural origin (Jeník 1961) and its area is 1,048 ha (Treml and Banaš 2000). Nevertheless, medieval colonization caused a change in the alpine timberline and influenced the species composition of vegetation by mowing and grazing (Novák et al. 2010). This alpine timberline ecotone naturally lacked dwarf pine (*Pinus mugo*), which is a non-native species in the Hrubý Jeseník Mts. (Rybníček and Rybníčková 2004). However, dwarf pine was planted near the alpine timberline in the second half of the nineteenth century, covering today large areas above it. In the second half of the twentieth century, there was a sharp increase in the area covered by dwarf pine. In the period 1973–2003, it increased by 63 % (Treml et al. 2010a).

In this mountain range there are 1,200 species of plants, which account for more than one third of the total number of plant species growing in the Czech Republic. The Velká kotlina Cirque is the richest locality in the whole mountain range as regards the number of plant species, with over 500 species of vascular plants and mosses (Jeník et al. 1980). In addition, there are five species that grow only in the Hrubý Jeseník Mts. and nowhere else in the world. These are *Poa riphaea*, *Campanula gelida*, *Plantago atrata sudetica*, *Dianthus carthusianorum sudeticus* and *Carlina biebersteinii sudetica* (Bureš 2013). Several glacial relicts have survived, e.g. *Salix herbacea lapponum*, *Bartsia alpina* and *Carex rupestris* (Bureš 2013), animals include *Sicista betulina*, *Charadrius morinellus*, *Xestia speciosa*, *Aphodius limbolarius*, *Aeshna subarctica elisabethae*, *Somatochlora arctica* and *Somatochlora alpestris* (Kočí 2007). In summit areas and mountain saddles there are peat bogs of different sizes. They began to form in the middle Holocene (during the Subboreal period, ca 4,700 BP, uncal. age ^{14}C, Rybníček and Rybníčková 2004; 4,180 BC, Dudová et al. 2013). Rare animals, which can be found there, include the lynx (*Lynx lynx*), lesser horseshoe bat (*Rhinolopus hipposideros*), peregrine falcon (*Falco peregrinus*), capercaillie (*Tetrao urogallus*) and water pipit (*Anthus spinoletta*). Sometimes wolves (*Canis lupus*) or bears (*Ursus arctos*) come to the range from the nearby Carpathians.

The great wealth of animate and inanimate nature was the reason for the founding of the Jeseníky Protected Landscape Area (the CHKO Jeseníky) in 1969 (covering an area of 740 km^2). The area is part of the Natura 2000 EU-wide network of protected areas with 13 different Sites of Community Importance and Special Protection Areas.

22.1.2 Main Landforms and Their Evolution

The main landforms of the Hrubý Jeseník Mts—steep slopes, deep valleys and summit plateaus—were formed during the Cenozoic (Czudek 1997). Tectonic movements, which were a response to the pressure of the Alpine-Carpathian system, started to intensify from the Oligocene/Miocene boundary (Demek 1985). However, the main phase of tectonic uplift took place at the end of the Pliocene and Quaternary (Kopecký 1986). The total Cenozoic uplift of the upper parts of the Hrubý Jeseník Mts. is estimated at up to 1,200 m. Tectonic uplift reached 60–70 m in the northern foreland of the mountain range during the Early Pleistocene (Badura et al. 2007). Krzyszkowski and Pijet (1993) estimated tectonic uplift along the Sudetic Marginal Fault at up to 60–80 m from the Middle to the Late Pleistocene. During the Late Pleistocene, the maximum uplift is estimated at 20–35 m. The post-Saalian uplift reached a maximum of 25 m in the nearby valley of Nysa Kłodzka river (Krzyszkowski et al. 2000). The mountain range has well-developed summit planation surfaces (Fig. 22.2), concentrated at four basic altitudinal levels. The highest one is situated at 1,300–1,463 m a.s.l. (Křížek and Jablonská, unpublished). The planation surfaces were very important for the formation of glacial and periglacial landforms during the last Ice Age and the Holocene. Snow that was blown from the summit plateaus of planation surfaces by prevailing westerly winds (Jeník 1961) helped to create the only one cirque of the Hrubý Jeseník Mts. in the headwater part of the Moravice valley, known as Velká kotlina, and many nivation hollows on leeward slopes and valleys. On the other hand, blown snow and thinner snow cover on wind-exposed plateaus brought specific environmental conditions allowing for more intensive frost weathering and sorting (sensu Křížek and Uxa 2013), which are necessary for the development of sorted polygons and nets.

During cold periods of the Pleistocene, the Hrubý Jeseník Mts. were located in the periglacial zone in a foreland area of continental glaciation (Czudek 1997). Mountain glaciation was nevertheless very limited. There was only one short cirque glacier in the range (Fig. 22.3), 600 m long (Prosová 1973), with the terminal moraine situated at 1,060 m a.s.l. The cirque is ca 300 metres long and 400 metres wide, its headwall is 195 metres high, and its cirque overdeepening is 0.82 (Křížek et al. 2012). The average altitude of the cirque floor is 1,170 m a.s.l., thus the snowline (TPW-ELA) altitude during the last glaciation was 1,170 m a.s.l. A high degree of continentality with low precipitation was the reason for weak mountain glaciation in the Hrubý Jeseník Mts. in the Last Glacial period.

Fluvial erosion was the main geomorphological process which formed and remodelled valleys in the Hrubý Jeseník Mts. after the end of the Last Glaciation. Longitudinal profiles of streams are steep and stepped. Material from the slopes and transported by streams or debris flows has accumulated at the bottom of the valleys and on weakly developed floodplains. Less resistant metamorphic rocks,

Fig. 22.2 Summit planation surface on the main ridge of the Hrubý Jeseník Mts (*Photo* M. Křížek)

mainly phyllites and mica schists, support a weathered mantle up to several metres thick. Debris flows occurred where the products of weathering were saturated with water from melting snow or rainfall, e.g. in 1893, 1903, 1907, 1938, 1940, 1951, 1965, 1991, 1994, 1997 and 2004 (Gába 1992; Polách and Gába 1998; Hrádek and Malik 2007; Malik and Owczarek 2009; Krause and Křížek unpublished). The largest debris flows occurred after intense rainfall in 1880 and 1921. For example, debris flows involving a total of 50,000 cubic metres destroyed an area of 16 ha on the slopes of Mt. Červená hora in 1921 (Polách and Gába 1998). Landslides into Hučivá Desná River caused large-scale devastation over a stretch of almost 25 km. Relicts of landslides at the valley bottoms are evidence of former debris flow activity. Some mounds of debris flow deposits are so large that they were previously interpreted incorrectly as relicts of glacial moraines (see Prosová 1973).

Some debris flow paths are also avalanche paths. Kříž (1995) described a total of 22 avalanche paths in eight areas of the mountain range, but the actual number is now lower (about 16 avalanche paths in six areas). This is because slopes prone to avalanches were afforested and overgrown with dwarf pine or trees, which prevent avalanches. Current avalanche activity in the Hrubý Jeseník Mts. is minimal—events occur once every few years.

22.2 Periglacial Landforms

The redistribution of snow during the Last Glacial period, the cold climate (periglacial) and extensive planation surfaces at high elevations of the Hrubý Jeseník Mts. have created favourable conditions for the formation and evolution of periglacial landforms, some of which are preserved to this day. Most of these landforms—tors, frost-riven cliffs, cryoplanation terraces, blockfields, sorted polygons and nets, and large solifluction steps—are inactive and relict, and only a small part of climatically less demanding periglacial landforms are active. These include ploughing blocks, earth hummocks, small sorted circles, nivation hollows and small

Fig. 22.3 Velká kotlina Cirque (*Photo* M. Křížek)

solifluction lobes. Likewise, some geomorphological processes, such as frost heaving, frost sorting, needle-ice activity (Fig. 22.4) and solifluction, which are necessary for the formation and evolution of periglacial landforms, can be observed on wind-exposed sites (Křížek et al. 2010). Although there is no evidence of recent permafrost occurrence in the Hrubý Jeseník Mts., selected summit parts of this range can be regarded as mountain periglacial zones according to certain criteria, e.g. mean annual temperature below +3 °C (sensu French 2007), position above the alpine timberline (sensu Ballantyne and Harris 1994), solifluction activity (sensu Williams 1961, Leser in Embleton 1984) or activity of ploughing blocks (Furrer in Washburn 1979).

22.2.1 Cryoplanation Terraces and Blockfields

Cryoplanation terraces in the Hrubý Jeseník Mts. are related to old planation surfaces (Demek 1969) and often occur as multiple steps on slopes. Their size varies from several tens to several hundreds of square metres. The cryoplanation flats are several metres wide and separated by distinct edges (accompanied by a significant change of slopes) from cryoplanation steps, whose slope mostly varies between 10 and 35°. Inclinations of the flats themselves vary from 3 to 12° (Křížek 2007). In some locations patterned ground has developed on cryoplanation terraces, above all sorted polygons, nets and stripes in steeper slopes, and there are tors. These have the character of isolated rock outcrops, such as Petrovy kameny (Fig. 22.5, 1,446 m a.s.l.), Ztracené kameny (1,245 m a.s.l.) and Vozka (1,377 m a.s.l.), and are often surrounded by a mass of collapsed blocks. These tors on summit plateaus, unlike the frost-riven cliffs below them, represent a more advanced stage of slope development, i.e. slope retreat (sensu Demek 1969). The heights of tors and frost-riven cliffs are limited to 10 m and their horizontal dimensions reach a maximum of a few tens of metres.

Blockfields and block streams in the Hrubý Jeseník Mts. create contiguous areas covered with angular blocks, produced by cryogenic disintegration of bedrock outcrops (tors

Fig. 22.4 Needle ice pushing up soil particles and clasts up to several tens of centimetres in diameter on Mt. Keprník (1,423 m a.s.l.) on the 20th April 2007 (*Photo* M. Křížek)

and frost-riven cliffs). The dimensions of these blocks are from tens of centimetres up to several metres. These blockfields and blockstreams are mainly developed on slopes with an inclination of between 10 and 50°. The essential difference between blockfields and blockstreams is in their shape, the latter being significantly elongated in the direction of the slope. Large blockfields occur on Mt. Břidličná hora (1,358 m a.s.l.), Mt. Ztracené kameny (1,245 m a.s.l.), Mt. Zelené kameny (1,178 m a.s.l.) and Mt. Suť (1,224 m a.s.l.). The blockfield on Mt. Suť is the largest, covering 4.13 ha. It has been gradually covered by forest growth like most other blockfields situated below the alpine timberline. In today´s conditions, mechanical weathering is of lower intensity, so the blockfields are not being enriched with new material, and movement of the blocks, formerly caused by solifluction, has ceased (Prosová 1954) or is very small (Demek et al. 2011).

22.2.2 Patterned Ground

Patterned ground covering an area of 98 ha above the alpine timberline is the most common periglacial landform in the Hrubý Jeseník Mts. (Křížek et al. 2007). There are two main genetic types of patterned ground on the high elevated planation surfaces, namely sorted and non-sorted variants. On flat planation surfaces (inclination of 0–3°) sorted polygons and sorted nets occur. They elongate gradually with increasing slope inclination (3–7°) and then change to sorted stripes on the slopes with an inclination of 7–12°. The case of non-sorted patterned ground is similar, where earth hummocks have been turned to non-sorted stripes as a result of solifluction on steeper slopes.

The best developed sorted polygons with straight sides occur on Mt. Břidličná hora (1,358 m a.s.l., Figure 22.6). Their average size is 489 × 377 cm (Table 22.1). This type

Fig. 22.5 Tor of Mt. Petrovy kameny (1,446 m a.s.l.) in 1907 and 2005 (*Photo* M. Křížek). At the back the highest point of the Hrubý Jeseník Mts.—Mt. Praděd (1,492 m a.s.l.) with an old watchtower (in construction) and the present transmitter tower, which is 162 m high

of patterned ground has the best developed frost sorting of clasts (Křížek and Uxa 2013). It can therefore be considered as the climactic stage of sorted patterned ground. While the borders of sorted polygons consist of coarse-grained clasts, the finer-grained and domed centres of the polygons are now covered with vegetation. The occurrence of these sorted polygons is associated with quartzite, which facilitated frost sorting by its characteristic disintegration.

Sorted nets surrounding areas of sorted polygons are the most common patterned ground in the Hrubý Jeseník Mts. They have an irregular shape and are smaller (their average size is 308 × 242 cm) than sorted polygons. Today, these landforms are covered with vegetation. In some places (e.g. on Mt. Vysoká Hole, 1,464 m a.s.l.), the relief of this vegetated surface does not always correspond to the real structure of buried sorted nets, i.e. the coarse clast rim of sorted nets is under the domed surface, and the centre is developed under the furrows of the hummocky surface (Křížek 2007). Sorted stripes occur on slopes steeper than 7° and adjoin sorted polygons and sorted nets areas. Sorted stripes are parallel lines of coarse-grained clasts elongated down the slopes. Small clasts and finer sediment situated between these stripes are domed. The length of sorted stripes ranges from a few metres to several tens of metres.

Sorted circles are found at only one site in the Hrubý Jeseník Mts.—on the wind-exposed edge of the summit plateau on Mt. Keprník. Clasts of this type of patterned ground are arranged in regular circles with a diameter of about 20 cm. They arise regularly in spring, but are very soon destroyed by the feet of tourists.

Earth hummocks occur at four sites: at the top of Mt. Keprník (1,423 m a.s.l.), on the northern plateau of Mt.

Fig. 22.6 Sorted polygons on Mt. Břidličná hora (*Photo* M. Křížek)

Table 22.1 Mean morphometric characteristics of patterned ground at selected sites of the Hrubý Jeseník Mts. (Krížek et al. 2007, modified)

Locality	Type of patterned ground	Altitude (m a.s. l.)	Length (cm)	Width (cm)	Height (cm)	Relative height (Height/Width)
Břidličná hora	Sorted polygons	1,355	489	377	18	0.05
Praděd	Earth hummocks	1,446	185	142	32	0.24
Keprník	Earth hummocks	1,420	166	126	40	0.34
Vysoká hole	Sorted nets	1,460	335	280	17	0.06
Velký Máj	Sorted nets	1,385	313	246	24	0.1
Kamzičník	Sorted nets	1,419	349	264	20	0.08
Mravenečník	Sorted nets	1,340	232	179	21	0.12

Červená hora (1,333 m a.s.l.), on the north-west plateau (1,435 m a.s.l.) of Mt. Petrovy kameny and on Větrná louka (=meadow; 1,400 m a.s.l.). The best developed earth hummocks are located on Mt. Kepník, where their average size reaches 166 × 126 cm and the average height is 40 cm (Treml et al. 2010b). Compared with the sorted polygons and nets they are higher and more domed, i.e. earth hummocks have a greater height/width ratio. These earth hummocks are characterized by a high proportion of the fine-grained fraction (particle size median of A_h − B/C horizons is between 0.1 and 0.2 mm) and a high content of organic matter (i.e. content of combustible soil organic matter is about 30 % down to the depth of 60 cm of the soil profile). Thus, the organic matter of the earth hummocks is susceptible to frost action (sensu Ballantyne 1996). The age of the earth hummocks has been identified as going back to the Subboreal/Subatlantic boundary. This is based on a pollen analysis and ^{14}C radiometric dating (2,090 ± 35 uncal. years BP; Křížek 2007). Evidence of present-day activity of earth hummocks is provided by soil horizons distorted by cryoturbation (Fig. 22.7) and vertical and horizontal displacement of clasts, indicating movement within the earth hummocks, cryoexpulsion features and cracks on crests of earth hummocks with disturbed vegetation cover, frequent freeze-thaw cycles and long-term freezing. Cores of segregation ice created inside the earth hummocks (Fig. 22.8) during the winter remain until the end of May or June, depending on the severity of winter. The freeze-thaw season for the earth hummocks starts in the second half of November and terminates at the end of spring (Table 22.2). In 2003 a core of segregation ice inside an earth hummock on Mt. Keprník was detected even on 7 July (Křížek 2007). The occurrence and activity of earth hummocks, sorted circles and ploughing blocks on Mt. Keprník allows us to regard this area as being part of a mountain periglacial zone (sensu Leser in Embleton 1984), located at its lower limit (sensu Furrer in Washburn 1979), which is defined by ploughing blocks activity. However, the top of Mt. Keprník and similar isolated sites in the Hrubý Jeseník Mts. are among the climatically most exposed areas in the Czech Republic (Křížek et al. 2010), which is why they are unique. The closest equivalent locations can be found in the significantly higher Alps, the High Tatras or in the distant Scandinavian mountains.

Non-sorted stripes (or hummocky stripes) in the Hrubý Jeseník Mts. can be described as being an elongated form of earth hummocks. This type of patterned ground occurs mainly on gentle slopes (typical inclination is 3–12°) around

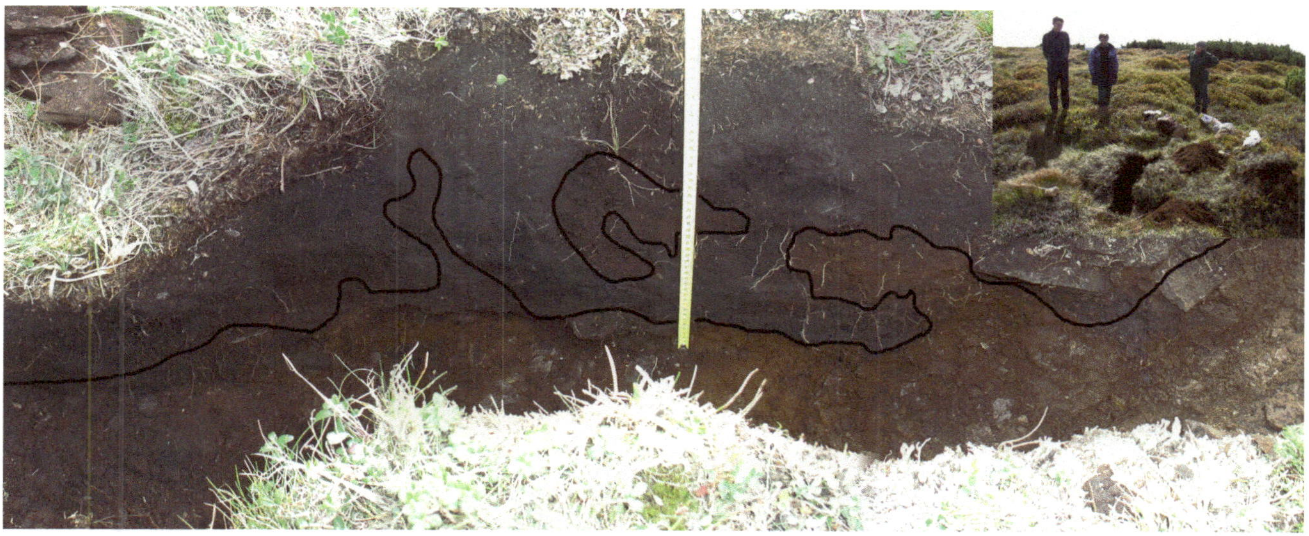

Fig. 22.7 Cross-section profile through the earth hummock on Mt. Keprník. There are cryoturbation pockets. Note: *solid black line* represents an interface between A and B horizon twisted by cryoturbation. (*Photo* Z. Engel)

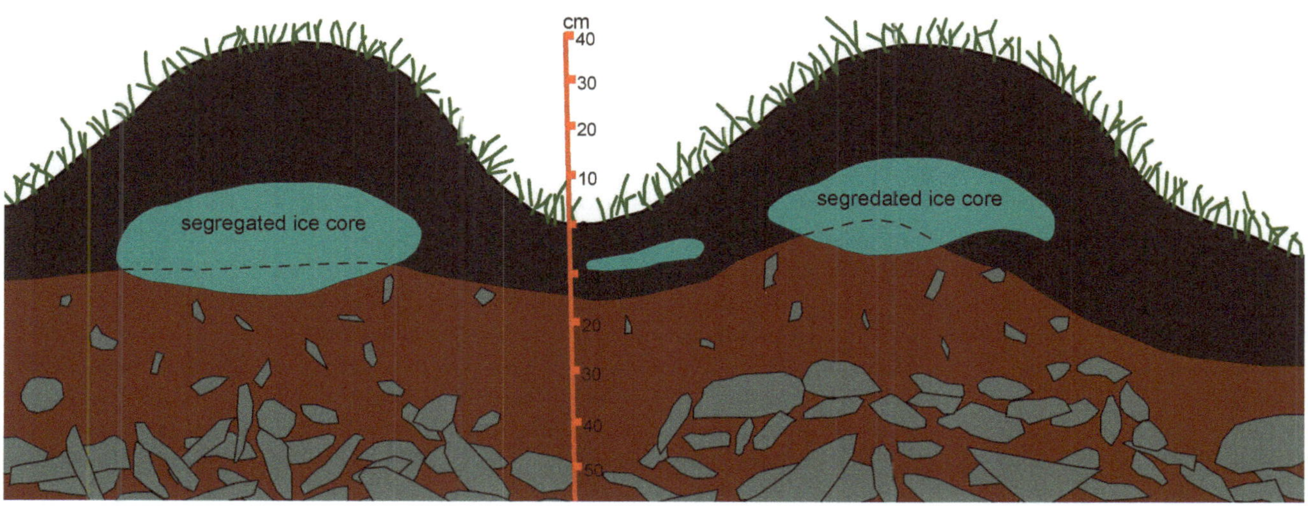

Fig. 22.8 Segregation ice forming lenses within the earth hummock on Mt. Keprník on 7th June 2003

Table 22.2 Selected characteristics of the freeze-thaw action of an earth hummock on Mt. Keprník (temperature measured at the depth of 5 cm)

Freeze-thaw season	Measured object	Start of feeze-thaw season	End of freeze-thaw season	Number of days of freeze-thaw season	Minimum temperature during freeze-thaw season (°C)	Number of drops below 0 °C
2008/2009	Earth hummock	18.9.2008, 14.00	14.5.2009, 9.00	238	−5.5	32
	Control measurement outside of earth hummock	17.9.2008, 9.00	13.4.2009, 11.00	208	−1.6	14
2009/2010	Earth hummock	22.10.2009, 14.00	20.5.2010, 10.00	210	−4.3	18
	Control measurement outside of earth hummock	3.11.2009, 11.00	21.5.2010, 11.00	199	−0.9	14
2010/2011	Earth hummock	30.10.2010, 10.00	5.5.2011, 9.00	187	−8.1	12
	Control measurement outside of earth hummock	30.10.2010, 9.00	9.5.2011, 8.00	191	−3.6	18
2011/2012	Earth hummock	17.11.2011, 5.00	25.4.2012, 8.00	161	−7.9	15
	Control measurement outside of earth hummock	21.11.2011, 8.00	26.4.2012, 11.00	157	−2.7	10
2012/2013	Earth hummock	3.12.2012, 6.00	21.4.2013, 8.00	139	−6.1	9
	Control measurement outside of earth hummock	7.12.2012, 16.00	22.4.2013, 10.00	136	−2.0	8

summit plateaus. Hummocky stripes have a domed centre (between 15 and 40 cm) like earth hummocks and are elongated down the steepest available slope. These stripes are parallel. The width of stripes is between 45 and 150 cm and they are usually several tens of metres long.

22.2.3 Nivation Hollows, Solifluction Phenomena and Ploughing Blocks

Nivation hollows are shallow depressions with a steep scarp (20–35°) and a gently inclined bottom (5–12°). Their dimensions range from tens (these are the most common) to hundreds of metres. The occurrence of nivation hollows is linked to eastern leeward slopes, where snow accumulates (Křížek 2007). Long-lying snow patches cause solifluction phenomena to concentrate near nivation hollows. The largest nivation hollow lies in the Mezikotlí locality (1,275 m a.s.l., situated south-west of the Velká kotlina cirque, in the neighbouring valley) and has two well-developed pronival ramparts. The first one is located on the edge of the nivation hollow and has blocked its outlet. The bottom of this hollow is covered by a 3–4 m thick layer of sediments and weathered rock, and a 1.3 m layer of peat (Křížek et al. 2010). The age of this peat bog is 1,520 ± 39 ^{14}C uncal. BP, and run-off sediments at the bottom of the peat bog are at least 1,000 years older (2,696 ± 40 ^{14}C uncal. BP; Křížek et al. 2010). The second pronival rampart is smaller and younger than the first one and occurs below the 40° steep headwall of the nivation hollow.

Hundreds of ploughing blocks are located on the slopes of Mt. Keprník (1,423 m a.s.l.), Mt. Vysoká hole (1,464 m a.s.l., Fig. 22.9), Mt. Jelení hřbet (1,367 m a.s.l.) and Mt. Břidličná hora (1,358 m a.s.l.). The boulders range in size from tens of centimetres to several metres. The basic feature of the ploughing blocks is the furrow behind them—a linear depression elongated in the direction of the slope gradient. Furrows are several metres long and 10–35 cm deep. Some of the ploughing blocks have 10–50 cm niches in front of their furrows. Almost 75 % of blocks have a mound, usually 20–40 cm high, developed in front of them (Křížek 2007). The lithology of the ploughing blocks usually corresponds to the basement rock in the area, with the exception of the Mt. Vysoká Hole—Mt. Břidličná hora ridge, where quartzite blocks originating from the upper parts of the slopes dominate. This fact, together with the developed furrows, proves that ploughing blocks in the mountain range are moving, and are thus active.

Fig. 22.9 Ploughing block with a developed mound on the north-west slope of Mt. Vysoká hole (*Photo* M. Křížek)

The solifluction steps in the Hrubý Jeseník Mts. are landforms with morphologically distinct short and high (up to 4 m) steps. Well-developed solifluction steps occur on the south-east slope of Mt. Keprník. The bodies of these landforms are consolidated, without disturbance to turf cover. Creep probes did not record any movement during the period of 2005–2010. The solifluction steps have been inactive in recent times, unlike solifluction lobes. Solifluction lobes are elongated in the direction of greater slope inclination and are between 5 and 20 cm high. Some of them have a disturbed turf cover. The rates of their movement ranged between 0 and 17 mm in 3 years (Křížek 2007). Solifluction movements usually reach the maximum depth of 20 cm and the rate of movement decreases with increasing depth. Further evidence of solifluction lobe activity is provided by an increased amount of fine-grained material and humus in the A_h horizon in the faces of the lobes (Treml et al. 2003).

22.3 Periglacial Landforms and Human Activity

When dealing with periglacial phenomena in the Hrubý Jeseník Mts., mention must be made of human encroachment into the highest localities of this range. In the second half of the twelfth century, prospectors began to operate, looking for and then mining iron ore and non-ferrous and precious metals, including gold. Locally, Střední Opava river is called the Golden Opava river, recalling the times when gold was panned. In the thirteenth century, the first settlements began to appear and the first castles were built in the middle of thick forests. The end of the sixteenth century saw the building of iron works and mills making use of the local supply of iron ore and charcoal. In the seventeenth century, people began to make use of mountain meadows above the alpine timberline to graze livestock (cattle grazed there even after the Second World War). Grazing, mowing and wood extraction led to the lowering of the alpine timberline. At this time, local settlers came into direct contact with periglacial phenomena, which they did not understand. Bizarrely shaped tors became the subject of myths and legends. According to one legend, the tor called Vozka (waggoner) is a petrified wagon that was carrying bread. God punished the driver, because he demonstrated his disrespect for bread by putting it under the wheels of his wagon on a broken forest road. Today, the silhouette of this well-known periglacial landform is used to promote beer from the local brewery. The strange and exotic shapes of a number of periglacial landforms were also associated with black magic and witchcraft. The tor at Mt. Petrovy kameny played a sad role in the seventeenth century as the location where witches' Sabbaths were allegedly held. As a consequence, 81 "witches and sorcerers" were tortured and burned to death in the period of 1679–1696. The mysterious nature of this tor was certainly enhanced by the surrounding regular patterns of sorted polygons.

The first scientists came to the Hrubý Jeseník Mts. and started to reveal the real secrets behind the individual periglacial landforms in the late nineteenth and early twentieth centuries. However, these periglacial landforms were not left in peace for very long. Following the foundation of Czechoslovakia, the summit planation surface of Mt. Vysoká hole and its surroundings became a military shooting range for the mountain artillery. Unfortunately, some of the local sorted nets and nivation hollows situated below the eastern edge of the summit plateau were destroyed by artillery shells that created craters of up to several metres in diameter. In addition, in 1944, construction of a mountain airport was commenced on the summit plateau of the planation surface of Mt. Vysoká hole. With the approaching end of the Second World War, the airport was not built, with the exception of concrete foundations for radar installations. The Jeseníky Protected Landscape Area (CHKO Jeseníky) was founded in 1969, and today the periglacial phenomena are protected along with other components of nature in the area. From the viewpoint of periglacial phenomena protection, the biggest current problem is presented by the invasive dwarf pine, which is not native to the Hrubý Jeseník Mts. (see Sect. 22.1.1). Earth hummocks covered by dwarf pine are destroyed mechanically by its roots and thermally by the increased accumulation of snow. This prevents deep freezing, which is necessary for the development of earth hummocks (Treml and Křížek 2006). For this reason, dwarf pines were removed in the location with the best developed earth hummocks, on the top of Mt. Keprník, in 2009–2010.

22.4 Conclusion

The presence of high elevation planation surfaces of the Hrubý Jeseník Mts. was the main preconditioning factor of formation of periglacial landforms that are good evidence of climate variability in Central Europe during the Quaternary. The local periglacial environment has created a wide variety of mountain cold climate landforms. There is very well-developed patterned ground, which reflects microclimatically more exposed sites (in the sense of intensity of freeze-thaw action). The age of patterned ground in the mountain range is estimated for the end of the Last Glacial period (sorted polygons and nets) or for during colder periods of the Holocene (sorted circles, earth hummocks, non-sorted stripes). Earth hummocks are unique landforms that are found at wind-swept sites on frost-susceptible, fine-grained regoliths. These landforms—together with ploughing blocks, nivation hollows and small solifluction lobes—are still active. Other periglacial landforms such as sorted polygons and nets, cryoplanation terraces, tors and frost-riven cliffs were active during the Last Glacial period, but are now inactive.

Acknowledgments I would like to dedicate this chapter to my wife, daughter and son. I acknowledge the administration of the CHKO Jeseníky for permitting me to conduct research within the protected landscape area. I wish to thank Tomáš Uxa for field assistance and Frederick Rooks for language editing.

References

Badura J, Zuchiewicz W, Štěpančíková P, Przybylski B, Kontny B, Canoń S (2007) The sudetic marginal fault: a young morphotectonic feature at the NE margin of the Bohemian Massif, Central Europe. Acta Geodynamica et Geomaterialia 4(4):7–29

Ballantyne CK, Harris C (1994) The periglaciation of Great Britain. Cambridge University Press, Cambridge, 330 pp

Ballantyne CK (1996) Formation of miniature sorted patterns by shallow ground freezing: a field experiment. Permafrost Periglac Process 7:409–424

Bureš L (2013) Chráněné a ohrožené rostliny CHKO Jeseníky. Rubico, Olomouc 314 pp

Coufal L, Langová P, Míková T (1992) Meteorologická data na území ČR za období 1961-1990. ČHMÚ, Praha 160 pp

Czudek T (1997) Reliéf Moravy a Slezska v kvartéru. Sursum, Tišnov 213 pp

Demek J (1969) Cryoplanation terraces, their geographical distribution, genesis and development. Nakladatelství ČSAV, Praha 80 pp

Demek J (1985) Morfogeneze epiplatformních pohoří České vysočiny (na příkladu Hrubého Jeseníku). Geografický časopis 37(2–3):303–313

Demek J, Havlíček M, Mackovčin P (2011) Quantitative monitoring of slope movements at the Břidličná hora Mt. (Hrubý Jeseník Mts., Czech Republic, EU). Acta Universitatis Carolinae Geographica 45 (2):31–45

Dudová L, Hájková P, Buchtová H, Opravilová V (2013) Formation, succesion and landscape history of Central-European summit raised bogs: a multiproxy study from the Hrubý Jeseník Mountains. Holocene 23(2):230–242

Embleton C (ed) (1984) Geomorphology of Europe. Verlag, Weinheim 465 pp

French HM (2007) Periglacial environment. Wiley, Chichester 458 pp

Gába Z (1992) Mury pod Keprníkem v červenci 1991. Severní Morava 64:43–49

Hrádek M, Malik I (2007) Dendrochronological records of the floodplain morphology transformation of Desná River Valley in the last 150 years, the Hrubý Jeseník Mts. (Czech Republic). Moravian Geographical Reports 15(3):2–10

Jeník (1961) Alpinská vegetace Krkonoš, Králického Sněžníku a Hrubého Jeseníku: Teorie anemo-orografických systémů. Academia, Praha, 407 pp

Jeník J, Bureš L, Burešová Z (1980) Syntaxonomic study of vegetation in Velká kotlina cirque, the Sudeten Mountains. Folia Geobotanica 15:1–28

Kočí K (ed) (2007) Jeseníky. Actea, Karlovice 220 pp

Kopecký A (1986) Neotektonika Hrubého Jeseníku a východní části Orlických hor. Časopis Slezského muzea—Vědy přírodní (A) 35 (2):117–141

Krzyszkowski D, Pijet E (1993) Morphological effects of Pleistocene fault activity in the Sowie Mts., southwestern Poland. Zeitschrift für Geomorphologie Supplementband 94:243–259

Krzyszkowski D, Przybylski B, Badura J (2000) The role of neotectonics and glaciation on terrace formation along the Nysa Kłodzka River in the Sudeten Mountains (southwestern Poland). Geomorphology 33:149–166

Kříž V (1995) Laviny Hrubého Jeseníku, Králického Sněžníku a Moravskoslezských Beskyd. Sborník Přírodovědecké fakulty Ostravské univerzity—geografie, geologie. 149(3):69–86

Křížek M (2007) Periglacial landforms above alpine timberline in the High Sudetes. In: Goudie AS, Kalvoda J (eds) Geomorphological variations. P3 K, Praha, pp 313–337

Křížek M, Treml V, Engel Z (2007) Litologická predispozice, morfologie a rozmístění strukturních půd alpinského bezlesí Vysokých Sudet. Geografie—Sborník České geografické společnosti 112(4):373–387

Křížek M, Treml V, Engel Z (2010) Czy najwyższe partie Sudetów powyżej górnej granicy lasu są domeną peryglacjalną? Czasopismo Geograficzne 81(1–2):75–102

Křížek M, Vočadlová K, Engel Z (2012) Cirque overdeepening and their relationship to morphometry. Geomorphology 139–140:495–505

Křížek M, Uxa T (2013) Morphology, sorting and microclimates of relict sorted polygons, Krkonoše Mountains, Czech Republic. Permafrost Periglac Processes 24(4):313–321

Malik I, Owczarek P (2009) Dendrochronological records of debris flow and avalanche activity in a mid-mountain forest zone (Eastern Sudetes—Central Europe). Geochronometria 34:57–66

Neuhäuslová Z et al (1998) Mapa potenciální přirozené vegetace České republiky. Academia, Praha 341 pp

Novák J, Petr L, Treml V (2010) Late-Holocene human-induced changes to the extent of alpine areas in the East Sudetes, Central Europe. Holocene 20(6):895–905

Polách D, Gába Z (1998) Historie povodní na šumperském a jesenickém okrese. Severní Morava 75:3–29

Prosová M (1954) Studie o periglaciálních zjevech v Hrubém Jeseníku. Přírodovědecký sborník Ostravského kraje 15(1):1–15

Prosová M (1973) Zalednění Hrubého Jeseníku. Campanula 4:115–123

Quitt E (1971) Klimatické oblasti Československa. Nakladatelství ČSAV, Praha, 73 pp

Rybníček K, Rybníčková E (2004) Pollen analyses of sediments from the summit of the Praděd range in the Hrubý Jeseník Mts. (Eastern Sudetes). Preslia 76:331–347

Tomášek M (2003) Půdy České republiky. Česká geologická služba, Praha 68 pp

Treml V, Banaš M (2000) Alpine timberline in the High Sudetes. Acta Univesitatis Carolinae—Geographica 15(2):83–99

Treml V, Engel Z, Křížek M (2003) Periglaciální tvary v alpinském bezlesí Vysokých Sudet. Geografie—Sborník české geografické společnosti 108(4):304–305

Treml V, Křížek M (2006) Vliv borovice kleče (Pinus mugo) na strukturní půdy české části Vysokých Sudet. Opera Concortica 43:45–56

Treml V, Wild J, Chuman T, Potůčková M (2010a) Assessing the change in cover of non-indigenous dwarf-pine using aerial photographs, a case study from the Hrubý Jeseník Mts., the Sudetes. J Landscape Ecol 4(2):90–104

Treml V, Křížek M, Engel Z (2010b) Classification of patterned ground based on morphometry and site characteristics: a case study from the High Sudetes, Central Europe. Permafrost Periglac Process 21:67–77

Washburn AL (1979) Geocryology. Edward Arnold, London 406 pp

Williams PJ (1961) Climatic factors controlling the distribution of certain frozen ground phenomena. Geografiska Annaler—Series A 43:339–347

Litovelské Pomoraví—Landscape Around Anastomosing River Pattern of Morava

23

Zdeněk Máčka

Abstract

The Morava River, rising on the slopes of the Králický Sněžník Mts. at the border with Poland, enters the vast tectonic depression of Hornomoravský úval after ca. 80 km of its course. Here it begins to deposit much of the sediments that were eroded in the mountainous part of the catchment. On the flat bottom of the Hornomoravský úval Basin, the Morava is branching to numerous wide as well as narrow channels creating the dense network of "arteries" similar to branches of a river in delta at the seashore. Modern fluvial geomorphology recognises this unusual river style to be an anastomosed channel pattern. The present-day channel network consists of the dominant Morava channel, from which numerous side channels are branching fed with water raised in the main channel by weirs. This basic pattern of channels is supplemented by a dense network of smaller, ephemeral channels that are filled with water only during the floods. Anastomosed channels, in many cases intensively meandering, are surrounded by large areas of natural floodplain forests. The landscape interwoven with numerous river arms and covered with fertile soils was ideal for construction of mills and establishing permanent settlements. Medieval economical activities stand at the beginning of change of the natural anastomosed fluvial system to the present-day harmonic cultural riverscape.

Keywords

The Morava River floodplain • Anastamosing channel pattern • Channel changes • Floodplain forest • In-channel wood • Water mills • Nature protection

23.1 Introduction

The Morava River is a left-side tributary of the Danube, with the drainage area spreading over the eastern part of the Czech Republic. The Morava River is a natural backbone of the historical land of Moravia, whose statehood roots date back to the first state of western Slavs, Great Moravia, founded in the ninth century. After the Morava leaves the Sudetes Mountains, it enters the system of Fore-Carpathian depressions in the contact zone between the Bohemian Massif and the Western Carpathians. Although the Morava drainage basin is densely inhabited, intensively cultivated from ancient times, dotted with small and large settlements and industrial centres, and intersected with important transport infrastructures, surprisingly, areas of natural value only marginally touched by human impact are still present. One of these few unspoiled regions is the riverscape of the Morava River with an extensive floodplain forest, spread between the towns of Mohelnice and Olomouc. The region is registered as an internationally important wetland on the list of The Ramsar Convention on Wetlands since 1993.

Z. Máčka (✉)
Department of Geography, Masaryk University, Brno, Czech Republic
e-mail: macka@sci.muni.cz

T. Pánek and J. Hradecký (eds.), *Landscapes and Landforms of the Czech Republic*,
World Geomorphological Landscapes, DOI 10.1007/978-3-319-27537-6_23

23.2 The River and Its Setting

23.2.1 Inland Delta of the Morava

When leaving the mountains, the Morava River enters the narrow tectonic depression of Mohelnická brázda (width 3–5 km, length 20 km; brázda = furrow). Then, it continues via the short wind gap valley at the Třesín Threshold to the extensive depression of Hornomoravský úval (25 × 60 km; elongated in NNW–SSE direction; úval = basin), that belongs to the system of Fore-Carpathian depressions. The river drops to altitudes of only 250–270 m in these basins, losing its stream power, and starts to branch and meander in its own alluvial sediments (Fig. 23.1). The most marked feature of the river is the dense network of parallel, interconnected channels, forming the so-called anastomosed river pattern. To past researchers, multiple river channels resembled river branching in deltas and that is why this region was once called "inner delta" or "inland delta".

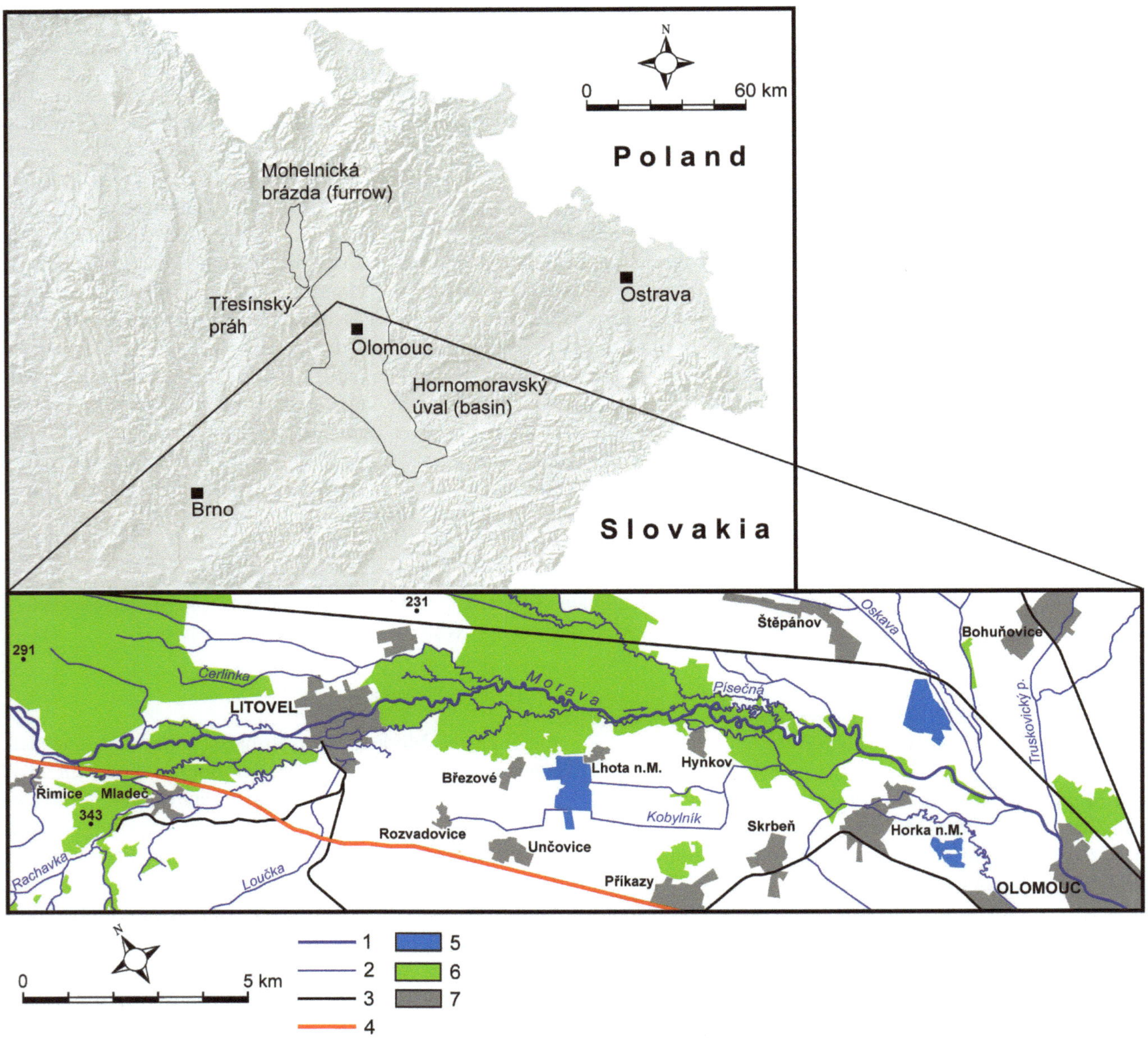

Fig. 23.1 Anastomosed channel pattern of the Morava River in the Litovelské Pomoraví region. Legend: *1* the dominant Morava River channel; *2* side channels and tributaries with permanent flow; *3* railways; *4* motorway; *5* artificial lake; *6* forest; *7* settlements

23.2.2 Geological and Hydrological Setting

The northern part of the Hornomoravský úval between Olomouc and Litovel, where the Litovelské Pomoraví area is located, was probably formed as early as in the Miocene (late Badenian) due to tectonic subsidence that continued through the Pliocene and Quaternary. Prolonged subsidence tendencies caused gradual filling of the depression with deposits of variegated character. Genetically these deposits originated mainly in shallow lacustrine and fluvial–lacustrine environment, and fluvial sequences began to prevail since the late Pliocene. The Pliocene-Pleistocene sedimentary complex is built of deposits ranging from clays to gravels, reaching the maximum thickness of 200–250 m (Růžička 1989). Quaternary fluvial sediments form a terraced assemblage that is traditionally divided into three stratigraphic levels. The highest level of the Kralice terrace also involves the alluvial fans of the Morava tributaries, the middle level is called Nenakonice terrace, whilst the lowest level is represented by the present-day floodplain. The basement of the Hornomoravský úval and slopes of the adjacent highlands are built of Paleozoic rocks—flysch rocks of Lower Carboniferous (shales, greywackes, sandstones) and limestones and metasediments of Devonian age are present (Otava et al. 1996).

Highest discharges in spring and lowest water stages at the break of summer and autumn are typical for the Morava River, however, floods may occur all over the year. More extensive inundations of the floodplain occur with discharges of 5–10 years recurrence intervals due to excessive incision of the main Morava channel. Two hydrological stations are operating on the Morava in Litovelské Pomoraví. The first one is located upstream in Moravičany (drainage area 1,561 km^2; Qa = 17.1 $m^3 s^{-1}$; Q_{100} = 394 $m^3 s^{-1}$), the second one in Nové Sady–Olomouc 41 km farther downstream (drainage area 3,324 km^2; Qa = 26.4 $m^3 s^{-1}$; Q_{100} = 551 $m^3 s^{-1}$). Arable land is protected by flood defence dykes; however, dykes are set far from the river itself, so that most of floodplain forests are spread in between the dykes and prone to regular flooding. Mean annual precipitation at the close rain-gauge stations falls between 570 and 650 mm (Hošek and Winkler 1993).

23.2.3 Anastomosed Channel Pattern as the Phenomenon of the Nature/Landscape Protection

The Morava River and its tributaries form a 4.2 km wide floodplain, where a richness of aquatic and semi-terrestrial habitats may be found. The axis of the area is the main channel of the Morava surrounded by the floodplain forest (Figs. 23.2 and 23.3). The main channel is accompanied by numerous oxbow lakes, network of side channels, periodical pools, tributary channels and many manmade water bodies—ponds and flooded gravel pits. The area was declared as landscape protected area in 1990. The protected area covers 3–8 km wide and 27 km long strip of the landscape, comprising mainly the floodplain, but partly also the higher land of the Třesín Threshold, where the Mladeč Karst is developed in Devonian limestones. The primary object of protection is a mosaic of aquatic and semi-terrestrial habitats in a lowland floodplain—bare sand and gravel bars with nesting wadable birds, periodical pools with rare crustaceans, wetland meadows, floodplain forest, but also water filled gravel pits inhabited by water birds.

23.3 Historical Changes of Anastomosed Channel Pattern Witnessed by Old Maps

The most recent period of channel network development may be traced on old maps that date back almost 200 years. Quite a lot of old maps are available for the Litovelské Pomoraví region, although with uneven coverage of the whole area. The richness of map sources is, among others, caused by the fact that large areas of floodplain forest were a property of Lichtenstein aristocratic family. The Lichtensteins let compile the maps of their forest properties (forest stands and management maps); frequently in detailed scales around 1:10,000. Some of those maps depict the anastomosed channel pattern with unprecedented planimetric accuracy. The first detailed map of a small area of floodplain forest comes from 1829. The aerial images are available from the 1930s. Thus, in some places the changes of channel pattern may be traced as long as 180 years back into the past.

23.3.1 Transformation (Degradation) of Anastomosed Channel Pattern

The most marked feature of Litovelské Pomoraví is a dense network of mutually interconnected, partly meandering channels. Kirchner and Ivan (1999) first identified this channel pattern as an anastomosing one. Branching of the Morava is depicted with variable degree of reliability already on the oldest maps of Moravia. Maps by P. Fabricius (1569; scale 1:288,000) and J.A. Komenský (1624; scale 1:520,000) show the Morava River mostly with a single channel, only in the surroundings of towns Litovel and Olomouc branching to two arms is shown. Among the older maps, the one by J.C. Müller (1716; scale 1:180,000) brings the most valuable information, showing branching of the Morava to three channels that may be identified with three

Fig. 23.2 Aerial view of the Morava River floodplain in the national nature reserve Ramena řeky Moravy (Branches of the Morava River) (*Photo* P. Holub)

recent river arms—Zámecká Morava (the Castle Morava), Městská Morava (the Town Morava) and Malá voda (the Little Water). Müller's map shows the northernmost Zámecká Morava as a main channel, albeit it is an inactive, ephemeral channel today.

Large scale maps dating back to the second half of eighteenth and the whole nineteenth century show progressive decline of the complicated system of river branches and concentration of flow in a single channel, specifically to the former Městská Morava. The acceleration of lateral channel migration and development of meanders is characteristic for the Městská Morava branch in that period. Fully functional anastomosed system changed gradually into a system with one dominant channel, dynamically meandering, which is accompanied with several smaller side channels with permanent flow. The lateral activity of the side channels was inhibited due to flow regulation by weirs. The remaining part of an anastomosed network became mostly disconnected from the dominant channel, only with ephemeral flows during floods and prevailing aggradation. Hydraulic connection of many side channels with the Městská Morava branch was substantially weakened, because the dominant channel has excessively incised and inlets of smaller river branches remained hanging above the level of the mean water in the main channel.

Societal economical activities are a suspected reason of metamorphosis of the channel pattern. Larger river branches were used as a source of energy for water mills since the thirteenth century—an activity that was accompanied with armouring of their banks in some channel reaches. Truly profound influence was rafting of wood that was carried out from sixteenth to eighteenth century. The floating of wood rafts necessitated the maintenance of the channel to make it navigable, removing the in-channel large woody debris at first place. Only a single river arm was probably chosen for this purpose, where the least number of weirs and mills was located. At the beginning of the nineteenth century the era of systematic river regulations began. First, river banks stabilisation by simple wooden constructions was carried out, followed by more demanding regulations including relocation of river channel (channel straightening, enlarging channel capacity, bank armouring) in order to protect forestry and agricultural production from flooding (Máčka et al. 2000, Fig. 23.4). In the present time, ca. 20 % of the main

Fig. 23.3 Morphologically diverse meandering channel of the Morava River with erosional banks, point-bars and in-channel woody debris surrounded by floodplain forest (*Photo* O. Žerníčková)

Morava channel in Litovelské Pomoraví is engineered in some way.

An activity that probably triggered degradation of the anastomosed channel pattern was the massive extraction of gravel and sand directly from the river bed. Mining of river gravels has expanded since the nineteenth century, when the noble family of Lichtensteins supplied gravel for the construction of embankments of the Olomouc–Prague railway. Originally localised mining was gradually replaced by dredging of the whole river reaches. The effect of "hungry water" (the flow lacking bedload for transport) that has arisen triggered river bed incision and accelerated bank erosion. River incision was further enhanced by channelization works in the years 1909–1940 (Kirchner et al. 1999).

In contrast to the dominant Morava channel, the plan form of the smaller as well as larger side river branches has remained basically unchanged. For example, the large southern branch called Malá voda (the Little Water), that branches off the dominant channel at the Řimice weir (in the wind gap section at the Třesín Hill), has retained exactly the same plan form of its meanders since 1834 and roughly the same outline is already visible on the map from 1774 (Fig. 23.5). The conservation of the plan form of side channels due to cessation of lateral erosion may be attributed to flow regulation by weirs that were built to supply the side channels with stable discharges necessary for driving water mills. Manipulation with water inflow substantially reduced the flow extremes at side channels and the rates of bank erosion (Kirchner et al. 1999).

23.3.2 The Kenický Meander—The Life Cycle of One Hydrological Node

In some cases it was possible to document the detail changes of the channel network in small areas. One of these is the area of the so-called Kenický meander (Gross Kenitzky, i.e. forest in ownership of the Kinský family) (Fig. 23.4), for

Fig. 23.4 Channel development of the Kenický meander hydrological node in the nineteenth century on old maps. **a** management map of forest district Štěpánov and Střeň from 1829 (original scale 1:5,200); **b** cadastral map of the Střeň settlement from 1870 showing the artificial shortening of the Morava channel (original scale 1:2,880); **c** management map of forest district Střeň from 1878 (original scale 1:11,520); **d** management map of forest district Střeň from 1894 (original scale 1:7,200) (Map collection of the provincial archive in Opava, branch Olomouc)

which old maps as well as present-day geodetic measurements are available. Thus, it was possible to track the development of the meander from the formation of the meander loop until the recent cut-off. Old maps from the break of eighteenth and nineteenth century show the water courses in this area as more straight than today; acceleration of lateral channel shifts may be traced on old maps from the nineteenth century. Several river branches connect with the dominant channel in this place, the position and discharge of which have changed during the time; thus, hydrological node (multiple bifurcation) with interesting historical development has arisen here.

In the first half of the nineteenth century the largest discharge flowed through the richly meandering Štěpánovská smuha channel. However, it has lost its dominance since that time, and today its remnants may be traced in the meadows eastward from the Kenický meander. The present-day dominant channel did not exist or was so small that it was not depicted on contemporaneous maps. The Štěpánovská smuha begin to lose discharge some 140 years ago and the development of the present-day dominant channel commenced further to the south. The abandoning of the Štěpánovská smuha was likely quite sudden as suggested by the plan form of its meanders which is exactly the same as on

Fig. 23.5 Cadastral map from the year 1834 showing the Římice weir close to Mladeč settlement and well developed meanders of the side channel called the Little water. The plan form of meanders has remained unchanged until the present (Map collection of the provincial archive in Opava, branch Olomouc)

maps from the first half of the nineteenth century. The changes in discharge of the Štěpánovská smuha were probably influenced by river regulations upstream that are visible on the cadastral map from 1870. There, three meanders were artificially cut-off that inevitably affected the lateral development of the channel downstream. The basic outline of the Kenický meander arose in the place of the former bifurcation of an old and new channel in 1890s. The radius of the meander was growing until 1980s when it reached its maximum. Later the development of meander's plan form continued only by narrowing of its neck.

The development of the Kenický meander has been geodetically surveyed since the year 2000. The channel width in the time of first measurement varied between 14 and 16 m, but locally the channel was widened up to 21 m due to the presence of large wood jam in the inlet arm of the meander. The length of the meander loop was almost 200 m, the radius was 60 m and the width of the meander neck only

8 m (Máčka et al. 2000). Resurveying of the meander plan form in 2006 showed that the meander neck narrowed to only 2.05 m (Máčka and Krejčí 2006). The Kenický meander natural cut-off occurred during spring flood in 2012 when the extremely narrow (ca. 0.5 m) neck collapsed. After the cut-off, a new channel with the width of 18 m arose. The life cycle of the Kenický meander was therefore closed, after ca. 150 years of its existence.

23.4 The Secret World of River Branches and Pools

The anastomosed network of side channels has been preserved in three forms on the territory of Litovelské Pomoraví (Fig. 23.6). River branches with permanent flow, fed by water from weirs built on the dominant Morava channel, represent the first type. Water levels in these permanent river branches fluctuate little and their morphology is rather stable (lateral stability). The second type is represented by channels with dynamic morphological changes, with the bottoms moulded into riffle and pool associations, locally with intensive bank erosion, however, with prevailing planform stability. The third type is represented by narrow channels, often discontinuous and blindly ending, where the dominant process is deposition of fine grained mineral and organic sediments. Sedimentary fills of these river arms are of variable thickness and grain size—clays, silts, silty sands and gravels—locally affected by gleyic process. A common fill of these narrow channels is black-brown to black loam with high organic content and frequent occurrence of large plant remnants (e.g. hazelnuts) (Máčka et al. 2000). The onset of inactivation of these small anastomosed channels in the nineteenth century is witnessed by the presence of Indian balsam (*Impatiens glandulifera*) pollen at the base of their sedimentary infills (Břízová 2002). This invasion species

Fig. 23.6 Examples of various types of side channels of the Morava River. **a** side channel with a permanent flow; **b** inactive (degrading) ephemeral channel with the periodical pool and the fill of fine grained mineral and organic sediments; **c**, **d** active ephemeral channel with gravel bed, shown during spring flood (**c**) and dry later in summer (**d**) (*Photo* L. Krejčí (**a**, **c**, **d**) and O. Žerníčková (**b**))

was introduced in Europe as decorative and honey-bearing plant as late as in 1840s.

A characteristic feature of the floodplain forest surrounding the Morava River are periodical pools. Periodical pools arise during spring floods that fill ephemeral anastomosed channels with water or even inundate the whole floodplain surface. After withdrawal of flood waters, the depressions in the floodplain surface remain filled with water that gradually dries out until summer. The periodic pools may be classified into two types, according to the hydraulic connectivity with shallow ground waters. The first type arises in the deeper ephemeral river branches with gravelly bed, where riffle and pool sequences are developed. These permanent pools may be filled with groundwater infiltrating through permeable sandy gravels underlying the floodplain surface during high-water stages in the Morava River. The second type is represented by pools located in shallow remnants of anastomosed channels, often disconnected from the active anastomosed network, which are filled only during large floods that inundate the whole surface of the alluvial plain. The colour of water in long-lived pools is often painted dark by organic substances deposited in water (mainly leaf litter). Periodical pools are known as a habitat of rare crustaceans, fairy shrimp (*Eubranchipus grubel*) and tadpole shrimp (*Lepidurus apus*).

The common phenomenon of the dominant Morava channel as well as of side river branches is the presence of large woody debris recruited from the surrounding floodplain forest. Available data suggests that the quantity of large woody debris (with minimal diameter 10 cm and length 1 m) varies between 5 and 67 pieces per 100 m of channel length. Prevailing mechanism of large wood input is bank erosion, thus, the largest quantities of large wood may be either found on the outer banks of meanders or they are transported to mid-channel positions, forming large wood accumulations (Fig. 23.7). Along smaller side river branches that are more laterally stable wood is supplied by wind throws, natural mortality (diseases, parasites, ageing), and floating from the dominant channel; bank erosion is less important. Litovelské Pomoraví is one of the last regions in the Czech Republic that is characterised by a high number of aquatic habitats linked to the in-stream woody debris. The formation of mid-channel gravel bars around large fallen trees, steps behind trees that transversally dam the channel, and widening of channels where bank erosion is enhanced due to acceleration of flow around mid-channel wood jams are noticeable examples. Blockage of flow by wood jams is one of factors conditioning faster spilling of flood waters from channels and the probable mechanism of channel avulsions (branching) in the past.

23.5 Floodplain Forest and Its Connection with the Fluvial System

The courses of the Morava River and its side channels are inseparably connected with the floodplain forest. Litovelské Pomoraví is a region where one of the last extensive complexes of the relatively preserved ecosystem of floodplain forest with the area of 2,435 ha may be found. A continuous complex of floodplain forest spreading along the Morava River between the towns of Mohelnice and Olomouc, surrounded by the fertile agricultural land of the Haná region, has been under strong human influence from the medieval times. The marginal, less frequently flooded parts of the forest were logged in the period between the twelfth and

Fig. 23.7 An example of large woody debris accumulation in the dominant channel of the Morava River originating from the erosion of floodplain covered by forest (*Photo* L. Krejčí)

fifteenth century, and initially continuous forest became fragmented. Relatively large stands of floodplain forest have remained preserved until recently mainly due to unmanageable flooding or thanks to preservation as game hunting territories (AOPK 2008).

Vegetation of the floodplain forest is dependent on geomorphological and hydrological regime of the anastomosed channel pattern that governs the height of the ground water level and the frequency of flooding (flood pulses). Side channels function as an effective distributional network for flood waters that spill out from the channels over the alluvial plain. Places that are regularly flooded for prolonged periods are overgrown with willow, ash, poplar and alder (so-called soft-wood floodplain forest). Places that are flooded only occasionally and for shorter periods, are covered with oak, lime, maple and elm (so-called hard-wood floodplain forest), with dense shrub understorey. Many transitional types of floodplain forest are present in which the species of soft-wood and hard-wood forest are mixed (Správa CHKO ČR 1998).

The beginning of systematic forest management in the floodplain forest may be traced back to the half of the eighteenth century. Until that time, forests were used only for so-called "wandering logging" (cutting only of individual trees as opposite to logging large clearings) or as an important source of acorns for feeding livestock. Regularly felled managed oak forest is first reported at the end of eighteenth century. Archival evidence suggests that large cleared plots were absent and the forest was regularly flooded in its entirety (AOPK 2008). Forest management maps were compiled for the forest plots of the Lichtenstein noble family dominion since nineteenth century; the maps are also valuable source of information about the state of the channel network of this period (Fig. 23.4).

The alteration of vegetation types (aspects) with seasons of the year is characteristic for floodplain forests. Enough sun radiation penetrates to the ground early in spring when trees are still lacking leaves. The sufficient amount of sunlight is utilised by species of spring vegetation aspect for blooming (*Galanthus nivalis*, *Leucojum vernum*) creating virtually the "white carpet" on the ground (Fig. 23.8). As soon as the sunlight is reduced by foliated trees the period of summer vegetation aspect comes in, with dominant *Allium ursinum*, *Urtica dioica* and lianas of *Clematis* and wild hops. Uncountable number of insects and other invertebrates living in the floodplain forest forms a food base for numerous bird

Fig. 23.8 Spring vegetation aspect of the floodplain forest with the dense undergrowth of snowdrops (*Galanthus nivalis*) (*Photo* O. Žerníčková)

species. Unique members of bird fauna are *Charadrius dubius* and *Actitis hypoleucos*, both nesting on the bare surfaces of gravelly channel bars.

23.6 The Morava River as a Mean of Subsistence and Relaxation—Making Use of River in the Past and Today

Most economic activities that took place in the territory of Litovelské Pomoraví from the early historical times were linked in some way with the interventions into the hydrological regime of the fluvial system. Anastomosed branches of the Morava were used for the propulsion of mills, fulling machines and lumber mills. Founding of water mills is evidenced in this region by archival materials from the thirteenth century, in connection with the development of agriculture and crafts in medieval Přemyslids royal dynasty medieval state (Semotanová 1998).

Founding of water constructions was connected with the use of water energy, among them weirs and mill drives were most important. Many weirs in Litovelské Pomoraví have already existed for more than 500 years; these old weirs often form important hydrological bifurcations in the channel network (Fig. 23.5). Water mills, which ceased their operation during the twentieth century, were in many cases rebuilt into small water power stations that are the continuation of an ancient tradition of water energy utilisation.

The River Morava was also used as a waterway for transport of various commodities, among others for floating of wood from the mountains in the north to the regions further to the south. For example, the written report from the year 1542 mentions that the Moravian Council ordered the local millers not to hinder the floating of wood rafts down the river. Since the sixteenth century, ambitious plans of making the Morava River navigable had repeatedly appeared. However, none of these plans and projects was implemented (Horák 1911).

Even the harmonic landscape of the Litovelské Pomoraví region was not saved from the large scale extraction of fluvial gravels from the bottom of opencast pits filled with groundwater that substituted the small-scale mining directly from the river channel. Farther upstream three extensive mining lakes arose in the floodplain between the settlements of Mohelnice and Moravičany, still active is a mining pit at the settlement of Náklo (Fig. 23.9). Nevertheless, opencast mining is not necessarily only a devastating activity for the fluvial landscape. An example is the Chomoutov Lake that is an extensive flooded mining pit with several islands and

Fig. 23.9 Gravel extraction by dredging from the bottom of mining pit filled with ground water near the settlement of Náklo (*Photo* O. Žerníčková)

densely vegetated littoral zone that is an important nesting site and migration stopping of water birds.

Flat alluvial landscape surrounding the lowland river, so monotonous at the first glance, has surprisingly considerable potential for tourism. The whole region is well suited for cycling thanks to low vertical differences and the dense network of asphalted pathways running through the floodplain forest. Almost intact meandering river is tempting people for canoeing. However, the number of boats passing the river is regulated due to the conservational status of the area. The river also offers high diversity of freshwater fishes (ca. 35 species), thus, not surprisingly much popular leisure activity is also angling.

23.7 Important Historical Localities Linked with the Morava River

The Morava River with its surrounding landscape has attracted humans since prehistoric times. The limestone karst of the Třesín Hill with caves, rising above the Morava floodplain, was inhabited by Paleolithic hunters some 30,000 years ago. In the middle ages, the river provided energy for water mills along its branches, whereas the richness of fishes, good harvest from soils developed on flood-deposited loam and wind-blown loess, and hospitable landscape in river's surrounding tempted to found towns.

23.7.1 Mladeč Cave—Karst Phenomenon and Its Prehistoric Settlement

The Mladeč Karst, located west of the town of Litovel, is developed in three isolated islands of Devonian limestone at Třesín Hill (345 m a.s.l.) and Skalka Hill (335 m a.s.l.), where tributaries of the Morava River created several caves. The largest one is the system of the Mladečská Cave; the cave originated as sub-horizontal phreatic cavern eroded by an underground stream of the Hradečka brook. The development of three-storey labyrinth of parallel as well as transverse corridors oriented in three main directions was influenced by the course of folded strata and faults; the cave system comprises 1,250 m of corridors with the vertical span of 30 m. The corridors, often blocked with roof-falls, bear remarkable modelling of walls and roofs and rich dripstone ornamentation.

The cave belongs among archaeological sites of the world importance. It is one of the oldest known settlements of anatomically modern humans in Europe, who already dwelled in the cave before 31,000 years. Remnants of modern humans comprise several skulls and numerous skeleton remains of men, women and children, who demonstrably practised cannibalism. They used tools made of stone and bone, weapons and decorations of the Aurignacian Paleolithic culture. A large number of bones of presently extinct vertebrate species was excavated here as well —mammoth, horse, cave bear, cave lion, hyena, bison, bovine, deer and reindeer. Most numerous are findings from the Dome of Dead, where original prehistoric fire places are preserved. The Mladeč Cave is supposed to be the oldest proved settlement of modern humans in the Central Europe, who migrated to the north to the regions still inhabited by Neanderthal humans (Wild et al. 2005).

23.7.2 Příkazy—An Open-Air Museum of Vernacular Architecture of the Haná Region

Litovelské Pomoraví lies within the distinctive Haná ethnographic region that spreads out in the middle part of the Morava River drainage basin. Haná is an important agricultural region with soils known for their fertility, characteristic dialect, folk costumes and customs. In the settlement of Příkazy, at the edge of the Morava floodplain, an open-air museum of vernacular architecture may be found with unique examples of clay buildings from the Haná region (Fig. 23.10). Represented mainly by four barns marked by window flannings from the first half of the nineteenth century and a complete Haná farm with its residential and workplace sections from 1875, it is accompanied with gardens planted with old fruit trees. The heritage buildings are equipped with rustic furniture of period 1799–1950. Traditional days of clay building or shows of traditional harvesting and threshing attract visitors who are encouraged to try out these previously common labours of the traditional countrymen.

23.7.3 The King's Town of Litovel—Venice of the Haná Region

Litovel is a historical town in the heart of the Litovelské Pomoraví region, placed on the seven river branches of the Morava, that was found by the Czech king Přemysl Otakar II in the thirteenth century. One of the river arms called Nečíz flows just underneath the city hall tower and through the town square. That is why the town is also called the Venice of Haná. Among other historical sights, the St. John′s stone bridge with the statute of St. John of Nepomuk is an outstanding example (Fig. 23.11). It is the third oldest bridge still in existence in the Czech Republic that has survived all the floods since the construction in 1592; the bridge is supported by six pillars and its length is almost 60 m. The

Fig. 23.10 Examples of vernacular architecture from the Haná region: clay barns (**a**) and roofed entrance to the residential house (so called žudr) (**b**) (*Photo* P. Holub)

Fig. 23.11 The bridge of St. John in the town of Litovel, the third oldest stone bridge in the Czech Republic (*Photo* O. Žerníčková)

town of Litovel was severely damaged during the "flood of the century" in July 1997 when its entire area was inundated and many houses had to be pulled down. Another natural hazard, rare in European conditions, was a tornado that struck the town on 9 June 2004 (third degree according to Fujita tornado intensity scale).

23.8 Conclusions

The Litovelské Pomoraví region with the king's town of Litovel as its cultural centre, which was founded in the middle of floodplain on the branches of the Morava River, lies at the northern border of the peculiar ethnographic region of Haná. The Morava River created at its entrance to the Fore-Carpathian depressions a biologically diverse landscape of multiple river branches (anastomosed pattern), vegetated by the well-preserved floodplain forest. Harmonic cultural landscape has arisen throughout the centuries of human interaction with the fluvial landscape, now dotted with mills, villages and towns seated on the river branches.

Nowhere else in the Czech Republic so well-preserved example of an anastomosed channel pattern accompanied with natural floodplain forest may be found. The natural value of the Litovelské Pomoraví landscape lies in the high diversity of aquatic and semi-terrestrial habitats that are the result of a unique set of hydrogeomorphological processes. Despite the human impact on the channel network and degradation of many anastomosed channels, the region still retains extraordinarily well-preserved dynamics of natural fluvial processes in the Central European context.

References

AOPK (2008) Rozbory Chráněné krajinné oblasti Litovelské Pomoraví (Analyses of the landscape protected area Litovelské Pomoraví). AOPK ČR, Správa CHKO Litovelské Pomoraví, Litovel 114 pp

Břízová E (2002) Records of changes of vegetation in sediments of the Morava River in the protected landscape area Litovelské Pomoraví. In: Kvítková L, Musil R (eds) Proceedings of the conference 8. Kvartér 2002, Masaryk University, Brno, pp 1–2

Horák J (1911) O regulaci řeky Moravy (On regulation of the Morava River). Zprávy Spolku českých inženýrů v markrabství Moravském, 1919–10, Brno, pp 131–173

Hošek A, Winkler I (1993) Hydrologické charakteristiky CHKO Litovelské Pomoraví (Hydrological characteristics of the landscape protected area Litovelské Pomoraví). Český hydrometeorologický ústav, Ostrava 16 pp

Kirchner K, Ivan A (1999) Anastomózní říční system v CHKO Litovelské Pomoraví (Anastamosed fluvial system in the landscape protected area Litovelské Pomoraví). Zprávy o geologických výzkumech na Moravě a ve Slezsku 6:19–20

Kirchner K, Lacina J, Máčka Z, Hrádek M, Ivan A, Hofírková S, Petrová A, Krejčí M (1999) Studium a modelování antropogenního ovlivnění říční sítě v NPR Vrapač (Investigation and modelling of the anthropogenic impact on the channel network in the national nature reserve Vrapač). Research report, Institute of Geonics AS CR, branch Brno, 59 pp

Máčka Z, Kirchner K, Hofírková S, Vašátko J, Voženílek V, Rulík M, Krejčí M, Roštínský P (2000) Studium a hodnocení vývoje říční sítě v Nárordní přírodní rezevaci Ramena řeky Moravy od Hynkovského jezu po ústí Cholinky (Investigation and evaluation of the channel network development in the national nature reserve "Branches of the Morava River" from the Hynkov weir to the mouth of the Cholinka River). Research report, Institute of Geonics AS CR, branch Brno, 82 pp

Máčka Z, Krejčí L (2006) Prognóza geomorfologického vývoje řeky Moravy v úseku od jezu Hynkov po Kenickou lávku (Prognosis of geomorphological development of the Morava River channel from the Hynkov weir to the Kenický footbridge). AOPK, Brno 62 pp

Otava J, Pospíšil Z, Krčmář Z, Kristen A (1996) Stanovení limitu ekologické únosnosti vlivů těžby nerostných surovin v CHKO Litovelské Pomoraví (Definition of limits for ecological acceptability of mineral mining in the landscape protected area Litovelské Pomoraví). Geologické výzkumy na Moravě a ve Slezsku 4:126

Růžička M (1989) Pliocén Hornomoravského úvalu a Mohelnické brázdy (Pliocene of the Hornomoravský úval Basin and the Mohelnická brázda Furrow). Sborník geologických věd—Anthropozoikum 19:129–151

Semotanová E (1998) Historická geografie českých zemí (Historical geography of Czech lands). Práce Historického ústavu AV ČR, series A, vol 16, Praha, 293 pp

Správa CHKO Litovelské Pomoraví (1998) Chráněná krajinná oblast Litovelské Pomoraví (Landscape protected area Litovelské Pomoraví). Invence, Litomyšl 27 pp

Wild EM, Teschler-Nicola M, Kutschera W, Steier P, Trinkaus E, Wanek W (2005) Direct dating of early upper palaeolithic human remains from Mladeč. Nature 435:332–335

The Nízký Jeseník—Highland with Abandoned Deep Mines

24

Jan Lenart

Abstract

The Nízký Jeseník is a large flat upland area situated in the north-eastern part of the Bohemian Massif. Despite being rather uniform geologically, its geomorphic diversity is very specific as it includes structural, depositional, erosional, volcanic, periglacial and man-made landforms. The Nízký Jeseník consists chiefly of Lower Carboniferous sedimentary flysch rocks: greywackes (sandstones and conglomerates) alternating with slate or siltstones, which were folded and thrusted during the Variscan orogeny. During the successive long-term evolution the whole massif was shaped by various processes with a long-standing geomorphic impact, including protracted denudation that led to the formation of a vast planation surface. Gradual tertiary tectonic uplift caused deep incision of rivers. Neogene–Pleistocene volcanic eruptions formed conical volcanoes, lava flows and a lava-dammed lake. Pleistocene evolution under the conditions of periglacial climate shaped asymmetric valleys–dells, whereas gullies and alluvial fans were formed in the Holocene. Finally, the area has been recently affected by human impact, as testified by diverse surface and underground mining and military landforms.

Keywords

Nízký Jeseník • Planation surface • Cenozoic volcanism • Slate mines • Military landforms

24.1 Introduction

The Nízký Jeseník is a large upland area of subdued topography, situated in the north-eastern part of the Bohemian Massif (Demek and Mackovčin 2006). Overall, it involves 2,894 km^2 and the mean altitude is 483 m (Demek and Mackovčin 2006). In the surroundings, lowlands mostly occur (Fig. 24.1). The whole upland is slightly inclined to the east and south-east. The highest point, Slunečná (800 m a.s.l.), is found in the west-central part of the area. From the point of view of geology, the area is generally composed of Paleozoic/Lower Carboniferous flysch rocks. Although the geological constitution is quite uniform, the story of the geomorphic evolution is specific in many respects. The Nízký Jeseník upland involves landforms of different origins and ages; they represent an uncommon mixture of a well-developed old planation surface dissected by deeply incised valleys that additionally contains young volcanic landforms and unique man-made geomorphic elements which have developed only recently. The aim of the chapter is to present specific geomorphology of the Nízký Jeseník area, starting with the formation of the old planation surface and finishing with a modern eventful period of the activity of man.

J. Lenart (✉)
Department of Physical Geography and Geoecology, Faculty of Science, University of Ostrava, Chittussiho 10, 710 00 Ostrava, Czech Republic
e-mail: jan.lenart@osu.cz

T. Pánek and J. Hradecký (eds.), *Landscapes and Landforms of the Czech Republic*, World Geomorphological Landscapes, DOI 10.1007/978-3-319-27537-6_24

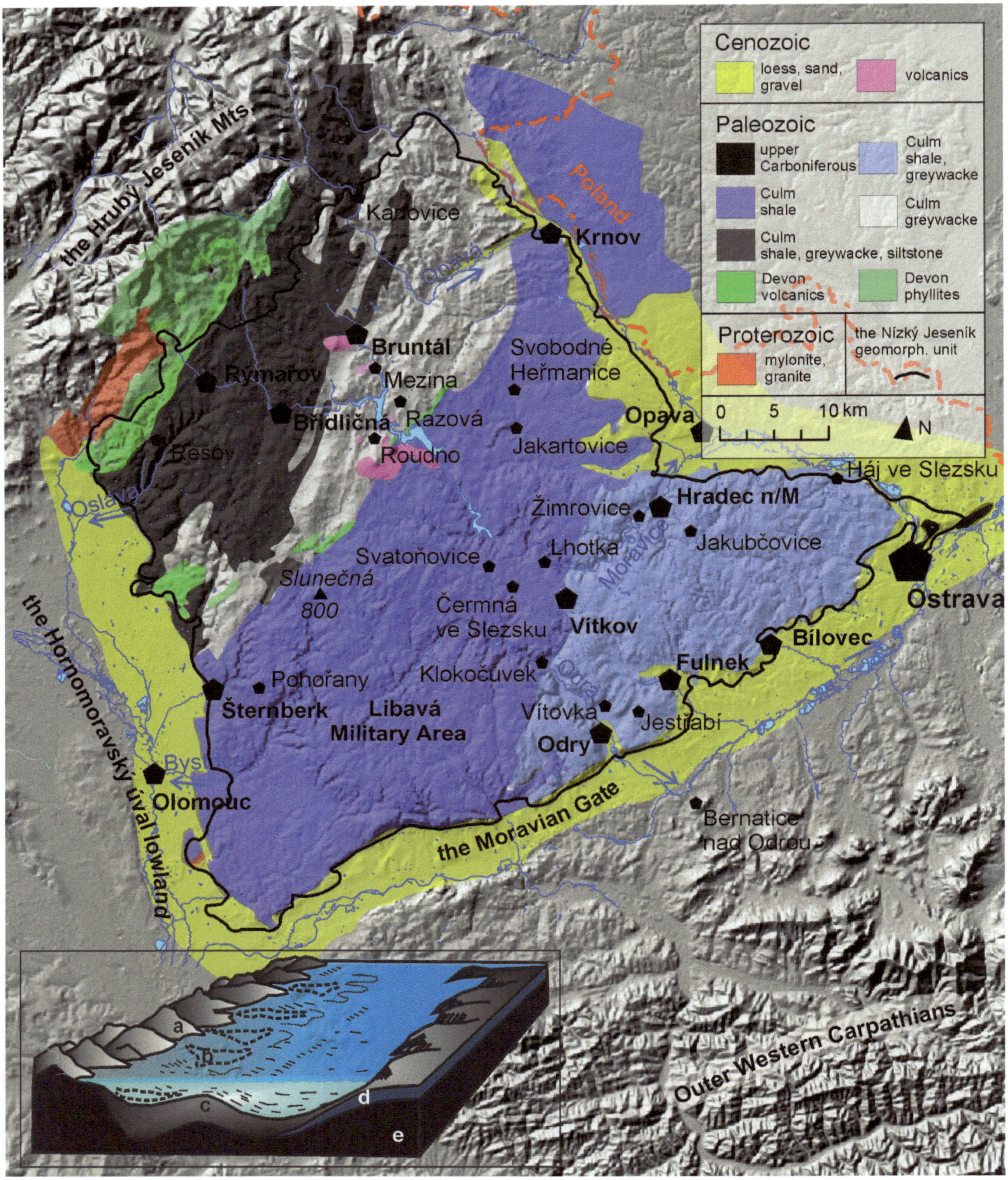

Fig. 24.1 Shaded relief model (SRTM) of the area with simplified geologic situation and the reconstruction of the sedimentary area of the Nízký Jeseník upland at the end of the Lower Carboniferous (after Chlupáč et al. 2011): **a** Variscan mountains of Silesicum; **b** sedimentary area; **c** slate and greywackes; **d** Devonian carbonates; **e** Brunovistulicum basement

24.2 Throughout the Geologic and Geomorphologic History

The Lower Carboniferous rocks of The Nízký Jeseník were formed in conjunction with the Variscan orogeny during which huge turbidity currents brought a large amount of sediments that were deposited over older Devonian rocks (Chlupáč et al. 2011, Fig. 24.1). The age of these rocks decreases towards the east which was caused by a gradual shift of the area of accumulation (Chlupáč et al. 2011). The flysch rocks consist of greywackes (sandstones and conglomerates) alternating with slate or siltstones. The greywackes reach the thickness of up to 100 m in the central part of the area (Chlupáč et al. 2011). The strata were distinctly folded and thrusted during the Variscan orogeny. Then, the whole area was substantially weathered and flattened until Cretaceous. The Alpine orogeny caused the tectonic disruption of whole rock complex and deep incision of the valleys. Badenian sea transgressions stored sands and clays of the Lanzendorf Formation onto the plateau. These sediments are recently preserved chiefly as infills of valleys, whereas in the culmination parts of Nízký Jeseník were eroded. The follow-up tectonic uplift connected with sea regression caused the evacuation of older valleys, but it did not reach the pre-Badenian bottoms in many places. Several newly formed valleys were deeply incised. These vertical movements probably continued until the Late Pleistocene (Czudek 1971). During different phases of the Cenozoic, several volcanoes, which belong to the youngest ones in the Czech Republic, broken through the Lower Carboniferous rocks. Quaternary glacial deposits are presented by diamict located close to the northern and directly in the eastern peripheral parts of the Nízký Jeseník. These sediments are relics of the Elsterian and Saalian ice sheets that leaned against steep marginal slopes (Czudek 1971; Nývlt et al. 2011) forming an ice-dammed lake in the Odra River basin (Nývlt et al. 2011). Recently, the landscape of the Nízký Jeseník has been shaped mainly by the sheet and gully erosion, shallow landslides, accumulation processes on alluvial fans and especially by human activities.

24.3 Planation Surface Dissected by Deeply Incised Valleys

The long period between Carboniferous and Cretaceous was characterised by deep weathering and planation (Czudek 1971). The planation culminated during the Paleogene and the weathered mantle was gradually eroded probably during the Neogene (Czudek 1971). Resulting planation surface with less than 1° slope gradient in some parts is a typical feature of the Nízký Jeseník area and it can be classified as etchplain (Czudek 1971). Mainly in the northern part of the area, the etchplain is remarkably undulated and punctuated by isolated hummocks (inselbergs) formed by more resistant and rigid greywackes that rise above the surrounding densely jointed surface by about 20–50 m. Even though the hummocks have evolved since the Tertiary, they were mainly exposed in the Pliocene as an effect of tectonic uplift together with the denudation of regolith (Czudek 1995). The evolution of etchplain continued by denudation through the Quaternary/Pleistocene (Czudek 1971). From the north-east, east and south-west the etchplain is laterally limited by up to 100 m high and steep fault-generated escarpment slopes that are very straight and follow the directions of main regional faults. During the tectonic uplift, streams substantially incised into a homogenous upland to form deep but wide valleys which cut the etchplain into particular blocks (Fig. 24.2). The tectonic movements caused the differences in elevation of particular blocks which are most pronounced in the eastern and southern part of the Nízký Jeseník. Impressive scenery contrasts are enhanced by vegetation cover: windy plateaus are used as fields and meadows, whereas steep valley slopes are densely forested. The valleys are poorly accessible, which explains why villages were historically established on the etchplain, while the valley slopes and bottoms remained rather wild and not settled. The valleys are flowed through by rivers such as Odra, Moravice, Bystřice and Oslava whose courses follow rectangular patterns, predisposed by the directions of main regional fault systems (NW–SE, SW–NE). Gradual erosion of rivers exposed the most resistant rocks, which nowadays create impressive rock cliffs with talus accumulations below (Fig. 24.3). Moreover, cross sections of dells within the upper parts of the valleys are asymmetrical, which was caused by solifluction, irregular melting of permafrost and snow accumulation and melting during the cold phases of the Pleistocene (Czudek 1971). As a result, western slopes of the dells are now steeper than the southern ones. During the Holocene, the valley slopes were dissected by gullies with alluvial fans formed at their foot.

24.4 Moravice River as an Axis of the Nízký Jeseník Upland

The most beautiful valley in the region is the deeply incised valley of the Moravice river. The river is 100 km long (Kříž 2004) and the headspring is located in the Hrubý Jeseník Mountains built of metamorphic rocks which supply gravel to the river sediments. The river continues through the northern part of the Nízký Jeseník, where it meets Cenozoic volcanic rocks and incorporates them into the sediment load. The rest of the river course intersects the sedimentary rocks of the Nízký Jeseník area, which means that the composition of the river sediments is very heterogeneous. In the central

Fig. 24.2 View from the village of Bernartice nad Odrou towards the high and steep SE marginal fault scarp of the Nízký Jeseník with the Odra River valley. (*Photo* J. Lenart)

Fig. 24.3 Slate rock cliff above the incised Odra river valley—the Orlí skála Cliff near the village of Klokočůvek. (*Photo* J. Lenart)

Fig. 24.4 Deeply-incised Moravice River valley: **a** Shaded relief of incised meanders near Lhotka Village; **b** remnants of the Vikštejn Castle; **c** Weisshuhn's Water Channel near the village of Žimrovice. (*Photo* J. Lenart)

part of the Nízký Jeseník upland, 30 and 50 km from the spring, the river continuity is interrupted by two huge river dams which have made it impossible for the sediments to go further for the past tens of years (the Kružberk dam was built between 1949 and 1955; Brosch 2005). The deluvial sediments are then the only recent sources of river bedload. For example, one of the tributary, the Melčský potok Creek, collects the sediments from stone accumulations and creates gravel bars and rapids at the point of junction. Downstream from the dams, the remaining, nearly 50 km long river reach occupies the bottom of a deep valley, characterised by incised meanders cutting through the northern part of the Nízký Jeseník area as a consequence of the Neogene uplift (Fig. 24.4a). The direction of the valley with several rectangular bends follows the direction of regional faults. Vertical differences between the valley bottom and the surrounding plateau reach nearly 200 m within a relatively short distance. Steep slopes are covered by debris coming from exposed rock walls, protruded by small rock towers or structurally predisposed rock ridges. These conditions led to the formation of old talus forests that are nowadays protected within the Natural Protected Areas of Nové Těchanovice, Valach, Kaluža and Údolí Moravice. The resulting morphology brought about by the deep incision has been used by humans many times in the history. For example, at the end of the 13th century, the Vikštejn Castle was founded on a distinct ridge above one of the steepest slopes of the Moravice River valley (Kouřil et al. 2000, Fig. 24.4b), while one of its foremost bastions was built on the rocky cliff above the Moravice River. Several incised meanders were exploited for industrial purposes in order to build cowsheds or paper mills. Some meander necks were dug through to drain the water via tunnels, a good example of which is a 3.5 km long artificial Weisshuhn's water channel from the late nineteenth century. The channel starts with a 45 m long tunnel within an incised meander neck, whose creation enabled obtaining useful superelevation. It further continues with a series of aqueducts and other smaller tunnels along steep and rocky slopes and finally enters the paper mill. The channel that is still in service is listed as a technical monument (Fig. 24.4c).

24.5 Rešov Waterfalls

After the Neogene uplift of the western part of the Nízký Jeseník plateau, the Huntava River, situated in the western part of the Nízký Jeseník, incised into the basement rocks. This incision was slowed down by a barrier of very resistant Devonian metamorphic rocks: the porphyroids. The river was forced to cut the barrier shaping a deep, 200 m long gorge with 25 m high rock walls (Janoška 2001). At the end of the gorge, where resistant porphyroids are bordering less resistant greenschists, greywackes and slate, we can observe several rocky steps with waterfalls. The last step is more than 3 m high (Fig. 24.5). Inaccessibility of the gorge was made use of to build a sentry castle in the Middle Ages (Fišera 2004). Outcrops of greenschists near the waterfalls are rich in magnetite which used to be mined in an old mine as a source of iron ore. Nowadays, a spectacular hiking trail leads tourists over the viewpoint, while the waterfalls with the surrounded rocks are protected as a national natural reserve.

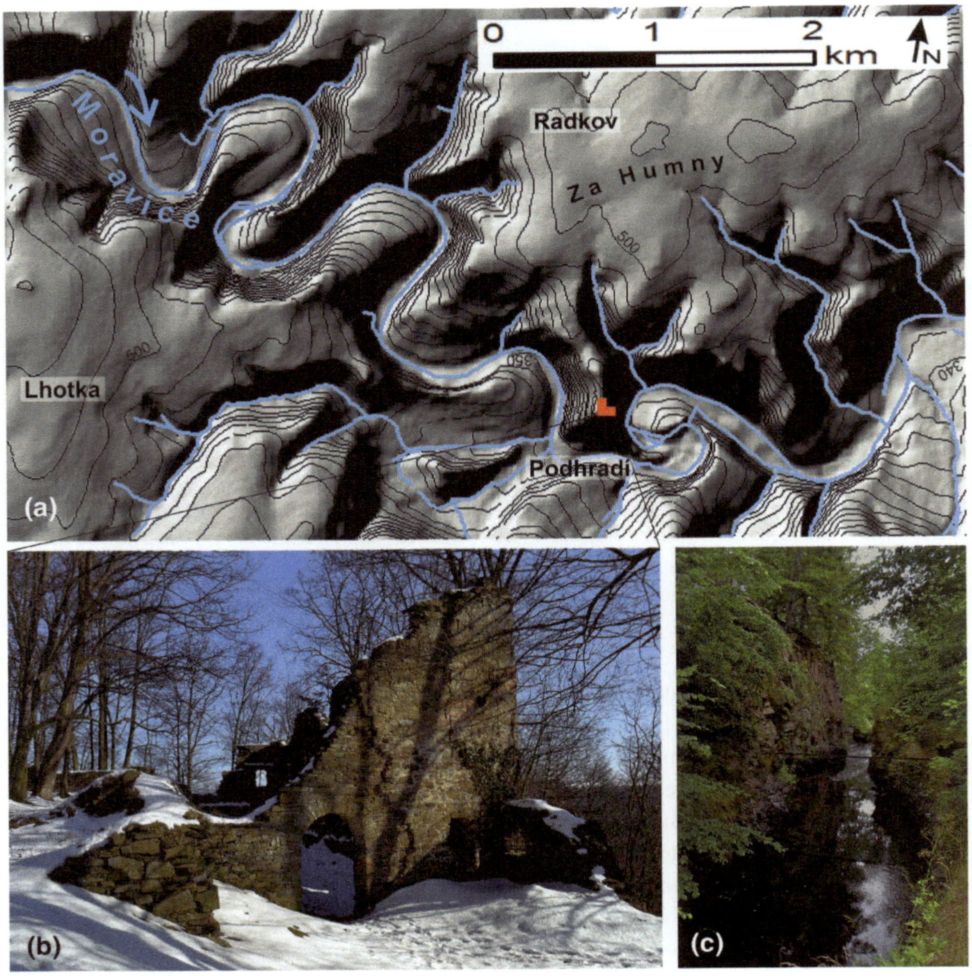

Fig. 24.5 The lowermost, more than 3 m high step of the Rešov Waterfalls. (*Photo* P. Migoń)

24.6 Relics of Cenozoic Volcanism

The relics of volcanoes, lava flows and tuffaceous sediments testify to young Cenozoic volcanism which took place in the central part of the Nízký Jeseník upland, along the SE continuation of the Marginal Sudetic Fault (Chlupáč et al. 2011). Cajz et al. (2013) studied contemporary samples of volcanic rocks from several localities in the Nízký Jeseník area. They used the K/Ar and paleomagnetic dating in order to determine the phases of volcanic activity. Strombolian-type eruptions started in the Pliocene near the town of Břidličná. The volcano formed a huge cinder cone and there were probably also lava flows leading to the valley of the Moravice River (Cajz et al. 2013). During the Early Pleistocene, a huge eruption formed the massif of the Roudný Hill. Vast lava flows of the Roudný Volcano covered a large area around. The longest lava flow extended most likely 13 km to the south, terminating near the Červená Hora Mt (Cajz et al. 2013). At that time, large lava flows filled the valley of the Moravice River and caused its damming. The river was pushed to the north where it incised into less resistant sedimentary rocks. Being the remains of this event, tuffaceous sediments are preserved on contemporary river banks. The strata contain volcanic material from the Roudný Volcano as well as the material from that time eroded lava flows of the Břidličná Volcano mixed with redeposited clasts of sedimentary and metamorphic rocks from the upper catchment. Nowadays, the tuffaceous strata are exposed within several quarries (Fig. 24.6a). Naturally dammed segment of the valley was used for the construction of the Slezská Harta Dam between the years 1987 and 1997. The natural reservoir was thus superseded by an artificial one. The original volcano was worn down during the rest of the Quaternary and divided by erosion into two separated elevations: the Velký Roudný Hill (780 m a.s.l., Fig. 24.6c) and the Malý Roudný Hill (771 m a.s.l.). The remnants of lava flows are also exposed in several quarries. Minor eruptions of at least two other volcanoes (Venušina sopka, 654 m a.s.l. and Uhlířský vrch, 672 m a.s.l.) occurred in the Early Pleistocene. Both the elevations are formed by volcanic cones with pyroclastic upper part. The eruption of Uhlířský vrch was rather calm, whereas the eruption of the Venušina sopka was quite strong, as demonstrated by the findings of sizable volcanic bombs (Ø

Fig. 24.6 The relics of Cenozoic volcanism: **a** tuffaceous strata exposed in the quarry: the Razová tuffs near the village of Razová; **b** hexagonal jointing of basalt in the old quarry: lava flow near the village of Mezina; **c** the Velký Roudný volcano near the village of Roudno. (*Photo* J. Miklín)

1 m) (Chlupáč et al. 2011; Cajz et al. 2013). One of the lava flows of the Venušina sopka is exposed in the quarry near Mezina Village (Fig. 24.6b), where pentagonal and hexagonal jointing of basalt is protected as a natural monument. The top of Uhlířský vrch is known for the baroque church and 1 km long alley, both from eighteenth century. Southern slopes are exposed by tuff quarries from nineteenth century. There are also other localities with the occurrence of older volcanic rocks such as veins (e.g. Chlupáč et al. 2011; Šešulka et al. 2012); however, these are not geomorphologically remarkable.

24.7 Anthropogenic Landforms

24.7.1 Abandoned Mines: Silent and Dark Witnesses of the Past

Due to its geological composition, the area of the Nízký Jeseník hosts more than one hundred presently abandoned mines of slate. Mining activity started in the region in the second half of the eighteenth century, while the greatest expansion occurred in the nineteenth century. Solid slate slabs were used as a building material and for roofing (e.g. the roof of the National Theatre in Prague). Besides quarries, there were numerous underground mining works. Many of them are only short exploratory adits or small mines with the overall length of up to several tens of metres. Yet, some of the mines hidden in forests are of a different character. They are large multilevel underground complexes, which are tens of metres deep and contain many kilometres of passages, beautiful galleries and vertical shafts. While some of them are still active, others have been abandoned and left dark and silent.

Miners used the Rhenish mining method within which the slate was extracted using the system of chambers with pillars between them. A typical activity was to store the slate in the form of mine fill within the chambers. Except for the handling passages, the mines were sometimes completely filled with them (Fig. 24.7a). By piling up the mine fill at the bottom of the mining chambers, these were shifted upward the massif. This has led to the creation of large underground

Fig. 24.7 Abandoned slate mines: **a** slate mine fill in the mine near the town of Odry; **b** extensive dumps in front of the old mine near the village of Svatoňovice; **c** The Atlantis in the Black Mine near the village of Čermná ve Slezsku. (*Photo* **a**, **c** J. Wagner; **b** J. Lenart)

chambers interconnected via chimney-like shafts. The mine fill also played a stabilising role underpinning sub-vertically deposited and exposed slate beds. Unusable rocks were deposited on dumps in front of the mine entrances (Fig. 24.7b).

24.7.2 The Black Mine: The Lost Atlantis

The remains of the Black Mine represent one of the largest, deepest and also the oldest underground works in the Nízký Jeseník. Mining activities were terminated before the First World War (Wagner 2010). The present entrance is situated near the village of Čermná ve Slezsku and it is created by a huge, 60 m deep vertical shaft. A large area around the entrance is covered by slate dumps with unstable slopes formed by slipping slate plates. The entrance leads into horizontal corridors with a tens of metres deep shaft which is the only connecting space with the second and third underground level. The second level is characterised by huge chambers of enormous dimensions (100 × 20 × 10 m). The corridors have a regular character with flat walls and ceilings. Sometimes the chambers are filled up with mine fill. The most spectacular part of the complex is the lowest level. The corridors are built inside voluminous slate mine fill with preserved arched passages. Because of mine excavation just under the groundwater level, the corridors are partly flooded. This part is called the Atlantis and it is one of the most beautiful places within the whole underground area of the Nízký Jeseník (Fig. 24.7c). The mine is protected as a natural monument, namely due to the occurrence of the endangered species of bats, especially *Barbastella barbastellus,* which reaches hundreds of specimens that use gaps in the slate mine fill as a hibernaculum during winter.

24.7.3 The Nittmann's Mine: Passages with an Underground Stream

The 3-storey Nittmann's Mine is located in the Moravice River valley in the central part of the Nízký Jeseník. Mining activities started in the eighteenth century and finished before the First World War. The corridors of the lower level collect underground water streams which are further joined and the resulting stream flows down narrow drainage passage to the lowermost place in the mine. There, tens of metres under the ground, the small waterfall marks the entrance to the Chamber of Lake. The follow-up passages are partly flooded creating an anthropogenic-induced underground lake. The Nittmann's Mine has a very unique microclimatic regime. In accordance with the outdoor thermal conditions, the underground microclimate reacts with a delay creating a thermal gradient with sequential chimney-like effect in the winter, when warm air rises up from the shafts. On the contrary, cold air sinks into the lower level where it interacts with the water of underground lake. These therefore freeze over every year. When the thermal gradient is very high, the cold air can freeze almost everywhere, while the ice crystals cover the walls and ceilings (Fig. 24.8a). The mine environment is very suitable for the hibernation of endangered bats: *Myotis myotis*, *Myotis emarginatus*, *Myotis nattereri*, *Myotis daubentonii*, *Myotis mystacinus*, *Rhinolophus hipposideros*, *Plecotus auritus*, *Eptesicus nilssonii* and *Barbastella barbastellus* (Fig. 24.8b). In the case of *Barbastella barbastellus*, the Nittmann's Mine represents one of five most inhabited hibernacula in the Czech Republic. The Nittmann's Mine is protected within the NATURA 2000 network. Unfortunately, there are also some negative effects and threats which make the locality very vulnerable. In the past centuries, the entrances were used as a waste material dump. To make things worse, there are occasional efforts to bury one of the shafts by backfilling material. These activities are justified in the name of the safety of the inhabitants that have built houses just alongside the shafts. However, the backfilling of the shaft may cause irreversible changes in the microclimatic regime and thus possible disappearance of the bat hibernaculum.

Fig. 24.8 The Nittmann's mine: **a** ice crystals covering the bottom of the lower passages during winter and spring; **b** colony of *Barbastella barbastellus* species. (*Photo* J. Wagner)

24.7.4 Hazards of Mines

The underground world is endangered by poor stability of vertically inclined slate layers that tend to exfoliate and leave their in situ position. This process is most accelerated within the mine's entrances due to the effect of frost weathering. A different situation is observed deep underground. In conjunction with the destruction of old slate mine fill, the exposed slate beds exfoliate and dip into mined space. Tens of years after the termination of mining activities, many places underground are affected by the collapse of huge chambers. Although the largest mines are still accessible, many shorter mines have been buried completely by downfallen rock blocks.

An extreme situation developed in the village of Čermná ve Slezsku near the town of Vítkov. The passages of the deep mine with an old vertical shaft are located under cottages that are permanently endangered by episodic collapses. Therefore, the underground passages are very unstable (Fig. 24.9a, b). Subsequent slope movements have become the main geomorphic factor involving and endangering the settled area. Up the slope, there are typical landforms connected with landslide development, namely tension cracks and trenches. Although the existence of the mine can represent a threat for people, the underground complex with one of the largest and deepest underground lakes is protected by the NATURA 2000 network.

24.7.5 Svobodné Heřmanice Quarry and Other Flooded Quarries

In the northern part of the Nízký Jeseník, near the village of Svobodné Heřmanice, an old opencast mine called Šífr represents a well-preserved relic of surface slate quarrying. In the present time the mine has respectable planform dimensions (500 × 70 m). Moreover, the whole quarry is completely flooded and the depth of the man-induced lake reaches 36 m. It is one of the deepest flooded quarries in the Czech Republic. Nowadays, the area is profusely used by divers. Mining activities were terminated during the 20th century. There are also other flooded quarries nicely framed by the landscape, e.g. west from the Jakartovice Village (Fig. 24.10). Other similar anthropogenic lakes can be found near the towns of Bílovec and Břidličná and near exploratory adits of the Čermná ve Slezsku, Jestřabí, Pohořany and Vítovka villages.

24.7.6 Military Landforms in the Nízký Jeseník: From the Old History to the Present

The list of sites of geomorphic interest in the Nízký Jeseník would not be complete without mentioning of landforms connected with military activities. The landscape of the

Fig. 24.9 Collapsed rocks within the old mine near the village of Čermná ve Slezsku: **a** the passage is shifted upwards due to collapses of slate plates; **b** collapsed walls and ceilings in the same mine. (*Photo* J. Wagner)

Fig. 24.10 One of the opencast mines situated west from the Jakartovice village. (*Photo* J. Lenart)

Nízký Jeseník is marked by the legacy of past war conflicts or recent military trainings. In the northern part of the mountains, the deeply incised Moravice river flows into a lowland and leaves the steep northern fault scarp. There, in the surrounding of the historical town of Hradec nad Moravicí, the northern state border between Austrian Habsburg Monarchy and the Kingdom of Prussia had to be guarded in the late eighteenth century. During the Silesian Wars in the second half of the eighteenth century Austria had to protect the borders against Prussian attack. Several fortifications with trenches and firing positions were founded on the edges of plateaus. After the first battles, the fortifications were modernised and extended to the surrounding hills. The remnants of these fortifications can still be found on several hills such as Hanuše, Kalvárie, Nad Přiletem, Šance or Kamenný šanc that are situated near the town of Hradec nad Moravicí and the village of Jakubčovice. Large ramparts with trenches are preserved (Fig. 24.11a). Nowadays, the fortifications have become overgrown with forests, but they are accessible via hiking trails.

Between 1935 and 1938, the Czechoslovakia decided to build a border fortification against Nazi Germany. According to the model of French Maginot line, hundreds of infantry cabins were built and interconnected. The fortification system enters the northern part of the Nízký Jeseník near the village of Karlovice in the west and leaves it near the city of Opava in the east. The cabins were built of solid concrete and expertly set into the ground. Tens of years later, the cabins started to be a part of the surrounding terrain. Many of them form contemporary conspicuous landforms. Some are covered by soil on the upper side and create unobtrusive cuts in the slope. The underground concrete construction is not visible. In some places trenches have also been preserved. The largest fortresses were founded at strategically important positions. One of these is situated on the upper edge of the northern marginal fault scarp on the Padařov Hill (339 m a.s.l.), which is the northernmost promontory of this part of the Nízký Jeseník. The Smolkov artillery fortress consists of 5 concrete objects, partly covered by soil, partly embedded in the terrain. The objects are interconnected by a system of underground passages. Today, the fortress represents a peculiar example of a combination of a strictly technical building and terrain configuration. Throughout the time, the artificial fortress has become a part of the natural relief. Some of gradually eroded concrete walls resemble natural rock walls (Fig. 24.11b), while underground passages yield to karstification.

In the twentieth century, the relief of the Nízký Jeseník was shaped by other military activities. In 1946 the southern culmination part of the Nízký Jeseník called the Oderské vrchy Hills was declared to be a military area. The villages were destroyed and inhabitants forcibly displaced. Until now the area has been used as a military training ground, which has brought about extensive transformation of relief (Fig. 24.11c). In this respect, the most interesting objects are extensive rifle ranges with high sides and frontal mounds. The surface relief has also been shaped by the impact of heavy missiles. The landscape is scarred due to a number of craters, some of which are flooded. Similar features such as flooded depressions and transformation of bare hills can be observed in the training areas of heavy military vehicles. The landscape of the military area is open to public twice a year.

Fig. 24.11 Military landforms of the Nízký Jeseník: **a** remnants of Nad Přiletem Fortification near the village of Jakubčovice covered by vegetation growth; **b** Smolkov Artillery Fortress near the village of Háj ve Slezsku with spectacular concrete walls; **c** furrowed relief within the Libavá Military Area: view from the military lookout towards the shooting range. (*Photo* **a**, **b** J. Lenart; **c** T. Galia)

24.8 Conclusion

The Nízký Jeseník upland is among the most extensive geomorphic unit of the Czech Republic. Although its geological constitution is rather uniform, it includes unique landforms. The relief of the Nízký Jeseník has been shaped by many different processes. The main geomorphic factor was gradual denudation which led to the flattening of the relief. This process culminated in Neogene by formation of the etchplain. Miocene/Badenian sea transgressions stored sediments onto the plateau, whereas follow uplift caused the sea regression, gradual stripping of regolith and the incision of rivers deeply into the massif creating fluvial forms and waterfalls. The denudation continued through the Pleistocene. Late Cenozoic volcanism created typical forms in the central part of the Nízký Jeseník, including volcanic cones and radiating lava flows, later eroded in different degree. In the Pleistocene, the ice sheet leaned against the marginal slopes of the mountains. The relief was strongly affected by periglacial processes which led to the creation of asymmetric valleys–dells. The Holocene is then the period of formation of gullies, alluvial fans and human-induced landforms, such as mining and military ones. Consequently, the Nízký Jeseník represents a region with a long, complex and diverse geomorphic evolution at the eastern margin of the Bohemian Massif.

References

Brosch O (2005) Povodí Odry. Anagram, Ostrava, 323 pp (in Czech)

Cajz V, Skácelová Z, Schnabl P, Radoň M (2013) Svrchněkenozoický severomoravský vulkanismus: rekonstrukce činnosti, paleomagnetismus, geofyzikální obraz, návrh litostratigrafie. Zpr Geol Výzk v Roce 2012:20–25 (in Czech)

Chlupáč I, Brzobohatý R, Kovanda J, Stráník Z (2011) Geologická minulost České republiky, 2. opravené vydání, Academia, Praha, 436 pp (in Czech)

Czudek T (1971) Geomorfologie východní části Nízkého Jeseníku. Rozpravy ČSAV, ŘMPV, 81, 7, Academia, Praha, 80 pp (in Czech)

Czudek T (1995) Kupovitý reliéf v severní části Nízkého Jeseníku. Čas. Slez. Muz. Opava (A) 44(1):31–42 (in Czech)

Demek J, Mackovčin P (eds) (2006) Zeměpisný lexikon ČR: Hory a nížiny. AOPK ČR, Brno, 582 pp (in Czech)

Fišera Z (2004) Skalní hrady zemí koruny české. Libri, Praha, 328 pp (in Czech)

Janoška M (2001) Nízký Jeseník očima geologa. Univerzita Palackého v Olomouci, Olomouc, 64 pp (in Czech)

Kouřil P, Prix D, Wihoda M (2000) Hrady českého Slezska. Archeologický ústav AV ČR, Brno, 645 pp (in Czech)

Kříž V (2004 Moravskoslezský kraj—klimatické a hydrologické poměry. Ostravská univerzita v Ostravě, Ostrava, 43 pp (in Czech)

Nývlt D, Engel Z, Tyráček J (2011) Pleistocene Glaciations of Czechia. In: Ehlers J, Gibbard PL, Hughes PD (eds) Developments in quaternary science, vol 15. Amsterdam, The Netherlands, pp 37–46

Šešulka V, Drápalová R, Přichystal A, Všianský D (2012) Nové poznatky o neovulkanitu v Pohoři u Oder (okres Nový Jičín). Geol Výzk Mor Slez 19(1–2):48–52 (in Czech)

Wagner J (2010) Historie záchrany jednoho podzemí. Speleo 54:22–26 (in czech)

Black Land: The Mining Landscape of the Ostrava-Karviná Region

25

Monika Mulková, Petr Popelka, and Renata Popelková

Abstract

The Ostrava-Karviná Mining District is the most significant black coal mining area in the Czech Republic. The beginnings of coal extraction date back to the end of the eighteenth century while an intensive excavation took place here in the second half of the nineteenth century and the beginning of the twentieth century. Black coal excavation is taking place in the Karviná part of the region to date. Underground black coal mining significantly affected the physical landscape in the whole region. Waste heaps, submerged ground subsidence areas and tailings ponds have become the main anthropogenic landforms. Different geological conditions in the Ostrava and Karviná parts of the area led to different manifestations of mining activity in the landscape. In the Ostrava part of the district, where the thickness of coal seams was less, more extensive mining-related surface change did not occur to such an extent as it happened in the Karviná area. The Ostrava area can be described as a landscape affected by mining and industry but with maintained residential purpose. In the mining landscape of the Karviná area, original housing estates disappeared in many places to be replaced with mostly semi-natural areas of trees, scrub and/or herbaceous vegetation associations.

Keywords

Ostrava-Karviná Mining District • Anthropogenic landforms • Landscape development • Land cover • Industrialisation

25.1 Introduction

The Ostrava-Karviná Mining District (further as OKMD) represents the main black coal mining district in the Czech Republic (Fig. 25.1). In the human memory it remains recorded as a black land with winding towers, heavy industry and devastated landscape. A transition from an originally dull agricultural area with underdeveloped industry has been taking place together with an onset of the industrialisation process of the nineteenth century connected with a growing demand for black coal. Deep black coal mining which has been conducted since the end of the eighteenth century has resulted in dynamic transformations of the landscape structure. Anthropogenic landforms have changed the appearance of the original landscape. Waste heaps, inundated ground subsidence areas, tailings ponds, revitalised areas and embankments, etc., can be listed as examples of typical anthropogenic landforms. The impacts of deep black coal

M. Mulková (✉) · R. Popelková
Department of Physical Geography and Geoecology,
Faculty of Science, University of Ostrava,
Ostrava, Czech Republic
e-mail: monika.mulkova@osu.cz

R. Popelková
e-mail: renata.popelkova@osu.cz

P. Popelka
Department of History, Faculty of Arts, University of Ostrava,
Ostrava, Czech Republic
e-mail: petr.popelka@osu.cz

T. Pánek and J. Hradecký (eds.), *Landscapes and Landforms of the Czech Republic*,
World Geomorphological Landscapes, DOI 10.1007/978-3-319-27537-6_25

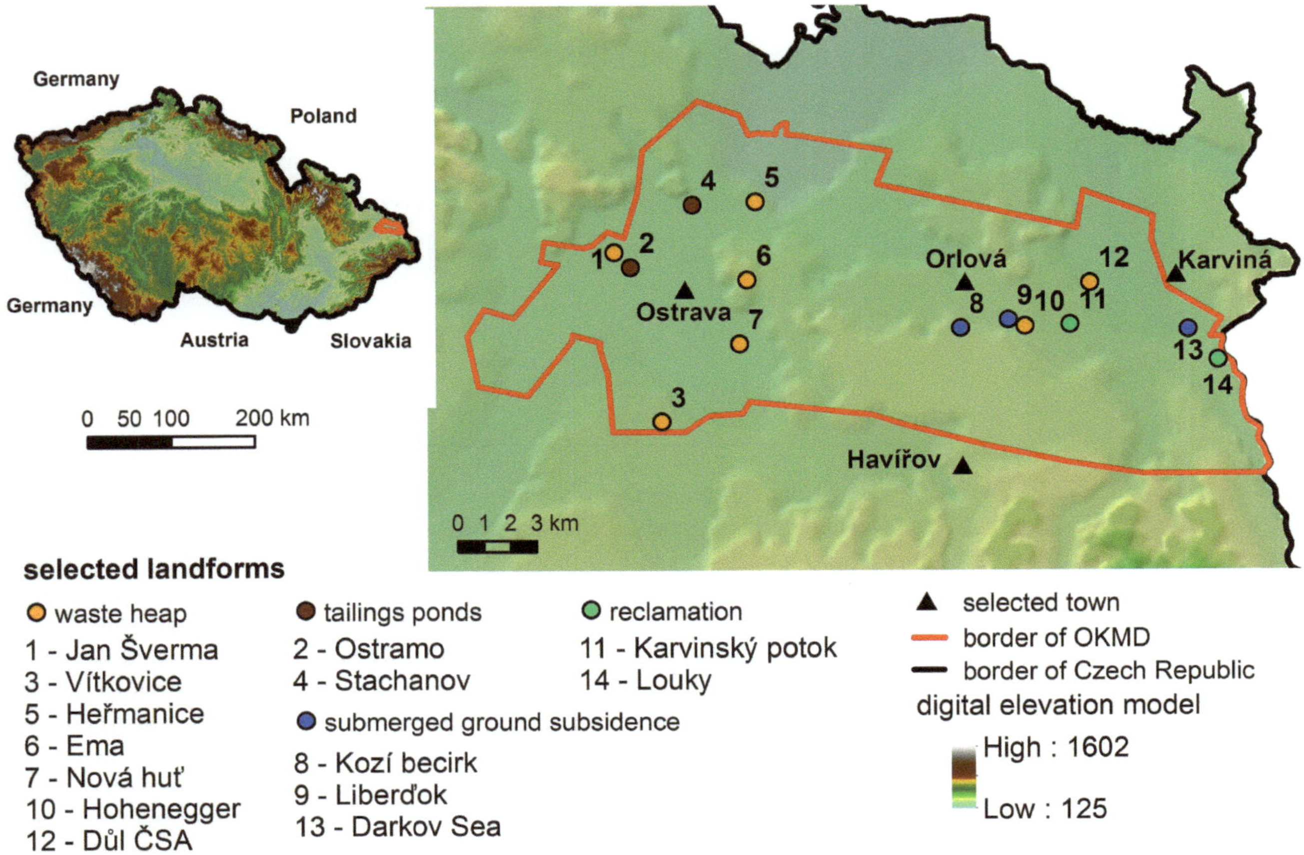

Fig. 25.1 Spatial delimitation of Ostrava-Karviná Mining District and selected land forms. *Data source* Portal of the Public Administration of the Czech Republic

mining were manifested in the landscape in an uneven way. Their biggest impact is observed in the Karviná area where the originally densely populated natural landscape has changed into a fully mining landscape in which the original residential role vanished. The coal mining activity was expressed in a different way in the western part of the district where we cannot observe such a high number of deserted and resettled areas. At the same time, the Ostrava part of the district is an industrial landscape with some areas of dense urban residential fabric. Currently, there are four active mines in the Karviná part of OKMD: ČSA, ČSM, Darkov and Lazy. Mining activity was discontinued in the Ostrava part in 1994—with the exception of the Paskov Mine. With respect to the landscape structure, the area of OKMD represents a specific type of the Czech Republic landscape in which human activity demonstrated itself in a very specific way over the last 200 years. The aim of the following chapter is to make the reader aware of a historical development within the local landscape and anthropogenic landforms, which is typical for a landscape affected by underground black coal mining.

25.2 The Basic Geological and Geomorphological Characteristics

With respect to the geomorphologic division of the Czech Republic, the area of interest belongs to the Ostrava Basin, where slightly undulated uplands and alluvial plains are typical (Bína and Demek 2012). Flat, erosion accumulation or accumulation landscape established on Quaternary sediments of various genesis (loess loams, fluvial gravels and sands) has been altered due to deep black coal mining, industry and urbanisation.

Regarding the regional geology, the Ostrava-Karviná Mining District belongs to the Bohemian Massif, particularly its Moravian—Silesian section. It is a part of the Upper Silesian Basin which enters the Czech Republic from Poland and occupies an area of approx. 1,600 km^2 (Mísař et al. 1983). The protrusion of the Upper Silesian Basin is in our territory is divided into the northern Ostrava-Karviná part and the more southerly located area at the foothills of the

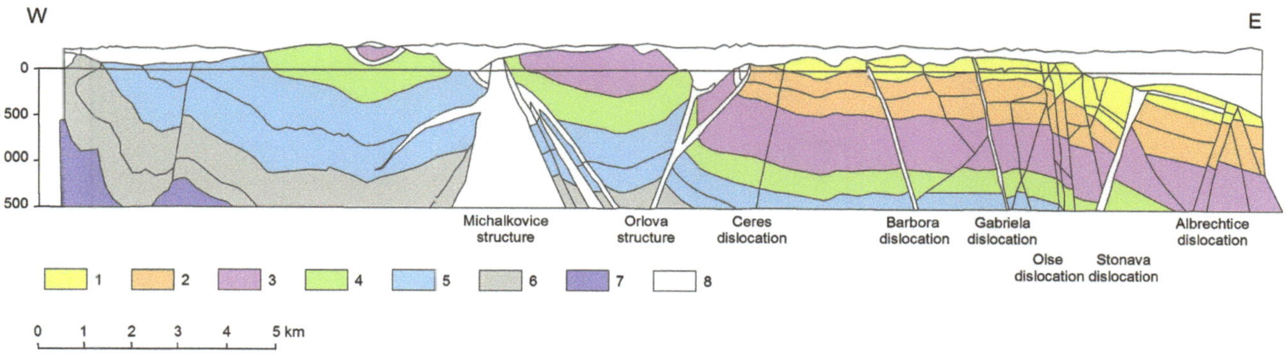

Fig. 25.2 Transversal geological section of Ostrava-Karviná Mining District (2x exaggeration, modified from Dopita et al. 1997). Karviná formation: (1) Doubrava beds and Sucha upper beds (*Westphalian A*), (2) Sucha lower beds and "sedlové" beds (*Middle* and *Upper Namurian*). Ostrava formation (*Lower Namurian*): (3) Poruba beds, (4) Jaklovec beds, (5) Hrušov beds, (6) Petřkovice beds, (7) Lower Carboniferous, (8) superficial deposits

Beskydy Mountains (Chlupáč 2002). Coal mining has not taken place in the latter area due to unfavourable local geological conditions. The bedrock of the basin consists of Brunovistulicum basement with a cover of Devonian and lower Carboniferous deposits. The basin is filled with Upper Carboniferous sediments with black coal seams. The Orlová tectonic structure divides the Ostrava-Karviná Mining District into the Ostrava part in the west and the Karviná part in the east. The Upper Carboniferous rocks are covered with Neogene deposits of the Carpathian Foredeep and with the Outer Carpathians nappes, therefore they emerge at the surface in only very limited exposures in Ostrava (Chlupáč et al. 2002).

Stratigraphically, the Upper Carboniferous of the Upper Silesian Basin is divided into the Ostrava formation which consists of Petřkovice, Hrušov, Jaklovec and Poruba beds and the Karviná formation, which comprises Sedlové, Suchá and Doubrava beds (Fig. 25.2). The Ostrava formation represents sediments of alternating continental and marine environment. With respect to the changeability of the facies, where marine, transitional and brackish facies alternate, the Ostrava formation probably belongs to the most varied sediment unit of the Bohemian Massif (Chlupáč et al. 2002). The thickness of the Ostrava formation is max. 3,200 m and a high number of coal seams of low thickness is typical. Less than ¼ of them can be excavated. In the superincumbent bed of the Ostrava formation there is the Karviná formation with the thickness over 1,000 m (Chlupáč et al. 2002). Marine deposits are missing in the Karviná formation and sedimentation took place in an endorheic basin. Coal seams are less frequent but they have a higher average thickness (180 cm). The coal bearing capacity of the Karviná formation is up to four times higher than that of the Ostrava formation. Interlayers of claystone—siltstone-sandstone type penetrate the seams (Dopita et al.1997). Dopita et al. (1997) stated that 1.419 billion tonnes of coal was excavated in the Ostrava-Karviná Mining District between 1782 and 1996. The volume of tailings was between 0.8 and 0.9 billion tonnes and the production of slurry coal was 40 million tonnes.

25.3 Agricultural Landscape and Its Transition into the Industrial Landscape

Up to the 1830–1840s the area of the OKMD represented a traditional agricultural landscape with high percentage of agricultural terrain and natural areas covered mainly by woods. When characterising the local landscape, one should not neglect the ubiquity of water courses and water bodies which created an extensive system of ponds established in previous centuries. A swift development of coal mining together with ironworks was set off from the 1840s, followed by expansion of chemical industry from the 1860s. This radical change significantly affected the local landscape and its use. The OKMD became a region with pronounced and at first glance clearly visible time and space changes in the landscape (Myška 1988; Matějček 1993; Popelka et al. 2013).

Throughout the nineteenth century considerable changes occurred both in the ways the landscape looked like and in the ways it was used. Primarily, the process of deforestation was launched. This was caused by the demand for timber from mining industry, by damage caused to woodlands by industrial exhalations, and also by demand for arable land and attempts to gain new areas where tailings heaps could be piled (Fig. 25.3). Contrary to this, arable land maintained its considerable representation in the landscape until the middle of the twentieth century. This was connected mainly with the efforts of miners' families to carry out small-scale farming. The hunger for land, together with mining activities instigated the end of the once extensive network of ponds which was—with only a few exceptions—wiped off from the

Fig. 25.3 Heaps in the centre of Moravian Ostrava in the second half of the 1930s when their liquidation started. The exhibition ground of Černá louka was founded in their original place. Archive of the City of Ostrava, collection of photographs, not categorised

OKMD landscape. The changes in farming in the second half of the nineteenth century resulted in a visible drop in the area of meadows and pastures in the landscape (Myška 1985; 1989; Mulková et al. 2012; Popelka et al. 2013).

25.4 Mining (Industrial) Landscape

25.4.1 Industrialisation and Landscape Development

The second half of thetwentieth century was characterised by accelerating changes in the OKMD landscape. It is obvious even from the first aerial photographs taken in 1949 that the process of industrialisation affected the appearance of the local landscape at that time. Compared to the beginning of the century, we can observe mainly a decrease in the area of agriculturally cultivated plots of land. The trend to suppress agricultural activity was considerably strengthened in connection with dramatic changes in agricultural production and with the preference given to mining industry in the 1950s. During a very short period of time, a total de-agrarisation of the landscape occurred. On the other hand, urbanised areas and areas affected by mining and industrial activities (Fig. 25.4) expanded noticeably. Semi-natural areas of trees, scrub and/or herbaceous vegetation associations started to appear in deserted agricultural or urbanised areas. The share of areas which were occupied by waste heaps (Fig. 25.5), inundated subsiding ground and tailings ponds increased too (Mulková et al. 2012; Popelka et al. 2013).

From the second half of the 1960s a new way how to solve environmental issues emerged, in connection with contemporaneous societal changes (Popelka 2013). The landscape—now strongly affected by industrial activity—was at least partly revitalised within a relatively short period. Thanks to reclamation and decreasing mining activity, the number of natural and semi-natural areas started to grow considerably, primarily in the Karviná part of the region. Forests together with woodland, scrub and/or herbaceous vegetation associations were now representing the prevailing element of the landscape cover. Between the 1960s and 1980s considerable relocations of residential areas took

Fig. 25.4 Industrial landscape of the central part of Ostrava. View of Vítkovice Steelworks and Karolina Coking Plant. Probably 1950s. Archive of the City of Ostrava, collection of photographs, sign. v 183-6/59

place too. On one hand, the mining activity took its toll in extinction of large inhabited areas where mining colonies were closed down; on the other hand new residential units were built.

The trend launched at the turn of the 1970s which aimed at making the landscape affected by industrialisation greener and more attractive has been continuing rather successfully to this date. The analysis of the landscape cover clearly shows differences between the Ostrava and Karviná parts of the area. The residential role of the landscape has been preserved in the Ostrava part and the total area of natural and semi-natural areas tends to be lower. Contrary to this, the Karviná landscape lost its original residential role in some areas (e.g. the towns of Orlová and Karviná). In abandoned areas there is a clearly higher total area of natural and semi-natural areas.

Due to the decrease of the mining activity in the 1990s, typical signs of deep coal mining, i.e. winding towers, bare waste heaps, tailings ponds and some water-filled subsidence troughs are quickly disappearing. The existing waste heaps were purposefully reclaimed or they have become covered with self-seeded trees, shrubs or grass (Fig. 25.6).

25.4.2 Main Mining Landforms

One can come across both primary (waste heaps, water-filled subsidence areas, tailings ponds, surface buildings belonging to mines) and secondary manifestations of deep coal mining (reclaimed areas, dry tailings ponds, communication landforms) in the OKMD mining landscape (Mulková and Popelková 2013). While the Karviná part of OKMD represents a typical mining landscape, mining, industrial and urban landscapes are interwoven in the Ostrava part of OKMD.

25.4.2.1 Waste Heaps

The most pronounced anthropogenic landforms in the mining landscape of OKMD are waste heaps. Waste heaps represent convex landforms whose area can reach from a few

Fig. 25.5 Heaps near the Ignat Mine in Ostrava-Mariánské Hory. Archive of the City of Ostrava, collection of photographs, sign. LX-10-1184

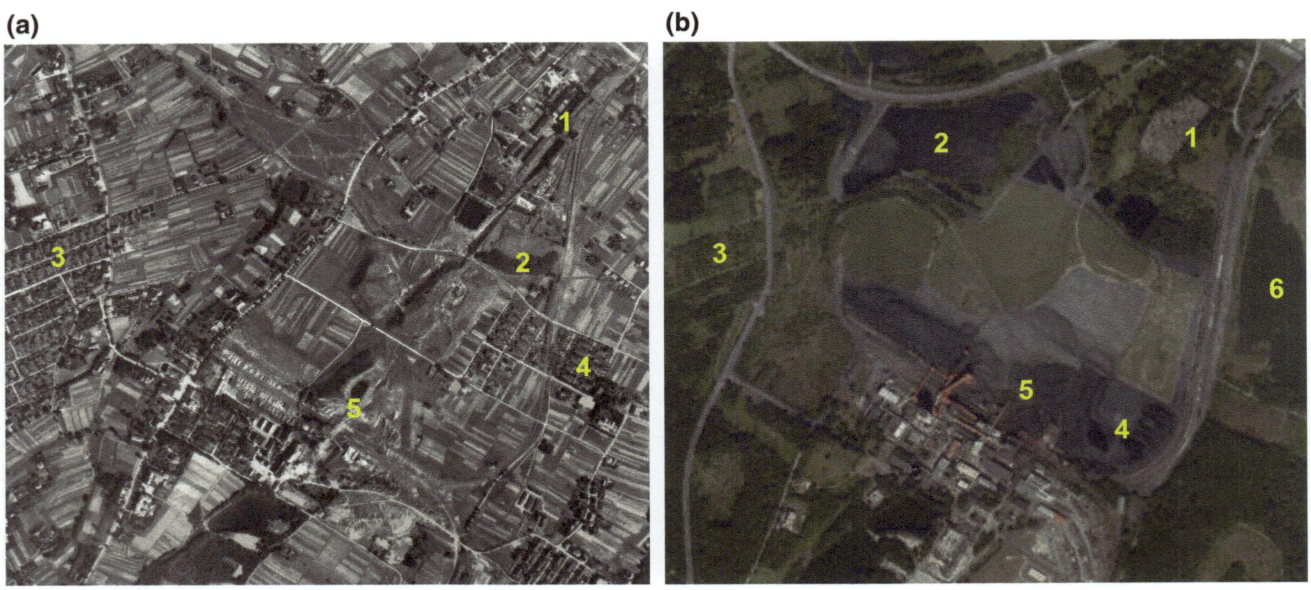

Fig. 25.6 **a** The Orlová region landscape heavily affected by industrialisation. A cut-out of an aerial photo from 1947. A body of Starý odval waste heap to the north of mine buildings (5), a waste heap in the N-E (2) and a part of a coke plant and associated land (1). Two mining colonies are seen: Liberdova colony (4) and Chobotova colony (3). The photo was provided by the Military Geography and Hydrometeorology Office in Dobruška. © MO ČR/GeoSI AČR. **b** The same area of Orlová landscape completely transformed by industrialization at the beginning of the twenty first century. A cut-out of an orthophoto from (2012) *1* area of a former coke plant, *2* final sedimentation tailings ponds, *3* area after the Chobotova colony, *4* tailings ponds, *5* Starý odval waste heap, *6* submerged ground subsidences Liberďok. Photo source: map services of ČÚZK (State Administration of Land Surveying and Cadastre)

Fig. 25.7 Heřmanice waste heap (*Photo* M. Mulkova)

ares to tens of hectares. They originate as a result of deposition of extracted coal waste during deep coal mining. The OKMD area includes the following basic types of waste heaps (Havrlant 1980): cone-shaped, waste piles, plate-shaped, and terrace-like waste heaps, flat waste piles or their combinations. Most waste heaps have been rehabilitated and covered with forest. A part of the waste heaps has been built over (communications, dams of tailings ponds). Some waste heaps are still burning inside. In later stages, the excavated matter was not stored only in waste heaps. Tailings were utilised also as a material for reclamation construction with the aim to conceal the impact of undermining.

There are a total of 50 waste heaps of different size and covering a total area of approx. 600 ha in the area of the city of Ostrava (Kozelská 2011). The largest waste heap complex is the central waste heap of the Heřmanice Mine which is still thermally active (Figs. 25.1 and 25.7). This is a plate-shaped feature with irregularly piled fills, and a tailings pond was created in its northern part (Havrlant 1980). The volume of the heap is 15,349,000 m^3 (Kozelská 2011), its relative elevation is between 26 m and 38 m (Havrlant 1980) and the area is 110.52 ha. We can also mention the complex of waste heaps next to the Jan Šverma Mine in Mariánské Hory with the volume of 6,040,000 m^3 and the waste heap complex in the vicinity of the Petr Bezruč Mine in Silesian Ostrava made from tailings and waste matter from the Trojice coking plant (Fig. 25.1). Piling of the central Jan Šverma waste heap commenced in 1930 and it was continuously topped up until 1970. The originally cone-shaped hill was remodelled into a flat-top cone. The elevation above the terrain was 21 m and the total area occupied was 52.45 ha (Havrlant 1980). The material was partly extracted for the construction of the road infrastructure. The Ema cone-shaped waste heap with a relative elevation of 80 m above the surrounding terrain and 110 m above the level of the city centre (Havrlant 1980) makes a dominant of the Petr Bezruč mine complex, with the area of 30.68 m (Kozelská, 2011). The waste heap complex started in 1920 and its filling was discontinued in 1950. The Ema waste heap, now protected as a cultural heritage, is made from burnt-through matter while it is still thermally active and releasing exhalations and water vapour.

The waste heaps for the Vítkovické železárny (Vítkovice Ironworks) and Nová huť Ironworks can serve as examples of ironworks waste heaps (Fig. 25.1). The waste heap for the Vítkovice Ironworks is located in the southern part of Ostrava not far from the settlement area. According to

Fig. 25.8 Waste heap for ČSA Mine. The tailings pond in the foreground of the photo, ČSA Mine on the right (*Photo* M. Mulkova)

Havrlant (1980), its area is 106.81 ha and it represents a plate-shaped heap with the elevation of 28 m. Slag and waste from the ironworks and rolling mills along with construction waste was piled up there (Pleva 2011). The piling of ironworks waste was discontinued after 1960 and the piling of dangerous waste (neutralising slurry and crude oil substances) in 1969. At present most of the waste heap is covered with greenery, the remaining part is used for extracting slag for construction purposes. The waste heap for Nová huť Ironworks has been used since 1956 for piling up slag from the ironworks. The area of this plate-shaped waste heap is 80 ha with the elevation of about 25 m above the surrounding terrain. The banks of the waste heap are covered with a tree-type vegetation.

In the Karviná part of OKMD, plate-shaped waste heaps prevail and their development was often morphologically changeable. Some were created from original waste piles or cone-shaped waste heaps. Their character can also be described as complex, where plate-shaped heaps, waste piles and terrace-like heaps are combined. We can list the waste heap of the ČSA Mine, which belongs to the largest waste heaps of the Karviná part, as an example of a varied waste heap with a changeable development (Figs. 25.1, 25.8). The height of the waste heap is approx. 20 m above the surrounding terrain and its area is 39 ha (Wróbel 2012). The waste heap of the former Hohenegger Mine, which was still being filled in the 1980s, can be listed as a representative of a plate-shaped waste heap (Fig. 25.1). According to Havrlant (1980), its area was 54.55 ha and it created a dominant structure with an elevation of approximately 25 m. At present the heap has been reclaimed. Its top is covered with grass and its slopes are overgrown with scrub and/or herbaceous vegetation associations.

25.4.2.2 Ground Subsidence Areas

Beside waste heaps, ground subsidence areas are other typical landform of the OKMD, although in this case they are concave. Ground subsidence originates as a result of surface subsidence above the mined-out space (Demek 1988). The hollows can be then filled with water. Surface deformations were incomparably smaller in the Ostrava part of OKMD compared to Karviná due to storage conditions, the width of seams and lithologic characteristics (Dopita et al. 1997). Kukal and Reichmann (2000) informed that in

Fig. 25.9 Contemporary view of the Church of St. Peter of Alcantara in Karviná-Doly mining area. The church subsided by 36 m. (*Photo* M. Mulkova)

the Karviná area only subsidence affected more than 5,000 ha and its magnitude can be as much as 20 m deep. In some locations subsidence can be even as much as 30 m deep (Fig. 25.9). The size of the affected area and the depth of subsidence will increase with respect to ongoing mining. Contrary to this, in the Ostrava part where seams with a low width and located rather deep under the surface were mined, inundates subsiding ground may be more extensive but more flat and shallow at the same time.

In the Ostrava part of the OKMD compared to that of Karviná, the depth of ground subsidence varies between 3 to 4 m, with which a smaller extent of mining damage and a lower density of submerged ground subsidence is related. In the past, larger submerged ground subsidence areas were located e.g. in the vicinity of the Zárubek Mine and the Silesian Ostrava Castle (Popelka 2013). Most of the inundated ground subsidence areas in the Ostrava region were filled with tailings. In 2009 their area was 8.32 ha, in comparison with 43 ha in 1964 (Kozelská 2011).

In the Karviná part of the OKMD mining of relatively shallow and wider seams soon started to manifest itself in the appearance of ground subsidence, with the affected areas often subject to inundation. Ground subsidence started to affect the stability of buildings (Fig. 25.10), which eventually resulted in the disappearance of original towns and villages (e.g. Karviná, Orlová, Doubrava, Darkov or Louky nad Olší). Related concave landforms can be found over the whole area. Their number, shape and area have altered as the mining process has continued. Some inundated pans have been filled in under the reclamation process of the area. The largest water-filled ground subsidence area is Darkovské moře (Darkov Sea), with the area of 31.4 ha (Figs. 25.1 and 25.11). The submerged ground subsidence in the northern part of the mining area in Karviná-Doly I (area of 16.3 ha), Liberďok (11.4 ha) and Kozí becirk (7 ha) in Orlová-Lazy can be listed as further examples. Altogether, in 2012 there were approximately 80 waterlogged ground subsidence areas, occupying the total area of 150 ha in the Karviná part of OKMD.

25.4.2.3 Other Anthropic Landforms

Tailings ponds are another anthropogenic concave elements of relief, connected with the following conditioning of extracted coal. The tailings pond, a natural or excavated

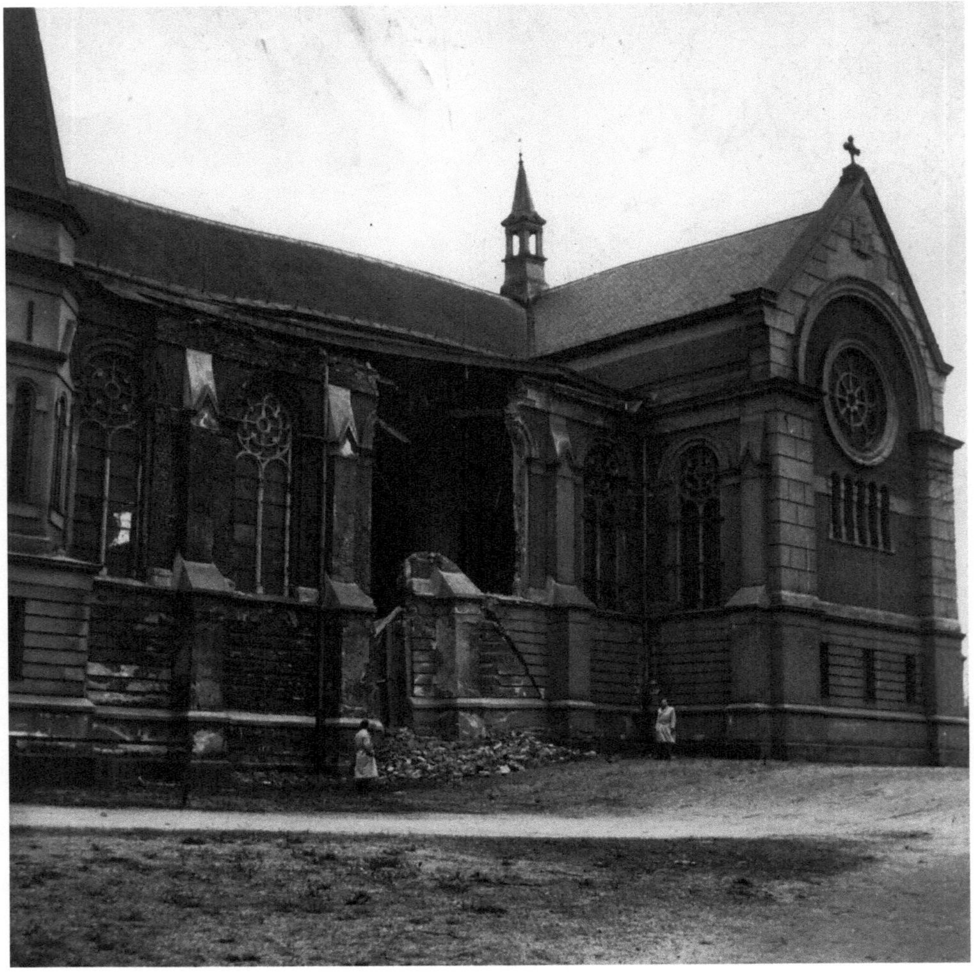

Fig. 25.10 St. Henry Church in Karviná Solca at the time of its downfall. The neo-renaissance building was constructed between 1894–1897 and it could hold up to 4,000 people. It was heavily damaged due to mining subsidence and was demolished in 1960. Archive of the City of Ostrava, collection of photographs, sign. LXXXV-5-1/5

basin, serves for permanent or temporary storing of hydraulically transported tailings (Kirchner and Smolová 2010). Final sedimentation ponds can have a character of natural water surfaces.

There are tailings ponds to store coal slurry and other waste (e.g. phenol–ammonia waste from coking process) in the landscape of Ostrava. The location of Stachanov (Fig. 25.1) in Ostrava-Přívoz with the area of 10 ha is an example of a reclaimed tailings pond (Kozelská 2011). Tailings ponds of the second type represent a risk for environment. As an example we can mention four Ostramo lagoons (Fig. 25.1), where refinery waste was stored from the end of the nineteenth century and waste from recuperation of used lubricants was stored from 1964 (Diamo 2014). Currently the lagoons are being revitalised. The total area of tailings ponds was 68.51 ha in 2009. Their area was largest in 1994, occupying 94.53 ha (Kozelská 2011).

In the Karviná part of OKMD the tailings ponds covered 384,4 ha in 2009. The majority of tailings ponds are usually reclaimed after they were filled or the tailings management discontinued. Some cleaning ponds remain preserved in the landscape.

Surface buildings belonging to mines are inseparably connected with the mining landscape and they help, together with winding towers, to create its typical appearance. Revitalisation of the area affected by undermining leads to the origin of secondary landforms, mainly connected with reclamation areas and communication. Reclamation areas concern temporary convex landforms in a shape of low flat waste heaps. These should be aligned with the surrounding landscape after the termination of reclamation works and are made of tailings. As an example of extensive reclamation area we can list the area of Louky (Fig. 25.1) where the subsided surface is being re-modelled on the area of 107 ha. 220, 000 m^3 of tailings were used for another reclamation construction near the Church of St. Peter of Alcantara (Fig. 25.9), where the centre of the already disappeared old town of Karviná used to be situated. The reclamation of the Karvinský potok valley (Fig. 25.1) covers the area of 22 ha (OKD 2010; Popelka 2013).

Fig. 25.11 The western part of the Darkov Sea submerged ground subsidence area. ČSM North Mine on the right. (*Photo* M. Mulkova)

The area of 470.52 ha, which makes 4.7 % of the total area of the Karviná part of OKMD, was reclaimed between 1970 and 1990. Between 1991 and 2011 the area amounted at 1,186.34 ha, i.e. nearly 12 % of the whole area (Priesnitz 2012). In the case of undermined area embankments are created to level surface deformations damaging communications. As a result, embankments reaching up to a few metres high are built. The types of embankments, which are characterised by linear shapes, include railway and road embankments as well as embankments of engineering networks.

25.5 Deindustrialisation of the Ostrava-Karviná Mining District and New Forms of Utilisation of the Landscape

Although a mining landscape is usually presented as a devastated one due to the impact of mining on the original cultural landscape, the reality is more complex. The presence of anthropogenic mining landforms creates favourable conditions for many rare flora and fauna species which were not present in the original landscape. Tailings ponds, which are not used any more, and also submerged ground subsidence terrains can positively affect ecological value of the landscape as they increase species diversity of the territory. Wetland fauna and flora species appear on the banks of these water bodies (Harabiš and Dolný 2012), including rare invertebrates (e.g. specially protected dragonflies, crayfishes, shells). Dolný and Ďuriš (2001) described the presence of e.g. the Scarce Chaser (*Libellula fulva*), the Turkish Narrow-clawed Crayfish (*Pontastacus leptodactylus*) and Swan Mussel (*Anodonta cygnea*) in these water bodies which were created as a result of anthropogenic activity. These are specially protected and highly endangered species. The Scarce Chaser was considered extinct species in the Czech Republic until recently.

Populations of rare thermophilic steppe beetle species, e.g. the very rare type of rove beetle *Staphylinus pedator* and the rare *Quedius curtipennis* (Dolný 2000), were created on unreclaimed waste heaps where specific microclimatic conditions typical for mining waste heaps with black tailings were preserved.

Reclamation of undermined areas is currently implemented not only due to agricultural or forestry renewal but

Fig. 25.12 Ferdiš Duša—Ostrava industrial landscape. Oil on canvass, 80 × 100 cm, without date. Gallery of fine art in Ostrava. Author of photographs BcA. Vladimír Šulc, GVUO

other ways of how to utilise the landscape occur too. Many submerged subsiding areas have not been dried out. They act as biocentres (e.g. Liberďok in the mining area of Lazy) or are reclaimed to be used for fishing or leisure time activities (e.g. Darkovské moře, Kozí becirk). An example of how mining waste heaps can be used for leisure time activities is the Dinopark founded on the waste heap in Doubrava or the planned reclamation of the industrial waste heap in Hrabůvka and its conversion into a golf course. A golf course was also created in the undermined area of Lipiny, not far from the town of Karviná. The location left after the Karolina coking plant in the centre of Ostrava is another example of a new use. After the contaminated area was revitalised and reclaimed, the multi-purpose and shopping centre of Nová Karolina was created.

25.6 The Ostrava-Karviná Mining District Landscape in Literature and Art

The transition of an agriculture landscape into one that was heavily affected by the process of industrialisation was strongly reflected in fine arts. The onset of the industrial revolution and its first scarce evidence in a landscape were portrayed by artists such as Jakob Alt (1789–1872) or Ernst Wilhelm Knippel (1811–1900) as early as in the middle of the nineteenth century. During this period, industrial expansion was portrayed in an idyllic and romantic way, where the industrial motive did not disturb the depicted countryside but in a way complemented it and made it special. Profound changes in the portrayal of the Ostrava region landscape can be observed in fine arts during the period after the World War I, when the authentic Ostrava art was born. Artists depicted the landscape in varied ways but they always referred to characteristic phenomena of industrial development. Artists' attention was focused primarily at mining activity and everything connected to it. Therefore, the Ostrava region was presented as an area of industrial locations, mining heaps, and also typical peripheries with mining colonies (Fig. 25.12). Several high quality local artists such as Vladimír Kristin (1894–1970), Bohumír Dvorský (1902–1976), Jan Sládek (1906–1982) or Josef Dvořák (1898–1942) dealt with this theme in their lifelong work. Some works of art reflect the transition of the local landscape in a particularly persuasive way and the landscape of the Ostrava region was shown in an almost apocalyptical way. This is particularly valid for unique and imaginative depicting of the Ostrava region landscape with heaps by Jan Zrzavý (1890–1977) (Fig. 25.13). A different portrayal of Ostrava can be seen in a unique painting of Ostrava landscape in the work of famous Oskar Kokoschka. His expressive painting called Moravská Ostrava from 1937 shows a contrast between a town hidden in smoke and remnants of natural countryside of the meandering Odra River. The region of the OKMD as an area strongly affected by human activity remains to be a popular theme for fine artists over the second half of the twentieth century (Holý 2003; Ivánek and Smolka 2013).

Czech literature, which has dealt with the industrially expanding Ostrava region from the beginning of the twentieth century, generally presents the local countryside as an opposite to a natural countryside. Agricultural rural areas are described as balanced and naturally working places connected with time cycles, traditions and rituals. Contrary to this, the industrial countryside is presented as a landscape of disharmony and restlessness. The industrial landscape therefore gains connotations of *locus terribilis* where destruction of natural countryside together with destruction of spiritual and moral values of those who live there take place. However, there is a major turn in the elucidation of the industrial landscape at the end of the 1950s, when the same motives which previously portrayed as disharmony now become signs of optimistic images. Literary works of the era when socialism was built interpret the signs of industrial countryside in the spirit of this building effort. The Ostrava region was still perceived as a "black" region but it became a region characterised by human effort, a progressive place where a new order is born in dust and sweat. This view of industrial landscape gradually weakened to be replaced by a new-and-old paradigm in the 1990s. The Ostrava region still remains to be

Fig. 25.13 Jan Zrzavý—Ostrava heaps II. Oil on canvass, 66 × 100 cm, 1933. Gallery of fine art in Ostrava. Author of photographs BcA. Vladimír Šulc, GVUO

depicted as an industrial landscape, while the previous accent on the destructive impact of human activity returns. The era of de-industrialisation of the last two decades is manifested in literature as an image of a countryside falling apart and dying —but its transition is indicated too (Málková 2012).

25.7 Conclusion

Over 200 years long history of black coal mining and a fast development of industrial production caused fundamental changes in the physical landscape of the Ostrava-Karviná Mining District. The area which originally used to represent the landscape with an agricultural character underwent transition into a mining landscape. Therefore we can observe a strong anthropogenic influence with specific anthropogenic landforms in the landscape of the Ostrava-Karviná mining district. The process of de-industrialisation, which brought a fast and massive recession of industrial production connected to mining industry after 1989, brings new impulses to the process of landscape formation. There is an ongoing effort both to revitalise the landscape affected by mining activity and to remove old ecological burden. The consequences of deep black coal mining resulted not only in the deterioration of conditions for existing ecosystems but they also created conditions for new ecosystems which would have not occurred without this anthropogenic activity in the original landscape at all. Older waste heaps, submerged ground subsidence areas and unused tailings ponds have become unique habitats for significant fauna and flora species. The current landscape of OKMD provides its inhabitants with new ways of how they can enjoy it for leisure time activities (e.g. golf courses, amusement parks, hiking or angling).

Acknowledgements The contribution was created on the basis of the grant project solution: GAČR P410/12/0487 "The process of industrialization and landscape changes in the Industrial Zone of Ostrava in the nineteenth and twentieth century".

References

Bína J, Demek J (2012) Z nížin do hor: geomorfologické jednotky České republiky. Academia, Praha 343 pp

Demek J (1988) Obecná geomorfologie. Academia, Praha 480 pp

Diamo (2014) Laguny Ostramo. http://www.diamo.cz/lokality-odra/laguny-ostramo

Dolný A (2000) Budou na odvalech chráněná území přírody? Živa 4:173–176

Dolný A, Ďuriš Z (2001) Výskyt ohrožených bezobratlých na důlních odkalištích v Karviné. Živa 49:268–270

Dopita M, Aust J, Brieda J, Černý I, Dvořák P, Fialová V, Foldyna J, Grmela A, Grygar R, Hoch I, Honěk J, Kaštovský V, Konečný P, Kožušníková A, Krejčí B, Kumpera O, Martinec P, Merenda M, Muller K, Novotná E, Ptáček J, Purkyňová E, Řehoř F, Strakoš Z, Tomis L, Tomšík J, Valterová P, Vašíček Z, Vencl J, Žídková S (1997) Geologie české části hornoslezské pánve. Ministerstvo životního prostředí České republiky, Praha 278 pp

Havrlant M (1980) Antropogenní formy reliéfu a životní prostředí v ostravské průmyslové oblasti. Spisy Pedagogické fakulty v Ostravě 41. SPN, Praha, 153 pp

Harabiš F, Dolný A (2012) Human altered ecosystems: suitable habitats as well as ecological traps for dragonflies (Odonata): the matter of scale. J Insect Conserv 3:1–10. doi:10.1007/s10841-011-9400-0

Holý P (2003) Hornické Ostravsko a výtvarné umění. In: Machač J, Langrová et al. (eds) Uhelné hornictví v ostravsko-karvinském revíru. Anagram, Ostrava, 564 pp

Chlupáč I, Brzobohatý R, Kovanda J (2002) Geologická minulost České republiky. Academia, Praha 436 pp

Ivánek J, Smolka Z (eds) (2013) Kulturně-historická encyklopedie českého Slezska a severovýchodní Moravy. Ostravská univerzita, Ostrava 1145 pp

Kirchner K, Smolová I (2010) Základy antropogenní geomorfologie. Palacký University, Olomouc 287 pp

Kozelská M (2011) Antropogenní tvary reliéfu v ostravské části OKR. University of Ostrava, Ostrava, MSc Theses 83 pp

Kukal Z, Reichmann F (2000) Horninové prostředí České republiky: jeho stav a ochrana. Český geologický ústav, Praha 189 pp

Málková I (2012) Industriální Ostravsko jako literární fenomén. In: Malura J, Tomášek M (eds) Krajina. Ostravská univerzita, Ostrava, Vytváření prostoru v literatuře a výtvarném umění, p 396

Matějček J (1993) Změny prostředí v hornických a hutnických oblastech českých zemí od počátku průmyslové revoluce do roku 1918. Studie k sociálním dějinám 19. století 2:266–283

Mísař Z et al (1983) Geologie ČSSR I. Český masív, SPN, Praha 336 pp

Mulková M, Popelková R (2013) Displays of hard coal deep mining in aerial photos. Acta Universitatis Carolinae—Geographica 1:25-39

Mulková M, Popelková R, Popelka P (2012) Landscape changes in the central part of the Karviná region from the first half of the nineteenth century to the beginning of the twenty first century Ekológia. Int J Ecol Prob Biosph 1:75–91

Myška M (1985) Uhelný průmysl a počátky devastace životního prostředí na Ostravsku. Rozpravy Národního technického muzea, Studie z dějin hornictví 17:105–116

Myška M (1988) Průmyslová oblast před průmyslovou oblastí. Slezský sborník 86:194–219

Myška M (1989) Průmyslová revoluce a proměny životního prostředí v ostravské aglomeraci. Časopis Slezského muzea 38:241–261

OKD (2010) Vracíme krajině život. Rekultivace krajiny na Ostravsko-Karvinsku. OKD, Ostrava, 52 pp. http://www.okd.cz/files/dokums_raw/okd_rekultivacni_brozura_cz.pdf

Pleva M (2011) Na haldě bude golf, ale i trasy na brusle a běžky. Moravskoslezský deník. http://moravskoslezsky.denik.cz/zpravy_region/serial_vitkovice_halda_golf20110529.html

Popelka P (2013) Nová krajina. Počátky rekultivace krajiny ostravsko-karvinského revíru (do konce 60. let 20. století). Časopis Matice moravské 132:445–476

Popelka P, Popelková R, Mulková M (2013) Vliv industrializace na změnu krajiny ostravsko-karvinského revíru. Příklad Slezské Ostravy a její proměny v 19. a 20. století. Historická geografie 1:49–84

Priesnitz V (2012) Rekultivace krajiny v karvinské části OKR. Bachelor thesis. University of Ostrava, Ostrava, 85 pp

Wróbel L (2012) Antropogenní tvary reliéfu karvinské části OKR (východní část). University of Ostrava, Ostrava, MSc Theses 99 pp

Poodří—Landscape of Ponds and a Preserved Meander Belt of the Odra River

26

Jan Hradecký, Radek Dušek, Marián Velešík, Monika Chudaničová, Václav Škarpich, Radim Jarošek, and Jan Lipina

Abstract

The Odra River is one of the important rivers of Central Europe. Within the Czech Republic it flows through the territory of the Moravian Gate that was glaciated by the continental ice sheet several times in the past (Elsterian and Saalian glaciations). The territory of the Moravian Gate makes a part of Subcarpathia and is characterised by graben-like morphostructure with Devonian and Carboniferous, Tertiary and Quaternary sediments. The relief was largely formed during the Quaternary when the terraced system of the Odra River originated. Flood river branches and ancient river branches were subsequently used by people to build water supply canals for fish pond systems. From the point of view of paleogeography, it is a significant territory as it enables understanding of the Quaternary evolution of Central European landscape. The Poodří Region is a territory of precious natural values that have been created by fluvial processes and also by humans who started to participate in landscape development in the Middle Ages. Since the medieval times several fish pond systems have constituted a fundamental characteristic feature of the local landscape, which is nowadays appreciated for numerous wetland ecosystems preserved, thanks to the conservation of the meandering stream with a natural flood regime. The Poodří landscape has mainly changed when the area of the floodplain forest was reduced and substituted by meadows. Unstabilised banks and ongoing fluvial processes led to faster retreat of banks and formation of the meander belt. At present the Poodří landscape is protected and it has also been included in the list of wetland ecosystems of the Ramsar Convention of Wetlands.

Keywords

Odra River • Quaternary history • Meander belt • Ponds • Cultural landscape

J. Hradecký (✉) · R. Dušek · M. Velešík · M. Chudaničová · V. Škarpich
Department of Physical Geography and Geoecology Faculty of Science, University of Ostrava, Ostrava, Czech Republic
e-mail: jan.hradecky@osu.cz

R. Jarošek
Agency of Landscape Protection of the Czech Republic, Ostrava Branch, Trocnovská 2, Ostrava, Czech Republic

J. Lipina
Jan Lipina, Jistebník 145, Czech Republic

T. Pánek and J. Hradecký (eds.), *Landscapes and Landforms of the Czech Republic*, World Geomorphological Landscapes, DOI 10.1007/978-3-319-27537-6_26

26.1 Introduction: Preserved Riverine Landscape

The riverine landscape of the Odra River (Fig. 26.1) represents a unique territory within Central Europe as it preserves a natural river bed pattern as well as the natural character of fluvial processes. Although man has exerted certain influence over the local landscape, the character of floods that belong to important landscape-forming processes has not been affected greatly here. The Odra River is a significant European river that starts in the Oderské vrchy Hills; however, its course within the Czech Republic is relatively short, with its 112 km out of the total 854 km. Minimal channel regulation (Fig. 26.2), absence of large valley dams and sensitive approach in river management have led to the conservation of a unique type of riverine landscape with a meandering stream. Since the Middle Ages a part of the valley floodplain has taken advantage of to establish ponds. Nevertheless, the area along the river had already been used by much older cultures inhabiting the territory of Czech lands and thus a harmonic cultural landscape came into being and has been preserved in the vicinity of the important Ostrava industrial agglomeration. A very interesting characteristic of this cultural landscape is fluvial relief (Fig. 26.2) that creates conditions for the development of unique floodplain forests and other wetland ecosystems. The status of Protected Landscape Area (PLA) was granted to the Poodří region in 1991, to the area covering 81.5 km^2 that in 1993 was included in the list of important wetland ecosystems of the Ramsar Convention and, finally, in 2005 the area was included in Special Protection Areas under Birds Directive of Natura 2000. The landscape of the Odra River valley is paleogeographically related to the so-called Moravian Gate, an important territory that contains a geological record of Quaternary evolution, particularly in connection with the intervention of the continental ice sheet into Central

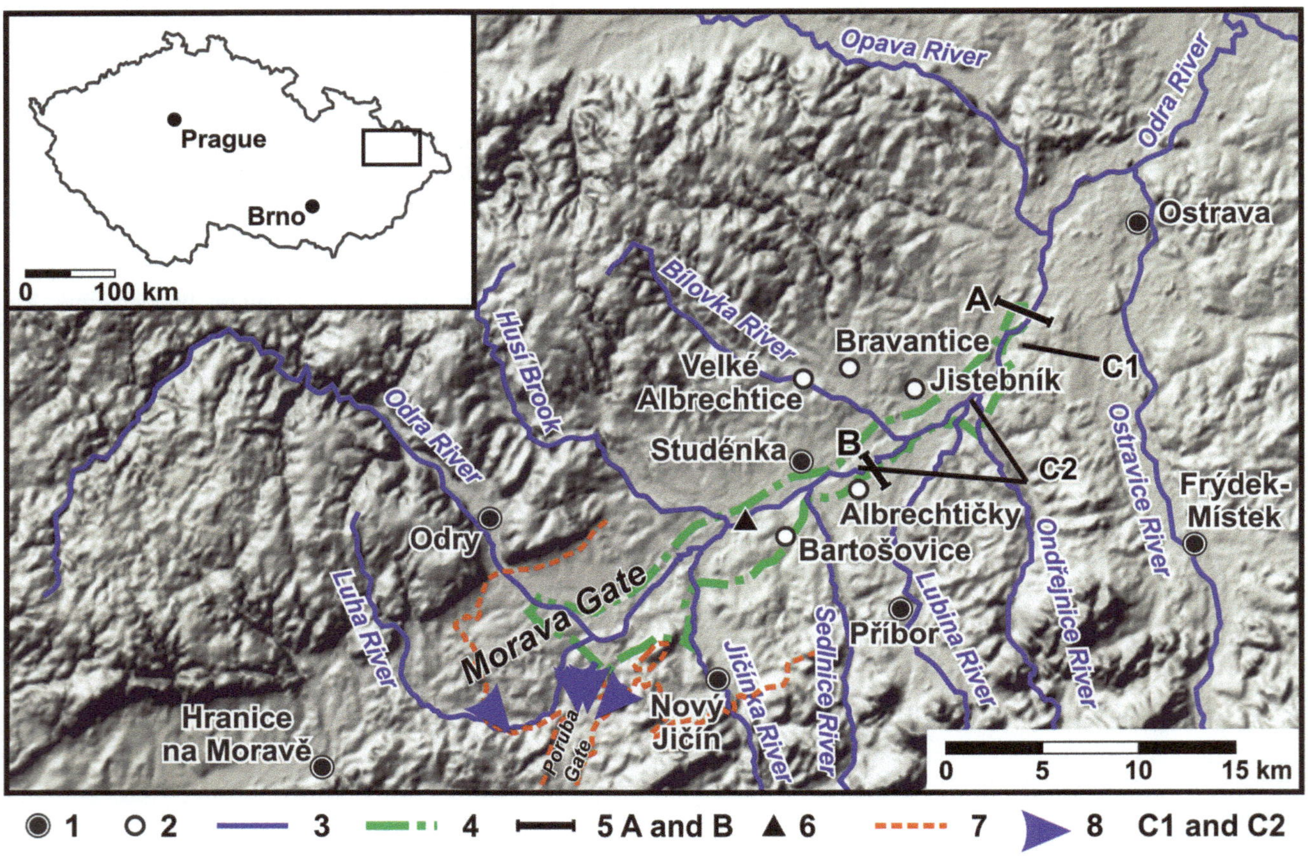

Fig. 26.1 The Poodří area—location of the preserved meander belt. *1* main cities, *2* villages, *3* rivers, *4* border of the Poodří Protected Landscape Area, *5* location of geological profiles, *6* limnigraphic station Bartošovice, *7* maximum limits of the continental glaciation sediments (Tyráček 2011), *8* direction of the meltwater drainage (Tyráček 2011), C1—intensity of lateral erosion (sites M1-M4), C2—historical changes in channel pattern

Fig. 26.2 Meander belt of the Odra River in the landscape of Poodří with extensive agricultural land use. (*Photo* J. Lipina)

Europe. The Moravian Gate is also the area that is passed through by the Main European Watershed between the drainage basins of the Odra and Danube Rivers.

26.2 Regional Settings

The Poodří landscape lies in the territory of the Moravian Gate (Moravská brána) that constitutes a part of Western Subcarpathia. The Moravian Gate stretches in the south-west north-east direction between Přerov and Ostrava cities. The Poodří region occupies the north-east part of the Moravian Gate that is referred to as the Odry part and follows the Bečva part behind (southwards) the Main European Watershed (Odra–Danube Rivers). This morphostructure represents a zone of the so-called Carpathian Foredeep of the Outer Western Carpathians that constitutes a geological border between the Outer Western Carpathians and the Bohemian Massif. The Moravian Gate is a tectonic graben whose geological structure is complex. The basement is made up of Proterozoic metamorphic rocks while above these there is a suite of Devonian and Carboniferous sediments, flysch formations of the Silesian Culm and, finally, Neogene and Quaternary sediments. Neogene sediments are represented by various rocks of Miocene marine transgression that started in the Karpatian. Carpathian nappes jut out above the surface in isolated forms that include mainly rocks of the partial Těšín Nappe. Quaternary sediments are connected with two advances of the Scandinavian ice sheet (Elsterian and Saalian Glacials), distinct aeolian activity during dry glacials, and fluvial activity of humid interglacials.

Although the Odra River can be considered the main axis of the territory, landscape formation and evolution are closely related to the activity of its many tributaries. Right-bank tributaries (Jičínka, Sedlnice, Lubina and Ondřejnice Rivers)

Table 26.1 N-year maximum peak discharges for the Bartošovice limnigraphic station (*source* Vodohospodářský portál)

N-year maximum peak discharge	Q_1	Q_2	Q_5	Q_{10}	Q_{20}	Q_{50}	Q_{100}
$m^3 \cdot s^{-1}$	78	110	157	196	239	299	349

(*source* http://app.pod.cz/portal/SaP/cz/PC/Mereni.aspx?id=300021344&oid=1)

start in the Carpathian part that is linked to the frontal foredeep in the south-east direction. An exception is the Luha River that starts in the Nízký Jeseník Upland but enters the Odra River from the right. Left-bank tributaries (e.g. Bílovka River and Husí potok Brook) connect the Moravian Gate with the Variscan area that is represented by the Nízký Jeseník Upland lying to the north-west. These tributaries have brought a large amount of sediments into the Odry part of the Moravian Gate, forming alluvial fans. Subsequently, these sedimentary complexes became a part of the Quaternary valley bottom infill.

From the hydrological point of view, the Odra River maintains a regime unaffected by any large hydraulic structures. It regularly experiences floods (Table 26.1) coming from spring snowmelt or related to intensive or long-lasting precipitation (e.g. floods of the year 1997, 2009 or 2010). The first historically recorded flood is that on 20 August 1501 (Brázdil et al. 2005). A total of 31 floods have been evidenced on the Odra River by historical records, many of which include the cause and exact date or at least the time of year. The period of instrumental measurement in the Bartošovice station revealed the following floods as three largest ones: 8 July 1997 (Qcul = 367 $m^3 \times s^{-1}$ more than N-100 maximum peak discharge), 25 June 2009 (Qcul = 126 $m^3 \times s^{-1}$ more than N-2 maximum peak discharge), and 18 May 2010 (Qcul = 76.2 $m^3 \times s^{-1}$ less than N-1 maximum peak discharge). The analysis of the twentieth century floods shows that the frequency of floods decreased considerably after the year 1950, namely 1–3 floods every 10 years. However, in the previous 10 year periods the Odra River experienced 8–9 floods (Brázdil et al. 2005).

26.3 Quaternary History of the Moravian Gate

The territory of the Moravian Gate has been glaciated twice, namely during the Elsterian glaciation and the Saalian glaciation. The fronts of the ice sheet crossed the Main European Watershed (Odra and Danube Rivers) at the Poruba Gate (Nývlt et al. 2011). Proglacial outwash drained into the Danube River basin (Bečva River) (Tyráček 2001). Due to this fact, the Moravian Gate is a very important area with respect to direct correlation of glacial sediments within the river terrace systems of the Odra and Danube Rivers (Macoun et al. 1965). The ice sheet reached the study area during the Elsterian stage twice (MIS 16/Elsterian 1 and MIS 12/Elsterian 2). The sediments of Elsterian glaciation were buried by deposits of the Saalian ice sheet (Nývlt et al. 2011). The advance of the Elsterian ice sheet to the Moravian Gate has long been subject to discussion. Tyráček (2011) stated that the lobe of the Elsterian ice sheet failed to reach the Main European Watershed; however, the existence of the northern termination in the valley of the Poruba Gate has not been satisfactorily explained (Tyráček 2011; Nývlt et al. 2011).

The area of the Moravian Gate was also glaciated during the Saalian (MIS 6) (e.g. Marks 2004, 2005; Nývlt et al. 2011). The North European ice sheet advanced into the Moravian Gate during the Saalian 1 (Macoun et al. 1965) and two oscillations were recorded (Růžička 2004). Saalian deposits represent the maximum Pleistocene glaciation in the Moravian Gate (Tyráček 2011). During the Saalian 1 the area of the Moravian Gate was drained through the Poruba Gate at the level of a 15-m-terrace to the south. There is evidence of glacifluvial and glaciolacustrine deposits overlying the main terrace in the Poruba Gate and at the same time Nordic rocks were found in the Bečva River equivalent terrace (Tyráček and Havlíček 2009). That means that at that time meltwaters overflowed into the Danube River basin (Nývlt et al. 2011). Nývlt et al. (2008) brought the first results of numerical dating of retreating proglacial glacifluvial sands in the Moravian Gate (close to Kunín municipality). These deposits, which directly overlie the main Odra River terrace, have been dated by optically stimulated luminescence to 162.0 ± 9.4 ka BP and they confirm the end of the first cool period of MIS 6 (Saalian 1).

There are two accumulation terraces in the area, that is the main terrace of the Saalian age and a valley terrace of the Weichselian age. The existence of older terraces has not been documented. Typical composition of valley deposits along the Odra River is presented by Fig. 26.3. The main terrace is the largest part of the depositional system within Northern Moravia and Czech Silesia but in the Odra valley it has developed in a specific way since the largest amount of deposits was accumulated by tributaries. From the stratigraphical point of view, the main terrace dates back to the phase between the Elsterian and the Saalian (Macoun et al. 1965). The main terrace is composed of two erosional levels too. The higher one represents the original depositional level and is covered by glaciolacustrine sands and gravel-sands of the Saalian glaciation. The lower level is a result of erosion

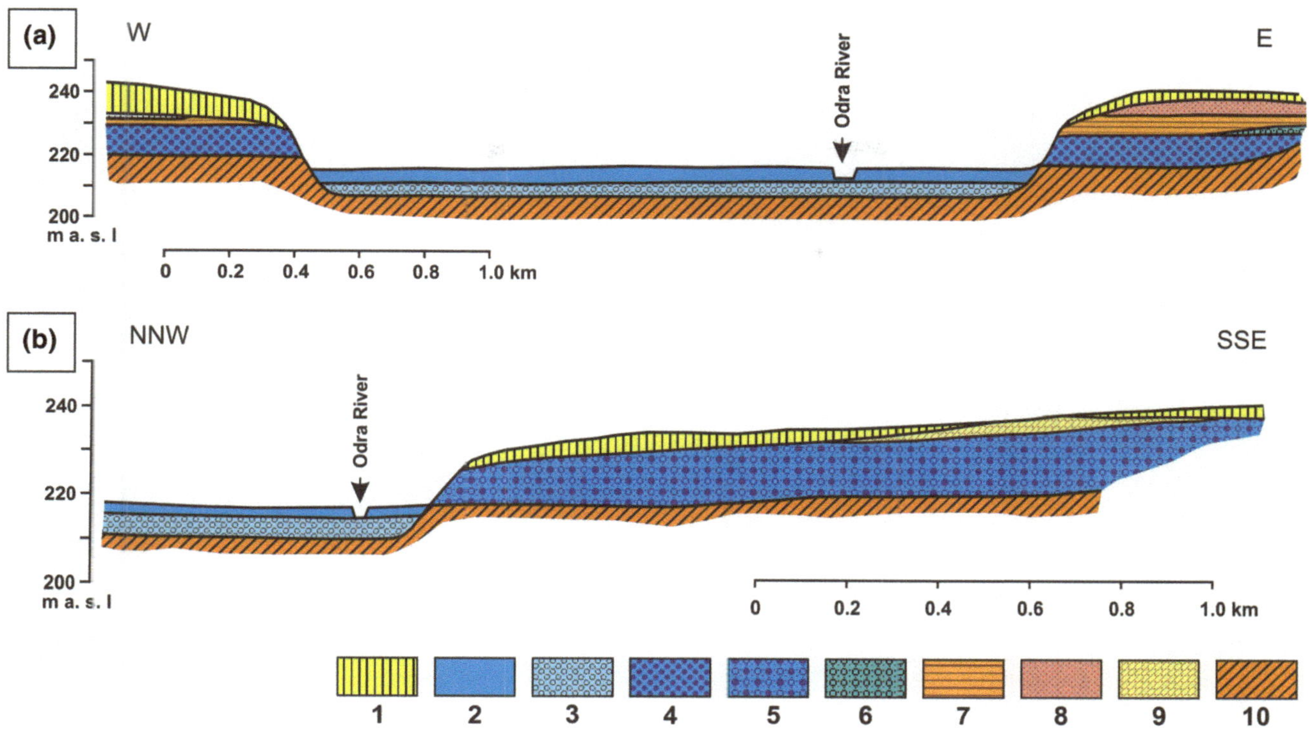

Fig. 26.3 Transects of the Odra River valley presented by Macoun et al. 1965 (for location see Fig. 26.1): **a** near Výškovice (Municipal District of Ostrava) and **b** near Albrechtičky municipality; legend: *1* loess, *2* Holocene flood loam, *3* sandy-gravel deposits of floodplain terrace, *4* sandy-gravel deposits of older terrace, *5* sandy-gravel deposits of an abandoned floodplain terrace (non-divided), *6* sandy-gravel deposits of younger terrace, *7* glaciolacustrine clay deposits, *8* glaciolacustrine sand deposits, *9* solifluction deposits, *10* Pre-quaternary basement

in the body of the main terrace and is covered by Late Pleistocene loess deposits. Relative height of the surface of the main terrace above the river oscillated between 5 and 17 m, while relative height of the base is between −1.5 and 6 m above the river. The valley terrace has a relatively regular structure. The base part is composed of fluvial sand and gravels with clayey interlayers (Weichselian age). The upper part is represented by 4–5 m thick deposition of flood loams (Atlantic and post-Atlantic age) within the reach of floods. Relative height of the surface of the valley terrace is of 2–3 m above the river, while relative height of its base is located from −4 to −6 m above the river level (Macoun et al. 1965).

Floodplain sediments of the Odra River are predominantly silty with a slightly increased proportion of sand within the bottom 40 cm of the profile (Fig. 26.4). The chronostratigraphy of the upper layers is enabled by the dating of ^{137}Cs radioisotope whose trace elements are first discovered in the depth of 50 cm. An increase in ^{137}Cs amount upwards starts in the depth of 37.5 cm, which can be related to the year 1953, whereas its peak then most likely indicates increased fallout after the Chernobyl explosion in 1986 (Walling and He 1997). Industrial age of the upper 50 cm is also confirmed by increased values of magnetic susceptlibility (χ), lead (Pb) and zinc (Zn). Anthropogenic origin of the increase of χ values is indicated by low frequency dependent magnetic susceptibility (χ_{fd}) and it is connected with intensive fossil fuel burning in the second half of the twentieth century (Thompson and Oldfied 1986). Higher concentration of Pb and Zn in upper sediments is also related to industrial production and fossil fuel burning. The increase in Pb is then primarily related to the worldwide introduction of leaded petrol in the first half of the twentieth century (Föstner and Wittmann 1983). These data point to gradual vertical aggradation of flood sediments in the Odra River floodplain.

26.4 Dynamics of the Meander Belt

The Odra belongs to those few rivers of the Czech Republic that have remained unaffected by any major channel regulation or construction of a large hydraulic structure. Almost natural state of the river, which is favourable to the natural course of floods, makes it possible to maintain the meander belt in a dynamic state characterised by bank erosion and aggradation. This fact has been validated by a few geomorphological studies that have provided information about

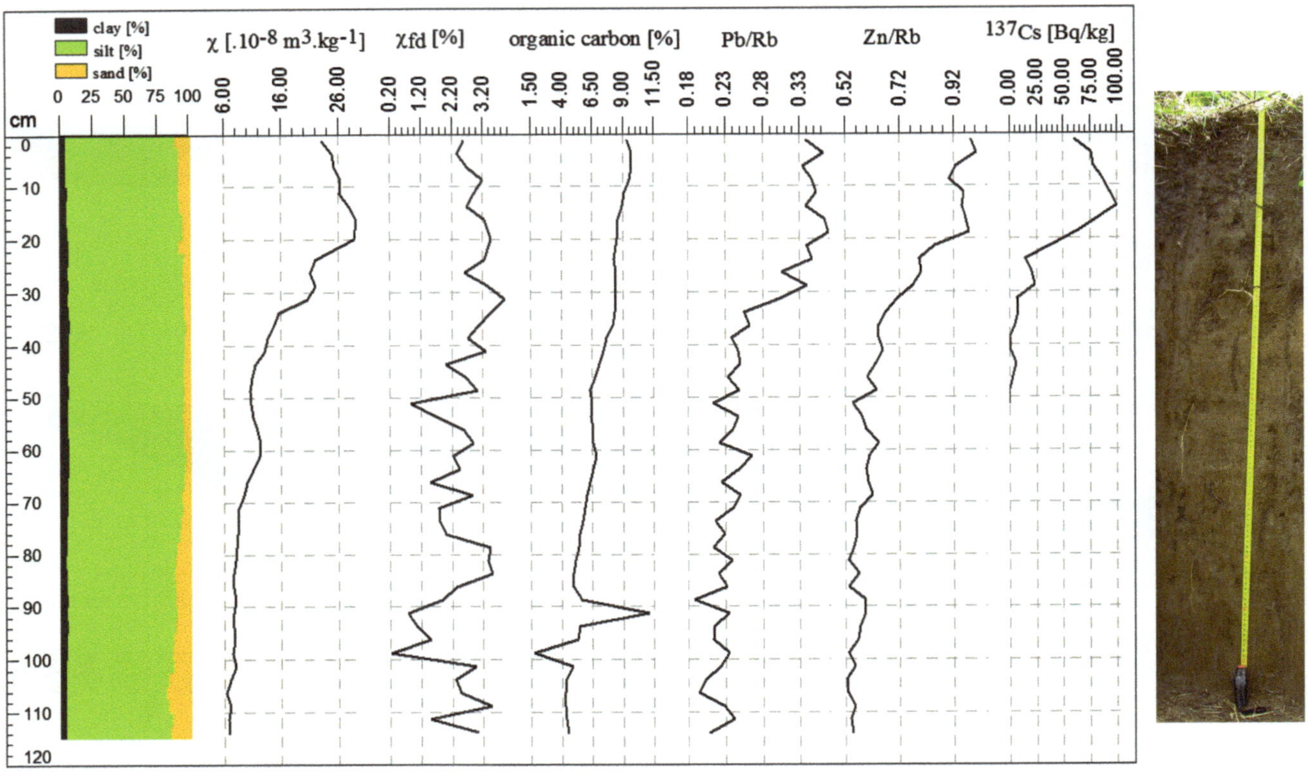

Fig. 26.4 Detailed sedimentological analyses of typical floodplain deposits of Odra River (*Note* Pb and Zn are normalised by rubidium-Rb, the so-called lithogenic element, which eliminates the impact of various sediment granularity and geology)

the speed of the evolution of the river pattern and the speed of the retreat of the bank zone.

From the point of view of morphology, the study area contains meso- and microforms that are typical of preserved parts of valley floodplains. The most conspicuous landform is the Odra River channel itself that, despite some local regulations, has largely preserved the character of a lowland middle-size stream with numerous meanders deeply incised into Holocene flood deposits. The total thickness of flood sediments of the Odra River and its tributaries ranges between 200 and 450 cm. They are made up mainly by sandy soils of light grey or yellow-brown colours; on the base there are clay or clay-sand sediments of grey, dark grey or black-brown colours. At present, the Odra River channel cuts into the whole formation of flood deposits and is incised into a layer of gravel sand bedrock. The profile of flood deposits becomes exposed at low water stages (Opravil 1999).

The study of the meander belt has been focused on the reach between the Oderská lávka Bridge close to Albrechtičky municipality (river km 44.2) and the confluence with the Lubina River (river km 31.4). The basic element of the meandering river pattern is a meander (bend) and its tendency towards lateral migration due to lateral erosion in poorly consolidated fluvial sediments (flood deposits). A shift in meanders can be observed within a zone in the valley bottom whose width varies and which is called the meander belt. This dynamic zone clearly reflects the processes that take place within the whole basin, i.e. climatic changes, transformation in the delivery of sediments, and influence on the hydrological balance. Also, it closely correlates with the frequency and magnitude of floods (Table 26.1). An important role in the evolution of the meander belt is played by vegetation, particularly the presence and character of the floodplain forest (Dušek and Hradecký 2011).

A complete overview of channel location and changes in the evolution of the river pattern with the position of individual meanders and the advancement of channel migration are given in Fig. 26.5. Seventy meanders were identified in the study area in the year 1839. Their number reached its maximum in 1937 (72) and has been decreasing since (Table 26.2). The 1955–1969 period was exceptional since an increase to 68 meanders was observed. At the end of the twentieth century the channel contained 57 meanders whose average radius reached 37 m, which is the maximum value in the whole studied period. It points to a trend of gradual joining of meander bends by which larger curves are created and the channel shortened. At the beginning the channel was 13.2 km long, but since 1969 it has shortened progressively

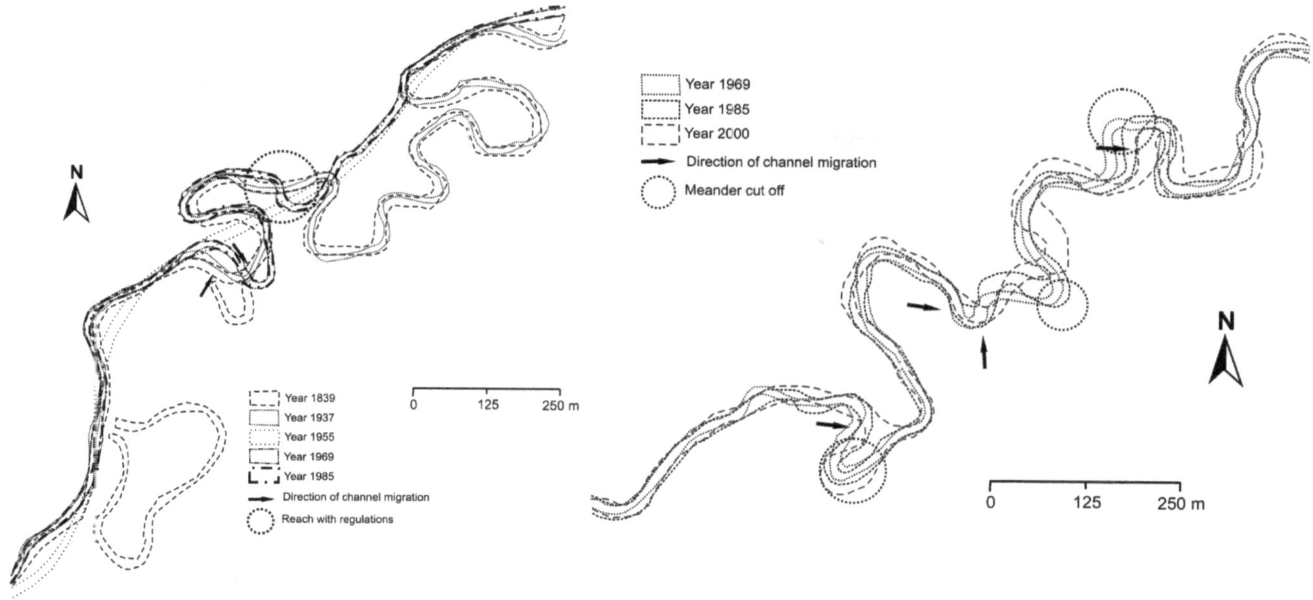

Fig. 26.5 Example of the channel pattern evolution in the channel reach C2—for location see Fig. 26.1 (Vávra 2006)

Table 26.2 Parameters of the meanders and the channel in the studied reach in the years 1839–2000 (after Vávra 2006)

Year/parameter	1839	1937	1955	1969	1985	2000
number of meanders	70	72	62	68	60	57
Radius—minimum (m)	16	11.4	13.6	15	14.7	15.5
Radius—maximum (m)	105	60.1	63	113.7	112.7	111.2
Average radius (m)	33.5	31.8	31.6	34.1	35.6	37.2
C_L (channel length) (m)	13262	12500	11303[a]	13221	12808	12669
V_z (direct distance) (m)	7421	6541	6056	7393	7401	7418
C_I (channel index)	1.79	1.91	–	1.79	1.73	1.7

([a]incomplete figure)

by c. 600 m to 12.6 km. The shortening is a result of water management regulations of the river channel and the fact that some meander necks were cut through and thus ceased to exist. Throughout the whole studied period the character of the river channel is that of a meandering stream with the channel sinuosity index ranging between 1.7 and 1.91 but a trend towards channel straightening can be recognized (Table 26.2).

A detailed study of changes in the position of the meandering channel was conducted in the years 2006 to 2012, via monitoring the Odra River meanders. The studied area (r. km 25.5–26.8) is located between the municipalities of Polanka nad Odrou and Stará Bělá. The monitoring was performed on the upper rim of four concave banks (M1-M4—see Fig. 26.1). The measurement of the position was complemented with a detailed study of spatial changes in a selected reach of the M3 bank using close-range photogrammetry. Six position monitoring measurements made it possible to observe meander changes in five phases (Fig. 26.6). An overview of all the shifts is given in Table 26.3 (Drozdek 2006; Klasová 2011; Hanzelková 2012; Černochová 2013). Originally curved bank was eroded mainly in the right, eastern part, which led to its straightening. Within 7 months the right edge of the monitored reach shifted by c. 7 m, i.e. 1 m per month. Even though the studied reach is relatively short (14 m), not only bank shifts but also changes in the bank shape can be observed. The results of close-range photogrammetry are presented by Fig. 26.7.

26.5 People and the River

The Poodří Region as a part of the elongated depression of the Moravian Gate has had a transport function since the prehistoric times. Hunters and fishermen passed through this region as early as in the Mesolithic period as it is documented

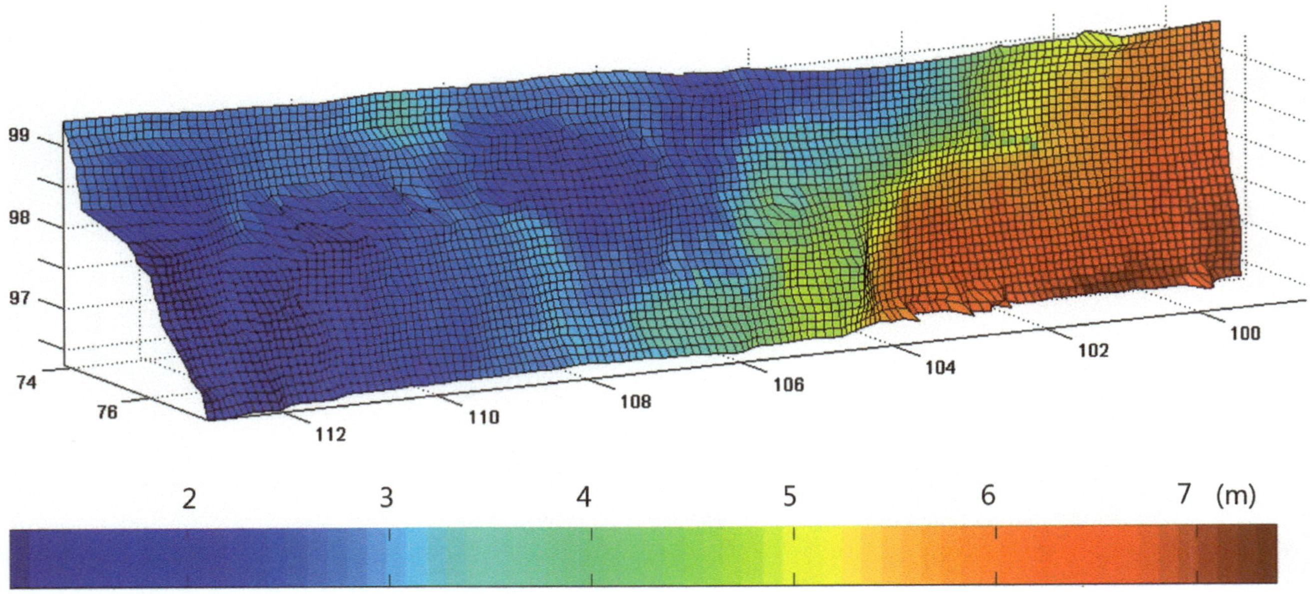

Fig. 26.6 Results of the lateral erosion measurement in the selected reach of the bank M1 the coloured bar represents the intensity of river bank retreat

Table 26.3 Overview of position monitoring results

Site	Bank length (m)	Average shift in individual phases (cm)					Total shift (cm)	Average shift (cm/month)
		A	B	C	D	E		
M1	336	187	13	107	145	0	452	6.3
M2	88	559	137	122	230	67	1115	15.5
M3	156	330	88	108	190	44	760	10.6
M4	140	53	0	30	155	49	287	4.0

by findings on river terraces along the alluvial plain. The Odra River alluvial plain connected the areas of fertile soils of the central Moravia with the northern territories of today's Poland and it made a logical corridor for people travelling between the Mediterranean and Baltic Seas (Knápek 2011). An important change came with the linear pottery culture in the Neolithic period (6,000 before AD) when the territory started to be deforested, the soil cultivated and settlements established on the Odra left-bank terraces, e.g. in today's Studénka and Hladké Životice municipalities. Archaeological findings include not only agricultural tools and quern-stones but also fishing tools. However, the settlements were of short duration and ceased to exist soon. As a result, the agricultural landscape largely returned to its original state. Other settlements appeared in the Eneolithic, yet until the Early Bronze Age the Poodří region was mainly used as a transit zone. The following Urnfield culture brought cattle breeding and grazing to the forests of the Poodří region. Anthracological analysis of Novák (2008) at the archaeological sites of Olbramice and Klimkovice municipalities made it possible to identify two dominant tree species, namely fir (*Abies alba*) and beech (*Fagus sylvatica).* Heliophytes such as hazel, birch, members of the genus of *Prunus* (*cherries*, *plums*, etc.), hornbeam and tilia were also identified in large numbers (Novák 2008). The Urnfield culture disappeared in the Iron Age, which left the landscape in a desolate state again. The settlements occupied mainly the eastern margin towards the Podbeskydská pahorkatina Hillyland. After a relatively long and calm period people came to resettle the area in the mid-thirteenth century. As Opravil (1974) states, the landscape until that time was characterised as "rather suitable for hunting". However, from that time, a medieval settlement structure started to appear along with a network of roads and typical structure of village patches of arable land (Hosák 1956). Forests containing cultural elements such as pastures, arable fields and villages represented dominant ecosystems. Other elements of the landscape included first ponds and canal pounds that drove water to the ponds and water wheels (Fig. 26.8). We can suppose that medieval colonisation led to massive soil erosion, material accumulation in river channels and changes in the hydrological regime of water streams. Apart from that, deforested floodplains may have experienced fast lateral

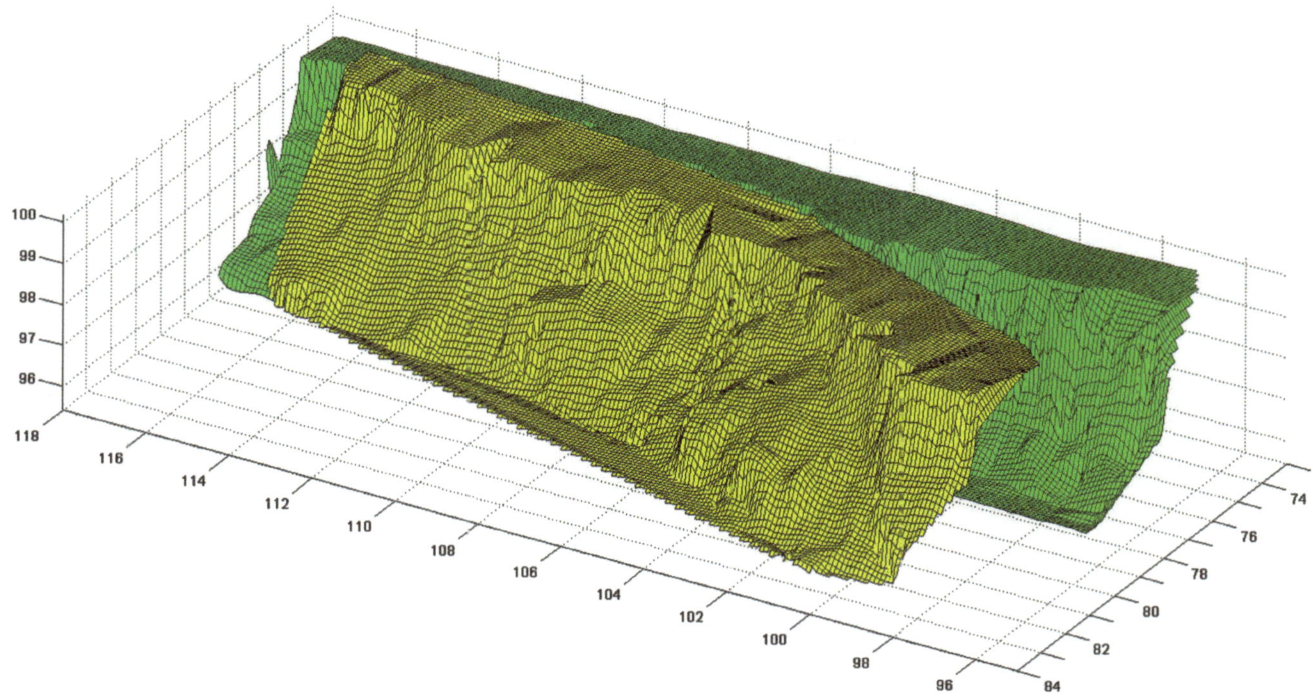

Fig. 26.7 Digital models of the bank from the 1st (*green coloured surface*) and 2nd (*yellow coloured surface*) photogrammetric measurement (local coordinate system, coordinates in meters)

erosion and massive migration of the meander belt. Generally, the historical phase of evolution of the local landscape is represented by the ponds themselves and a network of water-supplying canals.

26.6 Poodří Protected Landscape Area—Landforms and Biodiversity

The Poodří Protected Landscape Area contains mostly the flat wide floodplain of the Odra River and partly higher river terraces lining the river. The aim of the local nature protection is to combine the protection of biodiversity and ecosystems with economic activities and tourism development. The Poodří region is not only a large protected territory, but also a wetland (Fig. 26.9) of international importance (within the Ramsar Convention) and it is also included in the Natura 2000 network. Of particular importance with regard to valuable wetland ecosystems is the well-preserved state of its discharge regime that is unaffected by any larger technical structure in the upper part of the basin. Therefore, water stages within the year vary, while the maximal values are mainly measured in March and April. Several times a year the territory experiences large flooding, mainly as a result of spring snowmelt, when waters brought by right-sided Odra River tributaries from the Beskydy Mts meet the waters of its left-sided tributaries from the foothills of the Nízký Jeseník Upland. Other floods can occur in summer months as well as in autumn during extreme rainfalls. The flooding usually lasts several hours or days; some terrain depressions stay filled with water for a transitory period of time. Deeper periodic pools usually dry up during a period of several weeks to months. The extent of the annually flooded territory, which is approximately 1/5 to 1/4 of the extent of the Poodří PLA, comprises particularly meadows with numerous green patches and floodplain forests containing a network of mature distributary channels and ponds. Floods in the Poodří region represent a common natural phenomenon to which the landscape and local inhabitants have adapted very well.

In terms of areal extent, the most widespread in the Poodří region are meadows (Fig. 26.2) that create a unique continuous complex of more than 2,300 ha. Other elements of the floodplain are numerous microbodies in the form of periodic and permanent pools (oxbow lakes, deeper pools, shallow depressions) that have been left in the landscape by active streams. These locations host specific flora and fauna that have adjusted to water level oscillations throughout the year. Floodplain forests, which form natural vegetation cover in this altitude, have only been preserved in fragments occupying 10 % of the PLA, while the species composition is derived from concrete local conditions.

Fig. 26.8 Landscape of the Poodří area in the vicinity of Studénka municipality, old painting from the year 1737 (view to the south) with roads and important bridges. Selected items of the legend: *a* Studénka municipality, *b* Odra River, *c* bridge over the Odra River, *d* Sedlnice River, *e* bridge near Butovice quarter, *g* bridge over three rivers (in the direction towards Bílovec municipality), *h* bridge called Urbrucke over a canal pound and wetlands, *i* bridge called Ländebrücke over Little Odra River, *l* bridge in the direction towards Nová Horka and Hungary, *m* road to Frýdek-Místek through Paskov, *s* Moravian border. (*Source* Zemský archiv v Opavě, Hejtmanský úřad knížectví opavsko-krnovského, Opava, inv. č. 158, kart. 33)

Fig. 26.9 Velký roh weatland is typical biotope located in the floodplain of the Odra River with high biodiversity connected with evolution of fluvial landforms. (*Photo* J. Lipina)

Another important landscape element is extra-forest green vegetation with abundant solitary woody plants resembling a park in some places. There are also distinct linear elements such as bank vegetation, stretches of green vegetation on pond dikes and tree avenues along roads (Fig. 26.10). A dynamic element is the Odra River itself, characterised by numerous meanders that due to fluvial erosion and accumulation processes create biotopes of high clayey river cliffs and clayey-gravel point bars with the vegetation cover dominated by willow and poplar. From the point of view of economy, river terraces are exploited more intensively for being out of reach of floods. Slightly undulated terraces have been turned into fields and residential areas.

Ponds represent a significant ecological phenomenon in the cultural landscape of the Poodří Region (Fig. 26.10) since they offer home to a wide range of protected species of plants, invertebrates, amphibians and birds. Some of the ponds help to preserve the populations of critically endangered plant species such as *Salvinia natans, Najas minor, Nymphoides peltata* and *Elatine alsinastrum* and also the population of endangered fish *Missgurnus fossilis.* Littoral communities are inhabited by precious populations of endangered amphibians, e.g. *Rana ridibunda, Rana esculenta, Bombina bombina* or *Hyla arborea.* Also, the Poodří PLA was declared to be a Special Protection Area due to relatively rich avifauna and critically endangered Eurasian bittern (*Botaurus stellaris*), endangered species of western marsh harrier (*Circus aeruginosus*), common kingfisher (*Alcedo atthis*), gadwall (*Anas strepera*) and their biotopes. The ponds are extremely important for a vast number of migratory birds making

Fig. 26.10 Harmonious landscape in the area of Jistebnické rybníky ponds. The Odra River channel is in the close contact with the pond dikes—on the right side. (*Photo* J. Lipina)

long-distance flights across the territory (Sovíková 2011). Landscape protection management has long been trying to bring economic activities of fishermen into accord with the conservation of unique communities, which is not always easy and thus a decline in the number of some of the species has been recorded in the last tens of years.

26.7 Fish Farming in the Poodří: Landscape of Ponds

The beginnings of fish farming in the Odra River basin date back to medieval times. The oldest sources indicating local fish farming come from the mid-fifteenth century and mention ponds near Svinov municipality in the year 1431, Klimkovice municipality in 1447 and Polanka municipality in 1461. With an increasing interest in fish farming the number of ponds increased in the second half of the fifteenth century while the foundations of fish pond systems are laid. The documents mention, among others, the ponds near Studénka, Jistebník, Suchdol nad Odrou and Bartošovice municipalities (Hurt 1960a). The documents of Jan Starší from Žerotín (a noble) mention a pond in Výškovice and ponds of the nobility near the Odra River near Výškovice and Stará Bělá municipalities (ZA Opava—pobočka Olomouc). Even names of some ponds are given, such as Kukla, Kynovský and Podhorník ponds in Jistebník municipality and Vacek pond near Klimkovice and Polanka municipalities. Ponds and fish farming flourished in the period of the end of the fifteenth century and the first half of the sixteenth century. The ponds were no longer filled up by minor tributaries of the Odra River since they had stopped being sufficient as the source of water. Pond builders were forced to use water coming right from the Odra River itself. Using water from large streams was neither common, nor safe. Therefore, a canal pound called Oderská strouha, which belongs to the oldest constructions of this type in Czech lands, was constructed in order to fill up ponds in Studénka, Jistebník, Polanka n. Odrou and Klimkovice municipalities. Ponds built in the Poodří region were different from those in Bohemia and Moravia for the style of their construction since the dikes were built around the whole ponds whose depth was smaller, ranging around 1 m (Hurt 1960a). The depth affected farming as the ponds of small depth were more easily filled up with sediments and eutrophication made itself felt more intensively. The decline in fish farming during the 30 Years' War was followed by slow renovation of damages caused by the war due to lack of finance and general neglect. The second half of the seventeenth century witnessed a short-term period of recurrent development connected with the renovation of defunct ponds, modifications of the existing ponds and constructions of new ones, mainly small spawning and nursery ponds (Hurt 1960b).

The decline in profitability along with societal changes led to the destruction of many ponds and pond systems since the 1770s (situation of the landscape in the 1st half of the eighteenth century is presented by Fig. 26.8. The main reason was the need to obtain arable land for agriculture. The ponds in the Poodří region ceased to exist especially in places in which the state exercised its power, mainly at church farms, foundation farms and town farms. The need for agricultural land was only short-time and was soon substituted by a gradual demand for fish farming while fish farming management was modified. The management of ponds was observed, for example, near Studénka municipality (Fond ÚSBVs Bravantice).

At present, the Poodří PLA contains 60 ponds within the area of 700 ha, which represents 8.5 % of the total area of the Poodří PLA (Sovíková 2011). The pond founders made use of suitable conditions of the relief, particularly flat terrain of the Odra River floodplain and depressions in the terrain (oxbow lakes, avulsion channels, etc.) as well as many accumulation forms used for the construction of pond dikes. The natural network of minor river channels in the Odra River floodplain became the basis for building ditches and canal pounds that brought water into the ponds or from the ponds back into the main Odra River channel when the ponds were drained.

The ponds are grouped into eight smaller pond systems (Fig. 26.10). The largest ponds include the Bezruč Pond (75.1 ha) (Fig. 26.12), Bartošovický dolní pond (73.7 ha) and Kotvice pond (54.6 ha). The depth of the ponds is relatively small with the average value of 1 m, which is caused by relatively high sediment supply from the upstream courses of the Odra River and its tributaries. The ponds are mostly used for intensive fish farming (Fig. 26.11). They are fertilised and the fish are actively fed, which leads to the eutrophication and hypertrophication of the water (Fig. 26.10) in the ponds (Sovíková 2011). Typically farmed fish is carp, while secondary species are tench, bream, pike, silver carp, bighead carp and grass carp. Except for the hibernating ponds, the common ponds are emptied before the winter starts (the so-called wintering of ponds). In these phases of the year, it is possible to see the pond bottom with a high layer of accumulated fine sediment in which silt-clay fractions prevail. From the point of view of sustainable fish farming in the Odra River ponds, the hazards comprise not only sediment accumulation but also the burst of dikes that

Fig. 26.11 Atmosphere of autumn fishing out at the Křivý pond. (*Photo* J. Lipina)

Fig. 26.12 Bezruč pond with the front view of the Moravskoslezské Beskydy Mts and its highest peak Lysá hora Mt (1323 m a. s. l.). (*Photo* J. Lipina)

are in direct contact with freely meandering Odra River (Fig. 26.10). The dike burst hazard is normally eliminated via technical measures that, however, limit natural retreat of river cliffs. One of the possibilities how to prevent the dike burst is the monitoring of selected meanders.

26.8 Conclusion

The Poodří landscape is an example of a preserved harmonic cultural landscape in which an important role in its evolution and functioning was and is still played by the river and fluvial processes. The irreplaceable role in the development of the local landscape was that of man that has made use of the relief and hydrological conditions for establishing ponds since the medieval times. Thanks to minimal modification of the Odra River course and morphology and the conservation of its hydrological regime, extremely valuable wetland ecosystems have been preserved that quickly disappeared from other parts of the territory of present-day Czech Republic after the World War II. Quaternary evolution increased the value of this territory to an extremely precious area from the point of view of natural sciences as the research of local Quaternary sediments and landforms could significantly contribute to the study of history of Central European landscape.

Acknowledgement We acknowledge support by the Student Grant Competition of the University of Ostrava (SGS18/PřF/2015-2016). Authors would like to thank Monika Hradecká for the English style revision.

References

Brázdil R, Dobrovolný P, Elleder L, Kakos V, Kotyza O, Květoň V, Macková J, Müller M, Štekl J, Tolasz R, Valášek H (2005) Historical and Recent Floods in the Czech Republic. Masaryk University, Czech Hydrometeorological Institute, Brno, Praha, 370 p

Černochová K (2013) Geodetické zaměření meandrů Odry—V. etapa. Bakalářská práce na PřF OU

Drozdek M (2006) Aplikace metod pozemní digitální fotogrammetrie při sledování změn průběhu koryta Odry v CHKO Poodří. Diplomová práce na PřF OU

Dušek R, Hradecký J (2011) Geomorfologický výzkum fluviálních procesů v CHKO Poodří. Poodří 14:10–17

Fond ÚSBVs Bravantice inv. č. 2089, ÚSBVs Bravantice

Förstner U, Wittmann GTW (1983) Metal pollution in the aquatic environment. Springer-Verlag, Berlin, 486 p

Hanzelková H (2012) Geodetické zaměření meandrů Odry—IV. etapa. Bakalářská práce na PřF OU

Hosák L (1956) Středověké osídlení a kolonisace mezi Odrou, Ostravou a Beskydami. Časopis Společnosti přátel starožitností 64:17–22

Hurt R (1960a) Dějiny rybníkářství na Moravě a ve Slezsku. I. díl, Krajské nakladatelství, Ostrava, 274 p

Hurt R (1960b) Dějiny rybníkářství na Moravě a ve Slezsku. II. díl, Krajské nakladatelství, Ostrava, 323 p

Klasová P (2011) Monitoring říčních břehů pomocí pozemní stereofotogrammetrie—5. etapa. Bakalářská práce na PřF OU

Knápek A (2011) Vliv lidské přítomnosti na Poodří od pravěku do vrcholného středověku. Poodří 14:7–10

Kolektiv (2009) Plán péče o chráněnou krajinnou oblast Poodří na r. 2009–2018. Studénka, Depon. Správa CHKO Poodří, 221 p

Krejčí J (1971) Vývoj průchodní funkce Moravské brány. Folia Facultatis Scientiarum Naturalium Universitatis Purkynianae Brunensis, Geographia 5. Brno, Univerzita J.E. Purkyně, pp 5–63

Macoun J, Šibrava V, Tyráček J, Kneblová-Vodičková V (1965) Kvartér Ostravska a Moravské brány. Nakladatelství ČSAV, Praha, 419 p

Marks L (2004) Pleistocene glacial limits in Poland. In: Ehlers J, Gibbard PL (eds.) Quaternary Glaciations—Extent and Chronology, Part I: Europe. Developments in Quaternary Sciences 15, pp 295–300

Marks L (2005) Pleistocene glacial limits in the territory of Poland. Przegl Geol 53:988–993

Novák J (2008) Dřevinná skladba severní části Oderské brány ve starší době železné z pohledu antrakologické analýzy. Bioarcheologie v České republice, České Budějovice, Praha, pp 267–284

Nývlt D, Engel Z, Tyráček J (2011) Pleistocene glaciations of Czechia. In: Ehlers J, Gibbard PL, Hughes PD (eds.) Developments in quaternary sciences 15, pp 37–46

Nývlt D, Jankovská V, Víšek J, Franců E, Franců J (2008) Deglaciační faze prvního sálského zalednění v Moravské bráně. In: Rozsková A, Vlačiky M, Ivanov M (eds) 14. Kvartér, Brno, pp 14–15

Opravil E (1974) Moravskoslezský pomezní les od začátku kolonizace. Archeologický sborník, Ostravské museum, Ostrava, pp 113–133

Opravil E (1999) Z historie údolní nivy v CHKO Poodří a v přilehlém území. Poodří. Současné výsledky výzkumu v Chráněné krajinné oblasti Poodří. Společnost přátel Poodří, Ostrava, pp 23–26

Růžička M (2004) The Pleistocene glaciation of Czechia. In: Ehlers J, Gibbard PL (eds.) Quaternary Glaciations—Extent and Chronology, pp 27–34

Sovíková L (2011) Rybníky v CHKO Poodří. Poodří 14:38–42

Thompson R, Oldfield F (1986) Environmental magnetism. Allen & Unwin, London, 227 p

Tyráček J (2001) Upper Cenozoic fluvial history in the Bohemian Massif. Quat Int 79:37–53

Tyráček J (2011) Continental glaciation of the Moravian Gate (Czech Republic). Sbor. geol. Věd, Antropozoikum 27:39–49

Tyráček J, Havlíček P (2009) The fluvial record in the Czech Republic: a review in the context of IGCP 518. Glob. Planet. Change 68:311–325

Vávra J (2006) Změny průběhu koryta řeky Odry v CHKO Poodří (ř. km 44.2–31.4). Diplomová práce na PřF OU

Walling DE, He Q (1997) Use of fallout ^{137}Cs in investigation of overbank sediment deposition on river floodplains. Catena 29:263–282

Zemský archiv Opava – pobočka Olomouc (ZA Opava pobočka Olomouc) Fond Lenní dvůr Kroměříž inv. č. 321, Lenní dvůr Kroměříž – pergamenové listiny 1249–1836

Zemský archiv v Opavě, Hejtmanský úřad knížectví opavsko-krnovského, Opava, inv. č. 158, kart. 33

Landslide Landscape of the Moravskoslezské Beskydy Mountains and Their Surroundings

27

Tomáš Pánek and Jan Lenart

Abstract

Moravskoslezské Beskydy Mts represent the highest part of the Western Carpathians in the Czech Republic and could serve as an archetype of elevated flysch landscape that originated within the Alpine thrust-and-fold belt. Major landforms of the area originated as a consequence of Miocene thrusting, rock control (alternation of sandstone and shale layers of different rock strength), fluvial processes and mass movements. Of major importance are ancient and active landslides that play a crucial role in increasing geodiversity of the mountain slopes. The Moravskoslezské Beskydy Mts together with surrounding flysch highlands host the most frequent and developed crevice-type caves in the Czech Republic, which makes this area unique in Central European context. However, increasing tourism together with intensive forest management and other human activities pose a major threat to the landscape underlain by fragile flysch rocks.

Keywords

Western Carpathians • Moravskoslezské Beskydy Mts • Flysch rocks • Thrust-and-fold belt • Rock control • Landslides • Crevice-type caves

27.1 Introduction

Steeply rising above densely settled and industrialized piedmont in the north, Moravskoslezské Beskydy Mts are among the most contrasting landscapes in Central Europe (Fig. 27.1). The area represents the highest part of the Western Carpathians in the Czech Republic and its geomorphic diversity benefits mainly from the specific nature of flysch bedrock which was folded and thrust during Alpine orogeny. In such circumstances, this region is a part of the youngest geological domain in the Czech Republic. Due to the presence of hardly accessible forested steep slopes modelled by landslides and gully erosion, the Moravskoslezské Beskydy Mts and their surrounding highlands contain some of the last patches of wilderness in the Czech Republic, providing habitats for endangered and protected species (e.g. Eurasian lynx, brown bear, grey wolf). However, the immediate vicinity of the Ostrava industrial area (one of major economical hubs of the country) and related increasing demand for recreational activities represent major threat to the landscape and landforms in the westernmost promontory of "true" Carpathians.

In this chapter we will demonstrate how geological distinctiveness of flysch substratum together with mass movement processes influence diversity of landforms within a medium-high mountain belt. Besides the Moravskoslezské Beskydy Mts forming the highest part of the region, the chapter also covers adjacent mountains (Slezské Beskydy Mts, Slovenské Beskydy Mts, Vsetínské vrchy Mts and Javorníky Mts) that together form one compact region of flysch highlands (Fig. 27.2).

T. Pánek (✉) · J. Lenart
Department of Physical Geography and Geoecology Faculty of Science, University of Ostrava, Ostrava, Czech Republic
e-mail: tomas.panek@osu.cz

J. Lenart
e-mail: jan.lenart@osu.cz

T. Pánek and J. Hradecký (eds.), *Landscapes and Landforms of the Czech Republic*,
World Geomorphological Landscapes, DOI 10.1007/978-3-319-27537-6_27

Fig. 27.1 Steep northern slope of the Moravskoslezské Beskydy Mts seen from the top of the Lysá hora Mt (1323 m a.s.l.). Note overall asymmetry of the culmination ridge with steep escarpment at the front of the overthrust and gently inclined opposite slopes conforming to bedding planes (*Photo* T. Pánek)

27.2 Geological Setting and Morphostructure

As a part of the flysch belt of the Outer Western Carpathians, the mid-altitude mountain terrain of the Moravskoslezské Beskydy Mts and surrounding ranges (Fig. 27.2) are built up by a series of Tertiary nappe stacks (Menčík et al. 1983). Generally, the mountain ridges trend in the WSW–ENE direction, which reflects the major structural grain of thrust sheets and the strike of flysch beds. The elevation culminates at the northern edge of the mountain region in the area of the Moravskoslezské Beskydy Mts, in ridges exceeding 1,200 m a.s.l. (Lysá hora Mt reaches the altitude of 1323 m a.s.l.), while other mountains such as the Slezské Beskydy Mts, Slovenské Beskydy Mts, Javorníky Mts and Vsetínské vrchy Mts reach the maximum altitude around 1000 m a.s.l. (Fig 27.2).

27.2.1 Geological and Geomorphological Evolution

Lithology of the area consists of flysch sequences which originated as turbidites in the Tethys Sea between the Early Cretaceous and the Lower Miocene (Picha et al. 2006) (Fig. 27.2). Flysch beds involve the alternation of proximal turbidity facies represented by sandstones and conglomerates and distal facies formed by shales and claystones. Some of the stratigraphic units reach considerable thickness, e.g. >3 km thick Godula Formation of the Late Cretaceous age which underlies the highest ridges of the Moravskoslezské Beskydy Mts (Menčík et al. 1983).

Similarly, as for the other European Alpine mountain belts, a crucial milestone in the evolution of the flysch Carpathians was the compressional phase due to Africa-Eurasia collision in the Paleogene and Neogene leading to the evolution of nappe stacks, which were thrust northward onto the sediments of the Miocene foredeep and the basement of the marginal part of the Paleozoic Bohemian Massif (Picha et al. 2006). Thrusting in the studied area culminated in the Miocene before ~15 Ma. The region is divided into three major nappe groups, successively from the north: the lowermost Sub-Silesian nappe almost completely covered by higher units, Silesian nappe underlying the most elevated part of the mountains and Magura nappe with the highest structural position outcropping in the southern part of the area (Fig. 27.2). Structures related to nappe tectonics such as shear zones, thrust planes and folds are occasionally exposed within natural outcrops, forming sites of major geological interest. One of the most important geosites providing insight into the geological history of the region is an erosionally exposed contact between the Silesian and Sub-Silesian nappes in the incising channel of the Ostravice River (Fig. 27.3).

Post-collisional evolution of the mountain range in the Late Neogene and the Quaternary involved uplift and erosion accompanying the formation of contemporary surface. Geological constraints as well as apatite fission track and (U-Th)/He thermochronology point to >2 km-deep

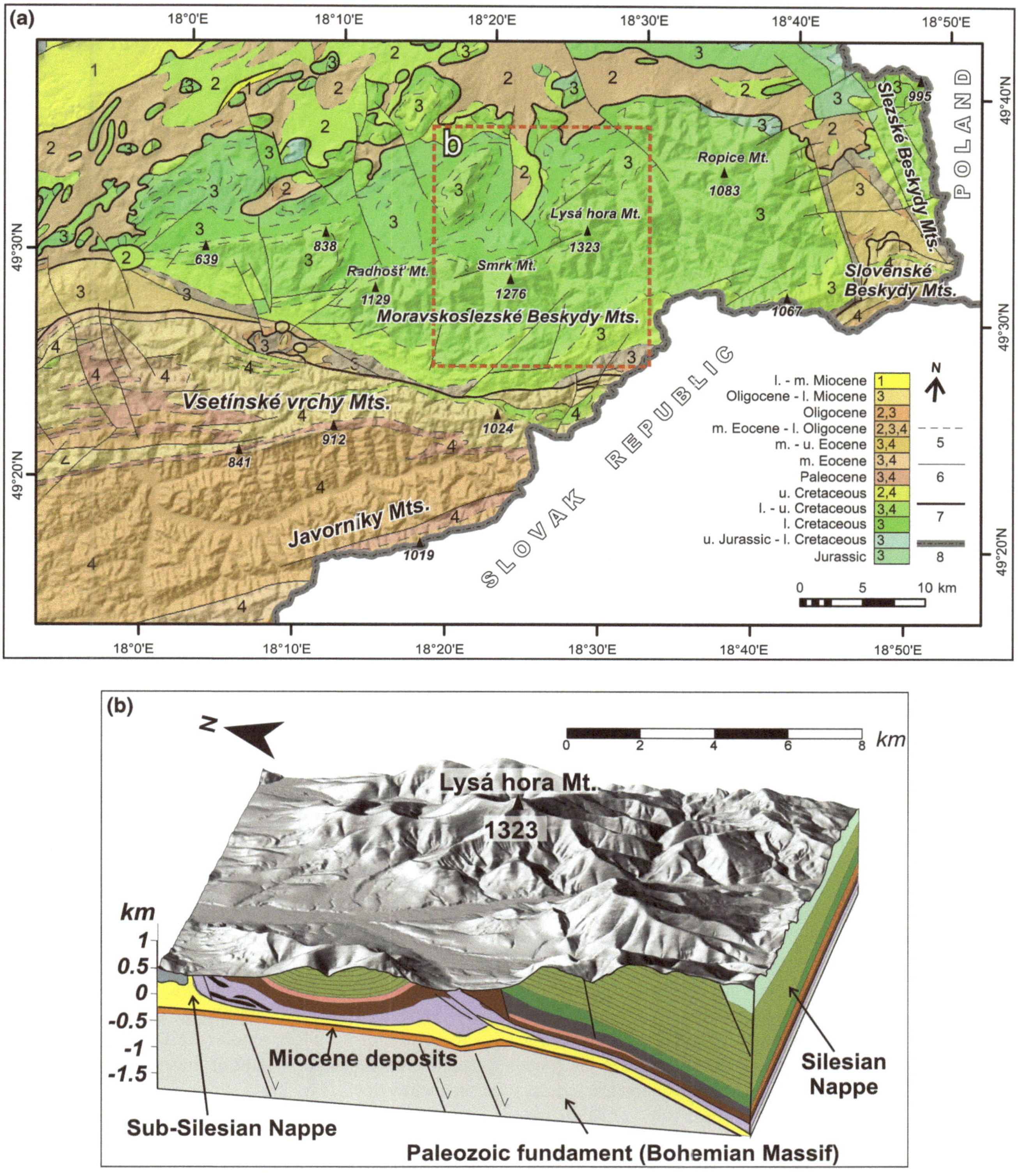

Fig. 27.2 **a** Geological sketch of the Moravskoslezské Beskydy Mts and surrounding highlands. Black dots mark the position of the most important crevice-type caves. *1* Sediments of the Carpathian Neogene foredeep (*lower* to *middle* Miocene), *2* flysch formations of the Sub-Silesian Unit (Outer Group of Nappes, *upper* Cretaceous–*lower* Miocene), *3* Flysch formations of the Silesian Unit (Outer Group of Nappes, *upper* Jurassic–lower Miocene), *4* flysch formations of the Raca Unit (Magura Group of Nappes, lower Cretaceous–lower Oligocene), *5* lithological boundary, *6* fault, *7* main thrust fault, *8* state boundary **b** Block diagram revealing the structure of the highest part of mountains (for location see the *red square* in the geological map)

Fig. 27.3 Natural outcrop within the Ostravice River channel (vicinity of the Ostravice village; Moravskoslezské Beskydy Mts) provides an insight into highly deformed basal sequence of the Silesian nappe thrust over the Sub-Silesian Unit (outcropping to the left from the photo) (*Photo* T. Pánek)

denudation since the Late Miocene (~9 Ma), i.e. defining the maximum age of gross landforms (Danišík et al. 2008). Since that time the main pattern of the mountain ridges and deeply incised river valleys has evolved. Climatic changes in the Quaternary dictated erosion-accumulation phases in the river valleys and at the foot of the mountain slopes. Despite the fact that the area is located at the southern limit of at least two advances of continental glaciation in the Middle Pleistocene, traces of mountain glaciation have not been identified (Nývlt et al. 2011).

27.2.2 Main Morphostructural Features

Main morphological features of the area (e.g. position and shape of main ridges and valleys) are influenced predominantly by (1) location of major thrust faults predisposing the main escarpments, (2) style of tectonic deformation of flysch beds (e.g. whether they form monoclinal structures or folds) and (3) thickness of individual flysch formations with distinct rock strength.

Morphostructural peculiarities are often closely related to individual nappes. The Silesian nappe, which is underlying the majority of the Moravskoslezské Beskydy Mts, is a large monoclinal structure, gently inclined (~5–20°) to the south and south-east. It is reflected in overall asymmetric, cuesta-like topography of the mountain range (Fig. 27.4). Monoclinal ridges with steep northward oriented escarpments and gently inclined opposite dip slopes are a characteristic feature of this area and we can observe this phenomenon in an excellent way along the highest part of the mountains, involving the ridges of Radhošť Mt (1,129 m a.s.l.), Kněhyně Mt (1,257 m a.s.l.), Smrk Mt (1,276 m a.s.l.), Lysá hora Mt (1,323 m a.s.l.) and Travný Mt (1,203 m a.s.l.). On the contrary, a higher degree of folding and the presence of anticlinal and synclinal belts within the Magura nappe led to the evolution of predominantly symmetrical ridges, sometimes revealing inverted topography with synclines forming the mountain ridges. Unlike in the Silesian nappe, the landscape within the Magura nappe is characterized by a dense trellis pattern of ridges and valleys, a typical feature of the Vsetínské vrchy Mts and Javorníky Mts (Fig. 27.4).

27.2.3 Rock Control

Rhythmicity and thickness of individual flysch formations of distinct rock strength are crucial factors influencing the morphology of both large- (e.g. ridges and valleys) and small-scale landforms (e.g. crags, minor structural escarpments and waterfalls). Massiveness of ridges is largely controlled by the rhythmicity of flysch layers and thickness of more competent formations involving sandstones and conglomerates. This phenomenon is particularly well developed within the Silesian nappe where sandstone-dominated, a few kilometres-thick Godula Formation forms long-wavelength topography with the highest local relief, whereas lithologically variegated Istebna Formation outcropping in the southern part of the Moravskoslezské Beskydy Mts predisposed short-wavelength system of lower, narrow ridges and subsequent valleys (Fig. 27.4).

Lithological variations are reflected in detail in the distribution of rock forms that mostly originated due to selective denudation of rigid sandstone and conglomerate beds. Step-like arrangement of rock faces with numerous outcrops of more resistant sandstones and conglomerates is a characteristic feature of the areas underlain by the Istebna Formation, especially along the border with Slovakia. Numerous small waterfalls throughout the entire mountain area are predisposed in a similar way. However, the most

Fig. 27.4 Landscape in the southern part of the Moravskoslezské Beskydy Mts (upper Morávka River catchment) with a characteristic dense network of structural ridges and valleys predisposed by alternations of competent sandstone/conglomerate and shale-dominated flysch layers within lithologically variegated Istebna Formation (*Photo* T. Pánek)

striking examples of rock control can be seen in the Javorníky Mts, at the SW margin of the area. Steeply inclined thick sandstone/conglomerate beds of the Luhačovice Member, forming anticline flanks within the Magura nappe, have been exposed whilst thin-bedded sandstones and shales were preferentially eroded away. One of the best examples of this phenomenon is the Pulčínské skály Rocks, the largest rock city in the Czech part of the Western Carpathians (Fig. 27.5). Uniqueness of this site attracts thousands of visitors every year and the whole locality has been protected as a National Natural Reserve since 1989.

27.3 Landslides: Deep-Seated Deformations and Catastrophic Events

Flysch rocks are extremely susceptible to various types of mass movements. In the described area, >1000 major landslides have been identified (Krejčí et al. 2002). Some mountain ridges are almost completely re-modelled by several generations of deep-seated landslides nesting numerous shallow instabilities such as debris slides and debris flows (Fig. 27.6). Main triggering factors for the

Fig. 27.5 Resistant sandstone layer of competent Luhačovice Member (Magura nappe) is a part of the SE limb of an anticline and has given rise to rock formations of the National Natural Reserve "Pulčínské skály Rocks" in the SW part of the Javorníky Mts (*Photo* T. Pánek)

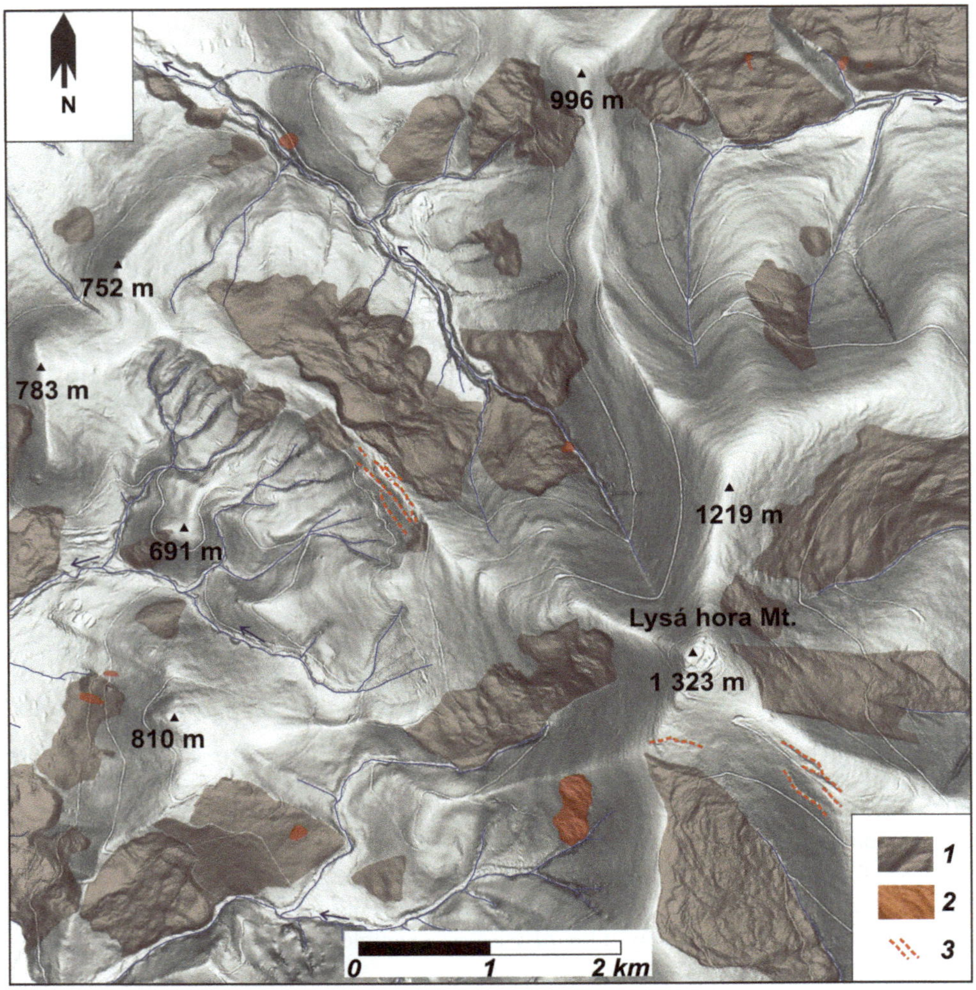

Fig. 27.6 LIDAR-based shaded relief (resampled to 5 × 5 m grid; provided by Czech Office for Surveying, Mapping and Cadastre) of the Lysá hora Mt revealing substantial modification of the topography by landslides. Note only a small proportion of the area affected by recent sliding activity. *1* ancient deep-seated landslides, *2* active landslides and landslides that originated since extreme rainfalls in July 1997, *3* Sackung features (tension cracks, double ridges and anti-slope scarps)

origin of landslide are long-lasting heavy rainfalls, occasionally combined with rapid snowmelt (Krejčí et al. 2002; Pánek et al. 2011). A majority of the landslides are ancient and stabilized, but several extreme hydrometeorological events during the last decades have revealed a great potential for reactivations of older instabilities as well as the occurrence of completely new slope failures. For instance, several hundreds of landslides originated during heavy rainfalls in July 1997 (maximum recorded five-day precipitation totals in the Moravskoslezské Beskydy Mts were 616 mm!; Brázdil et al. 2005); more than 300 mm of precipitation during three days in May 2010 led to the occurrence or reactivation of >150 slope failures (Pánek et al. 2011). As it will be demonstrated in the following chapters, landslides represent not only a major threat to local people in the landscape of flysch Carpathians, but they also contribute to geodiversity and overall natural beauty of the Moravskoslezské Beskydy Mts and surrounding highlands.

27.3.1 Landslides as a Major Landscape Features in the Region

Distribution, types and activity of landslides are controlled by the nature of flysch substratum. Thick-bedded and jointed sandstone-dominated layers of the Godula Formation in the Silesian nappe are affected by deep-seated landslides with sackung features such as ridge-top grabens, crevices and counter-slope scarps (Fig. 27.7). Monoclinal ridges within this geological domain are affected by vast translational rockslides affecting south-oriented dip slopes, whereas north-facing escarpments are deformed by rotational landslides with pronounced, amphitheatre-like headscarps. Recent landslide activity within the Silesian nappe is rather negligible. On the contrary, flysch rocks within the Magura nappe contain a higher percentage of shales and claystones producing regolith that is extremely prone to earthflows and debris slides and thus a majority of recent sliding activity occurs in this geological domain (Fig. 27.7).

Fig. 27.7 **a** Typical graben with rock outcrops in the upper part of an ancient deep-seated landslide affecting the sandstone-dominated Godula Formation in the Silesian nappe (the Čantoryje Mt/Slezské Beskydy Mts). **b** Recurrent landslide/earth flow in the Malá Brodská Valley (Vsetínské vrchy Mts; thin-bedded flysch of the Magura nappe) with the last major activity documented during extreme rainfalls in July 1997. As stated by radiocarbon dating of sediments from a nearby landslide-dammed lake, previous landslide originated in the same slope section ~1.1 cal ka BP (*Photo* T. Pánek)

At the slope scale, landslides represent major geomorphic features in the presented region (Fig. 27.6). However, the influence of some landslides extends well outside the limits of individual hillslopes. There are numerous examples of watershed lowering, narrowing and damming of valley floors, increasing steepness of valleys, or even major reorganization and piracy within the river valley network (Baroň et al. 2014).

Recent investigation aimed at radiocarbon dating of landslides has revealed that landslides have originated throughout the whole Holocene, with a few cases dating back to the Late Glacial. The majority of events is dated to the Late Holocene, i.e. Subboreal and Subatlantic chronozones (Pánek et al. 2013). Such a temporal bias reveals that younger landslide events and erosion often removed morphological and sedimentary legacy of older mass movements. The absence of data related to the Late Pleistocene landslides confirms this statement.

27.3.2 The May 2010 Girová Landslide: Example of a Recurrent Story

Morphological and sedimentological evidence shows that some ancient landslides in the region originated as catastrophic failures (rockslides, rock avalanches, etc.; Pánek et al. 2009). The May 2010 Girová landslide in the Slovenské Beskydy Mts (Magura nappe) provided an excellent contemporary lesson about such catastrophic events (Fig. 27.8).

The landslide originated on the southern slope of Mt Girová (839 m a.s.l.) during the night from 18 to 19 May 2010, at the termination of an intensive three-day rainfall event with precipitation totals >300 mm. According to an eyewitness account, the collapse started in the upper part of the slope where a pronounced, 25 m-high wedge-like rocky headscarp evolved within a few hours (Pánek et al. 2011). Translational displacement of this deep-seated "key block" significantly destabilized and unloaded the slope and caused other minor collapses. Frontal part of the slide subsequently turned into earthflow that was moving downslope for several following days and reached its contemporary position on 22 May 2010.

Detailed investigation involving geophysical measurements revealed that the landslide originated in shales along deeply weathered fault zone trending in an approximately north–south direction. Furthermore, the landslide is a part of an extensive deep-seated slope deformation of the area of ~1.5 km^2 that produced several slope failures in the Holocene history. Radiocarbon dating of these accumulations revealed that at least one Holocene long runout landslide (~7.5 cal ka BP) and several smaller slump-like failures (~1.5 and ~0.6 cal ka BP) had preceded this recent catastrophic failure (Fig. 27.8; Pánek et al. 2011).

With its area reaching 20 ha and length almost 1200 m, the May 2010 Girová landslide is one of the largest catastrophic landslides which originated in the Czech Republic during the last decades. Except for some economic losses (clearance of forest), it fortunately caused no fatalities and major damages on infrastructure, although the nearest inhabited houses are located just a few metres off the contemporary rocky headscarp. The existence of landslide of such dimensions, which is easily accessible to tourists, could

Fig. 27.8 The May 2010 Girová landslide (Slovenské Beskydy Mts). Forested slope on the right side of the photo is affected by a deep-seated landslide with numerous crevice-type caves (*Photo* T. Pánek)

be used for educational purposes as a clear demonstration of how local geological and geomorphological settings control high-magnitude recurrent slope failures.

27.3.3 Landslides and Geodiversity

Rock forms, caves, boulder accumulations, scree slopes, small lakes and peat bogs constitute a list of most typical habitats connected with landslides in the Moravskoslezské Beskydy Mts and their surroundings (Fig. 27.9). Mass movements significantly increase geodiversity of the flysch landscape (i.e. diversity of landforms, soil and water) that, in other respects, is characterized by smooth topography and absence of rock outcrops (Alexandrowicz and Margielewski 2010).

The majority of specially protected areas such as national natural reserves, natural reserves and natural monuments in the area is directly or indirectly connected with diverse topography of ancient landslide bodies. For instance, one reason for the establishment of the Ropice Natural Reserve (the Ropice Mt, Moravskoslezské Beskydy Mts) in 2010 was the existence of a highly dissected landslide terrain with numerous rock forms, boulders and scree slopes forming habitats for endangered species such as European lynx and Western Capercaillie. Other conservation areas protect aquatic habitats within landslides, e.g. peat bogs in near-scarp depressions and wetlands and small lakes behind landslide-blocked valley sections. Landslide peat bogs are also of major scientific importance, because they are often the only natural archives of Late Glacial–Holocene paleoenvironmental changes (Pánek et al. 2013).

Besides the specific topography, the reason why landslide terrains host some of the last wild areas within the Outer Western Carpathians is their inaccessibility. Remote position protects these places from massive forest clearance and thus allows the conservation of Carpathian Virgin Forests (e.g. acidophilic spruce forests and beech-fir forests) with some trees exceeding the age of 300 years. The best example is the Kněhyně-Čertův mlýn National Natural Reserve, with the last preserved ecosystem of an original spruce forest occupying the upper parts of an ancient deep-seated landslide.

27.4 Crevice-Type Caves: Specific Features of the Region

Among the most spectacular and fascinating phenomena connected with slope deformations in the area are crevice-type caves (Wagner et al. 1990; Lenart et al. 2014). These landslide elements significantly increase geodiversity of the landscape. Providing direct insight into the landslide bodies, they are also of major scientific importance for geoscientists. Crevice-type caves originate during gravitational widening (spreading, toppling, rotational movements, etc.) of discontinuities within flysch bedrock. The movements are usually very slow and reach the orders of 10^{-2}–10^{-1} mm/year (Klimeš et al. 2012). In contrast to the karst caves, the development of secondary speleothems in the

Fig. 27.9 Landslide elements increasing geodiversity of the area. **a** Rock form called "Devil's Table" situated in the depletion area (double ridge) of a deep-seated slope deformation (Čertův mlýn Mt; Moravskoslezské Beskydy Mts) **b** Blockfield in the accumulation part of the landslide in the Kobylík Valley (southern slopes of the Lysá hora Mt; Moravskoslezské Beskydy Mts). **c** The small landslide-dammed lake originated due to the July 1997 landslide in the Malá Brodská Valley (Vestínské vrchy Mts; photo taken in May 2006) (*Photo* T. Pánek)

crevice-type caves is very poor. Sporadically, there are small straw stalactites (10^{-1}–10^{1} cm) hanging on lower bedding planes or rare flowstones on the walls.

More than 100 crevice-type caves have been explored in the Moravskoslezské Beskydy Mts and the surrounding highlands (Fig. 27.2). Most of them do not reach the depth greater than 12 m and they are not longer than 50 m. These small caves are relatively frequent and may accompany the evolution of many slope failures. Crevice-type caves of greater dimensions are rare. Two most important ones of them, i.e. the Cyrilka Cave (longest) and the Kněhyňská Abyss (deepest), are described in the following text in detail.

27.4.1 Cyrilka Cave: Longest Crevice-Type Cave in the Czech Republic

The Cyrilka Cave forms a unique maze of subparallel underground crevices in three levels. The cave is 535 m long and its bottom is situated 16 m below the topographic surface. The entrance is situated at the bottom of a shallow trench under the landslide head scarp near the Pustevny settlement on the main ridge of the Moravskoslezské Beskydy Mts. The cave has been known in the community of local people for centuries. Many adventurers equipped only with a rope and candle tried to explore the cave, but they never reached the end. Old fables mention treasures guarded by a huge black dog (Wagner et al. 1990). Worshippers of the pagan god Radegast could hide religious documents in the cave and there are also stories about the Radegast's statue made from pure gold and hidden deep within the cave. Nevertheless, these treasures were never found. What is more plausible is the exploitation of the cave entrance by shepherds after the Wallachian colonization in the sixteenth century. Sheepfolds were built very close to the present entrance in the Pustevny settlement and there are still well-preserved pastures next to the cave. It is likely that old shepherds stored milk products in the cold entrance to the crevice. The oldest mention of the cave in the literature comes from 1639 (Skutil 1957), but modern exploration of the cave started in the nineteenth century. The first cave plan was published in 1953 (Tučník 1953) and the last discoveries in 2011 established the length of the system to 535 m (Wagner and Lenart 2012).

Narrow entrance leads into the entrance hall which is one of the most spacious rooms in the cave. The crevices are organized into three levels connected by small scuttles and up to a few metres deep abysses (Fig. 27.10). The underground space is horizontally dissected into three segments by two fault zones crossing the cave in the W–E direction.

Fig. 27.10 Narrow and high crevices within the old part of the Cyrilka cave are formed in the Pustevny sandstones (Silesian nappe, The Radhošť ridge, Moravskoslezské Beskydy Mts (*Photo* J. Lenart)

Each of these three segments has a slightly different genesis, dictated by the prevailing style of gravitational movement that influenced the origin of the crevices. Several small stalactites formed on the ceilings of the cave increase the uniqueness of the cave system. The crevices of the Cyrilka Cave are a hibernaculum for several species of bats (*Myotis myotis*, *M. mystacinus*, *M. emarginatus*, *Rhinolophus hipposideros*), i.e. species protected within the scope of the European Ecological Network NATURA 2000.

27.4.2 Kněhyňská Abyss: 57 m Deep into the Landslide

The spectacular system of deep underground crevices allows reaching 57 m under the surface of the landslide body. The cave entrance is situated at the bottom of the rocky trench in the central part of the vast landslide close to the Kněhyně Mt (1257 m a.s.l.) in the central part of the Moravskoslezské Beskydy Mts. Chambers and crevices lead the visitors to the bottom where continuing fissures are further inaccessible. The underground labyrinth of huge rock blocks broken by widened cracks forms a rugged abyss with very specific speleomorphology (Fig. 27.11). There is one of the biggest cave chambers in flysch Carpathians (11 × 2 × 8 m; Fig. 27.11), up to 12 m-deep abysses, passages filled with huge sandstone blocks and narrow but long crevices connected by small scuttles. The overall cave area is laterally predisposed by two distinctive fault zones which are detectable through slickensides and zones of tectonic breccia on lateral walls. The cave system was declared a natural monument in 1990. Tens of bats winter in the cave every year.

27.5 Human Activity and Mountains

In comparison with the highlands in the Bohemian part of the Czech Republic (e.g. Sudetes), which have been intensively settled and used by people at least since the medieval times, the highest parts of flysch Carpathians including the Moravskoslezské Beskydy Mts were inhabited by people several centuries later. Although Margielewski et al. (2010) noted locally important human influence on the mountain ridges in adjacent Polish Flysch Carpathians since the Neolithic, the majority of the mountainous landscape of the flysch Carpathians in the Czech Republic remained to be a pristine landscape until sixteenth–seventeenth centuries. Although most of the region has been a part of the Beskydy Protected Landscape Area since 1973, human activities still pose one of the threats to the landscape and landforms on the "roof" of the Czech Western Carpathians.

27.5.1 Man as a Geomorphological Agent

The beginnings of human-induced transformation of the natural landscape in the Moravskoslezské Beskydy Mts and surrounding highlands are dated back to the so-called Wallachian colonization in the sixteenth–seventeenth centuries. This colonization was a process that started in the thirteenth–fourteenth centuries in the Southern Carpathians (contemporary Romania) and terminated during the sixteenth–seventeenth centuries in the territory of the Czech Western Carpathians (Štika 2007). The population of the mountainous easternmost region of the Czech Republic increased during that period by 650 % and the area was markedly deforested (Jančík 1958). As a consequence of this dramatic change of the landscape, gully and sheet erosion as well as shallow landslides accelerated. Radiocarbon dating of floodplain sequences in the piedmont of the mountains revealed the onset of massive deposition of overbank facies

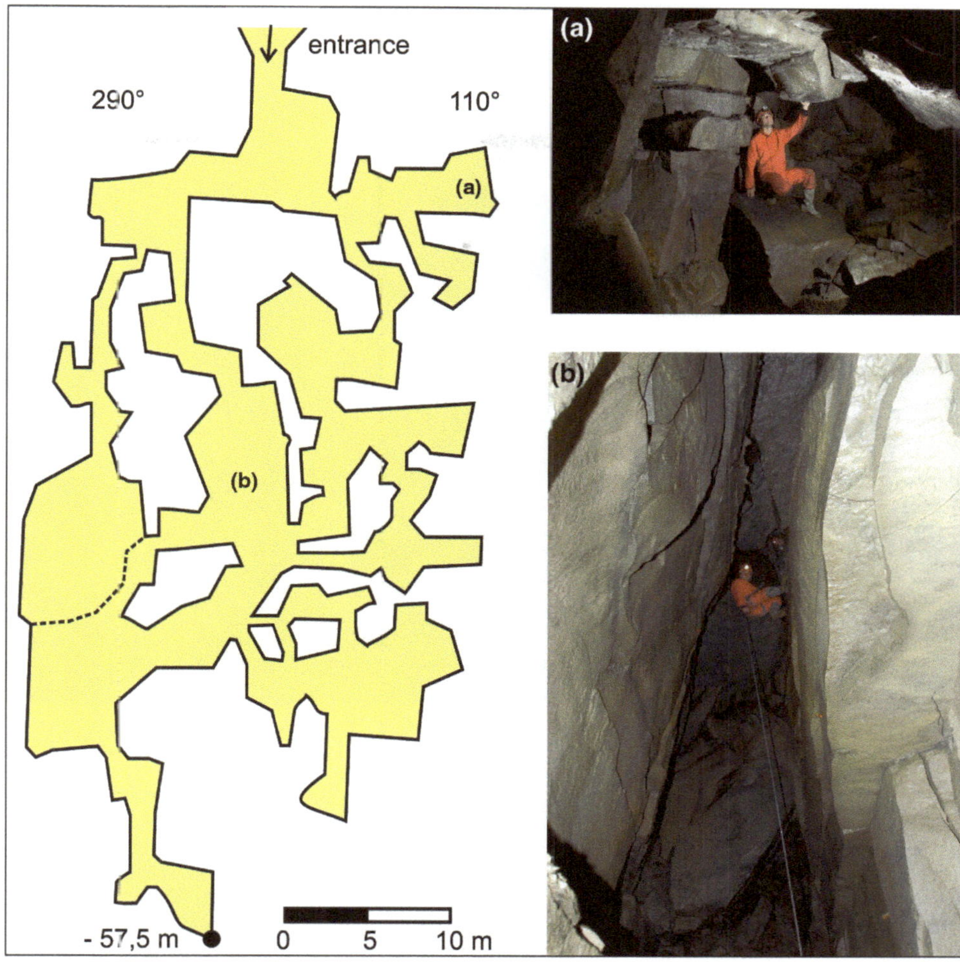

Fig. 27.11 Idealised cross-section of the Kněhyňská Abyss in the Moravskoslezské Beskydy Mts (Silesian nappe). **a** Upper chambers are formed within joint-guided crevices and partly between fallen boulders. **b** The Big Abyss in the central part of the cave: 12 m-deep abyss is the biggest vertical crevice in the Czech Outer Western Carpathians (*Photo* J. Wagner, cross-section after Wagner et al. 1990)

in the sixteenth–seventeenth centuries, corresponding to the deforestation of the mountain part of watersheds (Stacke et al. 2014).

Although the deforestation of the mountains terminated during the nineteenth century and the majority of slopes was afforested at the turn of the twentieth century, forest management, river regulations, building of dams and increasing tourism (e.g. construction of hotels and ski resorts) have brought additional negative impact on the landscape and landforms. For instance, channel straightening, reinforcement of river banks and construction of check dams led to dramatic incision of channels at the mountain piedmont. The incision of the Morávka River reached up to 8 m in the last 40 years, while accelerated erosion has recently been migrating upstream to the mountainous part of the watershed (Škarpich et al. 2013). Forest management connected with timber production contributes crucially to erosion and degradation of forest soils. As noted by Buzek (2007), the rate of water erosion has increased by 300 % since the network of forest dragging roads was built. Last but not least, geologically fragile flysch substratum is also degraded substantially by tourists, especially in the area of major touristic centres such as the Lysá hora Mt or the Radhošť ridge. For instance, the Lysá hora Mt is visited by more than 200,000 tourists every year, which has a negative effect on the face of the mountain (Fig. 27.12).

27.5.2 Cultural Value and Tourism Attractiveness

Elevated ridges with rock forms, waterfalls and crevice-type caves overgrown with dense forests have attracted tourists for more than one hundred years (Fig. 27.13). First cottages appeared at the turn of the twentieth century and since that time several tens of mountain resorts have been built. The whole area is interconnected with a well-marked network of tourist routes and some of them are used as educational trails. For instance, the educational trail around the Čertův Mlýn Mt (1206 m a.s.l.) explains the genesis of double ridges and other distinct landforms that originated due to ancient mass movement processes. Other trails with educational panels concerning geomorphic

Fig. 27.12 Hiking trail approaching the top of the Lysá hora Mt from the north is passed every year by approximately 100,000 tourists (*Photo* T. Pánek)

Fig. 27.13 Historical postcard of the Čeladná village dated back to the beginning of twentieth century. A small village at the foot of the highest part of the Moravskoslezské Beskydy Mts has grown into an important tourist resort. On the background large periglacial blockfields (nowadays densely forested) are visible covering slopes of the Malý Smrček Mt. Source: Beskydy Museum in the Frýdek-Místek town

phenomena can be found, e.g. on the Lysá hora Mt and around the Mionší National Natural Reserve.

Many geomorphic features played an important role in the regional mythology. Rock forms and cracks that originated on some mountains due to ancient landslides were considered by local people to be a result of supernatural forces. According to these stories, some curious rocks were created by devils (e.g. the Devil´s Table; Fig. 27.9), while crevice-type caves were shelters of legendary brigands or sleeping armies of knights (e.g. the Čantoryje Mt). Apart from the scientific values and natural beauties of the landslide landscape of the Moravskoslezské Beskydy Mts and surrounding flysch highlands, legends help to enhance its cultural value.

27.6 Conclusion

Although the area of the Moravskoslezské Beskydy Mts lacks the most evident geomorphic attractions such as rock cities in the north-eastern Bohemia or glacial cirques in the Krkonoše Mts, the region contains a particularly rich assemblage of landforms connected with mass movements. Such landforms provide a scientifically unique insight into the dynamics of Late Quaternary hillslope processes which have endangered human activities for several last centuries. The landforms also demonstrate that landslides belong to major agents diversifying flysch landscape, with rocky scarps, crevice-type caves, ancient landslide-dammed lakes and peat bogs. However, negative traces have been left on the face of the mountains by intensive human activity since the sixteenth–seventeenth centuries along with hundreds of thousands of visitors coming every year. It is of major importance—both for scientists and tourists—to conserve the landscape features that could be used as textbook examples for the understanding of dynamic geomorphic processes shaping the mountain slopes on mechanically weak flysch rocks.

References

Alexandrowicz Z, Margielewski W (2010) Impact of mass movements on geo- and biodiversity in the Polish Outer (Flysch) Carpathians. Geomorphology 123:290–304

Baroň I, Bíl M, Bábek O, Smolková V, Pánek T, Macur L (2014) The effect of slope failures on river-network pattern: a river piracy case study from the Flysch Belt of Outer Western Carpathians. Geomorphology 214:35–365

Brázdil R, Dobrovolný P, Elleder L, Kakos V, Kotyza O, Květoň V, Macková J, Müller M, Štekl J, Tolasz R, Valášek H (2005) Historical and Recent Floods in the Czech Republic. Masarykova univerzita and Český hydrometeorologický ústav, Brno-Praha (in Czech with English abstract)

Buzek L (2007) Water erosion in the watershed of the upper Ostravice river from 1976 to 2000. Moravian Geogr Rep 14:2–12

Danišík M, Pánek T, Matýsek D, Dunkl I, Frisch W (2008) Apatite fission track and (U-Th)/He dating of teschenite intrusions gives time constraints on accretionary processes and development of planation surfaces in the Outer Western Carpathians. Zeitschrift für Geomorphologie, NF Hauptbänd 52:273–289

Jančík A (1958) Odlesňování Těšínska v minulosti. Sborník Československé akademie zemědělských věd—Lesnictví 4:1019–1036 (In Czech)

Klimeš J, Rowberry MD, Blahůt J, Briestenský M, Hartwich F, Košťák B, Rybář J, Stemberk J, Štěpančíková P (2012) The monitoring of slow-moving landslides and assessment of stabilisation measures using an optical–mechanical crack gauge. Landslides 9(3):407–415

Krejčí O, Baroň I, Bíl M, Jurová Z, Hubatka F, Kirchner K (2002) Slope movements in the Flysch Carpathians of Eastern Czech Republic triggered by extreme rainfalls in 1997: A case study. Phys Chem Earth 27:1567–1576

Lenart J, Pánek T, Dušek R (2014) Genesis, types and evolution of crevice-type caves in the Flysch Belt of the Western Carpathians (Czech republic). Geomorphology 204:459–476

Margielewski W, Krąpiec M, Valde-Nowak P, Zernitskaya V (2010) A Neolithic yew bow in the Polish Carpathians: Evidence of the impact of human activity on mountainous palaeoenvironment from the Kamiennik landslide peat bog. Catena 80:141–153

Menčík E, Adamová M, Dvořák J, Dudek A, Jetel J, Jurková A, Hanzlíková E, Houša V, Peslová H, Rybářová L, Šmíd B, Šebesta J, Tyráček J, Vašíček Z (1983) Geologie Moravskoslezských Beskyd a Podbeskydské pahorkatiny (Geology of the Moravskoslezské Beskydy Mountains and Podbeskydská pahorkatina hilly country). Ústřední Ústav Geologický, Praha (in Czech, with English Summary)

Nývlt D, Engel Z, Tyráček J (2011) Pleistocene Glaciations of Czechia. In: Ehlers J, Gibbard PL, Hughes PD (eds) Developments in Quaternary Science 15, pp 37-46

Pánek T, Hradecký J, Minár J, Hungr O, Dušek R (2009) Late Holocene catastrophic slope collapse affected by deep-seated gravitational deformation in flysch: Ropice Mountain, Czech Republic. Geomorphology 103:414–429

Pánek T, Šilhán K, Tábořík P, Hradecký J, Smolková V, Lenart J, Brázdil R, Kašičková L, Pazdur A (2011) Catastrophic slope failure and its origins: Case of the May 2010 Girová Mountain flow-like rockslide (Czech Republic). Geomorphology 130:352–364

Pánek T, Smolková V, Hradecký J, Sedláček J, Zernitskaya V, Kadlec J, Pazdur A, Řehánek T (2013) Late-Holocene evolution of a floodplain impounded by the Smrdutá landslide, Carpathian Mountains (Czech Republic). Holocene 23:218–229

Picha FJ, Stráník Z, Krejčí O (2006) Geology and Hydrocarbon Resources of the Outer Western Carpathians and Their Foreland, Czech Republic. In: Golonka J, Picha FJ (eds) The Carpathians and Their Foreland: Geology and Hydrocarbon Resources. The American Association of Petroleum Geologists, Tulsa, Oklahoma, pp 49–175

Skutil J (1957) Radhošťské ďůry a přání s nimi spojené. Zprávy krajského muzea v Gottwaldově 3:2–3 (In Czech)

Stacke V, Pánek T, Sedláček J (2014) Late Holocene evolution of the Bečva River floodplain (Outer Western Carpathians, Czech Republic). Geomorphology 206:440–451

Škarpich V, Hradecký J, Dušek R (2013) Complex transformation of the geomorphic regime of channels in the forefield of the Moravskoslezské Beskydy Mts.: Case study of the Morávka River (Czech Republic). Catena 111:25–40

Štika J (2007) Valaši a Valašsko: o původu Valachů, valašské kolonizaci, vzniku a historii moravského Valašska a také o karpatských salaších. Valašské muzeum v přírodě, Rožnov pod Radhoštěm, 237 pp (In Czech)

Tučník D (1953) Radhošťské jeskyně. (Pukliny na Pustevnách na hoře Radhošti). Československý kras 6:185–186 (In Czech)

Wagner J, Lenart J (2012) V jeskyni Cyrilce překonána délka půl kilometru Speleofórum 31: 53–57 (In Czech)

Wagner J, Demek J, Stráník Z (1990) Jeskyně Moravskoslezských Beskyd a okolí. Knihovna České speleologické společnosti, Praha, 132 pp (In Czech, with English Summary)

Strážnické Pomoraví—Holocene Evolution of a Unique Floodplain and Aeolian Landforms

28

Zdeněk Máčka and Jaroslav Kadlec

Abstract

The Morava River drainage basin is the largest fluvial system of the eastern part of the Czech Republic. The Morava enters the Dolnomoravský úval Basin (northern tip of the Vienna Basin) at its lower course, where a 3.5 km wide floodplain is developed. The most interesting section of floodplain may be found between the towns Veselí nad Moravou and Hodonín, where Pleistocene and Holocene sediments of the Morava River are accompanied by the unique complex of lacustrine sediments remodelled by the wind action to the shape of up to 10 m high sand dunes. The Morava river was branching into many large as well as small arms in its floodplain, creating an anastomosed channel pattern. Diverse mosaic of aquatic and (semi)terrestrial habitats were present as it is displayed on old maps of the floodplain. The majority of small anastomosed channels vanished due to the river regulation works started in the nineteenth century and most of the river flow was concentrated into one dominant channel. This channel was affected by substantial deepening, widening and lateral migration in the second half of the twentieth century triggered by river regulation in 1930s. The aerial extent of floodplain inundation was reduced to approximately one-fourth of its original extent due to the construction of flood defence dykes. The Strážnické Pomoraví region is one of last remaining examples of a lowland meandering river with more or less preserved natural dynamics of fluvial processes in the Czech Republic.

Keywords

The Morava River • Anastomosing • Meandering • Sedimentary archive • Floods • Aeolian landforms • Human impact

28.1 Introduction

The section of the Morava River floodplain in Strážnické Pomoraví (named after the town of Strážnice) is one of last remaining regions where a large lowland river freely meanders in its several kilometres wide floodplain (Fig. 28.1). Sustained quasi-natural dynamics of fluvial processes of a lowland river is rather a unique phenomenon in Middle Europe due to the long tradition of river engineering spanning back to the nineteenth century. The rarity of the Strážnické Pomoraví landscape lies also in the connection of rich fluvial sedimentary archives with aeolian sands of "Moravian Sahara". In a small area, two different worlds are intimately interlinked—a flat alluvial plain of meandering river covered by arable land and managed forests meeting the rolling topography of sand dunes overgrown by pine forests. Though, Strážnické Pomoraví is also

Z. Máčka (✉)
Department of Geography, Faculty of Science, Masaryk University, Brno, Czech Republic
e-mail: macka@sci.muni.cz

J. Kadlec
Laboratory of Geomagnetism, Institute of Geophysics AS CR, v.v.i, Prague, Czech Republic
e-mail: kadlec@ig.cas.cz

T. Pánek and J. Hradecký (eds.), *Landscapes and Landforms of the Czech Republic*, World Geomorphological Landscapes, DOI 10.1007/978-3-319-27537-6_28

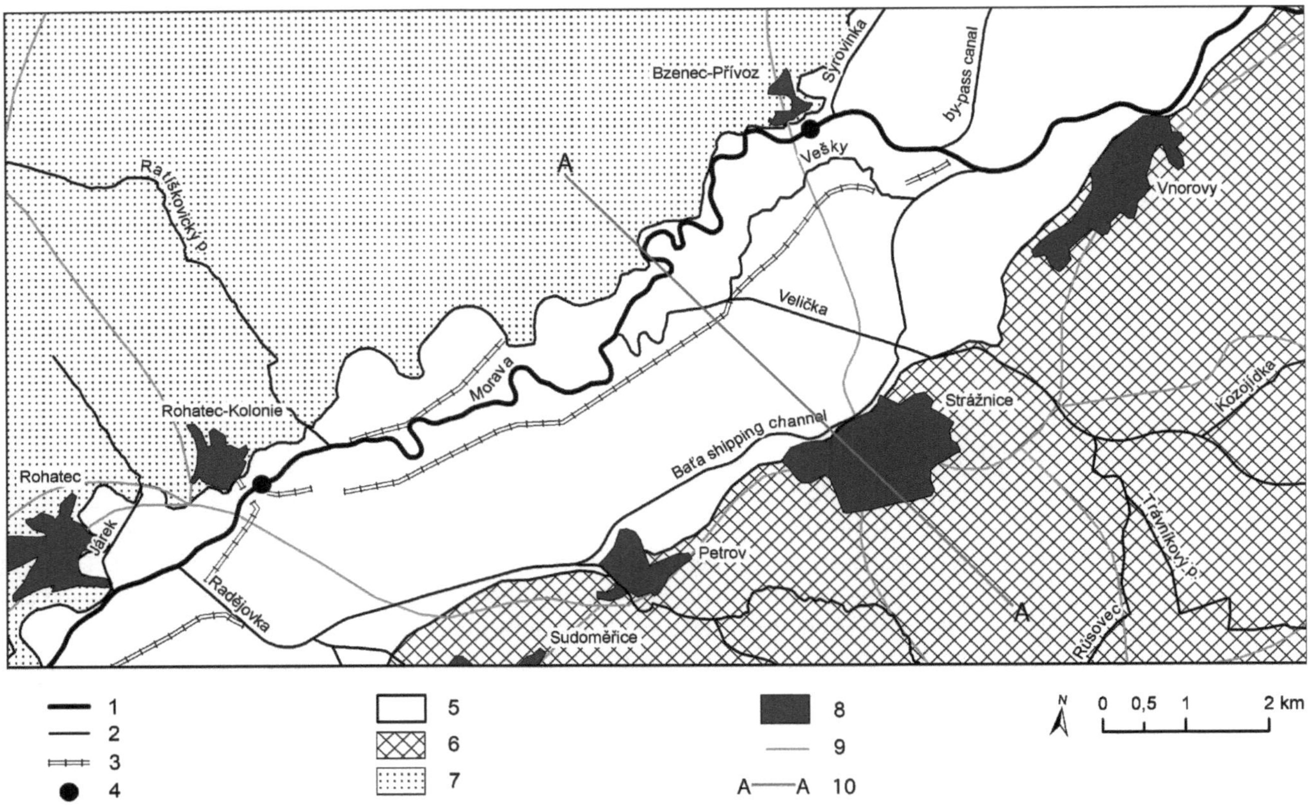

Fig. 28.1 The Morava River floodplain within the Strážnické Pomoraví region. Legend: *1* modern river bed; *2* side river branches and tributaries; *3* flood defence dyke; *4* hydrological station; *5* floodplain; *6* Mesozoic and Tertiary flysch rocks; *7* Upper Pleistocene (aeolian) sands; *8* settlement; *9* road; *10* line of geological section (see Fig. 28.3)

a landscape bearing the stigmas of human impact, a landscape profoundly moulded by humans since the early medieval times. The folk traditions of the past are still alive in the Strážnice region and the town hosts annually thousands of visitors of the International folklore festival. The slopes of flysch hills in the wider surrounding of Strážnice are covered with vineyards and wine production has given its unmistakable imprint into the local folk architecture (Fig. 28.2).

28.2 The Setting

28.2.1 The Morava River: Natural Backbone of the Strážnické Pomoraví Cultural Landscape

The Morava River drainage basin is the largest fluvial system in the eastern part of the Czech Republic. Above its confluence with the Danube, the Morava has a length of 353.1 km with a catchment area of 26,578 km^2. The Morava rises in the Králický Sněžník Mts. close to the Polish border at an altitude of 1,371 m and leaves the territory of the Czech Republic in the lowlands of southeastern Moravia, at an altitude of 148 m. The lower course of the river is located in the Dolnomoravský úval Basin (northern tip of the Vienna Basin, altitude between 150–200 m), where its floodplain is 3.5 km wide on average (Kirchner and Nováček 1991). Anastomosed channel pattern with meandering as well as straight branches was typical for the lower course of the Morava as long as to the first half of the twentieth century. As a result of river regulation, the flow in many anastomosed channels was reduced and discharge has concentrated primarily into a single channel. The area described in this chapter is, in fact, a small segment of the riverscape stretched along the 12.5 km long section of the Morava River between the settlements of Strážnice, Bzenec-Přívoz and Rohatec in the Czech historical land of Moravia. Mean annual discharge at the gauging station Strážnice is 59.6 $m^3 s^{-1}$. The Morava River is intimately woven into the history of western Slavic tribes that established their first empire "Great Moravia" (Moravia Magna) along its lower course in the ninth century.

Fig. 28.2 Painted facades of wine cellars in Petrov, the settlement placed at the margin of the Morava River floodplain (*Photo* V. Kubík)

28.2.2 Geological Setting

Geologically, Strážnické Pomoraví belongs to the Moravian part of the Vienna Basin. Sedimentary strata of the basin consist of Neogene claystones, siltstones and sandstones of Miocene age. Quaternary sediments on the right flank of the floodplain are represented by Upper Pleistocene aeolian sands of "Moravian Sahara". Aeolian sands cover the older Quaternary fluvial and Neogene marine sediments (Fig. 28.3). Aeolian sediments are undercut by lateral erosion of the river and locally the sprinkling sand banks up to 10 m high are developed (Fig. 28.4). (Hence the local name Osypané břehy, i.e. Sprinkling banks, comes) Aeolian sands are fine to coarse grained, yellow-brown in colour, with prevailing quartz (80–90 %), forming rolling topography of sand dunes up to 10 m high. Deposition of sand had not been continual, but interrupted by hiatuses of variable duration (Havlíček et al. 2008). Higher terrain in the wider

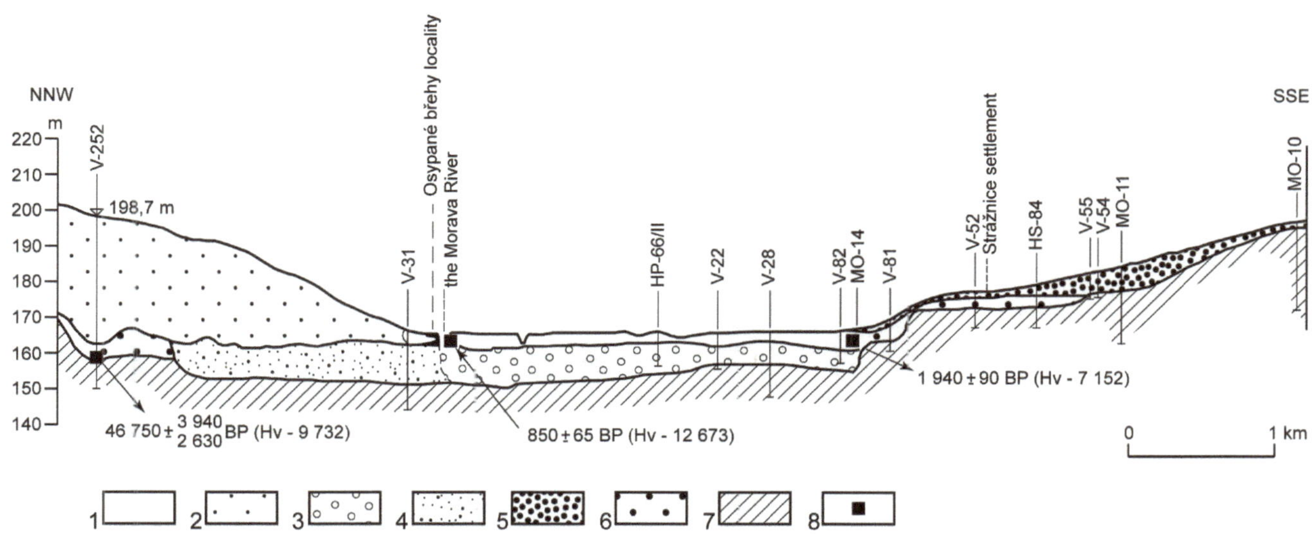

Fig. 28.3 Geological section across the Morava River floodplain between Bzenec-Přívoz and Strážnice. *1* flood loams. *2* aeolian sands; *3* fluvial sandy gravels (Holocene to Upper Pleistocene); *4* fluvial sandy gravels (Upper Pleistocene); *5* coarse grained sandy gravels of alluvial fans, locally with intensive cryoturbations; *6* loamy fluvial sandy gravels (Middle Pleistocene, two height levels); *7* Neogene sands and clays; *8* radiocarbon sample points with laboratory codes (Hv = Hannover). © Havlíček et al. (2008), Vydavatelství ČGS

Fig. 28.4 Eroded bank of the Morava River in the downstream part of the meander loop in the Osypané břehy locality. Bank erosion has exposed sediments of the sand dune of "Moravian Sahara" (*Photo* L. Krejčí)

surroundings is built by flysch sediments of Magura group of nappes of the Outer Western Carpathians (Bílé Karpaty and Rača units), that were thrust from SE to NW.

28.3 The Morava River Floodplain as a Unique Natural Archive of Fluvial Sedimentation

Alluvial fine-grained Holocene sediments of the Strážnické Pomoraví floodplain (overlying Upper Pleistocene sandy gravels) may serve as a natural archive. Overall thickness of fluvial sediments of Last Glacial and Holocene age, overlying the Tertiary clays, is up to 17 m (Fig. 28.3). Stehlík and Kadlec (2012) discriminated three distinct erosional phases separated by aggradation periods dating from the Late Glacial to the present.

The first erosional phase is dated to the transition from Allerød to Younger Dryas within the Late Glacial. The Morava created excessive channel with meander belt much wider than today and rapidly incised by 15 m into the neighbouring sandy sediments. River channels from that period (13.5 ky old) are still preserved at the NW margin of the floodplain in the form of large meander bends cut into the adjacent slopes built of aeolian sands (Fig. 28.5).

The second erosional phase was triggered by climatic changes at the glacial/interglacial transition, from Younger Dryas to Preboreal. An active meander belt shifted from NW margin of the floodplain to the present-day position and an early Holocene river incised deeply into its own deposits and locally even into underlying Tertiary clays. The incised valley of early Holocene was gradually filled with clayey sediments during the warm and wet Atlantic stage between 9 and 5.5 ky BP. The higher flanks of the floodplain were not affected by the Atlantic sedimentation and this type of deposits is lacking there. ^{14}C ages of sediments suggest that the sedimentation within the whole width of the floodplain was reactivated at the beginning of the Subboreal stage (approximately around 5 ky BP). At that time, the Morava formed an anastomosed channel pattern characteristic by interwoven channel branches, the winding relics of which are still noticeable in the arable fields far from the present-day river channel.

At the end of the Subboreal stage, a further climatically controlled change in the river regime resulted in a new erosional phase dated between 2500 and 0 BC. It is evident

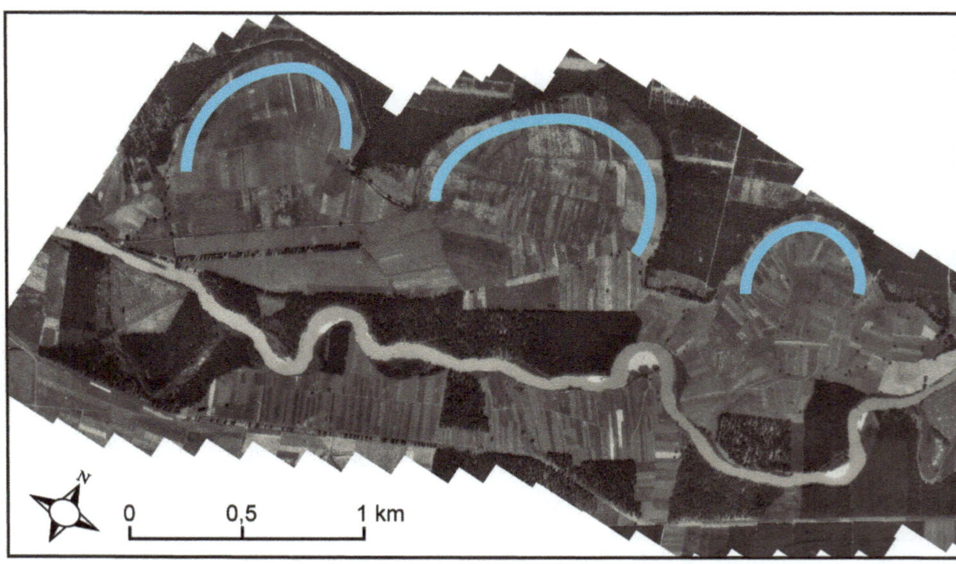

Fig. 28.5 Late Glacial paleomeanders cut into the sands of Moravian Sahara. The course of paleomeaders is accentuated by the boundary of arable land (alluvial deposits) and forest (sands reworked by wind). Note the markedly different size of present-day meanders and paleomeanders. Background image from 1954 © Vojenský geografický a hydrometeorologický úřad

that the river deepened its channel at least by 5 m below the present floodplain surface, incising into fine-grained deposits of the Atlantic and Subboreal age. The erosion trough that originated was filled with the deposits of meandering river in the following 1500 years. These sediments consist mainly of fine-grained flood loams deposited outside the channel (vertical accretion sediments) that buried the older river branches and point-bars deposits. The majority (70 %) of radiocarbon ages comes from this approximately 4 m thick sedimentary sequence, where the largest number of dates fall into period between the eleventh and seventeenth centuries. Stehlík and Kadlec (2012) infer that the dynamics of the Morava River increased at the turn of the second millennium. The reason was probably the combination of climatic influences (e.g. ending of Medieval Climatic Optimum and beginning of Little Ice Age) and rise in human activities in the drainage basin.

Most research attention was paid to the flood sediments of the youngest accumulation phase, from approximately the last millennium. Kadlec et al. (2009) refer to substantial change of the character of the flood sediments from more clayey to more sandy and silty at about AD 1550 on the basis of the study of five sections (thickness up to 5 m) exposed in erosional river banks. They assign the coarsening of floodplain fines to the onset of Little Ice Age. Despite the detected changes in river activity (reflected in coarsening of sediments), estimated mean sedimentation rates stayed rather uniform throughout the last millennium (around 0.2 cm $year^{-1}$).

The anthropogenic pollution has clearly affected the topmost 50 cm of flood sediments deposited over the last 50 years. The mean sedimentation rate based on this "industrial layer" is estimated at 0.8 cm $year^{-1}$. At a depth of 12 cm, a peak in ^{137}Cs activity is attributed to the Chernobyl nuclear power plant accident in April 1986. Kadlec et al. (2009) attribute the higher sedimentation rate in the topmost part of the floodplain sequence to increased erosion in response to land-use changes linked with the period of socialist collectivization of agriculture after the World War II.

Matys Grygar et al. (2011) refer that apart from past land-use changes and climatic forcing, sedimentation rates were also affected by building of flood defence dykes and downstream/upstream river channelization most recently. For proximal floodplain sediments, the mean deposition rate increased from 0.23 cm $year^{-1}$ in 700 AD to about 0.31 cm $year^{-1}$ at the end of the twentieth century. For distal floodplain sediments, the corresponding increase would be from 0.14 to 0.23 cm $year^{-1}$. In their previous paper, Grygar et al. (2010) reported higher depositional rates in the interval 0.2–0.6 cm $year^{-1}$, however, this data refer only to the faster deposited proximal floodplain sediments.

28.4 The Late Glacial Lacustrine, Fluvial and Aeolian Processes

The NW edge of the Morava River floodplain is bounded with a large sand body which is exposed in an extensive sand quarry and in the modern Morava River meander cut (Fig. 28.6). The sediments show laterally continuous beds ranging in thickness from 3 to 20 cm. The dominant feature of the beds is normal grading, in which sandy gravel to gravelly sand base fines upward to medium or fine sand. Grains are rounded, frosted and there is an absence of mud. The upper portion of the sections consists of ripple-to-dune scale cross-stratified sand with intercalated sandy gravel beds, and cross-stratified, fine-grained sand dominated by wind-ripple laminae. The dominant presence of graded beds and the paucity of sedimentary structures such as planar lamination or cross-strata argue that sedimentary-gravity flows within a quiet water setting were the primary depositional processes, and the stacked graded beds are interpreted as representing repeated deposition by turbidity currents. The frosted and rounded grains, and the absence of mud suggest an aeolian history to the sediment, and the turbidity currents are plausibly the result of dune slumping into a standing water body. The cross-stratified and sandy gravel beds overlying the turbidites, however, argue for a resumption of braided stream conditions followed by aeolian sand dune formation.

Incision of the modern meander bend of the Morava River gives rise to a natural outcrop in the Bzenec sand body. The outcrop called "Osypané břehy" is dramatically truncated by Holocene floodplains at both the upstream and downstream reaches of the meander bend. Bedding is similar to that observed in the sand quarries, showing laterally continuous beds, with normal grading in which sandy gravel or gravelly sand yields upward to medium sand, with an iron-stained fine-grained sand cap.

In terms of origin, this sedimentary sequence was interpreted as the wind-blown sediment body (e.g. Havlíček et al. 1996, 2007). However, recently conducted research has provided evidence allowing for a new interpretation and reconstruction of sedimentary processes controlled by the Late Glacial climatic oscillations (Kadlec et al. 2015). The Last Glacial stage was dominated by the braided river sediments which formed a terminal fan downstream from the Morava and Dyje river confluence during the MIS 3 and early MIS 2 periods. The subsequent MIS 2 stadial climatic

Fig. 28.6 The horizontally stratified lacustrine sand deposited by turbidity currents exposed in the Bzenec-Přívoz Sand Quarry (*Photo* J. Kadlec)

deterioration triggered an increase in aridity and freezing accompanied by fluvial activity decrease. The flow disappeared downstream, giving rise to aeolian dunes at the terminal fan top. Fluvial conditions withdrew to the area upstream, giving rise to a lake in the Dolnomoravský úval Basin. The OSL ages of the lacustrine sediments indicate the existence of the lake between 20 and 18 ka BP. Dam erosion and a subsequent lake cessation were driven by Late Glacial climatic amelioration, which increased the Morava River discharge. Fluvial systems changed from braided to meandering (Vanderberghe et al. 1994; Mol et al. 2000). The Morava River meandering channel incised into the lake deposits and formed an area for the Holocene floodplain processes.

The aeolian sand dunes have been formed on the surface of the Bzenec sand body after the lake cessation. These relic features trend roughly N–S and the forms are rounded and subdued, but most show asymmetry with a gentler W flank and a steeper E flank and range in height from 6 to 8 m. The internal stratal architecture, surveyed by a ground penetrating radar, shows steeply strata dipping to the east in agreement with the surface morphology. Overall dune migration was toward the east, but the poor quality of the exposures did not allow for a more detailed reconstruction of the constructive winds (e.g. Eastwood et al. 2012). The historic evidence of the sand dune reactivation is a reason why this area is called Moravian Sahara.

28.5 Fluvial Dynamics of the Morava River in the Nineteenth and Twentieth Centuries: Metamorphosis of Channel Pattern

The rate of geomorphological change of the Morava River may be studied not only from its sedimentary archive, but also from historical documents such as old maps and written records. For the most recent times, the evidence of old maps, photographs and chronicles may be combined with direct field measurements of fluvial processes.

28.5.1 The Testimony of Old Maps

Maps available for the studied area cover the period from the second half of the sixteenth century up to the present day (Fig. 28.7). The scale of maps varies from 1:530,000 (Comenius Map of Moravia, 1624) to 1:10,000 (present-day topographical maps). The planimetric accuracy of maps from the sixteenth to eighteenth century is not comparable to modern standards but they can be used for evaluation of the overall river pattern and its evolutionary tendencies over 100 years. It is evident that the Morava River had an anastomosed pattern through its middle and lower course as it flowed through the Outer Carpathian Depressions and the

Fig. 28.7 An example of an old map showing the anastomosed channel pattern of the Morava River. Müller's map of Moravia from 1730, original scale ca. 1:168,000 (Map collection of the Department of Geography, Masaryk University in Brno)

Vienna Basin. Taking a transect across the floodplain from Strážnice to Rohatec settlements, all the historical maps portray a channel pattern consisting of several parallel branches; from two to five parallel channels are shown depending on the map scale and cartographic generalisation.

Historical maps indicate that individual river branches varied in their width, course and sinuosity. Two large channels were located along the opposite margins of the floodplain and formed lateral boundaries of the channel network. From the second half of the eighteenth century, these two main channels were intensively meandering and were laterally active. The overall pattern of the channel network included numerous smaller meandering or straight river branches that diverted and re-joined the main channels, or in some cases crossing the floodplain and connecting the two main channels. The Morava River thus had an anastomosed character with meandering and straight individual channels. The sinuosity of the individual channels reflected the stream power; large branches (trunk channels) with a higher discharge had a meandering pattern and small branches (minor channels) were rather straight. Some of the minor channels are occasionally exposed in the cut banks of the present-day channel of the Morava River.

It can be inferred from historical maps that the arrangement of the channel network has not been affected by any dramatic changes at least since the first half of the seventeenth century. The overall pattern with two dominant channels can be traced back to the Comenius Map of Moravia which dates from 1624. The most valuable information on the dynamics and development of the channel network is provided by the maps of the First, Second and Third Austrian Military Surveys, which were conducted from the second half of the eighteenth to the second half of the nineteenth centuries. These maps provide the indirect and direct evidence of channel avulsions as well as the lateral shifts by gradual bank erosion. The relatively dense network of degrading, ephemeral watercourses, either still connected to the active channel network or preserved as isolated fragments within the floodplain, provides the indirect evidence of historical reorganisations of the channel network. Věšky and Medvídka channels are the current remnants of such abandoned channels (see Fig. 28.1). The direct evidence of channel relocation is depicted on maps from the first half of the nineteenth century, when the southernmost of the two trunk channels split into two branches. Old and new branches coexisted for several decades until the entire discharge eventually diverted into the new channel.

Cessation of natural avulsions, the abandonment of many smaller channels, and the concentration of discharge into one of two main channels are the main trends in channel network development during the second half of the nineteenth and the whole of the twentieth century. The transformation of the anastomosed system into a single channel meandering system was completed during the 1930s by the erection of flood defence dykes and local straightening of the dominant channel. Fluvial processes have been inhibited in the side channels but, on the other hand, the dominant meandering channel experienced profound morphological changes. Increased discharge has caused channel incision and widening and fast lateral migration has occured. Thus, the Morava River between Strážnice and Rohatec documents the transformation from a multichannel system with both sinuous and straight branches to a single channel meandering system with high lateral activity (Fig. 28.8).

28.5.2 Evolution of a Cut-off Meander: Case Study of the Osypané Břehy Locality

The spring flood of 2006 caused the meander cut-off upstream of the Osypané břehy locality, and the oxbow lake originated (Fig. 28.9). Many geomorphological and ecological changes may be observed at the oxbow lake since the cut-off. Meander cut-off occurred at the turn of April 2006 during a flood that culminated at the gauging station Strážnice on 29 March with the discharge of 733 m^3 s^{-1} (Q_{50}).

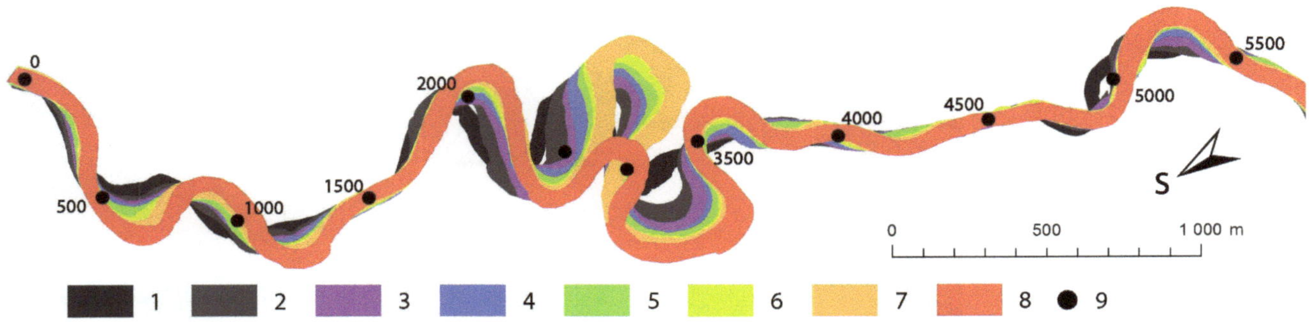

Fig. 28.8 Lateral shifts of the Morava River in the surroundings of the Osypané břehy locality in the period 1938–2012. The route of the channel in respective years: *1* 1938; *2* 1953; *3* 1963; *4* 1973; *5* 1982; *6* 1993; *7* 2003; *8* 2012; *9* distance from the beginning of analysed river reach (m). © Ondruch (2014), Masaryk University

Fig. 28.9 Oblique aerial image of the cut-off meander upstream to the Osypané břehy locality, photo taken in summer 2006. At the inflow and outflow parts of the oxbow lake are visible newly formed alluvial plugs (*Photo* J. Wenzel)

Cut-off was accomplished through bank erosion of opposite banks of the meander neck (neck cut-off type). The local sinuosity of the Morava channel dropped from 3.05 to 2.27 after the cut-off. The resulting oxbow lake had a length of 826 m.

The evolution (mainly infilling) of the oxbow lake was very dynamic in several years following the cut-off (Máčka et al. 2011; Ondruch 2014). One of the reasons was the low angle between the line of active and cut-off channel; thus, the suspended load from the active channel was easily transported into the oxbow lake. It resulted in the fast formation of alluvial plugs at inflow and outflow ends of the oxbow lake. Most marked was the growth of the inflow plug in first 6 months after the cut-off, when the area of the plug reached 10,000 m^2 (that is 22.5 % of the whole oxbow lake area), while the outflow plug reached the area of only 4,200 m^2. In the following period, the rate of oxbow filling decreased, but acceleration occurred again between 2009 and 2012. Faster infilling in this period may be attributed to hydrological situation, when a few larger floods brought larger quantities of sediment into the oxbow lake. The maximum thickness of deposited sediments reached 4–6 m until the year 2012. After 6 years of development the hydraulic connectivity between the active channel and the oxbow lake substantially decreased, with only higher flows penetrating into the lake.

Profound geomorphological changes affected close surrounding of the oxbow lake too. Due to the channel shortening by cut-off, a knickpoint originated in the active channel, and localised, accelerated channel erosion began. Around 2,000 m^2 of floodplain were eroded during 1 month after the cut-off and newly formed river bank retreated by 15 m. The destruction of further 9,600 m^2 of floodplains took place until 2012; the rate of bank erosion gradually declined from 7.92 m $year^{-1}$ (2006–2009) to 5.74 m $year^{-1}$ (2010–2012).

28.6 People and Floods

An inseparable feature of the Strážnické Pomoraví landscape were always the annual floods (floodplain inundations). Temporal changes in frequency and intensity of floods may be traced back to the end of the seventeenth century, thanks to archival sources and records of instrumental measurements. Among the most important documentary sources for flood data are monastery diaries, claims for tax relief, river channel engineering documentations and chronicles of settlements (Brázdil et al. 2005). Oldest instrumental hydrological measurements come from a water stages gauging station located at the mill weir in the settlement of Rohatec (since 1886) (Brázdil et al. 2011a). A sudden melting of accumulated snow reserves, mainly in the mountainous parts of the catchment, often accompanied by rainfall, is important for the generation of winter floods between November and April. Heavy rains, lasting for several days, cause summer floods between May and October.

A continuous flood chronology for the Morava River at the Rohatec/Strážnice stations over the 1886–2010 period was created based on peak water stages ($H_k \geq H_2$; Rohatec 1886–1920) and peak discharges ($Q_k \geq Q_2$; Rohatec 1921–1939, Strážnice 1940–2010) (Fig. 28.10). So far, the highest peak discharge at Strážnice, a flow of 901 $m^3\ s^{-1}$, was recorded on 10 July 1997, during the "flood of the 20th century" on the Morava River and significantly exceeded the value of Q_{100}. During the 90 years of discharge measurement, the annual peak discharge was $\geq Q_2$ in 39 years (i.e. 43.3 % of all years) with an accumulation of these cases

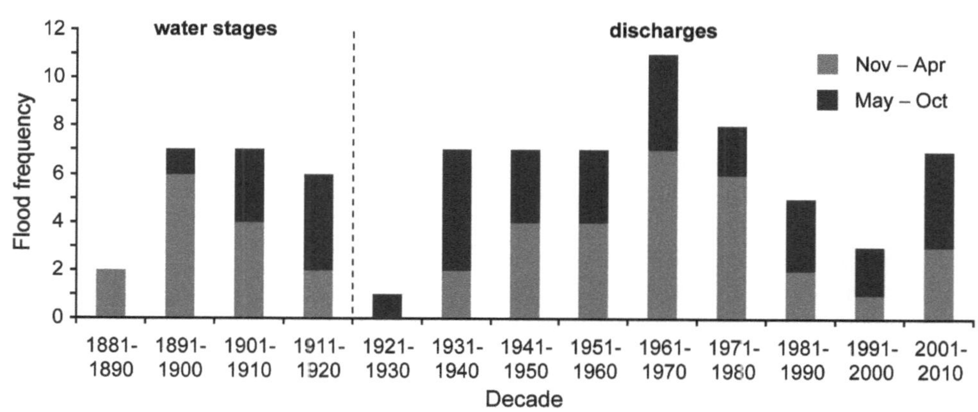

Fig. 28.10 Synthesis series of decadal frequencies of floods of the Morava River at the Rohatec/Strážnice stations in the 1881–2010 period with respect to their occurrence in the winter (November–April) and summer (May–October) hydrological half-years. Flood frequencies were derived from annual peak water stages ($H_k \geq H_2$) of Uherský Brod and Uherské Hradiště (1881–1885), peak water stages ($H_k \geq H_2$) of Rohatec (1886–1920) and peak discharges ($Q_k \geq Q_2$) of Rohatec (1921–1939) and Strážnice (1940–2010). © Brázdil et al. (2011b), Taylor&Francis

evident between 1937 and 1987. In terms of seasons, the occurrence of annual peak discharges exhibits a dominant concentration in March (31.1 %). The highest frequency of floods (seven or more floods per decade) was recorded in the periods 1891–1910, 1931–1980 and 2001–2010, with a clear predominance in the 1961–1970 decade, with 11 floods. In contrast, only one flood was recorded in 1921–1930, two occurred in 1881–1890 and three in 1991–2000. There is a higher proportion of winter floods as compared to summer floods (55.1 % cf. 44.9 %) among the 78 floods in this compiled series, with summer floods prevailing in the years 1911–1940 and 1981–2010.

Although, the dimensions of the Morava channel are not known for the break of nineteenth and twentieth century, it is evident that overbank flows occurred annually, in some years even twice. On the other hand, in the more recent times, the overbank flow of the Morava River at the Strážnice gauging station occurs at the value greater than ~520 $m^3\ s^{-1}$, which corresponds to a discharge with a reoccurrence frequency of 5 years. The inundation frequency of the floodplain decreased dramatically since the 1930s, when extensive river engineering works were accomplished. The reason lies in the progressive enlargement of the channel cross-section arising out of accelerated incision and bank instability triggered by river training. In the presently inundated part of the floodplain, i.e. only within flood defence dykes, rather coarser sediments are deposited in comparison to the state before regulation work (Grygar et al. 2010).

28.7 Conclusions

Strážnické Pomoraví is a traditional settlement area intensively moulded by human activities for more than 1000 years, since western Slavs founded their first state of Great Moravia there. Folk traditions of the past are still alive in the Strážnice region with richly decorated costumes, moving songs, "ride of kings" and others that are presented every June at the International folklore festival. Strážnice region is also one of Moravian wine production areas that invites the visitors to rest in wine cellars, to cycle along the Moravian wine trails or to sail along the Baťa shipping channel.

The natural backbone of the region is the River Morava with its floodplain that touches the area of sand dunes called "Moravian Sahara". River was branching and meandering in its 3.5 km wide floodplain creating anastomosed channel pattern as long as to the beginning of the twentieth century. The period of great river management projects just before the World War II caused the cessation of anastomosed river system functions and the transition towards a single channel meandering pattern. The unique combination of fluvial and lacustrine sediments, the latter reworked by wind action, creates a remarkable natural archive suitable for the investigation of environmental changes since the Late Glacial. Fluvial sediments with thickness up to 17 m conceal the evidence of alternating phases of floodplain erosion and aggradation from Late Glacial and Holocene, but also bear the testimony of industrial and agricultural pollution in the upper 50 cm of floodplain soils (DDT, PCB, lead contamination).

References

Brázdil R, Dobrovolný P, Elleder L, Kakos V, Kotyza O, Květoň V, Macková J, Müller M, Štekl J, Tolasz R, Valášek H (2005) Historické a současné povodně v České republice (Historical and recent floods in the Czech Republic). Masarykova univerzita, Český hydrometeorologický ústav, Brno 370 pp

Brázdil R, Řezníčková L, Valášek H, Havlíček M, Dobrovolný P, Soukalová E, Řehánek T, Skokanová H (2011a) Fluctuations of floods of the River Morava (Czech Republic) in the 1691–2009 period: interactions of natural and anthropogenic factors. Hydrol Sci J 56:468–485

Brázdil R, Máčka Z, Řezníčková L, Soukalová E, Dobrovolný P, Matys Grygar T (2011b) Floods and floodplain changes of the River Morava, the Strážnické Pomoraví region (Czech Republic) over the past 130 years. Hydrol Sci J 56:1166–1185

Eastwood EN, Kocurek G, Mohrig D, Swanson T (2012) Methodology for reconstructing wind direction, wind speed and duration of wind events from aeolian cross-strata. J Geophys Res Earth Surf (2003–2012), 117(F03035):1 20

Grygar T, Světlík I, Lisá L, Koptíková L, Bajer A, Wray DS, Ettler V, Mihaljevič M, Nováková T, Koubová M, Novák J, Máčka Z, Smetana M (2010) Geochemical tools for the stratigraphic correlation of floodplain deposits of the Morava River in Strážnické Pomoraví, Czech Republic from the last millennium. Catena 80:106–121

Havlíček P, Krejčí O, Novák Z, Stráník Z, Uhlířová I (1996) Geological map of Czech Republic, 1:25,000, sheet 34-22 Hodonín. Czech Geological Survey, Prague

Havlíček P, Adámek J, Adamová M, Břízová E, Bubík M, Čtyroká J, Čtyroký P, Macek J, Nekovařík Č, Neudert O, Novák Z, Nováková D, Petrová P, Skácelová Z, Stráník Z, Šikula J, Švábenická L (2007) Explanation to the basic geological map of the Czech Republic 1:25 000, sheets 34-224 Strážnice and 34-242 Mlýnky. Czech Geological Survey, Prague 57 pp

Havlíček P, Kučera Z, Vachek M (2008) Přírodní park Strážnické Pomoraví—Osypané břehy: zkrácení toku Moravy (The Natural Park Strážnické Pomoraví—Osypané břehy: the shortening of the river course). Zprávy o geologických výzkumech v roce 2007: 91–92

Kadlec J, Grygar T, Světlík I, Ettler V, Mihaljevic M, Diehl JF, Beske-Diehl S, Svitavská-Svobodová H (2009) Morava River floodplain development during the last millennium, Strážnické Pomoraví, Czech Republic. The Holocene 19:499–509

Kadlec J, Kocurek G, Mohrig D, Shinde DP, Murari MK, Varma V, Stehlík F, Beneš V, Singhvi AK (2015) Response of fluvial, aeolian and lacustrine systems to Late Pleistocene to Holocene climate change, Lower Moravian Basin, Czech Republic. Geomorphology 232:193 208

Kirchner K, Nováček V (1991) Hodnocení fyzickogeografických poměrů údolní nivy Moravy u Strážnice (Evaluation of physical geographical setting of the Morava River floodplain by the town of

Strážnice). Geografie teorie a výzkum, vol 13. Geografický ústav ČSAV, Brno, 32 pp

Máčka Z, Ondruch J, Michalková M (2011) Geomorfologické a vegetační změny opuštěného meandru Moravy v oblasti Osypaných břehů pět let po odškrcení (Geomorphological and vegetation changes of the Morava River oxbow lake in the locality of Osypané břehy 5 years after the cut-off). Geologické výzkumy na Moravě a ve Slezsku 2:37–42

Matys Grygar T, Nováková T, Mihaljevič M, Strnad L, Světlík I, Koptíková L, Lisá L, Brázdil R, Máčka Z, Stachoň Z, Svitavská-Svobodová H, Wray DS (2011) Surprisingly small increase of the sedimentation rate in the floodplain of Morava River in the Strážnice area, Czech Republic, in the last 1300 years. Catena 86:192–207

Mol J, Vandenberghe J, Kasse C (2000) River response to variations of periglacial climate in mid-latitude Europe. Geomorphology 33: 131–148

Ondruch J (2014) Fluviální geomorfologie širšího okolí lokality Osypané břehy (Strážnické Pomoraví) (Fluvial geomorphology of the wider surroundings of the Osypané břehy locality (Strážnické Pomoraví)). Master thesis, Department of Geography, Faculty of Science, Masaryk University, Brno, 108 pp

Stehlík F, Kadlec J (2012) Dolní tok Moravy v holocénu aneb Co řeka napsala do svého archivu (Lower course of the Morava River in Holocene or What the river wrote to its archive). Vesmír 91: 100–102

Vanderberghe J, Kasse C, Bohncke S, Kozarski S (1994) Climate-related river activity at the Weichselian-Holocene transition: a comparative study of the Warta and Maas rivers. Terra Nova 6:476–485

Limestone Klippen of the Pavlov Hills

29

Tomáš Pánek, Jan Miklín, and Karel Kirchner

Abstract

The Pavlov Hills represent a distinct geomorphological landscape of the Outer Western Carpathians in the South Moravia. They comprise a string of limestone klippen which originated as an interplay of Lower Miocene nappe tectonics and later selective erosion that removed weak Tertiary flysch and morphologically enhanced limestone blocks. Major morphological features are attributed to the lithology and thrust-and-fold tectonics of Late Jurassic-Late Cretaceous limestones which form the core of individual klippen structures. Monoclinal structures with west-facing escarpments and rather gentle eastward-oriented dip slopes have been sculptured chiefly by mass movement and karstification. However, besides structural landforms and karst features related to bedrock geology, immediate surroundings of the Pavlov Hills offer some of the famous Late Quaternary localities in Europe involving worldwide known Upper Paleolithic (Gravettian) excavations. More than 25 ka of human inhabitation established a highly valuable cultural landscape with specific habitats of limestone hills resembling Mediterranean landscape, ruins of medieval castles and vineyards and picturesque villages in the piedmont of limestone hills.

Keywords

Western Carpathians • Pavlov Hills • Limestone klippen • Thrust-and-fold belt • Rock control • Karst • Upper Paleolithic • Loess

29.1 Introduction

Containing limestone cliffs rising above surrounding alluvial plains and low flysch piedmont, the Pavlov Hills form a prominent landscape in the South Moravia (Fig. 29.1). The exceptionally diverse landscape of the area is attributed to specific limestone geomorphology and position of the region within the warmest part of the Czech Republic. It is the reason why the Pavlov Hills host many endangered species of fauna and flora whose origin is in the Mediterranean region and why their foothills produce some of the best wines in Central Europe. In this respect, a string of white limestone klippen with ruins of castles and scenic villages on the hillslopes resembles a landscape of Southern Europe. In 1986, the uniqueness of the Pavlov Hills led to the integration of the area into the network of the UNESCO Biosphere Reserves—one of the six of these in the Czech Republic. Pronounced limestone terrain with many thrust-and-fold structures, karst phenomena and famous Paleolithic localities revealing more than 25 ka of human settlement makes the Pavlov Hills first-order geosite in the European context.

T. Pánek (✉) · J. Miklín
Department of Physical Geography and Geoecology, Faculty of Science, University of Ostrava, Ostrava, Czech Republic
e-mail: tomas.panek@osu.cz

J. Miklín
e-mail: jan miklin@osu.cz

K. Kirchner
Institute of Geonics, Academy of Sciences of the Czech Republic, v.v.i. Branch Brno, Brno, Czech Republic
e-mail: kirchner@geonika.cz

T. Pánek and J. Hradecký (eds.), *Landscapes and Landforms of the Czech Republic*, World Geomorphological Landscapes, DOI 10.1007/978-3-319-27537-6_29

Fig. 29.1 The highest part of the Pavlov Hills (with the Děvín Hill in the centre) rising above subdued flysch landscape with vineyards (*Photo* J. Miklín)

29.2 Geographical and Geological Setting

The Pavlov Hills (*Pavlovské vrchy* or *Pálava* in Czech) are an approximately 12-km-long range of klippen formed by Late Jurassic-Cretaceous limestones between the Dyje River and Czech-Austrian border near the town of Mikulov (Fig. 29.2). Isolated limestone hills named Leiser Berge continue to Austria having their southern limit at the town of Falkenstein, some 8 km beyond the Czech border. From the perspective of the whole Carpathian mountain arc, the Pavlov Hills represent its westernmost promontory or the transition zone between the Western Carpathians and the Eastern Alps.

Regarding the local relief and altitude, the height of the area rises in the northward direction and culminates with the Děvín Hill (549 m a.s.l.) (Fig. 29.2). However, the size (i.e. massiveness and height) of individual klippen varies. Looking from the north, the elevation of the Pavlov Hills with rocky escarpment at the NW slope of the Děvín Hill resembles true mountainous landscape with the local relief of more than 200 m. On the contrary, some of the klippen structures (e.g. Šibeniční vrch or Kočičí skála Hills) rise only a few metres above the surrounding landscape. The klippen structures are mostly overgrown with grasslands, forest steppes and forests (e.g. oak-hornbeam and shrubs), which contrast with cultivated gentle footslopes mostly used as vineyards and orchards (Fig. 29.1).

The klippen of the Pavlov Hill consist of Jurassic and Cretaceous limestones tectonically embedded within weak Tertiary flysch sequences (Fig. 29.2, Stráník et al. 1999; Poul et al. 2011). Mesozoic formations are arranged in four main Late Jurassic-Cretaceous lithostratigraphical units. The oldest is the Klentnice Formation (Oxfordian/Kimmeridgian to Lower Tithonian) consisting predominantly of claystones with some limestones in its topmost section. The basal part of this formation is of tectonic origin and corresponds to thrust plane (Poul et al. 2011). The Klentnice Formation passes (with intercalation of the so-called nodular limestones) to overlying massive Ernstbrunn Limestone (Upper Tithonian) which forms the core of individual klippen structures. An extremely long hiatus ($\sim$55 Ma) followed the sedimentation of the Ernstbrunn Limestone during which erosion and karstification prevailed. Stratigraphically, the uppermost Mesozoic sequences of the Pavlov Hills thus span the Late Cretaceous and involve predominantly claystones (and secondary limestones) of the Klement (Turonian to Coniacian) and Pálava (Coniacian to Campanian) Formations (Stráník et al. 1999; Poul et al. 2011). Tertiary flysch involving mainly claystones, sandstones and conglomerates of the Paleocene-Lower Miocene age are observed mostly in the piedmont positions and in some depressions between limestone klippen structures (Fig. 29.2).

From the tectonic point of view, the area is situated in the frontal part of the Carpathian accretionary wedge, more specifically in the Ždánice Nappe, which originated during the younger phase of the Alpine orogeny in the Lower Miocene (Stráník et al. 1999; Poul et al. 2011). The nappe is thrust in the W and NW direction and overlies another nappe stack (Pouzdřany Nappe) and clastic deposits of the Neogene Carpathian Foredeep. Thrust tectonics was responsible for uplift, dislocation and folding of the originally continuous Mesozoic limestone platform which formed the

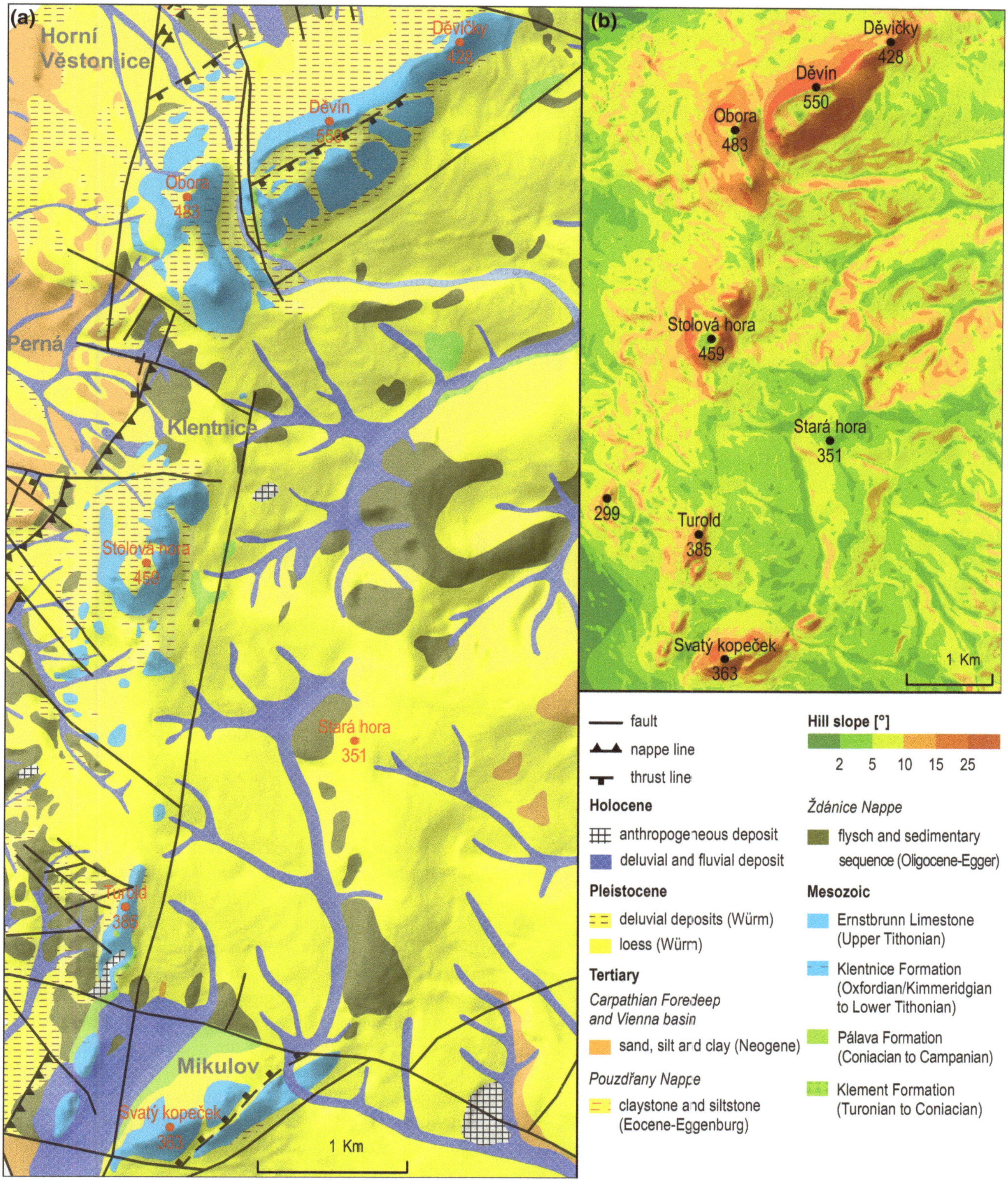

Fig. 29.2 Geology of the area placed over a shaded relief (**a**) and an inserted slope gradient map (**b**) revealing highly contrasting topography of limestone klippen structures in comparison with adjacent flysch foothills (*source* Czech Geological Survey)

basement for establishing nappes. According to a new tectonic model of the Pavlov Hills (Poul and Melichar 2009; Poul et al. 2011), klippen structures are formed by several thrust-related anticlines plunging slightly to the NE. The anticlines are interpreted as a part of flat-ramp-flat geometry which originated due to fault-bend fold mechanism (Poul et al. 2011). The whole area is cross cut by several generations of faults. Besides thrust faults striking in NE–SW and NNE–SSW directions in accordance with the front of the Ždánice Nappe, very important is a system of transverse "en echelon" sinistral strike-slip faults trending in NW–SW direction (Fig. 29.2, Poul and Melichar 2009). These faults dislocated Ernstbrunn Limestone to blocks and prepared weaker zones of rock massif for subsequent selective erosion and morphological separation of individual klippen structures (Poul and Melichar 2009).

Fig. 29.3 Aerial photography of the northern part of the Pavlov Hills showing cuesta-like klippen topography with slope deformations in the foreground (*Photo courtesy* T. Soudek)

29.3 Main Landforms

The Pavlov Hills are a unique example of a passive morphostructure. Although their geological structure resulted from Lower Miocene thrust tectonics and activity of numerous strike-slip faults, gross morphological features are attributed to Late Tertiary and Quaternary selective erosion of mechanically hard limestones enclosed within weaker flysch sequences. Generally, west- and north westward-oriented emplacement of the nappe stack caused the overall W–E asymmetry of the range and cuesta-like morphology of some klippen structures. Slopes oriented to the west are generally steeper and have a character of escarpments; those facing to the east follow bedding planes and form dip slopes, sometimes with characteristic flatirons (Fig. 29.3). Tectonically and lithologically anisotropic bedrock of the area is additionally prone to mass movements which furthermore increase morphological distinctiveness of the limestone hills.

29.3.1 Two Types of Limestone Klippen

The area involves two geomorphologically distinct types of limestone hills—(1) relatively high and massive structural ridges (cuesta or monoclinal ridges) and (2) small residual limestone hills completely surrounded by smooth flysch terrain.

The first category is represented especially by pronounced ridges of the Děvín (549 m), Kotel (483 m), Stolová hora (459 m), Turold (385 m) and Svatý kopeček (363 m) Hills. The Děvín Hill represents a true textbook example of a monoclinal ridge with steep NW escarpment and gentler dip slopes on the opposite side (Fig. 29.3). Different morphology reveals the Stolová hora Hill near the Klentnice village. English equivalent of its name is a "Table mountain", owing to extensive (~300 × 300 m) flat surface on its top. Some of the authors (e.g. Ivan 1973) interpreted this flat area as a planation surface which originated due to marine abrasion during the Upper Badenian sea transgression in the Middle Miocene. Although this transgression post-dated thrusting in the area and thus should leave some morphological signatures in the Pavlov Hills, further research is needed for the evaluation of its geomorphic importance. Common features of all "high" klippen structures are rock forms, sometimes with a character of continuous vertical cliffs, rock towers and pinnacles. Some of these rocks (e.g. on the NW escarpment of the Děvín Hill, in the Soutěska Gorge or rock group around the Martinka Rock) exceed 50 m in height.

End members in the geomorphic decay of tectonically fragmented Upper Jurassic-Cretaceous limestone body are small residual limestone hills and knobs. Although such features rise only a few metres above the landscape, it is

Fig. 29.4 Small limestone residual hill (Kočičí skála Rock) protruding only a few metres above a rounded flysch ridge (*Photo* J. Miklín)

their presence that definitely completes the picturesque scenery of the Pavlov Hills (Fig. 29.4). Like tors in granite landscapes, they attract visitors and form local geodiversity "hot spots". Most expressive examples of such small klippen structures are the Kočičí skála Rock (~10 m high) and Kočičí kámen Stone (~5 m high) between the Mikulov town and Klentnice village and the Šibeniční vrch Hill (~20 m high) in the vicinity of the Austrian border.

29.3.2 Tectonic Landforms

Landforms related to active tectonics are not presented in the landscape of the Pavlov Hills, yet there are numerous features which resulted from selective denudation of Alpine tectonic structures—especially those associated with Lower Miocene thrust tectonics and activity of transverse sinistral strike-slip faults. Selectively denuded fronts of thrust structures are expressed in the contemporary landscape as pronounced escarpments with vertical cliffs, particularly at the western slopes of klippen. The thrust fault cutting the Děvín Hill contributes to its double-ridge character with the main ridgeline situated in the western part of the elevation and the secondary one in the eastern part of the elevation (Fig. 29.3).

An important role in the morphological individualisation of klippen structures was played by transversal sinistral strike-slip faults cutting the elevation of the Pavlov Hills in NW–SE direction on several sites (Poul and Melichar 2009). These faults led to the occurrence of deep corridors between particular limestone hills and associated fault-line slopes. Some rock outcrops on slopes reveal polished surfaces with sub-horizontal striations associated with strike-slip tectonics. The most expressive of these corridors is more than 100 m deep "Soutěska" depression (Eng. "Gorge") dividing the Děvín and Kotel elevations (Fig. 29.5). It consists of two opposing valleys trending in NNW–SSE direction and a saddle in the central part. The eastern part of the "Soutěska" is flanked by a 400-m-long and more than 20-m-high cliff on the Ernstbrunn Limestone which represents one of the most important rock forms in the area.

29.3.3 Mass Movement-Induced Landforms

Klippen structures are found in an advanced stage of gravitational disintegration. Many fascinating landscape sceneries such as vertical limestone cliffs, rock towers and blocks originated as a result of mass movement—mainly rotational

Fig. 29.5 Fault-related escarpment forms of the pronounced NW slope of the Soutěska Valley (*Photo* J. Miklín)

landslides, toppling and rockfalls. The occurrence of landslides in the Pavlov Hills is facilitated by (1) an abundance of clay-rich rocks (e.g. Klentnice, Klement and Pálava Formations together with Tertiary flysch), (2) position of competent beds easily breaking up overlying plastic claystones (e.g. rigid Ernstbrunn Limestone overlying Klentnice Formation) and (3) existence of a thick cover of limestone debris resting on impermeable flysch layers at the foot of klippen. Rockfalls together with toppling contribute to the morphology of limestone cliffs (e.g. on the Stolová hora Hill and Obora and Děvín Hills) and lead to the evolution of vast scree slopes, which can be seen in their typical form along the NW slopes of the Děvín Hill (Fig. 29.6).

Spectacular landslide morphology with numerous rock forms is revealed by the NW part of the Kotel klippe, ~1 km SE from the Horní Věstonice village. There, an approx. 70-m-high limestone block of the local name "Martinka" was gravitationally detached from the massif and rotated backward by 20–25° (Fig. 29.7, Poul et al. 2010). The frontal (NW) face of the rotated block was additionally affected by toppling along the bedding planes, which formed a 3–4-m-wide and ~20-m-deep "gorge", partly roofed by collapsed boulders. Martinka Rock and adjacent cliffs demonstrate how mass movement increases the geodiversity of limestone terrain. Besides its scientific value, the site has been a popular target for climbers since the mid-twentieth century.

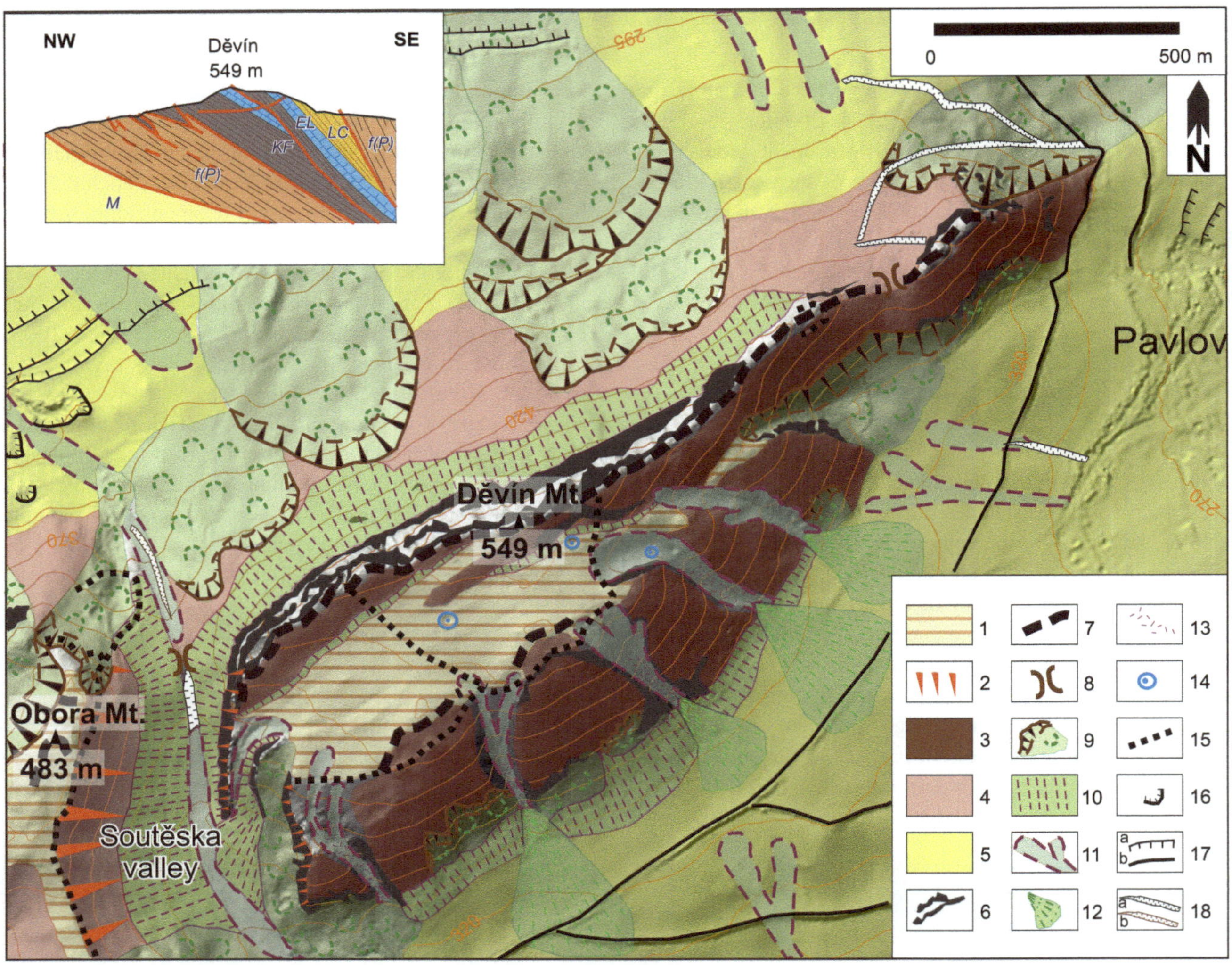

Fig. 29.6 Geomorphological map of the Děvín Hill (LIDAR-derived shaded relief used as a background). *1* Planation surface (Miocene?), *2* fault-line scarp, *3* dip slope ("flatiron"), *4* erosion-denudation slope, *5* loess slope, *6* rock wall, *7* structural ridge, *8* saddle, *9* landslide, *10* scree slope, *11* dell and dry valley, *12* dejection cone, *13* boulder deposit infilling valley floor, *14* sinkhole, *15* late Bronze Age fortification, *16* abandoned brickyard, *17a* agricultural terrace, *17b* road, *18a* hollow way, *18b* gully (inserted cross-section of the Děvín Ridge is modified after Poul 2004; *M* Miocene (Eggenburg-Baden) deposits, *f(P)* flysch deposits of the Paleogene age, *KF* Klentnice Formation, *EL* Ernstbrunn Limestone, *LC* Late Cretaceous deposits)

Fig. 29.7 Martinka Rock (*left side of the photo*) is a huge detached and backward-rotated limestone block (*Photo* T. Pánek)

However, mass movements also pose a significant hazard in the region, especially for the villages (e.g. Klentnice and Pavlov) situated at the footslopes of klippen. Several rainfall-triggered catastrophic failures (esp. shallow landslides) were recorded here in the years 1900, 1906, 1910, 1911, 1915, 1916, 1917 and 1919 and during a humid period in the first half of the 1940s (Špůrek 1972).

29.4 Karst Phenomena

Although not being so expressive as in the Bohemian Karst or Moravian Karst (see Chaps. 6 and 20 in this book), karst features play an important role in the geodiversity of the Pavlov Hills. Despite the fact that exo-and-endokarst landforms are not presented in their whole spectrum and some characteristic landforms are almost missing, for instance, sinkholes that are only represented by a few individuals on the Turold and Děvín Hills (Fig. 29.6), the area contains one of the most developed karren structures in the Czech Republic and the "Na Turoldu" Cave at the town of Mikulov points to extremely long evolution, which started minimally in the Miocene epoch.

29.4.1 Karren

According to Vítek (2013), the diversity of karren structures that can be found in the Pavlov Hills is pronounced with at least 13 different types of the structures. The most frequent ones include grikes (joint karren), pit karren, slaggy karren, rillenkarren, solution pans and especially cavernous karren. Some of the most expressive forms are also karren rock ridges, several metres high rock forms occurring at the upper edge of vertical cliffs (e.g. Děvín and Soutěska) or on the top of rock towers (e.g. Tři panny Rocks near the Děvičky Castle). Extensive areas are covered by karrenfields, which are documented especially on inclined and jointed bedding planes of the Ernstbrunn Limestone forming "flatirons" sloping to the SE from the Děvín Ridge (Vítek 2013).

By far the most interesting and scientifically explored exokarst features are cavernous karren forms that evolved on the eastern rock wall of the Soutěska Valley (Vítek 2013). The southern part of the 8–15-m-high cliff is sculptured by several tens of such microforms (Fig. 29.8). The microforms whose shape varies between nearly perfectly circular to "egg-like" are elongated sub-horizontally in accordance with the bedding planes or sub-vertically along the joints. They occur in various stages of evolution (e.g. isolated versus intersected composite features) and diverse morphologies (hollows, niches etc.) (Fig. 29.8). Most developed forms reach about 1 m in diameter and 0.5 m in depth (Vítek 2013). During almost 100 years of research (e.g. Jüttner 1922; Demek and Macka 1953; Ivan and Kirchner 1996), several interpretations of their genesis have been proposed (e.g. turbulent action of the sea water during the upper Badenian transgression; Jüttner 1922), but the most convincing is the corrosion of limestone by seeping water. This idea is supported by the existence of small channels (playing a role of subsurface conduits) which intersect the surface of the rock wall just at the position of individual hollows (Vítek 2013).

A majority of karren structures evolved on jointed surfaces of the Ernstbrunn Limestone by selective dissolution. An important factor in the distribution of karren structures is the tectonic and lithological anisotropy of limestone massif (joints, faults, bedding planes and microstructures), its textural properties and removal of the micrite component from the crystals of dolomitic rhombohedrons (Bosák et al. 1984).

Fig. 29.8 Cavernous karren (rock pits) on the rock wall in the Soutěska Valley (*Photo* T. Pánek)

29.4.2 "Na Turoldu" Cave: Evolution Since the Mesozoic?

"Na Turoldu" Cave is the most important endokarst geosite within the Pavlov Hills (Fig. 29.9). Its entrance is situated in an abandoned quarry below the Turold Hill, in the northern suburb of the Mikulov Town. The cave was explored in the early 1950s and the total length of all accessible passages is 2200 m with the depth reaching 50 m below the surface. Contemporary investigations (e.g. Kolařík 2000) have found several interconnections between adjacent smaller caves suggesting a more extensive cave system. The cave is of a multi-level type and it has probably also experienced a prolonged multi-stage evolution (Bosák et al. 1984; Poul 2004). The upper levels are mostly of tectonic/gravitational origin (e.g. "Old" or "Boulder" chambers), while the lower

Fig. 29.9 Lake's chamber situated in the deepest part of the "Na Turoldu" Cave (*Photo* J. Kolařík)

ones (e.g. Lake's chamber) originated predominantly due to corrosion (Poul 2004). Some parts of the cave have even been modified by anthropogenic activity, especially by explosions in the quarry during the first half of the twentieth century (Poul 2004).

What is most important from the geomorphic point of view is the longevity of the cave's evolution (Bosák et al. 1984; Poul 2004). One of possible interpretations is that the onset of the cave evolution was as early as in the Cretaceous period, i.e., during the stratigraphical hiatus between the Late Jurassic and Late Cretaceous, which was connected with subaerial conditions leading to significant corrosion of the Ernstbrunn Limestone. However, there is no convincing sedimentological and geochronological evidence for this theory within the cave system (Poul 2004). The most probable explanation is that the cave originated just after the Late Cretaceous epoch, and mainly during the Miocene (Eggenburg-Badenian), in the period of nappe thrusting and repeated marine transgressions (Poul 2004). A significant marker for this interpretation is the fact that several originally continuous passages and chambers are truncated by faults which originated during the Miocene orogenesis. Some of these faults are of the same type as strike-slip faults which separate the Pavlov Hills into individual klippen structures. The upper levels of the cave have been evolving since the end of nappe emplacements (<15 Ma) in connection with gravitational disintegration of the limestone klippen (Poul 2004).

"Na Turoldu" Cave is open for public and visitors can observe a great diversity of endokarst features (e.g. various types of cave passages related to corrosion, tectonics or gravitational disintegration) and touch polished surfaces of fault planes that are responsible for the genesis of the Pavlov Hills.

29.5 Late Quaternary Landscape Evolution: An Interplay of Climate Changes and Human Activity

The Pavlov Hills and especially their immediate surroundings contain some of the most important European localities showing a nearly continuous record of the Late Quaternary environmental changes. Besides demonstrating paleoclimatic variations and associated changes in the vegetation cover, they reveal a fascinating long-term history of human settlement in the area which started with the Gravettian culture (~28–22 ka) in the Upper Paleolithic and has continued up to the modern times.

29.5.1 Dolní Věstonice Loess Sequence: Late Pleistocene Geosite

Situated in a former brickyard at the NW footslope of the Děvín Hill, the loess sequence of the Dolní Věstonice village is among the most studied in the world. It was uncovered in the 1960s during archaeological excavations and it has been a focus of numerous investigations since (e.g. Kukla 1961; Demek and Kukla 1969; Musson and Wintle 1994; Frechen et al. 1999; Fuchs et al. 2013; Antoine et al. 2013 etc.). The profile shows a nearly continuous record of the Last Interglacial-Glacial cycle and is among key natural archives of climatic variations in Eurasia. Thanks to an excellent insight into the history of landscape changes in the Late Pleistocene, the site is locally called "Calendar of Ages" and has been protected as a National Natural Monument since 2005.

The last detailed OSL and radiocarbon dating of the sequence shows that the loess-paleoseol sequence was deposited between ~110–120 ka (Fuchs et al. 2013) and reveals 22 sedimentary units divided into four sub-sequences (Fig. 29.10, Antoine et al. 2013). The lowest, ~5-m-thick sub-sequence (humid soil complex) is one of the most complete pedo-sedimentary records of the transitional phase between the Last (Eemian) Interglacial and Weichselian Early glacial (~110 and 70 ka) (Antoine et al. 2013). Chernozem soils with intercalated aeolian horizons within this sequence reveal seven interstadial events and six cold phases. The conditions in the site were generally more continental and drier than those of similar sequences in the Western Europe (Antoine et al. 2013). The second sub-sequence (~2-m-thick) is represented by sandy loess and indicates the presence of a typical periglacial landscape of the Lower Pleniglacial (~70–50 ka), with an extremely arid climate and strong winds which drifted fine particles away from sandy and gravely bars of the adjacent braided Dyje River (Antoine et al. 2013). The third, about 1-m-thick sub-sequence is represented by a brown soil complex with evidence of solifluction in the more humid Middle Pleniglacial conditions (~55–40 ka). The topmost sub-sequence involves ~6 m of Upper Pleniglacial (~30–40 ka) sandy loess evidencing the return to extremely arid periglacial conditions with very rapid loess accumulation. Tundra gley soil associated with Gravettian archaeological

Fig. 29.10 Loess paleoseol sequence of the "Calendar of Ages". The black arrow shows the position of the Gravettian archaeological layer (*Photo courtesy* J. Svoboda)

layer just at the base of the sub-sequence revealed remnants of a mammoth bone, while charcoals within this unit returned the radiocarbon age of 25,760 ± 190 BP (Antoine et al. 2013).

Besides the "Calendar of Ages" loess sequence, the whole surrounding area between the Dolní Věstonice and Pavlov villages is a worldwide-known archaeological locality (protected as a Natural Cultural Monument). Particularly, the investigations performed by Karel Absolon between 1924–1938 brought some of the famous findings such as a ceramic statuette of Venus (The Venus of Dolní Věstonice) or one of the most complete skeletons of early *Homo sapiens* (Klíma 1983).

29.5.2 Limestone Klippen as Human Refuges Since the Bronze Age

Despite the fact that the region of the Pavlov Hills has continuously been inhabited by humans since at least the Late Paleolithic (~30 ka BP), there are two historical periods that influenced the face of the limestone klippen in particular: the Bronze Age and Medieval Period.

The Late Bronze Age (1300–950 BC) was a period in which klippen structures of the Pavlov Hills were substantially deforested by man. Furthermore, fortified settlements were established on some elevations—e.g. on the Stolová

Fig. 29.11 Ruins of the Sirotčí hrádek Castle from thirteenth century represent a landscape dominant in the central part of the Pavlov Hills (*Photo* J. Miklín)

hora and Děvín Hills (Poborský et al. 1993). The position of fortified lines was hidden for a long time, but especially winter aerial photographs show continuous fortification on some hills. For instance, Late Bronze Age settlement on the Děvín Klippe covers 22–26 ha, i.e. a substantial part of the ridge area (Čižmář 2004). New high-resolution LIDAR-based digital elevation model performed by the Czech Office for Surveying, Mapping and Cadastre reveals fortification in a surprising detail and it can be assumed that additional parts of the system will be explored (Fig. 29.6).

However, the scenic appearance of the Pavlov Hills was not completed until the Medieval Period, when a series of castles was built on tops of some klippen structures during the thirteenth and fifteenth centuries. Ruins of some castles (e.g. Děvičky and Sirotčí hrádek) visually enhance the morphology of limestone klippen structures that offered an excellent strategic position for such buildings (Fig. 29.11). The castles survived until the modern times as ruins in various stages of decay—e.g. while the Děvičky Castle was partly restored, the Neuhaus Castle has only been preserved as a few walls on the rock cliff of the Kotel Hill.

29.6 Cultural Value and Touristic Promotion

From the point of view of natural and cultural values, the Pavlov Hills and their immediate surroundings represents one of the richest landscapes in the Czech Republic. Approximately, 50 km^2 of the landscape contain nine specially protected reserves and monuments and one National Cultural Monument. Partly due to its picturesque position, the centre of the Mikulov Town has been declared an Urban Monument Reserve, which is a status given to the most important historic towns in the Czech Republic. The whole territory of the Pavlov Hills is protected both as a part of the Pálava Protected Landscape Area and the UNESCO Biosphere Reserve.

The area is a focus of increasing tourism activities and, without any doubt, a part of this interest is attributed to its unique geological and geomorphological settings. Some of the visitors are motivated by trekking on the limestone hills, some of them admire Upper Paleolithic heritage of "mammoth hunters" in some of the local museums. However, one of the major symbols of the Pavlov Hills is vineyards, which are largest in Moravia, and production of delicious wines, out of which white varieties of Grüner Veltliner, Welshriesling and Pinot Blanc predominate. Wine production in this region has a very long tradition since it dates back to the Roman times. Vineyards are situated predominantly on the piedmont slopes which are formed by flysch formations containing claystones and marls, i.e. a substratum very suitable for wine production. Wine tradition attracts many tourists every year. For instance, bikers may use the famous "Mikulov Wine Trail", a route around the whole Pavlov Hills with plenty of educational panels and many opportunities to taste local wines.

29.7 Conclusion

Limestone geomorphology with numerous karst features makes the landscape of the Pavlov Hills highly different from other flysch uplands of the Outer Western Carpathians. The area originated as a result of Lower Miocene nappe tectonics which detached the autochthonous limestone body of the Mesozoic (Late Jurassic-Late Cretaceous) age thrusting it together with Tertiary flysch formations over deposits of the Carpathian Foredeep. Exceptionally diverse topography characterised by numerous limestone klippen structures coincides with highly contrasting geology. It was mainly selective erosion that has morphologically enhanced individual limestone klippen structures by the removal of less-resistant flysch and tectonically weakened rocks. However, the area would never get its picturesque scenery without human interference. Deforestation of limestone hills during the Late Bronze Age established the main image of the recent landscape resembling Mediterranean regions and the Medieval Period decorated several klippen structures with castles that survived until modern days in the form of scenic ruins. These attributes together with the presence of worldwide-known loess sequences and famous Paleolithic sites make the Pavlov Hills one of the major geomorphological landscapes in the Czech Republic.

References

Antoine P, Rousseau DD, Degeai JP, Moine O, Lagroix F, Kreutzer S, Fuch M, Hatté C, Gauthier C, Svoboda J, Lisá L (2013) High-resolution record of the environmental response to climatic variations during the Last Interglacial-Glacial cycle in Central Europe: the loess-palaeosol sequence of Dolní Věstonice (Czech Republic). Quatern Sci Rev 67:17–38

Bosák P, Čadek J, Horáček I, Ložek V, Tůma S, Ulrych J (1984) Krasové jevy vrchu Turold u Mikulova. Studie ČSAV, 5, Praha, 108 pp (In Czech)

Čižmář M (2004) Encyklopedie hradišť na Moravě a ve Slezsku. Libri, Praha, 304 pp (In Czech)

Demek J, Kukla J (1969) Periglazialzone Löss und Paläolithikum der Tschechoslowakei. Czechoslovak Academy of Science, Institut of Geography, Czechoslovakia, Brno, 157 pp

Demek J, Macka M (1953) Příspěvek k otázce mísovitých prohlubní ve vápencích Pavlovských vrchů. Sbor. Čs. Spol. Zem 58:54–56 (In Czech)

Frechen M, Zander A, Cílek V, Ložek V (1999) Loess chronology of the LastInterglacial/Glacial cycle in Bohemia and Moravia, Czech Republic. Quatern Sci Rev 18:1467–1493

Fuchs M, Kreutzer S, Rousseau DD, Antoine P, Hatté C, Lagroix F, Moine O, Gauthier C, Svoboda J, Lisá L (2013) The loess sequence of Dolní Věstonice (Czech Republic): a new OSL based chronology of the Last Climatic Cycle. Boreas 42:664–677

Ivan A (1973) Outline of denudation chronology of the Mikulovská vrchovina (Highland). Folia, Facultatis Scientiarum Naturalium Universitatis Purkynianea Brunensis—Geographia 13:35–43

Ivan A, Kirchner K (1996) Nové poznatky o geomorfologii Pavlovských vrchů. Geol. výzk. Mor. Slez 2:11–13 (In Czech)

Jüttner K (1922) Entstehung und Bau der Pollauer Berge. A. Bartosch, Mikulov, 68 pp

Klíma B (1983) Dolní Věstonice, tábořiště lovců mamutů. Academia, Praha, 176 pp (In Czech)

Kolařík J (2000) Nové objevy v jeskyni Na Turoldu. Speleofórum 19:18–20 (In Czech)

Kukla G (1961) Quaternary sedimentation cycle. In: Survey of the Czechoslovak Quaternary. Czwartorzed Europy srodkovej wschodniej. INQUA 6th International Congress, vol 34. Inst. Geol Prace, Warszawa, pp 145–154

Musson F, Wintle AG (1994) Luminescence dating of the loess profile at Dolní Věstonice, Czech Republic. Quat Geochronol 13:411–416

Poborský V, Čižmář M, Dvořák P, Erhart A, Janaák V, Medunová-Benešová A, Nekvasil J, Ondráček J, Pavelčík J, Salaš M, Stuchlík S, Stuchlíková J, Šebela L, Šmíd M, Štrof A, Tejral J, Valoch K (1993) Pravěké dějiny Moravy. Vlastivěda moravská—Země a lid. Nová řada sv. 3. Muzejní a vlastivědná společnost, Brno 543 pp (In Czech)

Poul I (2004) Názor na stáří jeskyně Na Turoldu na základě paleonapjatostní analýzy. In: Bosák P, Novotná J (eds) Český speleologický kongres, ČSS, Sloup 2004. Praha, pp 85–89 (In Czech)

Poul I, Melichar R (2009) Orientace příčných zlomů v Pavlovských vrchách na jižní Moravě (Západní Karpaty). Geol. výzk. Mor. Slez 16:70–74 (In Czech)

Poul I, Bubík M, Krejčí O, Švábenická L (2010) Strukturní interpretace vmístění svrchnokřídových sedimentů do svrchnojurských vápenců skalní stěny Martinka (Pavlovské vrchy). Geol. výzk. Mor. Slez 17:126–128 (In Czech)

Poul I, Melichar R, Janečka J (2011) Thrust tectonics of the Upper Jurassic limestones in the Pavlov Hills (outer Western Carpathians, Czech Republic). Geol Soc Spec Publ 349:237–248

Špůrek M (1972) Historical catalogue of slide phenomena. Studia Geographica 19. Geografický ústav ČSAV, Brno, 180 pp

Stráník Z, Čtyroký P, Havlíček P (1999) Geologická minulost Pavlovských vrchů. Sbor. geol. Věd, Geol. 49:5–32 (In Czech)

Vítek J (2013) Škrapy ve vápencích Pavlovských vrchů. Acta Mus. Moraviae. Sci geol 98:91–109 (In Czech)

Part III

Geoheritage and Geotourism

Geomorphological Heritage and Geoconservation in the Czech Republic

30

Lucie Kubalíková

Abstract

The high lithological and morphological diversity of the Czech Republic presents a basis for geoconservation activities. The conservation of geomorphological heritage has usually been included in general nature conservation, so it should be analysed within its context. Nature conservation and protection have a long tradition here; since the Middle Ages, the forests have been protected, primarily for the hunting reasons, and then for the aesthetic reasons. During the first half of the nineteenth century, first efforts of legal protection appeared (Velký a Malý Bezděz, established around 1833, Žofín forest, established in 1838), and during the second half of the nineteenth century and the first half of the twentieth century, conservation of abiotic features of the landscape became more important, but still unsystematic. The first law on the nature protection was adopted in 1956—at that time it was considered progressive legislative tool, but it allowed a lot of exceptions in the interest of mining or the construction of transport infrastructure. Today, nature conservation is covered by the Act 114/1992 Coll. that enables landscapes, karst features, minerals and fossils to be protected as well as the establishment of protected areas. In addition, there are many other geoconservation activities: the database of geosites, the database of karst forms or the network of national geoparks. Thanks to these activities, the development of geotourism and cooperation between scientists, authorities and local people, and the geomorphological heritage in the Czech Republic become more and more important and appreciated by the public.

Keywords

Geoconservation • Geodiversity • Geomorphological heritage • Geosites • Legal protection

30.1 Introduction

People have always been fascinated by lithological and morphological diversity (or geodiversity) of the landscape and already in prehistoric time they exploited mineral resources (stone, gems, metals), used the suitable landforms as shelters or as the important communication paths. In the Czech Republic there are a number of examples of such exploitation of geodiversity values and resources: the Paleolithic extraction and treatment of the hornblende in the region of Krumlov forest or at Stránská Rock in Brno (Přichystal 2002) or the extraction of clays and clay schist and its use for the production of ceramics, cult items and jewellery (Mrázek 1996). There are also some examples of the prehistoric use of landforms: the caves used as shelters (e.g. Býčí skála Cave in Moravian Karst, Klokočské skály in the sandstones of Bohemian Paradise), the significant elevations used for defence or sacral reasons (e.g. Velký Blaník in Blaník Knight's County geopark) or the Moravian Gateway, which was used as an important communication link and trade path

L. Kubalíková (✉)
Czech Academy of Sciences, Institute of Geonics, Brno, Czech Republic
e-mail: LucieKubalikova@seznam.cz

T. Pánek and J. Hradecký (eds.), *Landscapes and Landforms of the Czech Republic*, World Geomorphological Landscapes, DOI 10.1007/978-3-319-27537-6_30

between the Baltic and Mediterranean (the Amber road led through). Later, geodiversity resources were being used more and more often and during the nineteenth century the need for the conservation of these resources arose. Today there are many activities which help to protect geodiversity and geoheritage, but that journey was not always easy.

Before presenting the overview of geoconservation in the Czech Republic and the protection of geomorphological heritage, it would be appropriate to mention general concepts and issues related to the geoconservation activities and to explain the terms geodiversity, geoconservation and geoheritage.

30.2 Geodiversity

Geodiversity concept appeared in the early 1990 and was defined as "the diversity of Earth features and systems" (Sharples 1993). Later, the definition was extended (Dixon 1996; Eberhard 1997; Australian Heritage Commission 2002; Gray 2004) and today, the following broader definition is usually accepted: geodiversity is the natural range (diversity) of geological (rocks, minerals, fossils), geomorphological (landforms, topography, physical processes), soil and hydrological features, including their assemblages, structures, systems and contribution to landscapes (Gray 2013). Ibáñez et al. (1995) introduced the term "pedodiversity" to describe the diversity of soils as an abiotic component of global geodiversity and Panizza (2009) presented the term "geomorphodiversity" which refers to the diversity of landforms and processes. Both "diversities" are considered as a subset of geodiversity (Gray 2014, oral communication).

In the Czech ambient, the first who probably used the concept of geodiversity were Ložek (2000) and Cílek (2000, 2002) who discussed the relationships between geodiversity and biodiversity and the loss of geodiversity in the Czech landscape. Both authors agree that geodiversity is a base for biodiversity; they focus on its values and emphasize the importance of its protection.

30.2.1 Geoconservation

It is obvious that geodiversity, thanks to its values, has to be conserved and preserved. Geoconservation can be described as an activity of humans, which is oriented to the conservation of geoheritage and which aims to preserve the natural diversity of significant geological (bedrock), geomorphological (landform) and soil features and processes, and to maintain natural rates and magnitudes of change in those features and processes (Sharples 1995). Later, the principles of geoconservation were redefined by Australian Heritage Commission (1996), Sharples (2002), Australian Heritage Commission (2002), Dingwall (2005) and ProGEO (2011). The long history of geoconservation is described by Burek and Prosser (2008) and Wimbledon and Smith-Meyer (2012) summarized the geoconservation activities in Europe, where a particular chapter is dedicated to the geoconservation in the Czech Republic (Budil et al. 2012).

With regard to the needs of present-day society, it is not possible to conserve all geodiversity. Probably, the easiest way to protect and conserve geodiversity effectively is to protect and conserve its most valuable part—geoheritage (geological and geomorphological heritage) that can be represented by significant geological and geomorphological sites. Therefore, geoconservation is oriented towards a site-based focus, justifying protection of individual sites at different levels of significance, according to their relative scientific importance (Cleal 2007).

30.2.2 Geoheritage, Geosites and Geomorphosites

The concept of geoheritage is based on the definition of natural heritage which was presented already in 1972 (UNESCO 1972). The term geoheritage was later defined as those components of natural geodiversity of significant value to humans, including scientific research, education, aesthetics and inspiration, cultural development, and a sense of place experienced by communities (Dixon 1996). Sharples (2002) strictly distinguishes between the term "geodiversity" as a value-free quality, and "geoheritage" as those elements of geodiversity that are seen as significant according to particular subjective values. Shortly, geoheritage comprises examples of geodiversity that are represented by geosites and geomorphosites. According to Reynard (2004), geosites are defined as portions of the geosphere that present a particular importance for the comprehension of Earth history, geological or geomorphological objects that have acquired a scientific, cultural/historical, aesthetic and/or social/economic value due to human perception or exploitation; geomorphosites are landforms to which a value can be attributed and which can be used by society as a geomorphological resource (Panizza 2001; Reynard et al. 2009).

30.3 Geodiversity Values as a Basis for Geoconservation

It is evident that geodiversity plays a key role both in environment and human activities. Geodiversity is fundamental to the distribution of habitats and species. It has always provided and still provides many essential natural

Fig. 30.1 The basaltic knob of Říp **a** rises above the flat landscape of the central part of Bohemia. The site has a mythological value and it is related to the arrival of the old Bohemians. The Romanic rotunda of Saint George **b** on the top is protected as a national cultural monument (*Photo* Michaela Myklínová (**a**), Personal collection (**b**))

resources that societal and economic growth depends on (Gray 2004). The cultural influence of geodiversity on people is also extremely strong and it can be seen as a resource for tourism, respectively, geotourism activities, which may affect local development (Panizza and Piacente 2008). The importance of geodiversity can be represented by multiple values which represent a basis for geoconservation activities. Gray (2013) analyses these values in an "abiotic system services" context. Here, the brief overview of the values with examples is presented.

(1) *Intrinsic value* is independent on the human's evaluation and it refers to the ethical belief that some things are of value simply for what they are (Gray 2013).

(2) *Regulating and supporting value* plays an important role in the atmospheric processes or flood control, and it influences habitats and communities, e.g. acid sandstone (respectively, its structure, texture and chemical composition) in the Bohemian Paradise predetermines the growth of acidophile and psamophile plants, whilst in the deeply incised river valleys in the Moravian Karst, an inversion of the vegetation can occur; the bottoms of the valleys are wet and colder that allows microclimatic change and subsequent change of the vegetation, while the top parts of the slopes are dry and hot (websites of the Bohemian Paradise PLA and Moravian Karst PLA Administrations).

(3) *Cultural value* includes a wide range of meanings—spiritual, historical, aesthetic, artistic, etc. The *spiritual value* is connected to geomythology—an explanation of geological and geomorphological features using the supernatural forces and beings (Vitaliano 2007). In the Czech Republic there are a lot of places of geological and geomorphological interests connected to the myths. Probably, the most distinctive "magic" site is the Říp hill (Fig. 30.1), a basaltic knob with the Romanic rotunda of Saint George on the top. The site is usually linked to the legend about the arrival of the old Bohemians to the Czech lands. The genesis of the specific rock formations was often explained as the work of the devil which is reflected in the toponyms; there are a considerable number of geological and geomorphological sites bearing the name of the devil (e.g. Čertovy skály in Vizovice highland, eastern part of the Moravia). The *archaeological and historical value* of geodiversity is supported by the fact that fortification systems, castles or historical settlements were usually built on the elevated landforms for defence reasons and later, for aesthetic reasons, e.g. Trosky castle in northern Bohemia or Vranov castle in southern Moravia. *The aesthetic and artistic value* refers to visual (and non-visual) appeal provided by geodiversity including a psychological impact on human beings (Gray 2004). The aesthetic value of geodiversity and landscape was especially appreciated by the romantic artists; artworks of Antonín Balzer (e.g. Devil's Garden, Fig. 30.2), Karel Postl or Antonín Mánes which are considered the founders of Czech landscape painting, are a good illustration of the strong influence of the landscape and landforms on the artist. Other Czech painters, e.g. Antonín Chittussi, Antonín Slavíček or Josef Jambor, also tried to

Fig. 30.2 Antonín Balzer: Devil's Garden (1794)—a part of the series of coloured etchings depicting mainly the landscape of Krkonoše Mountains. The artwork is rather idealized, but it reflects the fact that geodiversity and landscape were inspiring and highly appreciated by the romantic artists (Courtesy of Krkonoše Museum in Vrchlabí, Czech Republic)

catch the atmosphere of rural areas and they also included abiotic features of the landscape into their pieces of work. The perception of geodiversity was also reflected in folk poetry and music. The cultural value of geodiversity is an important precondition for geotourist activities (Panizza and Piacente 2008).

(4) *Economic and functional value* is represented by the use of mineral resources (fuels, metals and construction materials), utilization of landforms (e.g. platforms as a suitable place for airports) and utilization of geodiversity, respectively, geoheritage, for geotouristic activities (Pralong 2005; Dowling and Newsome 2010). In the Czech Republic, there is a long tradition of extraction of the precious metals (silver mines in Kutná Hora and Jihlava), construction material (limestone quarries in Bohemian Karst, granite in the north of Bohemia within Lužický and Krkonošsko-jizerský pluton) or digging gravels and sand (Morava flood plains). Landforms are also exploited for economic purposes; the typical example is the use of narrow, deeply incised valleys and gorges for dam construction (e.g. Vír dam, Vranov dam, Brno dam). Also, the passes and gaps in mountain areas are used as suitable sites where the railways and roads can pass more easily (e.g. Vlárský gap, Červenohorské pass). As mountains, rock cities, significant outcrops and other landforms have always attracted the visitors, geodiversity and its valuable part, geoheritage, can be seen as a resource for tourism that supports regional and local economic development, especially in rural areas.

(5) *Research and educational value* is related to the understanding of the origin of life and landforms, evolution of the landscape and climate, paleogeographic reconstructions, the records of sediments in lakes or bogs that allow tracing the human impact on the landscape, and it can help assessing the effects of current and potential future impacts (Gray 2013). Geodiversity has a huge educational value, which is usually developed in geoparks (see Chap. 31 in this book).

30.4 Past and Present of Geoconservation in the Czech Republic

The protection of geomorphological heritage in the Czech Republic has always been included into general protection of nature, so the analysis of the past and present activities should be attempted and described within this broader frame. The protection of nature, or natural heritage, has already been widely accepted, but the protection of wildlife still

Fig. 30.3 Panská skála near Kamenický Šenov. The old quarry with an exemplary columnar separation of the basalts has been protected since 1893. Today it is a National Natural Monument (*Photo* J. Miklín)

receives more attention than geoconservation. However, there are good examples of conservation of geological and geomorphological heritage both in the past and at present.

30.4.1 First Attempts: From the Middle Ages to the New Year's Eve Decree

The first attempts to protect the nature in the Czech countries were linked to the protection of property–forests, wood and game (Vrška and Hort 2008). Already during the Middle Ages, in the period of the rule of Charles IV , the proposal of the legal code *Maiestas Carolina* comprised some elements of nature protection including punishments when breaking the rules. Later, some specific sites were protected both for forestry and for historical, cultural and aesthetical reasons, e.g. Týřov rocks, Velký a Malý Bezděz (Pešout 2013). Although it is evident that some protected sites had existed already before the nineteenth century, the Žofín forest in the Novohradské Mountains (established in 1838) is considered to be the first "official" nature reserve just because of clear evidence and attempt of the establishment of a natural reserve for purely conservation reasons ("to keep it for the future generations"). An important stimulus to protect not only the forests but also geological and geomorphological sites was the Romantic Movement, within which some people aimed to protect "pure" nature (Vrška and Hort 2008).

The first "geological" protected site in the area of today's Czech Republic was the Barrand Rock in Prague declared in 1884 (now a part of the Barrand Rocks National Natural Monument), and in 1893 the Panská skála near Kamenický Šenov (Fig. 30.3) and Vrkoč near Ústí nad Labem were declared as nature reserves—the main reason of protection having been outstanding columnar separation of basalts (Vítek 2012). At the turn of the twentieth century, other sites were declared as protected: Höllengrund near Česká Lípa (1895, the gorge in the sandstones, today a part of the National Natural Monument Peklo), Šerák and Keprník in Jeseníky Mountains (1904, mountain ridges with tors and isolated rocks, the oldest reservation in the Moravia, today Šerák-Keprník National Natural Reserve), Černé and Čertovo lake in Šumava Mountains (1911, lakes of glacial origin, today Černé a Čertovo jezero National Natural Reserve) and Šibeničník hill near Mikulov (around 1917, Jurassic limestone klippe, today Šibeničník Natural Reserve) (Kamarád 1975; Vítek 2012).

In 1919, soon after the establishment of the independent Czechoslovakia, the institute of "nature conservators" was established. These people were asked to look for the naturally important sites and to establish their protection. Usually, they were both university professors and local teachers interested in the nature conservation. The leading person was Rudolf Maximovič and nature conservation was developed under the Ministry of Education. In 1922, when he started to act, there were only 23 private nature reserves, and in 1938 there were already 113 reserves in Bohemia, 29 in Moravia and Silesia and 18 in Slovakia (Veselý et al. 1954; Čeřovský 2004). Since 1922, there had been some proposals for the law on the protection of nature, but none was accepted, so the nature reserves and other protected sites could have been declared only after an agreement with the owner of the land. These circumstances caused that protection was rather unsystematic (Vrška and Hort 2008); however, the importance of nature protection was already recognized as can be seen from the following Maximovič's definition which dates to the 1938 (Stejskal 2006):

> Nature conservation is an effort to implement principles of effective management of matter and energy of the nature, with respect to the needs and interests of future generations and to the preservation of preferably undisturbed nature, both biotic and abiotic, for reasons of public weal.

During the interwar period, the most important step forward was the *New Year's Eve Decree*, the first official list of the protected nature monuments. It was published in 1933 and comprised 108 sites from Czechia, 18 sites from Slovakia and 12 sites from Carpathian Ruthenia. Educational and protection values of the sites were presented and visits were also recommended. The list included especially forest areas, geological and geomorphological sites, e.g. Pravčická brána Arch (a rock arch in Bohemian Switzerland) or Tiské stěny (a gorge in Elbe sandstones) (Čeřovský 2004; Pešout 2013).

30.5 Legal Protection of Geological and Geomorphological Heritage

30.5.1 The Socialist Era: 1950s–1980s

In 1956, the first law of nature conservation (Act No. 40/1956 Coll.) was adopted. It defined several categories of territorial protection (National Park, Protected Landscape Area—PLA, State Natural Reservation, Protected Deposit, Protected Park or Garden, Protected Study Area, Protected Natural Monument) and enabled to protect minerals, fossils and rock formations. Although it was considered a progressive legislative tool at that time, it was based on the conservationist approach to the nature and already in the 1970s it was obvious that the philosophy of protection of isolated segments of the nature or endangered species is not sustainable and it cannot protect the nature effectively (Stejskal 2010). In addition, the law was very benevolent and allowed a number of exceptions in relation to agricultural, industrial, mining and construction activities. The cases of total destruction of natural phenomena were not rare. Some geologically and geomorphologically important sites were considerably damaged or even disappeared, e.g. in Lažánky area near Brno where a system of karst caves was destroyed by the limestone extraction (Himmel 1989), or the canyon-like valley of the Jihlava River (Fig. 30.4) which was flooded due to dam construction at Dalešice and Mohelno. The conflict between protection of geological and geomorphological heritage and economical activities was discussed in several papers. Skřivánek (1978), Müllerová and Vašíček (1988) examined the impacts of extraction of mineral resources on the geologically and geomorphologically important sites; Moldan (1972) discussed the negative influence of human activities on landforms in general.

Fig. 30.4 Jihlava valley at present. A considerable part of geomorphologically valuable canyon-like valley disappeared under the water; the remnants of the valley are protected as Natural Reserves or Natural Monuments (*Photo* J. Vítek)

In 1988, the Act No. 44/1988 Coll. on the protection of geological resources (Mining Act) was adopted, but its attempts were rather to protect the yields of mineral resources and to assure its economical and effective extraction.

30.5.2 From the Year 1989 to the Present

After the social and political changes in 1989, the situation changed and a new, quite progressive law (Act No. 114/1992 Coll.) was adopted. Currently, this law is the main legislative tool governing the protection of nature. It comprises territorial protection, species protection and general protection and enables landscapes, karst features, minerals and fossils to be protected. There are six levels of the territorial protection: National Park (NP), Protected Landscape Area (PLA), National Natural Reservation (NNR), National Natural Monument (NNM), Natural Reservation (NR) and Natural Monument (NM). In January 2015, 2481 sites and areas in total were protected. The number of protected sites is still increasing (for example, in 2005 there were 2210 protected

Table 30.1 Legal protection of nature in the Czech Republic (Source: http://www.ochranaprirody.cz/)

Category	Number	Area (ha)	% of the Czech Rep.	Geo(morpho) sites
National parks	4	119,489	1.51	–
Protected landscape areas	26	1,135,273	14.40	–
National natural monuments	114	5207	0.06	67
National natural reserves	110	28,319	0.35	97
Natural monuments	1420	28,984	0.36	507
Natural reserves	807	40,909	0.51	286
Total	2481	1,358,181	17.22	957

sites). In a considerable number of small-scale protected sites, geology and geomorphology are the reasons of protection or they form an important element/phenomenon within the protected site. General information about the number and area of them is given in Table 30.1.

Based on the websites of the Nature Conservation Agency of the Czech Republic and Czech Geological Survey (consulted in May 2014), the term "geo(morpho)sites" means that geology and geomorphology are the reasons of protection or it forms an important element/phenomenon within the protected site.

Important geological and geomorphological sites are usually protected as National Natural Monuments or Natural Monuments, although some NNR and NR with important geological or geomorphological phenomena were declared too. Natural Monuments, respectively, National Natural Monuments, are defined as a natural formation occupying a smaller area, in particular a geological or geomorphological formation, deposits of rare minerals or place of occurrence of endangered species in a fragment of ecosystems, having a regional, respectively, national or international environmental, scientific or aesthetic importance, including formations which were also formed by human activity (articles 35 and 36, Act No. 114/1992 Coll.).

Within the general nature conservation, a considerable number of natural outcrops, river valleys or alluvial flood plains are protected in the category of *Significant Landscape Component*; the landforms and landscapes at a larger scale are protected in the category of *Natural Parks*. In these sites and areas special conditions apply in relation to water management or construction activities.

According to the Act No. 114/1992 Coll., the limits of use of the protected territory are strictly specified and every protected area has a *care plan* which describes the conditions of protection and management of the site. These plans are prepared by conservation authorities and discussed with local authorities, organizations supporting nature protection, landowners, municipalities, non-governmental organizations and local residents. The proposed management measures are financed from public resources, financial support from the other parties being still relatively weak (Budil et al. 2012). There are several institutions governing nature conservation both on national, regional and local level (Table 30.2).

30.5.3 Other Geoconservation Activities

Probably, the most important geoconservation project in the Czech Republic in the last years has been the establishment of the National Geopark Network that is linked to the European Geopark Network. This will be discussed in following chapter.

A remarkable project is the Database of geological localities (more than 2,800 sites in May 2014), maintained and updated by the Czech Geological Survey. Geoscientific characteristics, the degree of protection and conflicts of interest are given for every site. The database is open and anyone can propose a new geosite or update information (via consultation with a specialist) (http://lokality.geology.cz/, Fig. 30.5). The Database of speleological objects is run by the Nature Conservation Agency of the Czech Republic and it is focused on karst phenomena (http://jeso.nature.cz/).

30.6 Geomorphological Heritage Within Protected Areas

The landscapes and landforms at a larger scale are protected as an important feature within protected landscape areas and national parks (Fig. 30.6).

Among the four Czech national parks, geological and geomorphological phenomena play a key role in the Bohemian Switzerland National Park established in 2000. The main reasons of protection are sandstone landforms (e.g. the Elbe canyon-like valley, the Jetřichovice rock cities, Pravčická brána Arch). The Elbe sandstone PLA immediately adjoins the Bohemian Switzerland National Park and also presents an important geomorphological heritage protected under the law. Moreover, rock cities in the Cretaceous sandstones of the Bohemian Cretaceous Basin are subject to nature protection in Bohemian Paradise PLA, Kokořínsko PLA (Fig. 30.7) and Broumovsko PLA.

Table 30.2 The state, regional and local institutions governing the nature conservation

Institution	Level	Main tasks and responsibilities
Ministry of environment	National	Central state administration authority for nature
		Protection preparation of the overall strategies and policies for the nature conservation
		Declaration of the NP, PLA, NNR, NNM
Nature conservation agency of the Czech Republic	National	Scientific and technical support for nature conservation
		Keeping the central register of nature conservation
		Keeping the database of speleological objects
Cave administration of the Czech Republic	National	Scientific and technical support for the speleology
		Surveys, inventories, registration and documentation on caves and other underground spaces
Czech Geological Survey	National	Expert services for the nature conservation authorities keeping the database of the geological localities
Administrations of protected landscape areas and national parks	Regional	Scientific and technical support for nature conservation in protected areas
		Administration of protected areas
Regional governments	Regional	Preparation of regional conservation strategies
		Declaration of the NR and NM
Municipalities	Local	Responsibility for the significant landscape components
		Close contact to landowners and stakeholders

Source Act No. 114/1992 Coll., Budil et al. (2012), the websites of the institutions. Abbreviations: National Park (NP), Protected Landscape Area (PLA), National Natural Reservation (NNR), National Natural Monument (NNM), Natural Reservation (NR) and Natural Monument (NM)

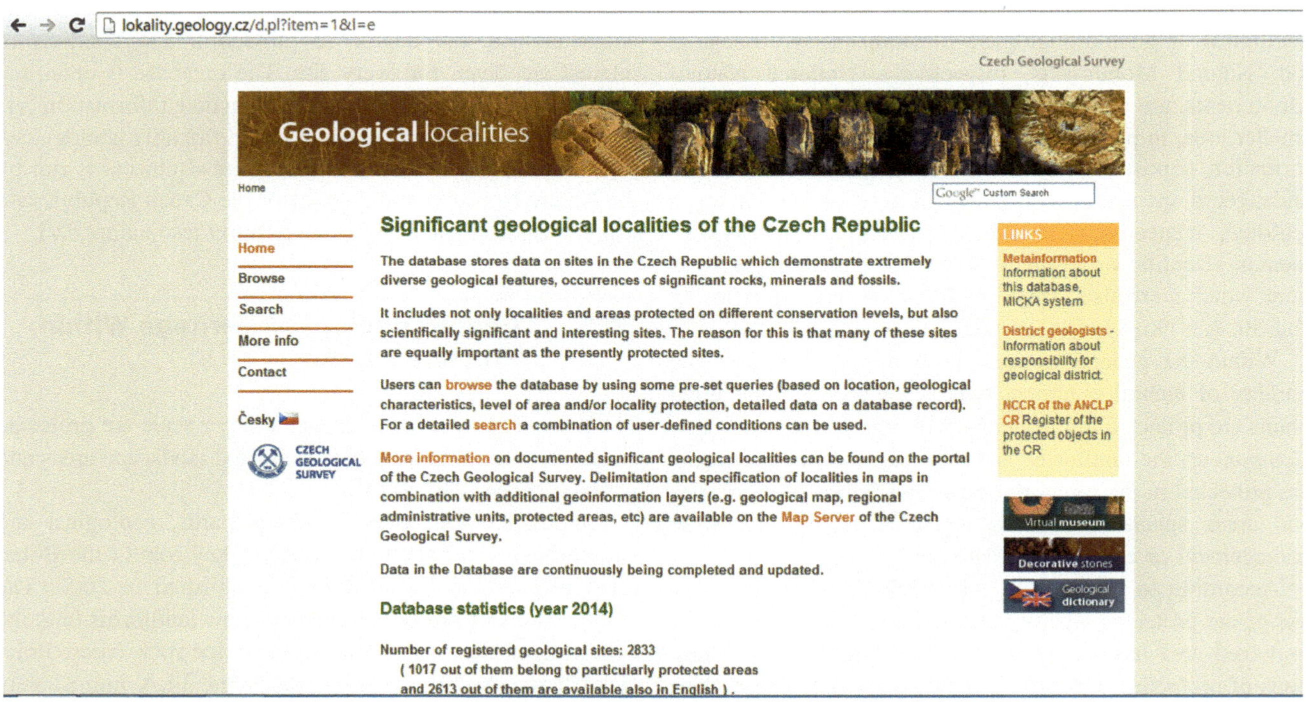

Fig. 30.5 The web database of the geological localities (Czech Geological Survey, http://lokality.geology.cz/)

The fluvial phenomenon is represented in Podyjí National Park (declared in 1991). The area is situated on the border with Austria, so that during the socialist era the access was strictly limited and human impact on the landscape was not very significant. The deeply incised valley of the Dyje River represents an exceptionally preserved illustration of how the

Fig. 30.6 Protected areas in the Czech Republic (Nature Conservation Agency of the Czech Republic, http://www.ochranaprirody.cz/), Data © ArcČR, ARCDATA PRAHA, ZÚ, ČSÚ, 2014

Fig. 30.7 Outcrops of Cretaceous sedimentary rocks (sandstones) attracted folk artist to create the "Devil's Heads" in Kokořínsko PLA (*Photo* J. Adamovič)

majority of the Czech river valleys would look like if they had not been flooded due to the dam constructions. The area is rich in fluvial, cryogenic, and gravitational landforms, especially incised meanders, rocky slopes, boulder fields and pseudokarst features. The unique geomorphological site of Ice Caves is situated within the first zone of the National

Park and consists of large crevice-type caves with specific climate. Natural fluvial processes and landforms (meanders, braided stream, oxbow lakes, etc.) can be also found in the PLAs Strážnické Pomoraví, Litovelské Pomoraví and Poodří. Some deeply incised valleys on the southeastern border of the Bohemian Massif are protected as Natural Parks, e.g. Střední Pojihlaví Natural Park, Oslava Natural Park, Svratecká hornatina Highland Natural Park. Within these Natural Parks, particular sites are protected as NNR, NNM, NR or NM. However, fluvial phenomena are the reason of protection throughout the Czech Republic, being represented not only by deep valleys, but also by bogs (e.g. Rejvíz in Jeseníky PLA, Boží dar in Krušné hory Mountains; as the biotic element is also very important, both sites are protected in the category of National Natural Reserve), fluvial landforms in the alluvial plains (e.g. Vrapač National Natural Reserve in the alluvial plain of the Morava River or Skalická Morávka National Natural Monument in Northern Moravia with braided stream) or waterfalls (e.g. Rešovské vodopády National Natural Reserve in Jeseníky PLA).

Krkonoše Mountains are the highest mountain terrain in the Czech Republic and they were declared as National Park already in 1963. Geomorphological phenomena are represented by numerous glacial and periglacial landforms such as cirques, trogs, tors, boulder fields and frost-riven cliffs. The typical geomorphological sites in Krkonoše are Dívčí kameny, Mužské kameny (examples of periglacial landforms—tors and adjacent boulder fields), Kotel or Obří důl (both with well-visible glacial phenomena). Glacial phenomenon is also important in the Šumava National Park (declared in 1994), the only place in the Czech Republic with glacial lakes (Černé, Čertovo, Plešné, Laka). Periglacial and cryogenic phenomenon is displayed practically in all mountain areas in the Czech Republic, which are declared as PLAs, e.g. Jeseníky, Žďárské vrchy, Beskydy and Jizerské hory. Within these protected areas, a considerable number of rock formations (boulder fields and streams, frost-riven cliffs, rock walls or tors) are protected in the categories of small-scale protected areas, e.g. Devět skal Natural Monument and Čtyři Palice Natural Reserve in Žďárské vrchy PLA.

Typical volcanic landforms are present in České Středohoří PLA and Lužické hory PLA and there are also many isolated volcanic sites that are protected as small-scale protected areas, e.g. Zlatý vrch in northern Bohemia, Železná hůrka in western Bohemia within the Egeria Geopark, Uhlířský vrch in northern Moravia—all these sites are protected in the category of National Natural Monument or National Monument.

Within the PLAs Bohemian Karst, Moravian Karst and Pálava, karst features are protected. These areas are partly influenced by quarrying which has a long tradition; the quarries Velká a Malá Amerika are also attractive as the exteriors for the movies.

Some anthropogenic landforms are also protected as small-scale sites; thanks to human activities as quarrying and mining, geoheritage which would be normally hidden under the surface, has become visible. Old quarries, clay or loess pits and mines are also important for the scientific purposes and some of them became the Global Boundary Stratotype Section and Point (GSSP), e.g. Požáry National Natural Monument where the boundary between Přídolí and Ludlow stages (Upper Silurian) is exposed. Such anthropogenic landforms are significant not only from the stratigraphic point of view, but they have also a historical value (important sites for the study of history of mining and quarrying), e.g. Vlčí Jámy Natural Monument (Fig. 30.8) in Krušné hory Mountains (the collapsed ceilings of the system of mediaeval galleries which were used for the extraction of tin).

30.7 Perspectives for the Protection of Geomorphological Heritage in the Czech Republic

The relatively long tradition and current satisfactory conditions of protection of geological and geomorphological heritage in the Czech Republic can be considered as a good starting point for future geoconservation activities. However, there are still some threats that can endanger the geoheritage and important geological and geomorphological sites. Especially changes in the water regime and ongoing mining can cause the loss of geodiversity, respectively, geological and geomorphological heritage. In addition, there is also a danger that is brought by mass tourism and overexploitation of the heritage from this point of view. One of the possibilities of how to avoid this risk is to accept the idea of Baba Dioum (1968):

> In the end we will conserve only what we love. We will love only what we understand. We will understand only what we are taught.

Education can lead to better understanding of the importance of conserving geological and geomorphological heritage. Another possibility of how to ensure good future of the protection of geological and geomorphological heritage is to support cooperation of institutions that deal with geoconservation and geoheritage activities (Czech Geological Survey, Czech Association of Geomorphologists, universities and high schools, administrations of national parks and protected landscape areas, regions, communities and civic association). The task is to propose them a space for exchanging information and experience and to get the

Fig. 30.8 Vlčí Jámy in the Krušné hory mountains–collapsed ceilings of the system of medieval galleries which were used for the extraction of tin (Personal collection)

influence on land planning and resources management. Other important task is rather professional: geoscientists that deal with geoconservation should cooperate with the scientists from other fields: biologists, historians, archaeologists, but also sociologists, economists or pedagogues. Geoconservation also should be a subject of discussion between the persons mentioned above and political entities.

References

Australian Heritage Commission (1996) Australian natural heritage charter for the conservation of the places of natural heritage significance: standards and principles. Australian heritage commission and Australian committee for the international union for the conservation of nature (ACIUCN), Sydney

Australian Heritage Commission (2002) Australian natural heritage charter for the conservation of the places of natural heritage significance, 2nd edn. Australian heritage commission and Australian committee for the international union for the conservation of nature (ACIUCN), Canberra

Burek CV and Prosser CD (eds) (2008) The history of geoconservation. The geological society of London, 305 pp

Budil P et al (2012) Czech Republic. In: Wimbledon WAP, Smith-Meyer S (eds) Geoheritage in Europe and its conservation. Pro GEO, Oslo, pp 92–99

Cílek V (2000) Geodiverzita—Geologická rozmanitost Čech. Vesmír 79(1):95–97

Cílek V (2002) Geodiverzita—opomíjený aspekt ochrany přírody a krajiny. Zprávy o geologických výzkumech v roce 2001:13–15

Cleal CJ (2007) Geoconservation—what on Earth are we doing? In: Hlad B, Herlec U (eds) Regional conference on geoconservation (2007, Ljubljana): geological heritage in the South-European Europe. Book of abstracts, p 25

Čeřovský J (2004) Vývoj hnutí dobrovolných konzervátorů, zpravodajů a strážců přírody na území České republiky. Veronica XVIII/16: 22–26

Dingwall P (2005) Geological world heritage: a global framework. A contribution to the global theme study of world heritage natural sites. IUCN, WCPA, UNESCO

Dixon G (1996) Geoconservation: an international review and strategy for Tasmania. occasional paper 35, Parks & Wildlife Service, Tasmania, 101 pp

Dowling R, Newsome D (eds) (2010) Geotourism. The tourism of geology and landscape. Goodfellow Publishers Ltd., 246 pp

Eberhard R (ed) (1997) Pattern and process: towards a regional approach to national estate assessment of geodiversity. technical series no. 2, Australian heritage commission and environment forest taskforce. Environment Australia, Canberra

Gray M (2004) Geodiversity: valuing and conserving abiotic nature. Wiley, Chichester 448 pp

Gray M (2013) Geodiversity: valuing and conserving abiotic nature, 2nd Edition. Wiley Blackwell, 508 pp

Himmel J (1989) Záchranný výzkum dvou jeskyní v Lažánecko-heroltickém krasu. Československý kras XL:122–128

Ibáñez JJ et al (1995) Pedodiversity concepts and tools. Catena 24: 215–232

Kamarád L (1975) Chráněná území a ochrana reliefu krajiny 1945–1975. Ochrana přírody 5–6:157–167

Ložek V (2000) Biodiverzita, ekofenomény a geodiverzita—Bohatství živé přírody je chráněno rozmanitostí terénu. Vesmír 79(1):97–98

Moldan B (ed) (1972) Geologie a životní prostředí. Ústřední ústav geologický, Praha 141 pp

Mrázek I (1996) Drahé kameny v pravěku Moravy a Slezska. Moravské zemské museum Brno, 204 pp

Müllerová J, Vašíček Z (1988) Geologie v tvorbě a ochraně životního prostředí. Vysoká škola báňská, Ostrava 157 pp

Panizza M (2001) Geomorphosites: concepts, methods and example of geomorphological survey. Chin Sci Bull 46:4–6
Panizza M (2009) The geomorphodiversity of the dolomites (Italy): a key of geoheritage assessment. Geoheritage 1(1):33–42
Panizza M, Piacente S (2008) Geomorphosites and geotourism. Rev Geogr Acadêmica 2(1):5–9
Pešout P (2013) Silvestrovský výnos—80 let od vydání. Ochrana přírody 6:8–11
Pralong JP (2005) A method for assessing tourist potential and use of geomorphological sites. Géomorphologie: relief, processus, environnement 1(3):189–196
ProGEO (2011) Conserving our shared geoheritage—a protocol on geoconservation principles, sustainable site use, management, fieldwork, fossil and mineral collecting. http://www.progeo.se/progeo-protocol-definitions-20110915.pdf. Accessed 23 April 2013
Přichystal A (2002) Zdroje kamenných surovin. Paleolit Moravy a Slezska, Dolnověstonické studie 8(11):55–70
Reynard E (2004) Geosite. In: Goudie AS (ed) Encyclopedia of geomorphology. Routledge, London, pp 440
Reynard E, Coratza P, Regolini-Bissig G (eds) (2009) Geomorphosites. Verlag Dr. Friedrich Pfeil, Mnichov 240 pp
Sharples C (1993) A methodology for the identification of significant landforms and geological sites for geoconservation purposes. Report to Forestry Commission, Tasmania 23 pp
Sharples C (1995) Geoconservation in forest management—principles and procedures. Tasforests 7(12):37–50
Sharples C (2002) Concepts and principles of geoconservation. Published electronically on the Tasmanian Parks & Wildlife Service website, 79 pp. http://dpipwe.tas.gov.au/Documents/geoconservation.pdf. Accessed 29 May 2014
Skřivánek F (1978) Vztah ochrany přírody k vyuţívání nerostných zdrojů a k ochraně nerostů a zkamenělin. Památky a příroda 7: 417–423
Stejskal V (2006) Stodvacet let od narození Rudolfa Maximoviče, zakladatele moderní ochrany přírody v Československu. Ochrana přírody 61(6):170–172
Stejskal V (2010) Principy právní úpravy ochrany přírody a krajiny. In days of Law 1, Masaryk University Brno, pp 1974–1989. http://www.law.muni.cz/sborniky/dny_prava_2010/files/sbornik/sbornik.pdf/. Accessed 29 May 2014
UNESCO (1972) Convention concerning the protection of the world cultural and natural heritage. http://whc.unesco.org/archive/convention-en.pdf. Accessed 29 May 2014
Veselý J et al (1954) Ochrana československé přírody a krajiny. Díl I. Nakladatelství ČSAV, Praha, 356 pp
Vitaliano DB (2007) Geomythology: geological origins of myths and legends. Geol Soc London Spec Publ 273:1–7
Vítek J (2012) Význačné geologické útvary jako předmět zájmu ochrany přírody. In: Machar I, Drobilová L (eds) Ochrana přírody a krajiny v České republice—vybrané aktuální problémy a možnosti jejich řešení (the appendix). Univerzita Palackého Olomouc, pp 23–28
Vrška T, Hort L (2008) Historie vzniku lesních rezervací v ČR do r. 1945. Ochrana přírody 63 (1). http://www.casopis.ochranaprirody.cz/clanky/historie-vzniku-lesnich-rezervaci-v-cr-do-roku-1945.html . Accessed 29 May 2014
Wimbledon WAP, Smith-Meyer S (eds) (2012) Geoheritage in Europe and its conservation. Pro GEO, Oslo 405 pp
Act No. 40/1956 Collection of Laws of the Czechoslovac Republic
Act No. 44/1988 Collection of Laws of the Czechoslovac Republic
Act No. 114/1992 Collection of Laws of the Czech Republic

Promoting Geomorphological Heritage: Bringing Geomorphology to People

31

Lucie Kubalíková

Abstract

Landscapes and landforms have always attracted people's attention, so visiting geosites and geomorphosites is nothing new. In the last years, the number of visitors to the sites has been growing. Of course, the demand of supporting services is also rising. As a result, projects and products focused on promotion of geoheritage began to be more and more important and geotourism as an emerging form of sustainable tourism is gaining more significance. In the Czech Republic, geological and geomorphological heritage is promoted especially within the National Geoparks and protected areas, and there is also a considerable number of smaller projects (geological gardens or open-air expositions and museums, geo-trails, etc.). Institutions such as the Czech Geological Survey, the Nature Conservation Agency of the Czech Republic or the Cave Administration of the Czech Republic also focus on promoting geomorphological heritage; their financing is based mainly on public resources. In the future, sustainability of the support of geomorphological heritage and its conservation and promotion will probably become more and more dependent on the cooperation between various stakeholders—both public and private, both professional and amateur.

Keywords

Geotourism • Geoheritage • Geoparks • Geoeducation

31.1 Introduction

Visiting natural sites that are important from geological or geomorphological point of view has been practiced for a long time (Migoń 2009; Dowling 2013). Already in the nineteenth century, some geosites and geomorphosites were frequently visited and they became traditional touristic destinations (e.g. Pravčická brána Arch, a natural sandstone arch in the Bohemian Switzerland National Park in northern Bohemia). Today, the number of people who are looking for the deeper experience at the visited site or area is growing, and these visitors are not satisfied only with a "come and see" experience, but they want to know more about the site or area and related issues (in the case of geosites and geomorphosites, this information should cover not only Earth Sciences but also historical, archaeological, ecological or artistic aspects of the site) and in the case that an area or a site is protected, they wish to understand why. Of course, the visitors also expect a good quality of support services such as safe and well-marked paths, information services (leaflets, information panels, guided tours, etc.), accommodation, transport facilities and catering and they count with an opportunity to try or buy products that are typical for the area, whether it is a local food, drink or handicraft product. These factors contribute to the development of sustainable forms of tourism, especially ecotourism and geotourism.

L. Kubalíková (✉)
Czech Academy of Sciences, Institute of Geonics, Brno, Czech Republic
e-mail: luciekubalikova@seznam.cz

T. Pánek and J. Hradecký (eds.), *Landscapes and Landforms of the Czech Republic*, World Geomorphological Landscapes, DOI 10.1007/978-3-319-27537-6_31

31.2 Geotourism: For What and for Whom?

In the last few years, geotourism has shown a considerable expansion all over the world (Dowling 2011) and the Czech Republic is not an exception. Before giving particular examples of how geomorphology is presented to the visitors and tourists in the Czech Republic, the brief overview of basic principles and approaches of geotourism and related issues is presented.

According to Butler (1992) in Pásková 2012), the real ecotourism is rather biocentric than anthropocentric, its aim being to enforce the environmental ethic, and it does not degrade and disturb natural processes and issues. Then, from the ecotourism, the nature should have a profit. Primarily, the geotourism concept copied this approach, but then it developed into a holistic concept that is rather people-oriented (Martini et al. 2012), nevertheless it still respects environmental aspects (respectively, geological and geomorphological) and the need for the protection of them. Shortly, it can be said that geotourism is not primarily for the nature, respectively, geology and geomorphology, but for the people (Zouros 2013, oral communication).

Table 31.1 presents some definitions illustrating the development of geotourism concept during the last 20 years; the shift from the geologically and geomorphologically oriented definitions towards the holistic concepts of geotourism can be observed.

It is obvious that geology and geomorphology form an essential resource for geotourism activities. Reynard et al. (2003) and Reynard (2008) analysed the relationship between geomorphology and tourism: geomorphology may be a tourist resource as part of the *primary or original offer* (geomorphological site as an attraction or geomorphological site as a support for tourist activity, e.g. climbing, Fig. 31.1) and *secondary or derived offer*, when tourist infrastructures (e.g. didactic trails), instruments (e.g. educational booklets) or services (e.g. guided tours) are proposed for effective use of the original offer. Gray (2004) also stated that geodiversity is of great value for geotouristic and geoeducational

Table 31.1 The development of approaches to geotourism

Hose (1995)	The provision of interpretive and service facilities to enable tourists to acquire knowledge and understanding of the geology and geomorphology of a site (including its contribution to the development of the Earth sciences) beyond the level of mere aesthetic appreciation
Hose (2000)	The provision of interpretative facilities and services to promote the values and social benefit of geologic(al) and geomorphologic(al) sites and their materials, and to ensure their conservation for the use of students, tourists and casual recreationalists
Slomka and Kicinska-Swiderska (2004)	An offshoot of cognitive tourism and/or adventure tourism based upon visits to geological objects (geosites) and recognition of geological processes integrated with aesthetic experiences gained by the contact with a geosite
National Geographic (2005)	Tourism that sustains or enhances the geographical character of a place—its environment, culture, aesthetics, heritage and the well-being of its residents
Dowling and Newsome (2006)	Tourism relating specifically to geology and geomorphology and the natural resources of landscape, landforms, fossil beds, rocks and minerals, with an emphasis on appreciating the processes that are creating and created such features
Hose (2008)	The provision of interpretive and service facilities to enable tourists to acquire knowledge and the understanding of the geology and geomorphology of the site (including its contribution to the development of the Earth sciences) beyond the level of mere aesthetic appreciation
Dowling and Newsome (2010)	A form of nature tourism that specifically focuses on landscape and geology. It promotes tourism to the geosites and the conservation of geodiversity and an understanding of Earth sciences through appreciation and learning. This is achieved through independent visits to geological features, use of geotrails and view points, guided tours, geo-activities and patronage of geosite visitor centres. The geotourism should be geologically based, environmentally educative, sustainable and locally beneficial and it should ensure tourist satisfaction
Hose (2012)	The provision of interpretative and service facilities for the geosites and geomorphosites and their encompassing topography, together with their associated in situ and ex situ artifacts, to constituency build for their conservation by generating appreciation, learning and research by and for current and future generations
Martini et al. (2012)	Geotourism allow tourists to know the local geology but also to better understand that this geology is closely related with all the other assets of the territory, such as biodiversity, archaeological and cultural values, gastronomy, etc
Dowling (2013)	Geotourism is sustainable tourism with a primary focus on experiencing the earth's geologic features in a way that fosters environmental and cultural understanding, appreciation and conservation, and is locally beneficial. Geotourism product protects, communicates and promotes geoheritage, helps build communities and works with a wide range of different people

Based on Dowling and Newsome (2010) and Kubalíková (2013)

Fig. 31.1 Climbing on Dráteníčky in the Žďárské vrchy PLA—an example of geomorphological site as a support or resource for tourist activity (*Photo* J. Vítek)

activities—it is one of the functions of geodiversity; however, it is evident that geodiversity as a whole cannot be used for geotourism purposes; tourist use of geodiversity is generally made through the exploitation of geosites and geomorphosites (Pralong 2003; Pralong and Reynard 2005). The cultural issues also have a big influence on geotourism development (Panniza and Piacente 2008) as they often increase the attractiveness of the geosites and geomorphosites (Fig. 31.2).

As can be seen above, geotourism definitions always include educational aspect. This is achieved mainly by the geo-trails and view points, guided tours and visitor centres situated at or in the proximity of destinations, but the geoeducation "ex situ" such as museums, popularization books or journals should be also included. In the Czech Republic, there are several good examples of books that popularize Earth Sciences, e.g. by Kříž (1999), Chlupáč (2002), Gába et al. (2002). Cílek (2002a, b) discussed the influence of landscape and landforms on humans. Some regional governments in cooperation with universities and nature conservation institutions issued materials about geologically and geomorphologically important sites in particular regions (e. g. Kühn 2006—Liberec Region). The Nature Conservation Agency of the Czech Republic has published comprehensive review which contains 13 volumes and describes protected sites and areas in the Czech Republic (Mackovčin and Sedláček 1999–2008).

The Czech Geological Survey also published some geological maps for tourists, e.g. Geological map of Brno and its

Fig. 31.2 Trosky Castle in the Bohemian Paradise National Geopark: cultural issues have a great influence on geotourism as they often increase the attractiveness of geosites and geomorphosites (*Photo* L. Brázdilová)

surroundings where the major geologically and geomorphologically important sites are described (Hanžl 1999) or Geological map of the Podyjí National Park (Batík 1992). Interesting links between geosciences and history are presented by the reproductions of old maps of the České středohoří Mountains (Čejchanová and Cajz 2009). In cooperation with the Nature Conservation Agency of the Czech Republic, leaflets about geology and geomorphology of some National Parks and Protected Landscape Areas were published, e.g. for Jizerské hory Mountains PLA (Pošmourný and Vítek 2003) or Pálava PLA (Stráník 2006).

Another possibility of how to bring Earth Sciences to the public is exhibitions in museums and other similar institutions. In the Czech Republic, museums usually have geological sections where geology (especially fossils and minerals) of particular area is displayed; however, geomorphological processes and issues are usually presented indirectly via pictures, movies or models, so the direct contact and experience with a site is missing. There are also many online projects to popularize Earth Sciences—the most remarkable one is probably the Database of geological localities (over 2,800 sites, see Chap. 30) and the Virtual museum of the Czech Geological Survey which presents fossils, minerals and rocks. Other online popularization activities of the Czech Geological Survey include Geohazards portal or Geological encyclopaedia (http://www.geology.cz/extranet/popularizace).

31.3 Geoparks: The Focus of Geotourism Activities

According to the European Geopark Network Charter (The EGN Charter 2000), a geopark is a territory with clearly defined boundaries, which includes a particular geological heritage and has sustainable territorial development strategy. It must comprise a certain number of geological sites of particular importance in terms of their scientific quality, rarity, aesthetic appeal or educational value; their interest may also be archaeological, ecological, historical or cultural. Geopark plays an active role in the economic development of the territory and its objective is to enable the inhabitants to re-appropriate the values of the territory's heritage. Geopark has to support education about the environment, training and development of scientific research in the various disciplines of Earth Sciences.

The EGN was established in 2000 by four European areas which became the first European Geoparks aiming for protection and enhancement of the European Earth heritage and promotion of a sustainable local development through geotourism (Zouros 2004; Eder and Patzak 2004). Soon after its establishment, the Convention of cooperation between EGN and UNESCO was signed. In 2004, the Global UNESCO Network of geoparks was formed together with adoption of the document "Operation Guidelines for National Geoparks seeking UNESCO assistance" (Zouros 2009). Today, the GGN has over 100 members, whereas the majority of them are situated in Europe (63 geoparks) and China (31 geoparks). In November 2015, the Global Geoparks became UNESCO sites. The Czech Republic is represented by the one member of the GGN—Bohemian Paradise. This geopark and the Bohemian Paradise Protected Landscape Area partly overlap.

Generally, geoparks are not anchored in legislation (respectively, they have not a status of a protected area declared by the national authorities of nature conservation) and are formed on the basis of voluntary activity of various entities in a given area and in close cooperation with institutions of nature protection, university experts and municipalities.

31.3.1 Geopark Network in the Czech Republic

On the national level, the National Geoparks Networks (or NGN) are established. In the Czech Republic, the National Geoparks Network is coordinated by the National Geoparks Committee which consists of the experts from different branches (geoscientists, economists, pedagogues, etc.). Any territory associated in some way with significant geoheritage can become a member of the NGN. Within the network, the members can exchange experience, organize visits, seminars and conferences (Fig. 31.3). The most important documents are the Charter of National Geoparks and the Directive of the Ministry of Environment No. 07/2006 (updated in 2011) which outlines criteria for establishing a national geopark as well as the criteria for the temporary status of candidate geopark. The conditions for the emission of the Certificate of the National Geopark and the revalidation rules (every 4 years) are also present. The requirements are quite strict and actually they come out from the conditions for establishing a European Geopark (Pásková 2014, oral communication).

Today, there are six territories which bear the title of "National Geopark": Bohemian Paradise, Železné hory, Egeria, GeoLoci, Kraj Blanických rytířů and Podbeskydí. The candidate geoparks are Vysočina, Ralsko, Jeseníky and Geopark of J. Barrande. There are other two areas that may be considered (in June 2014): Krkonoše and Broumovsko.

31.3.2 Bohemian Paradise Geopark: A Geological Handbook

The Bohemian Paradise National Geopark became the first European Geopark in the Czech Republic in 2005. It has been a traditional tourist destination for a long time (Fig. 31.4), already since the first half of the nineteenth century, when the Sedmihorky Spa and the watchtower on

Fig. 31.3 The first conference on geoparks under the auspices of the National Geoparks Network of the Czech Republic was held in GeoLoci National Geopark in April 2013 (GeoLoci geopark, http://geoloci.webnode.cz/)

Fig. 31.4 The Bohemian Paradise has been a traditional tourist destination for a long time. A local tourist group also initiated the construction of a watchtower on Hlavatice Rock already at the end of the nineteenth century and soon it became a favourite place for the tourists (*Photo* J. Prousek, around 1900)

the Trosky Castle were built. Later, the area became the favourite place for hikers and climbers. In 1937, the "Golden trail of Bohemian Paradise", which connected cultural and natural monuments, inns and pubs, was open. This was the initiative of the Club of Czech Tourists, the voluntary organization of tourist which was established already in 1888 and constructed tourist chalets, lookouts and watch-towers. Later, it began to create a dense network of marked paths and trails. Geology and geomorphology of the area, represented by sandstone rock cities, volcanic forms, karst phenomenon and deposits of precious stones, together with rich cultural heritage (castles, chateaux, folk architecture), are the principal resources for geotouristic activities. The support services are represented by the dense network of the marked paths for hikers and cyclists, a wide range of the educational products (information panels, leaflets, offer of the guided excursions and school trips). There are also several educational paths; some of them focused on geology and geomorphology (e.g. Hruboskalsko educational path, Prachovské skály educational path). A trade mark "Regional product of the Bohemian Paradise" is used to promote local products (food, handicrafts, glassworks, etc.) of over 50 producers (website of Bohemian Paradise geopark http://geopark-ceskyraj.cz).

31.3.3 Egeria Geopark and GeoLoci Geopark: Volcanism, Mineral Water and Mining History

Egeria National Geopark and GeoLoci National Geopark are situated in the westernmost part of the Czech Republic and their heritage is represented by the long tradition of using natural resources by human society. Both geoparks form a part of the bilateral Czech-Bavarian Geopark which extends on the both sides of the boundary (western Bohemia and eastern Bavaria). In 2010, Egeria was declared as National Geopark and 2 years later, GeoLoci also became the National Geopark.

Egeria National Geopark stands on six pillars that attract visitors: volcanism, mineral waters and spas, geological and geomorphological sites, mining history, brown coal and mineralogical and paleontological sites. The area is unique within the Czech Republic especially due to its post-volcanic activity and volcanic landforms; the youngest volcanoes, respectively, cinder cones (Železná Hůrka, Komorní Hůrka), can be found here as well as the mud volcanoes and moffetas which are present in Soos National Natural Reserve (Fig. 31.5). Another manifestations of post-volcanic activity are hot springs. Exploitation and use of mineral water (e. g.

Fig. 31.5 Gas bubbling—one among post-volcanic phenomena in the Soos National Natural Monument, western Bohemia, Egeria National Geopark (Personal collection)

in Carlsbad, Františkovy spas, Mariánské spas) have a long tradition and these waters present a significant economical resource for local people and municipalities (website of Egeria geopark http://www.geopark.cz/). This area was also the object of interest of J.W. Goethe who was a frequent visitor, collected minerals and also often found inspiration there (Cílek 2002a, b).

The GeoLoci National Geopark is especially rich in mining history heritage, which dates back to the Middle Ages. The silver, iron ore and other materials were extracted here and this fact is reflected in the toponyms (e.g. town Stříbro which means 'silver' in Czech). Old galleries, pits and other mining features can be considered as the most important anthropogenic landforms of the area. There are several museums (e.g. in Planá) and educational paths dedicated to geosciences and mining history. In 2013, GeoLoci National Geopark organized the first Conference of geoparks which provided the platform for exchanging experience and knowledge not only between the members of the NGN, but also people who are somehow involved in geotourism activities and who try to promote geoheritage of particular regions to visitors (website of GeoLoci geopark http://geoloci.webnode.cz/).

31.3.4 Železné Hory Geopark: The History of the Earth in Three Days

The staff of Železné hory National Geopark chose the motto "the history of the Earth in three days" because it exactly describes the heritage of the area. A visitor, who comes to this area on the border of Bohemia and Moravia, can see specific issues from nearly every geological period from Precambrium metamorphic rocks, over Paleozoic granitic plutons, Silurian schist and Devonian limestone (Fig. 31.6), Cretaceous sandstones, Tertiary volcanism to Quaternary landforms. The area is also rich in natural rock outcrops and anthropogenic landforms. The geopark administration closely cooperates with the owners of quarries and other stakeholders to extend the offer for visitors, e.g. in the granite quarry of Ctětínek where the visitor can see a process of the production of granite cobbles from the extraction of granite to manufacturing the stone to the final shape (http://www.geoparkzh.cz/cs/). Železné hory obtained the certificate of National Geopark in 2012 and currently the application to the EGN is being prepared.

31.3.5 Blaník Knights' County National Geopark and Podbeskydí Geopark

In 2014, two new National Geoparks were declared. Blaník Knights' County geopark, as the name suggests, is based especially on the rich history that is closely related to the landscape. It is one of the emblematic sites of the Czech Republic and as well as Říp hill in the Central Bohemia, this area is connected to the myths (according to the legends, there are sleeping knights inside the Blaník mountain and they will rise up when the Czech country will need them most). Local geoheritage is represented by the oldest rocks in the Czech Republic (Moldanubian) and also by cryogenic landforms (e.g. boulder fields, frost cliffs) which can be

Fig. 31.6 Silurian schist and Devonian limestone displayed in the quarry of Prachovice in the Železné hory National Geopark—an example how the geopark cooperate with private enterprises (Personal collection)

Fig. 31.7 The trademark "Blaník Knights' County regional product" supports regional products of the Blaník Knights' County National Geopark and the surrounding area. Both products, e.g. wooden joys, bee honey or traditional pottery, and services, e.g. inns or restaurants, can be promoted using this trademark (Public domain)

found on the Velký and Malý Blaník. Local products are promoted with the trade mark "Blaník Knights' County regional product" (Fig. 31.7) which includes both products, e.g. wooden joys, bee honey or traditional pottery, and services, e.g. inns or restaurants (http://www.blanicti-rytiri.cz/).

Podbeskydí geopark is situated in the north-eastern part of Moravia. This geologically, geomorphologically and historically rich area (Carpathian flysch sediments, Mesozoic submarine volcanism, paleontologically important Jurassic limestone, prehistoric settlements and documents of the Paleolithic working of flints) is the only geopark in the Czech Republic in the Western Carpathians (http://www.geoparkpodbeskydi.cz).

31.4 Bringing Geosciences to People: Geo-Trails, Geo-Gardens, Geo-Excursions

Administrations of National Parks and Protected Landscape Areas propose a wide range of interpretation facilities and services concerning geoheritage. All administrations manage visitor's centres where visitors can get knowledge about nature of the area. It is common that administrations cooperate with local museums, local and regional authorities and other institutions and associations, both public and non-governmental (e.g. Czech Union for Nature Conservation).

Administration of the Podyjí National Park organizes walks and excursions for the public, which are focused mainly on living nature features; however, there are also geological walks and occasionally, excursions to the pseudokarst Ice Caves (Fig. 31.8) which are normally inaccessible for the public are held (e.g. in the occasion of the International Night of Bats at the beginning of September

Fig. 31.8 Pseudokarst Ice Caves (Podyjí National Park) are normally inaccessible for the public; however, National Park Administration organizes guided excursions to show this unique site (Personal collection)

(http://www.nppodyji.cz/). Administration of the Podyjí National Park closely cooperates with Administration of Thayatal National Park which lies on the right bank of Dyje/Thaya River in Austria.

Guided walks and excursions are also organized in the Krkonoše National Park, the majority is also focused on plants and ecosystems, although glacial and periglacial geomorphology is also a theme of several excursions. In addition, the Administration runs several information centres (5 permanent and 2 seasonal). Promotion of geomorphological heritage is also ensured by information panels along the educational paths. Since 1970s, the Krkonoše centre for Environmental Education offers educational programmes for both primary and secondary schools and the public (http://www.krnap.cz/).

In the Bohemian Switzerland National Park and Šumava National Park, as well as other Protected Landscape Areas, promotion of geomorphological heritage is accomplished through leaflets, guided walks and educational paths. Generally, promotion of abiotic nature is not so frequent in comparison with promotion of living nature features (e.g. according to the calendar of walks and excursion published on the website of the Podyjí National Park in 2014, there are two geologically and geomorphologically oriented walks in comparison with 11 biologically oriented walks).

In some protected areas as well as outside them, there are geological educational paths; the list of them (which is continuously updated) can be consulted on the website of the Czech Geological Survey (http://www.geology.cz/extranet/popularizace/naucne-stezky). Some of them are focused on mining history (e.g. Mining in Jihlava, Mining in Kutná Hora). Within the Bohemian Karst PLA, the 39-km-long geo-path which gives information especially about stratigraphy, paleontology and karst features in the western part of the PLA is built (both for hikers and cyclists). An interesting project is "Geological and water open-air museum of the region Třebovsko-Orlicko" which connect 15 geologically, geomorphologically and hydrologically interesting sites and information panels inform about slope processes, outcrops, springs or quarries. The proposals of geological educational trail are sometimes a topic of diploma thesis (e.g. at Mendel University in Brno, where geoheritage and geoconservation courses are organized).

In the last years, the number of geological gardens and open-air exhibitions is growing. The first one was built in 1991 in Karlova Studánka in Jeseníky PLA (Budil et al. 2012). These expositions are sometimes set up by universities, schools and museums and they serve mainly for educational purposes of the students (e.g. Geological garden of Chotěboř Grammar School, Geological garden within Botanical garden of the Faculty of Sciences, Charles University in Prague, etc.). An interesting geological garden is situated on the border of Mariánské spa and thanks to the easy terrain, they are suitable also for less mobile visitors; another outdoor geo-exposition is situated near the cave Na Turoldu (Pálava PLA) and it displays rocks from all over Moravia (Fig. 31.9).

Guided tours outside the protected areas are often organized by the Czech Geological Survey, regional and local museums and universities. Information materials about geomorphologic heritage (leaflets, booklets, etc.) are also issued by regional governments and municipalities.

31.5 Karst Heritage: A Significant Attraction for Visitors

In the Czech Republic, there are two main karst systems that form a core of two Protected Landscape Areas—Bohemian Karst and Moravian Karst. Other karst systems are of smaller dimensions and they are usually protected as National Natural Reserves or Monuments (e.g. The Bozkov Dolomite Caves NNM in the Northern Bohemia or Zbrašov Aragonite Caves NNM in the central Moravia). The Cave Administration of the Czech Republic (governmental organization

Fig. 31.9 Geological garden (open-air exposition) which displays rocks from all over Moravia is situated near the cave Na Turoldu (Pálava PLA) and was built by the Centre of the Ecological Education Pálava (part of the Czech Union for Nature Conservation) (*Photo* J. Miklín)

established by the Ministry of Environment) provides scientific and technical support for speleological surveys, inventories and documentation on caves and manages 14 cave systems accessible for visitors according to the nature conservation needs. Its main task is to balance the protection of cave systems and its rational use and promotion for the public (http://www.jeskynecr.cz/).

31.6 Cultural Events Related to Geomorphology

An interesting opportunity of how to bring geomorphological heritage to the public is to use it as a place for cultural events. A remarkable project of connecting geomorphology, respectively, sandstone rock city, and music was the unique open-air concert "Philharmonic in the rocks" that was organized by Agency for Development of the Broumovsko Region with Hradec Králové Philharmonic. Another interesting ways of how to join geomorphological features and cultural events are small open-air theatres or "forest theatres", e.g. Forest Theatre in Sloup in Northern Bohemia (Fig. 31.10) or Nýrsko Forest Theatre in an old quarry which lies close to the Šumava Protected Landscape Area in the south-western Bohemia. Some sandstone landforms in northern Bohemia inspired the artists who created sculptures and reliefs, e.g. Braun's statues in Kuks which are now protected as National Cultural Monument, or Devil's Heads in Kokořínsko PLA (Fig. 30.7 in Chap. 30).

31.7 Perspectives for the Future

Promotion of geomorphological heritage in the Czech Republic is managed by different organizations and institutions. The most important and the most active ones are the National Geopark Network of the Czech Republic, the Czech Geological Survey and the Cave Administration of the Czech Republic. In addition, administrations of National Parks and Protected Landscape Areas in cooperation with the Nature Conservation Agency of the Czech Republic provide a wide spectrum of promotion activities and products related to geomorphological heritage, especially guided walks, excursions and geological educational trails. However, in some cases, promotion of geoheritage if compared with promotion of living nature is still low. The reason why it is so can lie in the fact that living nature is considered more important than abiotic nature and that changes in geological and geomorphological heritage are not as quick and visible as those in the living nature (Cílek 2002a, b).

However, there are a growing number of people outside the nature conservation organizations who have already realized that the Czech Republic is very diverse in terms of geomorphological heritage. These enthusiastic people play also a very important role in the promotion of geoheritage—some small-scale local geoheritage projects which have been realized by schools or local governments may serve as evidence.

Probably, the best way of how to raise awareness of the importance of geoheritage is to support geoeducation activities and increase an attractiveness of Earth Sciences in primary and secondary schools and grammar schools and to offer attractive products and activities focused on geoheritage both within and outside the protected areas. Another possibility how to bring geomorphological heritage to people is to relate it with other attractive activities (e.g. cultural or artistic events) and to support the rational use of geomorphological heritage via engaging local and regional authorities together with landowners and stakeholders; conserving and rationally promoting geological and geomorphological heritage can also have beneficial influence on

Fig. 31.10 Forest theatres present an amazing way of how to join geomorphological features and cultural events. Forest Theatre in Sloup in northern Bohemia has been regularly used since 1920s, nowadays is used occasionally during summer holidays (*Photo* J. Kuhn)

the local and regional economical development, attracting visitors to the area; therefore, protection of geomorphological heritage and its promotion should be included in integrated and sustainable landscape management and development strategies (Gray 2013).

Nevertheless, the future of promotion and rational use of geomorphological heritage depends on the cooperation between various institutions and organizations on all levels (national, regional, local), both governmental and non-governmental, both professional and amateur. Cooperation between specialists of different branches is important too. All this could lead to better understanding, effective conservation and rational promotion of geomorphological heritage in the Czech Republic.

References

Batík P (1992) Geologická mapa Národního parku Podyjí, Český geologický ústav, Praha, 1 sheet

Butler RW (1992) Alternative tourism: The thin edge of the wedge. In: Smith V, Eadington W (eds) Tourism Alternatives. University of Pennsylvania Press, Philadelphia, pp 31–46

Budil P et al. (2012) Czech Republic. In Wimbledon WAP and Smith-Meyer s (eds) Geoheritage in Europe and its conservation. Pro GEO, Oslo, pp 92–99

Chlupáč I (2002) Geologická minulost České republiky. Academia, 436 pp

Čejchanová A and Cajz V (2009) Geologické mapy Českého středohoří Josefa Emanuela Hibsche, Česká geologická služba Praha, 22 sheets

Cílek V (2002a) Krajiny vnitřní a vnější. Dokořán, 231 pp

Cílek V (2002b) Geodiverzita - opomíjený aspekt ochrany přírody a krajiny. Zprávy o geologických výzkumech v roce 2001:13–15

Dowling R (2011) Geotourism's Global Growth. Geoheritage 3(1):1–13

Dowling R (2013) Global Geotourism—an Emerging Form of Sustainable Tourism. Czech J Tourism 2(2):59–79

Dowling R and Newsome D (eds) (2006) Geotourism. Elsevier. 260 pp

Dowling R, Newsome D (eds) (2010) Geotourism. The tourism of Geology and Landscape. Goodfellow Publishers Ltd., Oxford 246 pp

Eder FW, Patzak M (2004) Geoparks - geological attractions: a tool for public education, recreation and sustainable economic development. Episodes 27(3):162–164

European Geopark Charter, available on-line on http://www.europeangeoparks.org/?page_id=357

Gába Z (2002) Geologické vycházky Českou republikou. Karolinum Praha 493 pp

Gray M (2004) Geodiversity: Valuing and Conserving Abiotic Nature. Wiley, Chichester 448 pp

Gray M (2013) Geodiversity: Valuing and Conserving Abiotic Nature, 2nd edn. Wiley Blackwell, Chichester 508 pp

Hanžl P (1999) Geologická mapa Brna a okolí, Český geologický ústav Praha, 1 sheet

Hose TA (1995) Selling the Story of Britain's Stone. Environ Interpretation 10(2):16–17

Hose TA (2000) European Geotourism—Geological Interpretation and Geoconservation Promotion for Tourists. In: Barretino D, Wimbledon WAP, Gallego E (eds) Geological heritage: its conservation and management. Sociedad Geologica deEspaña/Instituto Technologico GeoMinero de España/ProGEO, Madrid, pp 127–146

Hose TA (2008) Towards a history of geotourism: definitions, antecedents and the future. In Burek CV, Prosser CD (eds) The History of Geoconservation. The Geological Society of London, 305 pp

Hose TA (2012) 3G's for Modern Geotourism. Geoheritage 4(1–2): 7–24

Kříž J (1999) Geologické památky Prahy. Česká geologická služba, Praha 206 p

Kubalíková (2013) Geomorphosite assessment for geotourism purposes. Czech Journal of Tourism 2(2):80–104

Kühn P (2006) Geologické zajímavosti Libereckého kraje. Liberecký kraj, resort rozvoje venkova, zemědělství, životního prostředí a informatiky, 120 pp
Mackovčin P and Sedláček M eds. (1999–2008) Chráněná území ČR. Agentura ochrany přírody a krajiny ČR a EkoCentrum Brno
Martini G et al. (2012) Reflections about the geotourism concept. In Sá AA, Rocha D, Paz A and Correia V (eds) Proceedings of the 11 European Geoparks Conference. AGA – Associação Geoparque Arouca, Arouca, pp 187–188
Migoń P (2009) Geomorphosites and the World Heritage List of UNESCO. In: Reynard E, Coratza P, Regolini-Bissig G (eds) Geomorphosites. Pfeil, Munchen, pp 119–130
National Geographic Society (2005) Geotourism Charter. http://travel.nationalgeographic.com/travel/sustainable/pdf/geotourism_charter_template.pdf. Accessed 13 Oct 2012
Panizza M, Piacente S (2008) Geomorphosites and geotourism. Rev Geogr Acadêmica 2(1):5–9
Pásková M (2012) Environmentalistika cestovního ruchu. Czech J Tourism 1(2):77–119
Pošmourný K and Vítek J (2003) Jizerské hory: geologie chráněných oblastí České republiky. Česká geologická služba Praha, 1 shee
Pralong JP (2003) Valorisation et vulgarisation des sciences de la Terre: les concepts de temps et d'espace. In: Holzmann C, Guex D, Summermatter N (eds) Reynard E. Géomorphologie et tourisme, Actes de la Réunion annuelle de la Société Suisse de Géomorphologie (SSGm), pp 115–127
Pralong JP, Reynard E (2005) A proposal for a classification of geomorphological sites depending on their tourist value. Il Quaternario - Ital J Quat Sci 18(1):315–321
Reynard E et al (2003) Géomorphologie et tourisme: quelles relations? In: Holzmann C, Guex D, Summermatter N (eds) Reynard E. Géomorphologie et tourisme, Actes de la Réunion annuelle de la Société Suisse de Géomorphologie (SSGm), pp 1–10
Reynard E (2008) Scientific research and tourist promotion of geomorphological heritage. Geogr Fis Dinam Quat 31:225–230
Slomka T, Kicinska-Swiderska A (2004) Geotourism—the basic concepts. Geoturystyka 1:2–5
Stráník Z (2006) Pálava: geologie chráněných oblastí České republiky. Česká geologická služba Praha, 1 sheet
Zouros N (2004) The European Geopark Network, Geological heritage protection and local development. Episodes 27(3):165–171
Zouros N (2009) Geomorphosites within geoparks. In Reynard E et al. (eds) Geomorphosites. Pfeil Munchen, pp 105–118

Websites

Bohemian Paradise geopark http://geopark-ceskyraj.cz
Česká geologická služba http://www.geology.cz/extranet/popularizace/naucne-stezky
Egeria geopark http://www.geopark.cz/
GeoLoci geopark http://geoloci.webnode.cz/
Geopark Kraj blanických rytířů http://www.blanicti-rytiri.cz/
Geopark Podbeskydí http://www.geoparkpodbeskydi.cz
Krkonošský národní park http://www.krnap.cz/
Národní park Podyjí http://www.nppodyji.cz/
Správa jeskyní Českké republiky http://www.jeskynecr.cz/
Železné hory geopark http://www.geoparkzh.cz/cs/

Index

T. Pánek and J. Hradecký (eds.), *Landscapes and Landforms of the Czech Republic*,
World Geomorphological Landscapes, DOI 10.1007/978-3-319-27537-6

D

E

Q

R

S

T

FSC
www.fsc.org
MIX
Papier | Fördert gute Waldnutzung
FSC® C083411

Kolibri 360 GmbH
Ferdinand-Jühlke-Straße 7
99095 Erfurt, Deutschland
produktsicherheit@kolibri360.de